Lecture Notes in Artificial Intelligence   11471

Subseries of Lecture Notes in Computer Science

More information about this series at http://www.springer.com/series/1244

Hirosato Seki · Canh Hao Nguyen ·
Van-Nam Huynh · Masahiro Inuiguchi (Eds.)

Integrated Uncertainty in Knowledge Modelling and Decision Making

7th International Symposium, IUKM 2019
Nara, Japan, March 27–29, 2019
Proceedings

 Springer

Editors
Hirosato Seki
Osaka University
Toyonaka, Osaka, Japan

Canh Hao Nguyen
Institute for Chemical Research
Kyoto University
Uji, Kyoto, Japan

Van-Nam Huynh
Japan Advanced Institute of Science
and Technology
Nomi, Ishikawa, Japan

Masahiro Inuiguchi
Osaka University
Toyonaka, Osaka, Japan

ISSN 0302-9743   ISSN 1611-3349 (electronic)
Lecture Notes in Artificial Intelligence
ISBN 978-3-030-14814-0   ISBN 978-3-030-14815-7 (eBook)
https://doi.org/10.1007/978-3-030-14815-7

Library of Congress Control Number: 2019932915

LNCS Sublibrary: SL7 – Artificial Intelligence

This Springer imprint is published by the registered company Springer Nature Switzerland AG
The registered company address is: Gewerbestrasse 11, 6330 Cham, Switzerland

Preface

This volume contains the papers that were presented at the 7th International Symposium on Integrated Uncertainty in Knowledge Modelling and Decision Making (IUKM 2019) held in Nara, Japan, March 27–29, 2019.

The IUKM symposia aim to provide a forum for the exchange of research results, ideas, and experience of application among researchers and practitioners involved with all aspects of uncertainty modelling and management. Previous editions of the conference were held in Ishikawa, Japan (IUM 2010), Hangzhou, China (IUKM 2011), Beijing, China (IUKM 2013), Nha Trang, Vietnam (IUKM 2015), Da Nang, Vietnam (IUKM 2016), and Ha Noi, Vietnam (2018) and their proceedings were published by Springer in AISC 68, LNAI 7027, LNAI 8032, LNAI 9376, LNAI 9978, and LNAI 10758, respectively.

The IUKM 2019 was jointly organized by Osaka University and Japan Advanced Institute of Science and Technology.

This year the conference received 93 submissions from 20 different countries. Each submission was peer reviewed by at least two members of the Program Committee. After a thorough review process, 71 papers were accepted for presentation at IUKM 2018, of which 36 papers (38.89%) were accepted for publication in the LNAI proceedings.

The IUKM 2019 symposium was partially supported by Tateisi Science and Technology Foundation, Japan, Nara Visitors Bureau and the U.S. Office of Naval Research Global (ONR Global). We are very thankful to the local organizing team from Osaka University, Osaka Prefecture University, and University of Hyogo for their hard work, efficient services, and wonderful local arrangements.

We would like to express our appreciation to the members of the Program Committee for their support and cooperation in this publication. We are also thankful to Alfred Hofmann, Anna Kramer, and their colleagues at Springer for providing a meticulous service for the timely production of this volume. Last, but certainly not the least, our special thanks go to all the authors who submitted papers and all the attendees for their contributions and fruitful discussions that made this conference a great success.

March 2018

Hirosato Seki
Canh Hao Nguyen
Van-Nam Huynh
Masahiro Inuiguchi

Organization

General Co-chairs

Masahiro Inuiguchi   Osaka University, Japan
Thierry Denoeux   University of Technology of Compiègne, France

Advisory Board

Michio Sugeno   European Center for Soft Computing, Spain
Hung T. Nguyen   New Mexico State University, USA;
    Chiang Mai University, Thailand
Sadaaki Miyamoto   University of Tsukuba, Japan
Akira Namatame   AOARD/AFRL and National Defense Academy
    of Japan

Program Co-chairs

Hirosato Seki   University of Osaka, Japan
Canh Hao Nguyen   Kyoto University, Japan

Local Arrangements Co-chairs

Katsuhiro Honda   Osaka Prefecture University, Japan
Tomoe Entani   University of Hyogo, Japan
Puchit Sariddichainunta   Osaka University, Japan

Publication and Financial Co-chairs

Van-Nam Huynh   Japan Advanced Institute of Science and Technology,
    Japan
Katsushige Fujimoto   Fukushima University, Japan

Publicity Co-chairs

Yoshifumi Kusunoki   Osaka University, Japan
Seiki Ubukata   Osaka Prefecture University, Japan

Program Committee

Yaxin Bi   University of Ulster, UK
Jurek Blaszczynski   Poznan University of Technology, Poland
Tru Cao   Ho Chi Minh City University of Technology, Vietnam

Dominik Slezak	University of Warsaw and Infobright Inc., Poland
Roman Slowinski	Poznan University of Technology, Poland
Martin Stepnicka	University of Ostrava, Czech Republic
Marcin Szelag	Poznan University of Technology, Poland
Eiichiro Takahagi	Senshu University, Japan
Kazuhiro Takeuchi	Osaka Electro-Communication University, Japan
Xijin Tang	CAS Academy of Mathematics and Systems Science, China
Yongchuan Tang	Zhejiang University, China
Khoat Than	Hanoi University of Science and Technology, Vietnam
Phantipa Thipwiwatpotjana	Chulalongkorn University, Thailand
Araki Tomoyuki	Hiroshima Institute of Technology, Japan
Vicenc Torra	University of Skovde, Sweden
Dang Hung Tran	Hanoi National University of Education, Vietnam
Seiki Ubukata	Osaka University, Japan
Guoyin Wang	Chongqing University of Posts and Telecommunications, China
Zeshui Xu	Sichuan University, China
Koichi Yamada	Nagaoka University of Technology, Japan
Chunlai Zhou	Renmin University of China

Additional Reviewers

Duy-Tai Dinh	JAIST, Japan
Anh Hoang	JAIST, Japan
Thanh Phu Nguyen	JAIST, Japan
Panchalee Praneetpholkrang	JAIST, Japan
Kwanjira Kaewfak	JAIST, Japan
Thananut Phiboonbanakit	JAIST, Japan

Sponsoring Organizations

Tateisi Science and Technology Foundation, Japan

Office of Naval Research Global (ONRG)

Invited Speakers

Choquet Integral in Decision Making and Metric Learning

Vicenc Torra

Hamilton Institute, Maynooth University, Ireland

Summary: Choquet integrals are an effective tool for aggregation of numerical data. From a mathematical point of view they generalize the Lebesgue integral when the measure is not additive. Non-additivity permits us to represent interactions that cannot be represented by additive measures. Choquet integrals have been used in a large variety of contexts that include decision making, computer vision, and economy.

In this talk we will illustrate their use in decision making. We will review some results on Choquet integral based probability-density functions, which can be used to model decision making and classification problems. This will lead us to consider distances based on the Choquet integral, and the problem of measure identification. This last problem corresponds to metric learning. We will show its use in risk assessment in data privacy.

Logic for Thinking – From Mathematical Logic to Grammatical Logic

Michio Sugeno

Tokyo Institute of Technology, Japan

Summary: We see outer world with language and also think in inner world with language. Thinking is something different from imagining or dreaming; it is viewed as 'a logical act' associated with language.

In this talk, we first characterize 'thinking' on the semantic level from the perspectives of Systemic Functional Linguistics and then investigate logic for thinking imbedded in language on the grammatical level. Conventional logic with relations such as AND, OR and IF-THEN was born from language; in fact, these are grammatical relations between clauses (sentences). Next we explore primary and higher-order logic for thinking. As a result, there are a variety of grammatical relations found between clauses.

In addition, we consider when infants begin to think, reviewing their language development studied by Halliday and point out that grammar is a driving force for thinking as higher-order consciousness developed together with language. In this sense, resources necessary for thinking are all prepared in language.

Robust Decisions Under Uncertainty with a Rule Preference Model of Multiple Decision Makers

Roman Slowinski

Poznan University of Technology and Polish Academy of Sciences, Poland

Summary: Decision under uncertainty concerns acts characterized by outcomes that can be achieved with some probabilities. Recommending the best decisions is challenging because aggregation of the outcomes over probabilistic states of the world needs to respect preferences of decision makers (DMs). The method used to assist the DMs has to: rely on realistically available preference information, handle a possible inconsistency of this information, aggregate the outcomes in an intelligible and non-compensatory way. To respond satisfactorily to these requirements, we propose a methodology that relies on preference information in the form of decision examples provided by DMs on a subset of reference acts. As this information may be inconsistent with respect to stochastic dominance, it is structured using Dominance-based Rough Set Approach, and then used for inducing a preference model composed of "if..., then..." decision rules. Decision rules constitute an intelligible and non-compensatory aggregation model able to represent complex interactions. We induce all different minimal-cover sets of rules, each one being compatible with the consistent part of the preference information. Applying such compatible instances of the preference model on all considered acts, we get robust recommendations. We also present some indicators for judging the spaces of consensus and disagreement between DMs.

Variations and Generalizations of Fuzzy c–Means Clustering

Sadaaki Miyamoto

University of Tsukuba, Japan

Summary: There are many variations and generalizations of the method of fuzzy c-means. In this talk some of them are overviewed from three viewpoints:

1. relations between fuzzy model and statistical model are discussed;
2. introduction of size variables and covariance variables are considered;
3. theoretical properties of fuzzy classifiers are shown and contrasted with those in statistical model.

Specifically, generalizations of fuzzy c-means include (i) generalized entropy methods, (ii) Yang's fuzzified maximum likelihoods, and (iii) Generalized Gustafson-Kessel method. Variations encompass the noise cluster model, and also rough c-means model. When we consider the properties of fuzzy classifiers of generalized fuzzy c-means, a fuzzy classifier is obtained by substituting an object symbol x_k by the generic variable symbol x in the solution of the belongingness, which is defined on the whole space instead of the set of objects. The behaviors of fuzzy classifiers when x goes to an infinity point are studied. In addition to the above generalizations, a more abstract form of generalizations is studied. Simple illustrative examples are shown.

Contents

Uncertainty Management and Decision Support

Machine Learning Applications

Statistical Methods

Uncertainty Management and Decision Support

Scientometric Indices Based on Integrals and Their Adaptation in Different Domains

Andrea Stupňanová[✉][iD]

Slovak University of Technology in Bratislava,
Radlinského 11, 810 05 Bratislava, Slovakia
andrea.stupnanova@stuba.sk

Abstract. Based on the earlier discussed link between h-index and Sugeno integral due to Torra and Narukawa, a new Sugeno integral - based representation of h-index is proposed. Similarly, some other bibliometric indices are represented by means of particular universal integrals, and some new indices are proposed. More, a proposal based on modification of considered monotone measures allowing to compensate differences between particular scientific domains is presented and discussed.

Keywords: Bibliometric database · h-index · Monotone measure · Sugeno integral · Universal integral

1 Introduction

Since Hirch's proposal of his famous h-index [6], there were proposed many other types of scientometric indices aiming to evaluate the scientometric performance of researchers. These indices were introduced mostly in a verbal form and only later they were related to the corresponding formulas. So for example, Hirch's h-index $H(X)$ of a researcher X was introduced as an integer $h \in \mathbb{N}_0$ such that there are h papers of X having each at least h citations (in the considered database, such as Web of Science, Scopus, Google Scholar, DBLP, etc.). Later, Torra and Narukawa [20] have introduced a Sugeno integral - based [19] representation of h-index. Similar is the case of MAXPROD index introduced by Kosmulski [11] and related to the Shilkret integral [17], or $h(2)$-index introduced by Kosmulski in [10] again related to the Sugeno integral. Seeing the links between some scientometric indices and some integrals, the aim of this contribution is to propose a framework for integral - based scientometric indices. Note that this idea appeared already in [5]. We propose two approaches, depending on the monotone measure to be considered in the integration step. Further, modifying the considered measures depending on the scientific domain where the

This work was supported by the Slovak Research and Development Agency under the contract no. APVV-17-0066 and grant VEGA 1/0682/16.

H. Seki et al. (Eds.): IUKM 2019, LNAI 11471, pp. 3–12, 2019.
https://doi.org/10.1007/978-3-030-14815-7_1

evaluated researcher X is working, we give a hint for a more fair comparison of researchers from different domain than we can obtain using the h-index, $h(2)$ or MAXPROD indices, etc.

The paper is organized as follows. In the next section, we recall the concept of universal integrals due to Klement et al. [8], as well as two views for representation of scientific performance of considered researchers and the related h-index. In Sect. 3, we illustrate some other scientometric indices, including their visualization. In Sect. 4, we suggest a hint how to compare researchers from different domains. Finally, some concluding remarks are added.

2   Universal Integrals, Scientometric Records and h-index

Universal integrals were introduced by Klement et al. [8] as a general framework for integrals based on monotone measures and acting on an arbitrary measurable space $(U, \mathcal{U})$, integrating non-negative $\mathcal{U}$-measurable functions. Having in mind that the number of papers (of a considered scientist X) and the number of citations of papers of X are non-negative integers, we restrict our considerations to the space $(\mathbb{N}, 2^{\mathbb{N}})$ only, where $\mathbb{N} = \{1, 2, \dots \}$. We denote by $\mathcal{M}$ the set of all monotone measures $m : 2^{\mathbb{N}} \to [0, \infty]$ (i.e., $m(\emptyset) = 0$ and $m(A) \leq m(B)$ whenever $A \subseteq B \subseteq \mathbb{N}$), and by $\mathcal{F}$ the set of all non-negative function $f : \mathbb{N} \to [0, \infty[$.

Definition 1 ([8]). *A mapping* $\mathbf{I} : \mathcal{M} \times \mathcal{F} \to [0, \infty]$ *is called a universal integral on* $\mathbb{N}$ *whenever it is increasing (not necessarily strictly) in both components, and* $\mathbf{I}(m_1, f_1) = \mathbf{I}(m_2, f_2)$ *whenever* $m_1(f_1 \geq t) = m_2(f_2 \geq t)$ *for all* $t > 0$.

The function $h_{m,f} :]0, \infty[\to [0, \infty]$ given by

$$h_{m,f}(t) = m(f \geq t)$$

is decreasing (not necessarily strictly), and we denote as

$$\mathcal{H} = \{h_{m,f} | m \in \mathcal{M}, f \in \mathcal{F}\}.$$

Clearly, each universal integral $\mathbf{I}$ is related to an increasing functional $\mathbf{J} : \mathcal{H} \to [0, \infty]$,

$$\mathbf{I}(m, f) = \mathbf{J}(h_{m,f}),$$

and

$$\mathbf{I}(m_1, c \cdot 1_{E_1}) = \mathbf{I}(m_2, c \cdot 1_{E_2}) = c \otimes u$$

for some pseudo-multiplication $\otimes : [0, \infty]^2 \to [0, \infty]$, whenever $m_1(E_1) = m_2(E_2) = u$. More details we can found in [18].

We recall some of universal integrals:

- Consider a pseudo-multiplication $\otimes : [0, \infty]^2 \to [0, \infty]$, $\otimes$ is increasing in both coordinates and 0 is its annihilator, $0 \otimes a = a \otimes 0$ for all $a \in [0, \infty]$. Then the mapping $\mathbf{I}_{\otimes} : \mathcal{M} \times \mathcal{F} \to [0, \infty]$ given by

$$\mathbf{I}_{\otimes}(m, f) = \bigvee_{a \in [0, \infty[} \left(a \otimes h_{m,f}(a) \right) \tag{1}$$

is a universal integral;

- if $\otimes = \wedge$ (minimum) then $\mathbf{I}_\wedge = \mathbf{Su}$ is the Sugeno integral [16, 19];
- if $\otimes = \cdot$ (standard product), then $\mathbf{I}. = \mathbf{Sh}$ is the Shilkret integral;
- if $a \otimes b = a \wedge \varphi(b)$ for some increasing function $\varphi : [0, \infty] \to [0, \infty], \varphi(0) = 0$, then $\mathbf{I}_\otimes$ is a transformed Sugeno integral, $\mathbf{I}_\otimes(m, f) = \mathbf{I}_\wedge(\varphi \circ m, f)$.

- Choquet integral [1] is a universal integral related to the Riemann integral,

$$\mathbf{Ch}(m, f) = \int_0^\infty h_{m,f}(t) \, dt. \tag{2}$$

- A hierarchical class $\left(\mathbf{I}_{(k)}\right)_{k \in \mathbb{N} \cup \{\infty\}}$ of universal integrals (which are the only integrals that are also decomposition integrals [2]) was introduced in [13],

$$\mathbf{I}_{(k)}(m, f) = \sup \left\{ \sum_{i=1}^k a_i \cdot h_{m,f}(a_1 + \cdots + a_i) | a_1, \ldots, a_i \in [0, \infty[\right\} \tag{3}$$

if $k \in \mathbb{N}$, and $\mathbf{I}_{(\infty)} = \bigvee_{k \in \mathbb{N}} \mathbf{I}_{(k)}$. Note that $\mathbf{I}_{(1)} = \mathbf{I}.$ and $\mathbf{I}_{(\infty)} = \mathbf{Ch}$.

For the mathematical evaluation of scientometric indices, we will consider framework used in [12]. For a considered scholar X having n papers in a considered fixed database (WOS, SCOPUS, Google Scholar, DBLP etc.), the numerical performance $\mathbf{x}$ of X is an n-tuple, $\mathbf{x} = (x_1, \ldots, x_n)$, where the value x_i denotes the number of citations received by the i-th most cited paper of the scholar X. Clearly, then $\mathbf{x} \in \mathcal{S}$, where $\mathcal{S}$ is the set of decreasingly ordered non-negative integer sequences of any finite length, i.e.,

$$\mathcal{S} = \{(x_1, \ldots, x_n) | n \in \mathbb{N}, x_1, \ldots, x_n \in \mathbb{N}_0, x_1 \geq \cdots \geq x_n\}. \tag{4}$$

Here $\mathbb{N}_0 = \mathbb{N} \cup \{0\}$. Observe that a scholar X with no relevant publication is not considered in any of the mentioned databases. Note that then the h-index h is a value of function, $H : \mathcal{S} \to \mathbb{N}_0$, defined by

$$H(X) = \bigvee_{i=1}^n (x_i \wedge i) = \max\{\min\{x_1, 1\}, \ldots, \min\{x_n, n\}\}. \tag{5}$$

It is not difficult to check that the h-index $H(X)$ is just the Sugeno integral of the extended record $\bar{\mathbf{x}} = (x_1, \ldots, x_n, 0, \ldots) \in \mathbb{N}_0^\mathbb{N}$ with respect to the counting measure $\mu : 2^\mathbb{N} \to [0, \infty], \mu(A) = \text{card}(A)$.

Example 1. Consider $\mathbf{x} = (5, 3, 3, 3, 1, 0)$. Then $H(X) = 3$, $\bar{\mathbf{x}} = (5, 3, 3, 3, 1, 0, 0, \ldots)$ and

$$h_{\mu, \bar{\mathbf{x}}}(t) = \begin{cases} 5 & \text{if } t \in]0, 1], \\ 4 & \text{if } t \in]1, 3], \\ 1 & \text{if } t \in]3, 5], \\ 0 & \text{it } t > 5. \end{cases}$$

For the geometric interpretation, see Fig. 1.

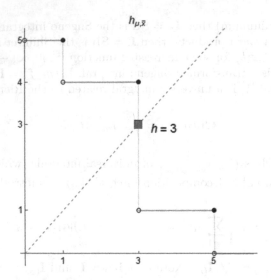

Fig. 1. Illustration of $H(X) = 3$ from Example 1

This representation was introduced already by Torra and Narukawa [20]. However, there is also a new alternative approach. In Torra and Narukawa approach, the considered scientist X is characterised by the couple $(\mu, \bar{\mathbf{x}}) \in \mathcal{M} \times \mathcal{F}$. In the new proposal, X is characterized by the couple $(m_{\mathbf{x}}, \mathrm{id}) \in \mathcal{M} \times \mathcal{F}$, where $\mathrm{id}(i) = i, i \in \mathbb{N}$, and $m_{\mathbf{x}}(\{i\}) = x_i - x_{i+1}, i \in \mathbb{N}$, with the convention $x_i = 0$ if $i > \mathrm{card}(\mathbf{x})$.

Theorem 1. *For a researcher X characterized by the record $\mathbf{x} \in \mathcal{S}$, it holds*

$$h = H(X) = \mathbf{I}_\wedge(m_{\mathbf{x}}, \mathrm{id}). \tag{6}$$

Proof: For $\mathbf{x} = (x_1, \ldots, x_n)$, it holds

$$h_{m_{\mathbf{x}}, \mathrm{id}}(t) = m_{\mathbf{x}}(\{i \in \mathbb{N} | i \geq t\}) = x_{\lceil t \rceil},$$

where $\lceil t \rceil$ is the ceiling of t, $\lceil t \rceil = \min\{j \in \mathbb{N} | j \geq t\}$. Then

$$\mathbf{I}_\wedge(m_{\mathbf{x}}, \mathrm{id}) = \bigvee_{t>0} \left(t \wedge x_{\lceil t \rceil}\right) = \bigvee_{i \in \mathbb{N}} \left(i \wedge x_i\right) = H(X). \qquad \square$$

The above result is illustrated in Fig. 2, using the same data as in Example 1. Now,

$$h_{m_{\mathbf{x}}, \mathrm{id}}(t) = \begin{cases} 5 & \text{if } t \in]0, 1], \\ 3 & \text{if } t \in]1, 4], \\ 1 & \text{if } t \in]4, 5], \\ 0 & \text{if } t > 5. \end{cases}$$

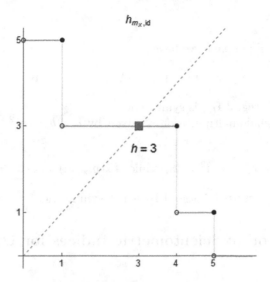

Fig. 2. Illustration of $h_{m_\mathbf{x},\mathrm{id}}$

3   Some Other Scientometric Indices

As observed in [20], when applying the Choquet integral, then $\mathbf{Ch}(\mu, \bar{\mathbf{x}}) = \sum_{i=1}^{n} x_i$ is the total number of citations of papers of researcher X. It is not difficult to check that also $\mathbf{Ch}(m_\mathbf{x}, \mathrm{id}) = \sum_{i=1}^{n} x_i$.

Recall that the Kosmulski MAXPROD [11] is given as the square root of a maximal value of $x_i \cdot i$,

$$\mathrm{MAXPROD}(X) = \left(\bigvee_{i=1}^{n} (x_i \cdot i) \right)^{1/2}.$$

Evidently, considering the Shilkret integral, we have

$$\mathbf{I}.(\mu, \bar{\mathbf{x}}) = \bigvee_{t>0} t \cdot \mathrm{card}\{j | x_j \ge t\} = \bigvee_{i=1}^{n} (i \cdot x_i) = \mathrm{MAXPROD}^2(X),$$

i.e., $\mathrm{MAXPROD} = \sqrt{\mathbf{I}.(\mu, \cdot)}$. Similarly, $\mathbf{I}.(m_\mathbf{x}, \mathrm{id}) = \mathrm{MAXPROD}^2(X)$. For $\mathbf{x} = (5, 3, 3, 3, 1, 0)$, it is easy to see that $\mathrm{MAXPROD}(X) = \sqrt{4 \cdot 3} = \sqrt{12}$.

In all till now discussed cases, the integrals related to couples $(\mu, \mathbf{x})$ and $(m_\mathbf{x}, \mathrm{id})$ had the same values. This fact is caused by the relation between the functions $h_{\mu,\mathbf{x}}$ and $h_{m_\mathbf{x},\mathrm{id}}$, which are linked by the pseudo-inversion [9], and a kind of symmetry of the considered integrals characterized by $\mathbf{I}(m_1, f_1) = \mathbf{I}(m_2, f_2)$ whenever h_{m_1,f_1} is a pseudo-inverse of h_{m_2,f_2}. This symmetry need not be valid for a universal integral $\mathbf{I}$, in general.

Example 2

(i) Continuing in Example 1, we have

$$\mathbf{I}_{(2)}(\mu, \mathbf{x}) = 14 \text{ and } \mathbf{I}_{(2)}(m_{\mathbf{x}}, \text{id}) = 14,$$

the universal integral $\mathbf{I}_{(2)}$ is symmetric.

(ii) Consider a pseudomultiplication $\otimes$ given by $a \otimes b = ab^2$. Then, for $\mathbf{x} = (5, 3, 3, 3, 1, 0)$,

$$\mathbf{I}_{\otimes}(\mu, \mathbf{x}) = 3 \cdot 4^2 = 48, \text{ while } \mathbf{I}_{\otimes}(m_{\mathbf{x}}, \text{id}) = 4 \cdot 3^2 = 36.$$

Evidently, the universal integral $\mathbf{I}_{\otimes}$ is not symmetric.

4 Modification of Scientometric Indices for Different Domains

For any increasing (not necessarily strictly increasing) mapping $\varphi : [0, \infty] \to [0, \infty], \varphi(0) = 0$, and a monotone measure $m : 2^{\mathbb{N}} \to [0, \infty]$, obviously also $\varphi \circ m : 2^{\mathbb{N}} \to [0, \infty]$, $\varphi \circ m(A) = \varphi(m(A))$, is a monotone measure. Therefore, for any $\mathbf{x} \in \mathcal{S}$, the Sugeno integral $\mathbf{Su}(\varphi \circ m, \text{id})$ is well defined. For $\alpha \in]0, \infty[$, consider $\varphi_{\alpha} : [0, \infty] \to [0, \infty]$,

$$\varphi_{\alpha}(t) = \frac{t}{\alpha}.$$

Then, modifying formula (6),

$$\mathbf{I}_{\wedge}(\varphi_{\alpha} \circ m_{\mathbf{x}}, \text{id}) = \bigvee_{i=1}^{n} \left(i \wedge \frac{x_i}{\alpha} \right).$$

Note that the above Sugeno integral need not have integer value. To ensure this, we propose the next H_{α} index, $H_{\alpha} : \mathcal{S} \to \mathbb{N}_0$,

$$H_{\alpha}(\mathbf{x}) = \bigvee_{i=1}^{n} \left\lfloor \min \left\{ \frac{x_i}{\alpha}, i \right\} \right\rfloor \tag{7}$$

(here $\lfloor \cdot \rfloor$ is the floor of a real number).

It is not difficult to check that $\lim_{\alpha \to 0^+} H_{\alpha}(\mathbf{x}) = j$ whenever $\mathbf{x} = (x_1, \ldots, x_n)$ is such that $x_j > 0$ and either $j = n$ or $x_{j+1} = 0$. However, related to the next verbal characterization of H_{α} we prefer to adopt a convention $H_0(\mathbf{x}) = n$ for any $\mathbf{x} \in \mathcal{S}, \mathbf{x} = (x_1, \ldots, x_n)$. On the other hand, obviously $\lim_{\alpha \to +\infty} H_{\alpha}(\mathbf{x}) = 0$ and $H_1(\mathbf{x}) = H(\mathbf{x})$ (i.e., H_1 is the standard h-index) for any $\mathbf{x} \in \mathcal{S}$.

Definition 2. *A scientist has index $H_{\alpha} = h$, if he or she has h papers, each having at least $\alpha \cdot h$ citations, but there are no $h + 1$ papers having each at least $\alpha \cdot (h + 1)$ citations.*

Considering formula (7), we see that $H_\alpha(\mathbf{x}) = k$ whenever the rectangle $]0, k] \times [0, \alpha \cdot k]$ is the greatest rectangle of type $]0, i] \times [0, \alpha \cdot i]$, $i \in \{0, 1, \ldots, n\}$, contained in the endograph of the function $f_\mathbf{x}$ given by $f_\mathbf{x}(t) = x_i$ for any $t \in]i - 1, i]$. (see Fig. 3). For this interpretation of H_α one can compare the approach discussed in [4]. The next figure is related to the record $\mathbf{x} = (10, 8, 7, 6, 3, 1, 0, 0)$, and $\alpha = 0.5$ is considered. Hence $H_{0.5}(\mathbf{x}) = 5$.

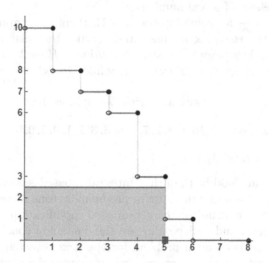

Fig. 3. Illustration of formula (7) for $H_{0.5}(\mathbf{x}) = 5$

Example 3. Consider, as above, $\mathbf{x} = (10, 8, 7, 6, 3, 1, 0, 0)$. Then

$$H_\alpha(\mathbf{x}) = \begin{cases} 0 & \text{if } \alpha > 10, \\ 1 & \text{if } \alpha \in]4, 10], \\ 2 & \text{if } \alpha \in]\frac{7}{3}, 4], \\ 3 & \text{if } \alpha \in]\frac{6}{5}, \frac{7}{3}], \\ 4 & \text{if } \alpha \in]\frac{3}{5}, \frac{6}{5}], \\ 5 & \text{if } \alpha \in]\frac{1}{6}, \frac{3}{5}], \\ 6 & \text{if } \alpha \in]0, \frac{1}{6}], \\ 8 & \text{if } \alpha = 0. \end{cases}$$

Remark 1. Note that the above mentioned formula $\mathbf{I}_\wedge(\varphi_\alpha \circ m_\mathbf{x}, \mathrm{id})$ is related to h_α-index (not integer valued, in general) proposed and discussed in [21]. The related graphical representation, can be found also in Gagolewski [3], see his Figure 2 left.

The index H_α is able to compensate a lower number of citations. Somewhat similarly, one can attempt to compensate a lower number of papers.

Definition 3. *A scientist has index $H^\beta = h$, if he or she has at least $\lfloor \beta \cdot h \rfloor \vee 1$ papers, each having at least h citations, but there are no $\lfloor \beta \cdot (h+1) \rfloor \vee 1$ papers having each at least $h + 1$ citations.*

Now, for $\beta \in]0, \infty[$, it holds

$$H^\beta(\mathbf{x}) = \lceil \mathbf{I}_\wedge(\varphi_\beta \circ \mu, \mathbf{x}) \rceil,$$

where $\lceil . \rceil$ is the ceiling of a real number.

Note that if $\beta = \frac{1}{m}$ for some integer $m \in \mathbb{N}$, then, to compute H^β, we have simply to repeat the record $\mathbf{x}$ m-times and reorder the related $n \cdot m$-tuple in decreasing sense, and then apply the standard h-index H to this extended record. It is not difficult to check that $H^0(\mathbf{x}) = x_1$, while $H^\infty(\mathbf{x}) = \lim\limits_{\beta \to +\infty} H^\beta(\mathbf{x}) = 0$.

Example 4. Continuing in Example 3, with $\beta = 0.5$, we have

$$\mathbf{x}^{0.5} = (10, 10, 8, 8, 7, 7, 6, 6, 3, 3, 1, 1, 0, 0, 0, 0)$$

and then $H^{0.5}(\mathbf{x}) = H(\mathbf{x}^{0.5}) = 6$.

Similarly, one can modify the other integral - based scientometric indices, modifying either the counting measure μ, or the monotone measure $m_\mathbf{x}$ induced by the record $\mathbf{x}$. For real use of these proposed modifications, an appropriate choice of parameters α and β, depending on the domain of considered researcher X, is necessary. This is an open problem for the next research in scientometry. Observe that several discussions concerning a quantitative distinction of some particular scientific domains were done and exemplified, e.g., in [7,14,15], etc. As an example, we add now Table 1 due to Podlubny [14].

Table 1. Comparison of the numbers of citations in different fields of science. Based on the data from Science and Engineering Indicators 2004. National Science Foundation, May 04, 2004, for the period 1992–2001, see [14,15].

Field	Average ratio of citation number to the number of citations in Mathematics
Clinical medicine	78
Biomedical research	78
Biology	8
Chemistry	15
Physics	19
Earth/space sciences	9
Engineering/technology	5
Mathematics	1
Social/behavioral sciences	13

5 Concluding Remarks

We have recalled some integral representation of well known scientometric indices. The first approach, based on the counting measure μ, was proposed already by Torra and Narukawa [20]. We have proposed a new alternative representation, where a monotone measure $m_{\mathbf{x}}$ is derived from the record $\mathbf{x}$ of a scientist X (and the integrand is identity function). More, we have proposed some new integral - based scientometric indices, e.g., those based on $\mathbf{I}_{(k)}$ integrals. Based on different performances (average number of papers and/or citations) in different scientometric domains, there is a need to modify the considered indices accordingly to the domain of considered researcher X. We have proposed two approaches allowing such a compensation, thus expanding the earlier work of Iglesias and Pecharromán [7]. As an open problem, an appropriate choice of modifiers α and β remains for the further study.

References

1. Choquet, G.: Theory of capacities. Ann. Inst. Fourier **5**, 131–295 (1953/1954)
2. Even, Y., Lehrer, E.: Decomposition-integral: unifying choquet and the concave integrals. Econ. Theory **56**(1), 1–26 (2014)
3. Gagolewski, M.: A remark on limit properties of generalized h- and g-indices. J. Inf. **3**, 367–368 (2009)
4. Gagolewski, M., Grzegorzewski, P.: A geometric approach to the construction of scientific impact indices. Scientometrics **81**(3), 617–634 (2009)
5. Gagolewski, M., Mesiar, R.: Monotone measures and universal integrals in a uniform framework for the scientific impact assessment problem. Inf. Sci. **263**, 166–174 (2014)
6. Hirsch, J.E.: An index to quantify an individual's scientific research output. Proc. Nat. Acad. Sci. **102**(4), 16569–16572 (2005). https://doi.org/10.1073/pnas.0507655102
7. Iglesias, J.E., Pecharromán, C.: Scaling the *h*-index for different scientific ISI fields. Scientometrics **73**(3), 303–320 (2007)
8. Klement, E.P., Mesiar, R., Pap, E.: A universal integral as common frame for Choquet and Sugeno integral. IEEE Trans. Fuzzy Syst. **18**, 178–187 (2010)
9. Klement, E.P., Mesiar, R., Pap, E.: Quasi- and pseudo-inverses of monotone functions, and the construction of t-norms. Fuzzy Sets Syst. **104**(1), 3–13 (1999)
10. Kosmulski, M.: A new Hirsch-type index saves time and works equally well as the original *h*-index. ISSI Newslett. **2**(3), 4–6 (2006)
11. Kosmulski, M.: MAXPROD - a new index for assessment of the scientific output of an individual, and a comparison with the *h* index. Cybermetrics **11**(1), paper 5 (2007)
12. Mesiar, R., Gagolewski, M.: H-index and other Sugeno integrals: some defects and their compensation. IEEE Trans. Fuzzy Syst. **24**(6), 1668–1672 (2016)
13. Mesiar, R., Stupňanová, A.: Decomposition integrals. Int. J. Approx. Reason. **54**, 1252–1259 (2013)
14. Podlubny, I.: Comparison of scientific impact expressed by the number of citations in different field of science. Scientometrics **64**(1), 95–99 (2005)

15. Podlubny, I., Kassayova, K.: Towards a better list of citation superstars: compiling a multidisciplinary list of highly cited researchers. Res. Eval. **15**(3), 154–162 (2006)
16. Ralescu, D., Adams, G.: The fuzzy integral. J. Math. Anal. Appl. **75**, 562–570 (1980)
17. Shilkret, N.: Maxitive measure and integration. Indag. Math. **33**, 109–116 (1971)
18. Struk, P.: Extremal fuzzy integrals. Soft Comput. **10**(6), 502–505 (2006)
19. Sugeno, M.: Theory of fuzzy integrals and its applications. Ph.D. thesis, Tokyo Institute of Technology (1974)
20. Torra, V., Narukawa, Y.: The h-index and the number of citations: two fuzzy integrals. IEEE Trans. Fuzzy Syst. **16**(3), 795–797 (2008)
21. van Eck, N.J., Waltman, L.: Generalizing the h- and g-indices. J. Inf. **2**, 263–271 (2008)

Normalization of Multiple Efficiency Intervals by Interval Data Envelopment Analysis from Different Frameworks

Tomoe Entani[1]([✉]) [ID] and Miho Isobe[2]

[1] University of Hyogo, Kobe, Japan
entani@ai.u-hyogo.ac.jp
[2] Shinshu University, Matsumoto, Japan
isobe@shinshu-u.ac.jp

Abstract. This paper proposes the method to derive the inner evaluation of a unit in a group of units in the sense of efficiency from the multiple perspectives. It aims to compare the multiple evaluations of a unit rather than the evaluations of the units. For a comprehensive analysis of a unit, its evaluations from the multiple perspectives are necessary, although it is not easy for us to compare them each other. Even from a specific perspective, we can have various evaluations depending on the viewpoints for the unit. In order to tackle these issues, first, we measure the efficiency of a unit from each perspective. We denote it as an efficiency interval considering various viewpoints based on interval data envelopment analysis. Then, we normalize the obtained multiple efficiency intervals with respect to the perspectives so that they can be compared. The normalized efficiency intervals are useful to know the characteristic of the unit in detail instead to rank the units.

Keywords: Interval data envelopment analysis · Efficiency ·
Interval probability · Normalization

1 Introduction

Digitalization of society makes us easier to collect various data than before. When we start looking for the data, we often collect the data which, we think, seems to be a little related to our purpose and then may have an inclusive data set. Hence, we may be able to analyze the collected data from various perspectives for our purpose other than one assumed at the beginning. It is sure that such a new analysis, in addition to the predetermined perspective analysis, is useful to confirm our purpose. Moreover, we prefer considering various perspectives into account for the better analysis rather than focusing on a specific perspective. Although the analysis from the multiple perspectives could derive the comprehensive conclusion than that from a specific one, in practice it is often difficult for the decision maker to compare multiple results in different frameworks of the perspectives and to derive his or her conclusion.

© Springer Nature Switzerland AG 2019
H. Seki et al. (Eds.): IUKM 2019, LNAI 11471, pp. 13–25, 2019.
https://doi.org/10.1007/978-3-030-14815-7_2

For instance, the approaches in multicriteria decision analysis (MCDA) are used to identify a preferred alternative, to rank the alternatives in decreasing order of preference, or to classify the alternatives into a small number of categories [10]. Most of them approach problems as if an undecided decision maker has to decide which course of action she or he should take. However, decision makers view the formal analysis as a kind of window dressing since they often already have something in their minds. Even if the decision maker has already decided what to do, there could be a legitimate purpose of doing careful analyses [8]. The decision maker might want to have the psychological comfort to secure and corroborate his or her decision. The formal analysis might help the communication process to convince the others. Furthermore, there is always the possibility these analyses will uncover new insights that result in a different alternative - one that is perceived as better than the original. The decision makers tend to be able to make up their decisions before or without any formal analyses, although they are welcome some supports by these analyses and also new additional insights. In this sense, the technology required in the analysis is not to find the best alternative but could be enough if it explains the alternatives from various perspectives in an organized manner. The analysis could be used to help and justify the decision maker's conclusion. Therefore, in this study, we aim to derive inner evaluation of the alternatives from the broadly collected data set by focusing more on each alternative rather than on comparing the alternatives.

As a tool to evaluate the alternatives, we use Data Envelopment Analysis (DEA). It is one of the non-parametric techniques and evaluates the relative efficiency of Decision Making Units (DMUs) [2]. In DEA, a DMU is considered as a system producing outputs from the inputs. The efficiency of the DMU is measured relative to the other DMUs which are also the systems with the same input and output terms. The DMU is more efficient if it produces more outputs from fewer inputs. DEA concerns with evaluations of performance and especially with evaluating the activities of organizations such as business firms, government agencies, hospitals, educational institutions, and so on [4]. The approaches other than DEA may have some difficulties of evaluating the activities with non-commensurable units since they have the complex and unknown nature of the relations between the multiple inputs and multiple outputs. Therefore, DEA is applicable to the performance evaluations in a broad sense, for instance, non-profitable activities and self-improvement activities.

In the efficiency literature, the efficiency of a unit is measured by a production function or a transformation function. In parametric methods, a production function is derivative so that it is straightforward and unambiguous. On the contrary, in non-parametric methods, such as DEA, it is no need to impose a specific function theoretically. It is an advantage not to require specification of function. In DEA, the production function is peculiar to the evaluated DMU and is determined from the optimistic viewpoint of the DMU. In other words, the efficiency value of the DMU is maximized under the condition that those of all DMUs are less than one for the normalization. The normalization has reconsidered so as to standardize the data with a different range and the best and worst normalized efficiency have been proposed by the relative positions from the best and worst DMUs, respectively [9]. The conventional DEA with precise data is generalized into that with uncertain data, and the efficiency score is defined so as to increase monotonically with uncertainty [5]. Instead of data uncertainty, Interval DEA concerns various viewpoints of the evaluation and defines the efficiency interval from the pessimistic viewpoint to the optimistic one [6]. The upper bound of the efficiency interval is the conventional efficiency value by DEA and its lower bound is obtained by minimizing the efficiency value. The efficiency interval in Interval DEA is obtained in the framework that the maximum efficiency value of those of all the DMUs is one for the normalization. Since it represents all the possible efficiency values of the DMU relative to the other DMUs, the decision maker understands the DMU comprehensively from a broad perspective. On the other hand, the possibility denoted as the width of the efficiency interval reflect the uncertainty of the evaluation and makes it difficult to order the DMUs. However, a ranking of the DMUs is not our primal goal. We focus more on analyzing each DMU with its composition than its rank in the group. Therefore, we use Interval DEA, which gives us comprehensive information on each DMU, to measure its efficiency from a perspective.

Considering multiple perspectives, each DMU has multiple efficiency intervals from the perspectives. The input and output terms selected from the inclusive data set depend on the perspectives so that each efficiency interval is obtained in a different framework. Our interest is in the inner evaluation of the DMU, such that its evaluation from the perspective is better or worse than those from other perspectives. Therefore, in order to compare the multiple efficiency intervals of a DMU, they are normalized. The multi-objective DEA approach has been proposed based on a variety of the projections of inefficient DMU to the production function [7] and it has been applied to the eco-efficiency problem [1]. On the other hand, the multiple evaluation contexts corresponding to production functions have been considered [3]. They include the methods to obtain the overall efficiency with respect to multiple objects. In the inner evaluation sense, we do not distinguish between two DMUs which are the best and the worst among all the DMUs from all the perspectives, since both DMUs are equally good from any perspective. For the normalization, the interval probability [11], which is one of the extensions of the conventional probability, is used. The interval probability is defined by two kinds of inequalities which make the sum of the values one and reduce the redundancy from the intervals. In other words, any value in the interval is necessary to make the sum of the crisp values in the other intervals be one. In order to generalize classic probability, the system of axiom for interval probability has been described [12].

This paper outlines as follows. In Sect. 2, we review Interval DEA from a single perspective. The input and output data corresponding to the perspective is selected from the inclusive data set and the efficiency intervals of all DMUs are obtained in the framework of the perspective. It is difficult for a decision maker to compare such efficiency intervals of a DMU in different frameworks. Hence, in Sect. 3, they are normalized to know the inner evaluation of the DMU and its characteristic is represented with the normalized efficiency intervals. Then, we show the numerical example in Sect. 4 to illustrate the proposed method and draw the conclusion in Sect. 5.

2 Interval Data Envelopment Analysis

Data Envelopment Analysis (DEA) is a nonparametric technique for measuring and evaluating the relative efficiencies of decision making units (DMUs) with common input and output terms. In DEA, the efficiency of a DMU is defined as the ratio of the weighted sum of outputs to that of inputs. In other words, the DMU which produces greater outputs from the smaller inputs is better than the others. Assume n DMUs. The problem to obtain the efficiency of DMU_o is formulated as the following LP problem [2].

$$
\begin{aligned}
\theta_o^* = \max_{v_o, u_o} \ & u_o^t y_o, \\
s.t. \ & v_o^t x_o = 1, \\
& u_o^t y_k \leq v_o^t x_k, k \in \{1, \ldots, n\}, \\
& u_o, v_o \geq 0,
\end{aligned}
\tag{1}
$$

where x_k and y_k are the given input and output vectors of DMU_k, respectively, and the variables are the input and output weight vectors, v_o and u_o, respectively. By the objective function with the first constraint, the efficiency of DMU_o, $u_o^t y_o / v_o^t x_o$, is maximized. Because of the maximization, this efficiency value is obtained from the most optimistic viewpoint for DMU_o. On the other hand, DMU_o can be evaluated from the pessimistic viewpoint and such a pessimistic efficiency value is explained later. DMUs other than DMU_o are taken into considerations by the second kind of constraints, where the maximum efficiency of all units with the weights is 1, i.e., the weighted sum of outputs is less than that of inputs with respect to all the DMUs. The obtained efficiency θ_o^* is an optimistic evaluation for DMU_o considering the other DMUs. The multiple inputs or outputs are synthesized into a pseudo input or output, respectively, although each input or output contribution to the pseudo one is not fixed. In this sense, DEA is a non-parametric technique so that the production function depends on the evaluated DMU. The optimal input and output weight vectors, v_o and u_o, depend on DMU_o and such a peculiar weight makes it possible to evaluate each DMU from its optimistic viewpoint.

On the other hand, there are the different weights peculiar to DMU_o from its pessimistic viewpoint. Then, the possible efficiency from various viewpoints from the optimistic one to the pessimistic one is denoted as efficiency interval and it could give us the observation on a DMU from a broad view. In interval DEA [6], the efficiency value by (1) is redefined as follows.

$$
\theta_o^* = \max_{v_o, u_o} \frac{\frac{u_o^t y_o}{v_o^t x_o}}{\max_k \frac{u_o^t y_k}{v_o^t x_k}},
\tag{2}
$$

$$
s.t. \ u_o, v_o \geq 0,
$$

where the relative efficiency of DMU_o comparing to the maximum of those of all the other DMUs is maximized. It is transformed into LP problem (1) by constraining the denominator is less than one and then by constraining the new denominator is one. Namely, it derives the optimistic evaluation for DMU_o in the same concept as that of (1). Instead of maximizing, the pessimistic efficiency value is defined as follows in Interval DEA [6].

$$\theta_{o*} = \min_{v_o, u_o} \frac{\frac{u_o^t y_o}{v_o^t x_o}}{\max_k \frac{u_o^t y_k}{v_o^t x_k}},$$
$$\text{s.t. } u_o, v_o \geq 0, \tag{3}$$

where the relative efficiency is minimized. The objective function value is obtained as

$$\theta_{o*} = \min_k \theta_o^k, \tag{4}$$

where

$$\theta_o^k = \min_{v_o, u_o} \frac{u_o^t y_o}{v_o^t x_o},$$
$$\text{s.t. } \frac{u_o^t y_k}{v_o^t x_k} = 1, \tag{5}$$
$$u_o, v_o \geq 0.$$

It is transformed into an LP problem by constraining the denominator of the objective function of (5) one in a similar way as transforming (2) to (1). The pessimistic efficiency θ_{o*} is obtained by solving at most n LP problems of (5), $k \in \{1, \ldots, n\}$. Thus, the efficiency interval of DMU_o, $[\theta_{o*}, \theta_o^*]$, with the input and output terms for the predetermined perspective is obtained by solving at most $(n + 1)$ LP problems: (1) for its upper bound and (5) for its lower bound. The efficiency intervals of all the DMUs are normalized so as to be less than one. They are obtained in the framework where the maximum efficiency in terms of the inputs and outputs of all the DMUs is one. The efficiency interval is comprehensive since it represents all the possible efficiency values from the pessimistic viewpoint for the DMU to the optimistic one.

3   Normalization of Efficiency Intervals

We often have an inclusive date set of a group of DMUs for their efficiency evaluations. For instance, when we are evaluating banks, we probably collect the numbers of their branches, employees, and customers, the prices of their net profits, deposits, and loans, and so on. Accordingly, in order to understand each bank in detail, we take several perspectives in efficiency evaluation such as technical efficiency, price efficiency, loan efficiency, and customer efficiency into consideration. In other words, the characteristic of a bank is represented with those efficiency values such that the bank is better from a technical perspective than a customer perspective. For the technical efficiency, the input terms are the number of branches and employees and the output terms the numbers of customers and the prices of deposits and loans. On the other hand, as for customer efficiency, the number of customers can be the input term, instead of the output term, and the output terms are the prices of deposits and loans. In this way, which kind of data in the inconclusive data set should be used depends on what kind of efficiency we are obtaining. In other words, each collected data can be used as the input or output term, or may not be used. As a result, corresponding to the evaluation perspectives, a DMU has multiple efficiency values obtained from some parts of a whole data. Moreover, each efficiency is obtained as an interval by Interval DEA in the former section. Namely, each of the efficiency intervals is relative to the other DMUs in the framework of each perspective. Then, we become interested in the relation among the efficiency intervals from various perspectives to understand a DMU. In other words, the issue is to compare the multiple perspectives of the DMU rather than to compare with the other DMUs from a specific perspective. Therefore, the next step is to compare multiple efficiency intervals of a DMU from various perspectives, each of which is obtained in the specific framework by comparing to all the DMUs.

The interval vector $([l_1, u_1], \ldots, [l_m, u_m])$ is normalized, if and only if

$$\sum_{i \neq j} l_i + u_j \leq 1, j \in \{1, \ldots, m\},$$
$$\sum_{i \neq j} u_i + l_j \geq 1, j \in \{1, \ldots, m\}. \tag{6}$$

These conditions are obtained in the study of interval probability [11]. In case of $l_i = u_i, \forall i$, they are equal to $\sum_i u_i = \sum_i l_i = 1$, which is the normalization of the real values. Hence, these conditions in (6) are one of the extensions of the normalization of the real numbers into that of the intervals. For any crisp value in an interval, $c_{i'} \in [l_{i'}, u_{i'}]$, there exist the crisp vales in the other intervals whose sum is 1, $c_{i'} + \sum_{i \neq i'} c_i = 1$, where $c_i \in [l_i, u_i], \forall i \neq i'$. In other words, the redundancy in each interval to make the sum be one is excluded.

Denote the number of the perspectives for efficiency is m and the efficiency interval of DMU_k on perspective $i \in \{1, \ldots, m\}$ as $[\theta_{ki*}, \theta_{ki}^*]$ so that there are m efficiency intervals of each DMU. In order to obtain the inner evaluation of a DMU, its multiple efficiency intervals need to be normalized since each is obtained dependently on the perspectives.

Assume two DMUs evaluated from three perspectives, $m = 3$, and three efficiency intervals of DMU_A and DMU_B as ([0.9,1], [0.9,1.0], [0.8,0.7]) and ([0.3,0.4], [0.3,0.4], [0.1,0.2]), respectively. Both DMUs look to be the worst from the third perspective among three perspectives, although DMU_A is superior to DMU_B from all the perspectives. From the viewpoint of the inner evaluation, which this paper concerns, the former finding of the relative evaluations among the perspectives of each DMU, is suitable. It is noted that the latter finding of the relative evaluations among the DMUs is necessary in case of ordering DMUs. When we give some feedback to the DMU for its improvement, there are also two approaches. One is followed by the latter finding and to encourage the DMU to reach the superior DMU. For DMU_B, DMU_A is one of the role models, while the better one, DMU_A, cannot find its role model. The other feedback is followed by the former findings and to show the DMU its strong or weak points by understanding its composition. Both DMUs find their weak point, which is the third perspective. Besides he worse one, DMU_B, the better one, DMU_A finds the weak point comparing its three efficiency values. However, it is noted that three efficiency intervals of each DMU are obtained in the different frameworks. This paper concerns the inner evaluation of each DMU instead of the ranking of the DMUs. We focus more on comparing the efficiency intervals of a DMU than comparing the efficiency intervals of all DMUs. Based on the interval normalization (6), the problem for the normalization of efficiency values of DMU_k is formulated as follows.

$$
\begin{aligned}
&\min \sum_i d_i - \varepsilon\alpha \\
\text{s.t. } &\sum_{i\neq j}(\alpha\theta_{ki*} + d_i) + (\alpha\theta_k^* - d_j) \leq 1, \forall j \in \{1,\ldots,m\} \\
&\sum_{i\neq j}(\alpha\theta_{ki}^* - d_i) + (\alpha\theta_{kj*} + d_j) \geq 1, \forall j \in \{1,\ldots,m\} \\
&\alpha \leq 1, \\
&d_i \geq 0, \forall i \in \{1,\ldots,m\},
\end{aligned} \tag{7}
$$

where ε is a small positive number and the efficiency intervals $[\theta_{ki*}, \theta_{ki}^*], \forall i$ by (1) and (5) are transformed into the normalized intervals $[\alpha\theta_{ki*}+d_i, \alpha\theta_{ki}^* - d_i], \forall i$ and the variables are α and $d_i, \forall i$. The upper and lower bounds of the efficiency intervals are reduced by α and then narrowed down by $d_i, \forall i$. It is noted that it is no need to solve (7) when $\sum_i \theta_{ki}^* - \max_j(\theta_{kj}^* - \theta_{kj*}) \leq \sum_i \theta_{ki*} + \max_j(\theta_{kj}^* - \theta_{kj*})$. We directly obtain $\alpha = \sum_i \theta_{ki*} + \max_j(\theta_{kj}^* - \theta_{kj*})$ and $d_i = 0, \forall i$.

By (7), the normalized efficiency intervals of DMU_A and DMU_B in the previous example are ([0.33,0.37], [0.33,0.37], [0.26,0.30]), and those of DMU_B are ([0.33,0.44], [0.33,0.44], [0.11,0.22]). By the normalization, both DMUs are in the same scale measurement. It is necessary for the inner evaluation to put away the effect by the absolute values such that DMU_A is superior to DMU_B. It is apparent that the weak point of both DMUs is the third perspective. Moreover, we find the degree of the weakness of the third perspective such that DMU_B is weaker than DMU_A, when we compare their normalized efficiency intervals.

The input and output terms correspond to the perspective of the efficiency evaluation. In the case of multiple perspectives, a DMU has multiple efficiency intervals in different frameworks. Therefore, normalizing the efficiency intervals makes us possible to compare the multiple perspectives with each other. Namely, the characteristic of the DMU is represented with the normalized efficiency intervals. Such an inner evaluation is different from the relative evaluation of the DMU by comparing it to the other DMUs and is helpful to understand the DMU in detail.

4   Numerical Example

We had 12 German texts written by non-native speakers, who were the students studying German for two to three years. They wrote their opinions about the media dependence of Japanese youth. For instance, the shortest text, text 7, is as follows: *Meiner Meinung nach ist der Medienabhängigkeit der japanischen Jugendlichen notwendig. Die Medien geben leicht uns viele Informationen und*

die interessiert uns. Wir leben in den viele Informationen, dass der Medien-
abhängigkeit der japanischen Jugendlichen notwendig ist. Die Hauptsache ist,
dass wir alle Informationen nicht müssen glauben. Wir müssen unsere Meinung
haben, indem wir andere Medien vergleichen.

The purpose was to evaluate the writing skill of 12 students from some parts
of the speech of these written texts. The writing skills could be evaluated from
various perspectives and we used four of them such as Vocabulary, and Subject, and Predicate, and Vorfeld (German prefield) based on a German expert's
knowledge and the structure is illustrated in Fig. 1. For instance, from the Vocabulary perspective, the numbers of verbs, nouns and so on with respect to the
numbers of the words in the text are considered. In other words, in Interval
DEA manner, the input from Vocabulary perspective is the number of words
and the outputs are the numbers of five parts of speech. Nouns have also used
the evaluation from Predicate perspective, which corresponds to the numbers
of sentences. Both noun and pronoun group are the outputs from Subject and
Predicate perspectives although the other outputs are different. DEA is applicable to such activity as writing a text and evaluates the writing skill from such a
perspective as Vocabulary. In addition, the perspectives to evaluate writing skill
need not be general, so the experts could set them based on their knowledge and
the types of writing beforehand.

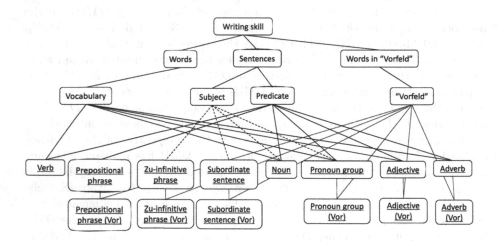

Fig. 1. Four perspectives to evaluate writing skills from written texts

In Table 1, the numbers of words, sentences, and parts of speech of 12 texts
are shown in the shortest order. It is natural that the longer the text is, the more
the sentence and each part of the speech in it become. It is noted that Vorfeld is
often a part of a sentence. The upper and lower parts are the inputs and outputs,
respectively, for the efficiency from four perspectives. The right four columns of
Table 1 show the parts of speech are marked corresponding to the perspectives
to evaluate the writing skill.

Table 1. Numbers of words, sentences, and parts of speech in whole text and "Vorfeld"

Text no.	7	8	5	2	1	6	3	4	10	11	9	12	Voc.	Sub.	Pre.	Vor.
Words	55	56	66	67	77	79	86	87	99	105	106	117	x			
Words (Vor.)	35	44	32	34	52	45	67	64	56	43	96	94		x	x	
Sentences	9	5	6	10	4	5	5	4	9	8	8	9				x
Verb	9	9	9	6	11	8	8	10	12	15	10	14	x		x	
Prepositional phrase	2	4	4	6	6	6	9	9	8	7	6	10			x	
Prepositional phrase (Vor.)	1	3	2	3	4	3	7	6	4	2	6	9				x
Zu-infinitive phrase	0	0	1	0	1	0	0	1	2	0	1	2		x		
Zu-infinitive phrase (Vor.)	0	0	0	0	1	0	0	0	2	0	1	2				x
Subordinate sentence	3	2	3	3	2	3	3	3	3	3	5	4		x		
Subordinate sentence (Vor.)	2	0	1	2	1	1	2	2	2	1	4	2				x
Noun	12	13	13	16	20	20	20	16	23	19	26	25	x	x	x	
Pronoun group	8	8	8	8	10	7	8	12	13	15	13	14	x	x	x	
Pronoun group (Vor.)	6	7	5	4	5	6	7	8	8	8	12	12				x
Adjective	5	4	8	8	6	7	9	6	7	15	10	9	x		x	
Adjective (Vor.)	2	2	2	3	3	3	7	4	3	5	8	6				x
Adverb	0	0	3	3	3	2	5	8	4	5	3	10	x		x	
Adverb (Vor.)	0	0	1	1	1	0	5	4	1	4	3	8				x

By (1) and (5), four kinds of efficiency intervals with respect to the perspectives are obtained and shown in Table 2. It is noted that we added 0.1 to each of the outputs in the bottom half of Table 1 to exclude 0 for a calculation. First, we focus on one of the four perspectives, Vocabulary perspective. The upper bounds of the efficiency intervals of 8 texts are one. Each of them can be efficient from its optimistic viewpoint with its strong points, which are represented as one or the combinations of input and output terms. The strong points are peculiar to the texts since Interval DEA is a non-parametric technique. However, the lower bounds of texts 7 and 8 are apparently small and those of text 6 is also smaller than those of the other texts, in addition. Since the lower bound of efficiency interval is obtained from the pessimistic viewpoints, these three texts are may be inferior to the other 5 texts. They have some apparent weak points comparing to the others as well as peculiar strong points. The evaluation of such a text that is outstanding in terms of the peculiar input and/or output terms is uncertain since these strong points boom up its evaluation and the rest boom it down. We can distinguish these three texts from the other 5 texts whose lower bounds are large enough, around 0.5. On the other hand, the text has balanced input and/or output terms comparing to the other texts, its evaluation tends to be certain. The efficiency intervals show the possible evaluations which help us to understand the texts from a broad view. They are more useful to understanding the detail of a text than the evaluation by a parametric method or from only the optimistic viewpoint.

Next, we consider all the four perspectives. As for text 4, its upper bounds of the efficiency intervals from four perspectives are one and it mentions that text 4 is possible to be the best among all the texts. However, the evaluations from Vorfeld and Vocabulary perspectives are less certain than those from the other perspectives because of their small lower bounds. As for texts 5, 6, and 9, we find that they cannot be the best from any perspectives since all the upper bounds are less than one. Moreover, the lower bounds of text 6 from all the four perspectives are apparently small. If we compare the texts at the column to rank them, we could conclude that these three texts are not superior to the others and text 6 is the worst among all the texts. It should be noted that this paper does not aim to find which text is good or bad, instead, it aims the inner evaluation of each text. In other words, we would like to know from which perspective the text is good or bad at the row. Hence, in the next step, we normalize four efficiency intervals from the perspectives of each text and compare the four each other.

Table 2. Efficiency intervals from four perspectives independently

Text no.	Vocabulary	Subject	Predicate	Vorfeld
7	[0.02,1]	[0.04,0.444]	[0.005,0.369]	[0.03,1]
8	[0.019,1]	[0.073,0.604]	[0.01,0.656]	[0.024,0.999]
5	[0.504,0.977]	[0.434,0.667]	[0.255,0.743]	[0.083,0.984]
2	[0.497,1]	[0.036,0.4]	[0.153,0.447]	[0.078,1]
1	[0.432,1]	[0.677,1]	[0.383,1]	[0.222,0.874]
6	[0.286,0.996]	[0.073,0.912]	[0.207,0.867]	[0.023,0.893]
3	[0.569,1]	[0.073,0.912]	[0.504,1]	[0.04,1]
4	[0.488,1]	[0.801,1]	[0.801,1]	[0.042,1]
10	[0.445,0.974]	[0.444,0.848]	[0.225,0.544]	[0.206,1]
11	[0.522,1]	[0.045,0.624]	[0.315,1]	[0.062,1]
9	[0.314,0.977]	[0.5,0.823]	[0.191,0.755]	[0.306,0.946]
12	[0.541,1]	[0.518,0.848]	[0.493,0.657]	[0.362,1]

By (7), four kinds of the efficiency intervals of each text are normalized and illustrated in Fig. 2. We solved 12 problems corresponding to 12 texts and as for 11 texts except text 4, the optimal values are $d_i = 0, \forall i$. In Fig. 2, the texts are ordered from the small number of the words to the large. The orders by the numbers of the sentences and words in Vorfeld are almost the same. These three kinds of numbers are the inputs to evaluate the writing skill. In Fig. 2, four lines of each text represent its inner evaluation and each line is based on the relative evaluation of the other texts. Comparing the vertical location of four lines of the text, we find that it is good or bad from each perspective. Moreover, the widths of the lines represent the possibility of the evaluations from the optimistic viewpoint to the pessimistic one. It is natural that some evaluations become uncertain from a broad view. The experts try to consider various viewpoints since they know it is necessary for the comprehensive evaluation, while they sometimes ignore some of them if it seems to be too difficult. The proposed method resulted in Fig. 2 shows the experts such a whole inner evaluation of each text numerically. From the left to the right, the widths of the normalized intervals become smaller, which mentions that the longer the text becomes the more precise the writing skill can be measured. It is reasonable that the evaluations of the writing skills are more varied as the text becomes shorter. Such a short text does not give enough information for the evaluation and can be evaluated both good and bad. The wide ranges increase and decrease the upper and lower bounds of the intervals, respectively, since four efficiency intervals are normalized so that the sum of the values in the intervals is one. Both of two shortest ones, texts 7 and 8, are better from Vocabulary and Vorfeld perspectives than from Subject and Predicate perspectives. It seems to mention that low writing skills of these perspectives tend to cause a short text.

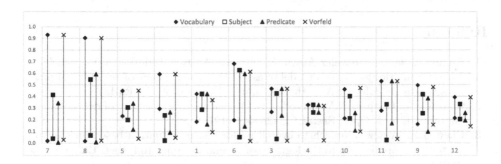

Fig. 2. Normalized efficiency intervals with respect to each text

As for feedback, we find that texts 1, 3, and 4 are equally good from all the four perspectives since their four lines are similar. Among these three texts, text 4, whose upper bounds of the efficiency intervals in Table 2 are 1, is located at the lowest because of less uncertain evaluations of Subject and Predicate perspectives. In other words, even from the pessimistic viewpoint, text 4 cannot be evaluated as bad from these two perspectives but can be evaluated as bad a little from the Vorfeld perspective. These findings help to understand the composition of text 4 in detail and could be useful feedback to the student of text 4. As for text 3, we may advise the student to rewrite Subjective and Vorfeld since the evaluations from these perspectives can be worse than the other perspectives. On the other hand, texts 10, 11, and 12 are bad from one of the four perspectives. For instance, text 12 is worse from Predicate perspective than from the other perspectives. Then, the feedback of their worst perspectives can be useful information to improve writing skill.

5   Conclusion

In order to understand a DMU comprehensively, we proposed the method to derive the normalized efficiency intervals from multiple perspectives. The efficiency interval from each perspective is relative to those of the other DMUs. It represents all the possible efficiency values from the pessimistic viewpoint to the optimistic one. A DMU has multiple efficiency intervals from the perspectives, each of which has an indigenous framework with respect to the input and output terms. Therefore, they are normalized so as to be compared to each other. The characteristic of a DMU is represented as the normalized efficiency intervals from the perspectives. It is useful for the decision maker to understand the DMU in detail. It is not easy for experts to simultaneously consider various viewpoints and various DMUs even from a single perspective. Furthermore, it is difficult to compare the multiple efficiency intervals from the perspectives in different frameworks. Hence, two advantages of the proposed method are summarized as follows: the consideration of various viewpoints and various DMUs for a DMU and the comparability of the efficiency intervals from various perspectives.

References

1. Angulo-Meza, L., González-Araya, M., Iriarte, A., Rebolledo-Leiva, R., de Mello, J.C.S.: A multiobjective dea model to assess the eco-efficiency of agricultural practices within the CF+DEA method. Comput. Electron. Agric. (2018)
2. Charnes, A., Cooper, W.W., Rhodes, E.: Measuring the efficiency of decision making units. Eur. J. Oper. Res. 2(6), 429–444 (1978)
3. Chen, J.X.: A new approach to overall performance evaluation based on multiple contexts: an application to the logistics of China. Comput. Ind. Eng. 122, 170–180 (2018)
4. Cooper, W.W., Seiford, L., Tone, K.: Data Envelopment Analysis: A Comprehensive Text with Models, Applications, References and DEA-Solver Software. Springer, New York (2007). https://doi.org/10.1007/978-0-387-45283-8
5. Ehrgott, M., Holder, A., Nohadani, O.: Uncertain data envelopment analysis. Eur. J. Oper. Res. 268(1), 231–242 (2018)
6. Entani, T., Maeda, Y.: HideoTanaka: dual models of interval DEA and its extension to interval data. Eur. J. Oper. Res. 136(1), 32–45 (2002)
7. Estellita Lins, M.P., Angulo-Meza, L., Moreira Da Silva, A.C.: A multi-objective approach to determine alternative targets in data envelopment analysis. J. Oper. Res. Soc. 55(10), 1090–1101 (2004)
8. Keeney, R.L., Raiffa, H.: Decisions with Multiple Objectives: Preferences and Value Trade-Offs. Cambridge University Press, New York (1993)

9. Liu, W., Wang, Y.M.: Ranking DMUs by using the upper and lower bounds of the normalized efficiency in data envelopment analysis. Comput. Ind. Eng. **125**, 135–143 (2018)
10. Ramanathan, R.: An Introduction to Data Envelopment Analysis: A Tool for Performance Measurement. Saga Publications, New Delhi (2003)
11. Tanaka, H., Sugihara, K., Maeda, Y.: Non-additive measures by interval probability functions. Inf. Sci. **164**, 209–227 (2004)
12. Weichselberger, K.: The theory of interval-probability as a unifying concept for uncertainty. Int. J. Approx. Reason. **24**(2), 149–170 (2000)

Overtime Assignment and Job Satisfaction in Noise-Safe Job Rotation Scheduling

Pavinee Rerkjirattikal[ID] and Sun Olapiriyakul[(⊠)][ID]

Department of Manufacturing Systems and Mechanical Engineering,
Sirindhorn International Institute of Technology,
Thammasat University, Pathum Thani, Thailand
suno@siit.ac.tu.th

Abstract. Under a harsh industrial environment, workforce safety, and satisfaction are key characteristics of a successful manufacturing operation. This paper proposes a job rotation approach that schedules workers based on NIOSH's noise-exposure criteria and worker satisfaction related to overtime assignments. Multi-period noise-safe job rotation models are developed to determine optimal job rotation schedules for a heterogeneous workforce with different skill levels and preferences for preferred overtime periods. Demand requirements and the effect of job rotation on process continuity are considered. A numerical example is used to validate the usefulness and effectiveness of the proposed models. The scheduling models are solved under two main objectives: minimizing the total labor cost and maximizing the minimum level of worker satisfaction. Then, a multi-objective optimization technique is applied to find the most suitable compromise solution between the two objectives. The ability of the proposed models to promote worker-satisfaction equality is discussed.

Keywords: Job rotation · Workforce scheduling · Noise exposure · Equality · Fairness · Satisfaction

1 Introduction

According to the Occupational Safety and Health Administration (OSHA), about 22 million workers in the U.S. are exposed to harmful noise levels each year [1]. Heavy industrial manufacturing processes usually generate high levels of noise from machinery. Excessive noise is the cause of headache, nervousness, stress, and speech interference among workers [2]. In the long term, repetitive exposure to a high noise level can cause noise-induced hearing loss and communication disorders in individuals [3]. The risk of hearing loss is doubled in workers who continuously worked under excessive noise level compared with those that did not [4]. In addition, the effects of repetitive noise exposure are cumulative and irreversible [5]. Job rotation is an administrative control method recommended by the National Institute of Occupational Health and Safety (NIOSH) [6] that is widely used in the manufacturing industry to mitigate workers' exposure to occupational risks. When implementing job rotation for noise exposure control, workers are rotated through tasks based on noise exposure criteria. Aside from the benefit of hazard control, proper design of a job rotation

© Springer Nature Switzerland AG 2019
H. Seki et al. (Eds.): IUKM 2019, LNAI 11471, pp. 26–37, 2019.
https://doi.org/10.1007/978-3-030-14815-7_3

schedule can help encourage job satisfaction [7] and motivation [8] of the workforce. To gain an insight into research trends and issues related to noise-safe job rotation modeling, a literature review is presented in the next section.

2 Literature Review

Job rotation schedules to prevent workers from excessive exposure to occupational noise have long been studied in the occupational safety and workforce scheduling literature. In general, noise-safe job rotation modeling is conducted using integer programming techniques, with an aim to reduce the noise exposure levels of workers to safer levels [9]. Alternatively, noise safety criteria can be included as part of the constraints, while formulating the objective function of job rotation models for process enhancements, such as improved productivity and optimal workforce size [10]. When considering productivity, the problem-solving process may involve the evaluation of worker limitations and task skill requirements due to the need to laterally transfer workers to jobs with different skill requirements. This is particularly important in the case of a heterogeneous workforce with varying levels of performance. In such a case, the ability of each worker to perform tasks can be specified [11]. Workers can also be classified into different skill levels, and each level can operate a different set of tasks [12, 13]. The classification of workers according to skill levels also enables job-rotation scheduling to be based on task preferences for employing workers with specific skill levels. As shown in a previous study, an ergonomic job rotation scheduling problem is solved to maximize the worker competency score [14]. The consideration of skills is also an important requirement in investigating the productivity-related performances of job rotation schedules, which include production cycle time [15] and overall production level [16]. Rotation frequency is another factor causing process discontinuity and productivity loss. Several research papers consider the adverse effects of job rotation by formulating noise-safe job rotation models to minimize the number of worker-location changeovers [17] or process setup time [18, 19]. In most of the aforementioned studies, the scheduling of workers is based on predefined task operation schedules. A more realistic assumption is to assume or use actual demand data. As in Niakan et al., noise-safe job rotation is scheduled according to demand requirements [20]. Their model evaluates the impacts of job rotation scheduling under the context of sustainability. The integration of actual demand data and overtime assignments in noise-safe job rotation scheduling problems is shown in a study by Rerkjirattikal et al. [21]. The inclusion of overtime shifts demonstrates the development of noise-safe job rotation schedules capable of achieving both occupational safety and production target objectives overextended work hours with a multi-period planning horizon. The inclusion of overtime helps improve the practicality of the job rotation model since overtime is generally used as a means for short-term adjustments to demand variation in the manufacturing industry.

   While extensive research efforts are devoted to improving the practicality of noise-safe job rotation models from a process performance perspective, there are a limited number of job rotation studies which attempt to investigate the impacts on worker job satisfaction. The factors affecting job satisfaction have a broad and varied interpretation, depending on the working conditions and the context in which the concept of job satisfaction is developed. Some researchers relate the perception of fairness in working conditions to job satisfaction. Adoly et al., identify that a balanced distribution of preferred work hours is key to improving job satisfaction among health care workers [22]. In an industrial setting with a harsh working environment, the distribution of workload in a workgroup is considered as the factor affecting workers' job satisfaction [23]. Aside from fairness, researchers also address worker's preference over different jobs and partners as the factor affecting job satisfaction [24].

The impacts of job rotation on worker job satisfaction can be more severe under an overtime schedule, as workers are associated with longer exposure to noise and other occupational hazards. According to a survey study, workers who work overtime are at higher risk of developing stress, fatigue, and sickness, due to insufficient recovery time [25]. Therefore, in industries with high demand fluctuation, it is crucial to investigate the impacts of job rotation scheduling and overtime assignment on both process performance and job satisfaction. By exploring the literature related to workforce scheduling and overtime assignments that focus on job satisfaction, fairness in overtime assignments is a key factor influencing worker job satisfaction, especially in health-care and service industries. The amount of overtime assigned to workers should be in line with what workers perceive to be a fair and neutral standard. As shown in a workforce scheduling study by Yang et al., workers decide their satisfaction level by comparing their overtime hours to the average overtime level of other outside firms [26]. Second, the distribution of overtime among workers with a similar job function should be fair. A previous workforce scheduling study by Agrali et al., suggests that management should focus on a worker's expectation about receiving a fair amount of overtime and day offs during holidays [27]. In this case, job satisfaction issues occur due to the disproportionate distribution of overtime during a worker's non-preferred period.

It is observed that the consideration of job satisfaction and fairness in overtime assignments has been based on the amount and distribution of overtime. In fact, the willingness of workers to work overtime during any specific time period is significantly related to their perception of the fairness of overtime and job satisfaction [28]. However, the consideration of worker preference over overtime hours and job satisfaction is still missing from the safe workforce scheduling literature. This study contributes to the existing literature by developing noise-safe job rotation models that can be used to improve worker's job satisfaction in a manufacturing context. Demand requirements and the use of overtime to fill the required operating capacity needed are considered in our study. Our proposed model is also capable of creating a multi-period job rotation plan. These factors are normally omitted by most of the previous job rotation studies.

3 Mathematical Model Formulation

In this study, three scenarios of noise-safe job rotation problems are constructed and solved based on integer programming models of a heterogeneous workforce with different skill levels and preferences on preferred overtime periods. The models used in these analyses are shown in this section of the paper. The modeling objective for the first scenario is to minimize the total labor cost. In the second scenario, a job rotation model is developed with an attempt to keep all workers equally satisfied with their overtime schedules. The modeling objective is to maximize the minimum satisfaction level of workers. In the third scenario, a multi-objective optimization model is developed and used to obtain a trade-off between the total cost and the fairness of satisfaction among workers. The planning period is 6 days. Each day has two 4-hour shifts, followed by a 2-hour overtime shift. To capture the general conditions of a manufacturing process, the following assumptions are made.

Assumptions

1. The skilled workers can perform all tasks, whereas the unskilled workers can only perform a subset of tasks at lower productivity levels.
2. Workers are able to perform tasks at higher levels of productivity (steady-state rate), when continuing to perform tasks at the same workstations for more than one consecutive shift.
3. Workers can perform one task at a time during each shift and can only be rotated to other workstations at the end of each shift.
4. Only workers who have performed tasks in both the two regular shifts are eligible to work overtime.
5. The daily noise exposure limit is 90 dBA (daily noise dose of 1.0) for an 8-hour working period and 88 dBA (daily noise dose of 0.75) for a 10-hour working period (with overtime).
6. During the planning period, the maximum number of overtime shifts each worker can obtain is set to 3.

All the parameters and indices used in this study are listed below. Then, the model formulation for the first scenario, which is the cost minimization case, is outlined.

Indices

i Workstation number ($i = 1, \dots , I$)
j Set of work shifts ($j = 1, \dots , J$), represented as morning shift (MS), afternoon shift (AS), and overtime shift (OS)
s Skill levels of workers ($s = 1, \dots , S$)
n Worker number ($n = 1, \dots , N$)
t Workday number ($t = 1, \dots , T$)

Parameters

D_i Noise level at workstation i
W_i Demand requirement of workstation i
$P_{i,s}$ Production rate of worker with skill level s, at workstation i
$SP_{i,s}$ Steady-state production rate achieved when worker with skill level s continues to perform task at workstation i over consecutive shifts
LD_s Daily wage of worker with skill level s
LO_s Overtime wage of worker with skill level s
OH Overhead cost of worker
MO Maximum allowable overtime shifts assigned to workers

Decision Variables

$X_{i,j,s,n,t}$ = 1 if worker n, with skill level s, works at workstation i, shift j, on day t; 0 otherwise

$Y_{i,s,n,t}$ = 1 if worker n, with skill level s, works both morning and afternoon shifts at workstation i; 0 otherwise

$Z_{i,s,n,t}$ = 1 if worker n, with skill level s, works both afternoon and overtime shifts at workstation i; 0 otherwise

$A_{s,n,t}$ = 1 if worker n, with skill level s, is scheduled to work on day t; 0 otherwise

$C_{s,n}$ = 1 if worker n, with skill level s, is selected under the job rotation plan; 0 otherwise

3.1 Scenario 1: Minimizing Cost

Minimize

$$\sum_{t=1}^{T} \sum_{n=1}^{N} \sum_{s=1}^{S} LD_s \cdot A_{s,n,t} + \sum_{t=1}^{T} \sum_{n=1}^{N} \sum_{s=1}^{S} \sum_{i=1}^{I} LO_s \cdot X_{i,j=OS,s,n,t} \\ + \sum_{n=1}^{N} \sum_{s=1}^{S} OH_s \cdot C_{s,n} \tag{1}$$

Subject to

$$\sum_{i=1}^{I} X_{i,j,s,n,t} \leq 1 \qquad\qquad \forall j,s,n,t \quad (2)$$

$$\sum_{i=1}^{I} \sum_{j=1}^{J} X_{i,j,s,n,t} \leq 3 \cdot A_{s,n,t} \qquad\qquad \forall s,n,t \quad (3)$$

$$\sum_{i=1}^{I} \sum_{j=1}^{J} D_i \cdot (X_{i,j=MS,AS,s,n,t} + 0.5 \cdot X_{i,j=OS,s,n,t}) \leq 1.0 - 0.25 \cdot \sum_{i=1}^{I} X_{i,j=OS,s,n,t} \qquad \forall s,n,t \quad (4)$$

$$2 \cdot \sum_{i=1}^{I} X_{i,j=OS,s,n,t} - \sum_{i=1}^{I} X_{i,j=MS,s,n,t} + \sum_{i=1}^{I} X_{i,j=AS,s,n,t} \leq 0 \qquad \forall s,n,t \quad (5)$$

$$\sum_{t=1}^{T} \sum_{n=1}^{N} \sum_{s=1}^{S} \sum_{j=1}^{J} P_{i,s} \cdot X_{i,j,s,n,t} \\ + \sum_{t=1}^{T} \sum_{n=1}^{N} \sum_{s=1}^{S} SP_{i,s} \cdot (Y_{i,s,n,t} + Z_{i,s,n,t}) \geq W_i \qquad \forall i \quad (6)$$

$$Y_{i,s,n,t} \leq \frac{X_{i,j=MS,s,n,t} + X_{i,j=AS,s,n,t}}{2} \qquad \forall i,s,n,t \quad (7)$$

$$Y_{i,s,n,t} + 1 \geq X_{i,j=MS,s,n,t} + X_{i,j=AS,s,n,t} \qquad \forall i,s,n,t \quad (8)$$

$$Z_{i,s,n,t} \leq \frac{X_{i,j=AS,s,n,t} + X_{i,j=OS,s,n,t}}{2} \qquad \forall i,s,n,t \quad (9)$$

$$Z_{i,s,n,t} + 1 \geq X_{i,j=AS,s,n,t} + X_{i,j=OS,s,n,t} \qquad \forall i,s,n,t \quad (10)$$

$$\sum_{t=1}^{T} A_{s,n,t} \leq Big M \cdot C_{s,n} \qquad \forall s,n \quad (11)$$

$$\sum_{t=1}^{T} \sum_{i=1}^{I} X_{i,j=OS,s,n,t} \leq MO \qquad\qquad \forall s, n \qquad (12)$$

Objective function (1) minimizes the total labor cost, which consists of daily wages, overtime wages, and overhead. Constraint (2) ensures that each worker is assigned to only one task during each time period. Constraint (3) keeps track of the assigned workdays of workers. Constraint (4) keeps the daily noise dose of all workers within the permissible exposure limit of 1.0. With overtime, the noise dose limit is 0.75. Constraint (5) specifies that only workers who perform both the morning and afternoon shifts are eligible to work overtime. Constraint (6) ensures that all demands are satisfied. Constraints (7) and (8) assign $Y_{i,s,n,t}$ as 1, when worker n works at the same workstation during both the morning and afternoon shifts, in workday t. Constraints (9) and (10) assign $Z_{i,s,n,t}$ as 1, when worker n works at the same workstation during the afternoon and overtime shifts, in workday t. Constraint (11) counts the number of workers to be employed under job rotation. Constraint (12) ensures that the total number of overtime shifts assigned to workers does not exceed the maximum allowable number of overtime shifts.

3.2 Scenario 2: Maximizing the Minimum Satisfaction Level

The model in Scenario 2 is developed to maximize the minimum satisfaction score of workers. The satisfaction score of each worker is evaluated based on the willingness of a worker to work overtime on a given day, which is identified on a scale from 0 to 5, with 5 representing the highest level of willingness. An example of overtime willingness values and the satisfaction score calculation is shown in Table 1. After that, the objective function and constraints are revised.

Table 1. Example of worker satisfaction score

	MON	TUE	WED	THU	FRI	SAT	
Overtime willingness	1	2	4	5	0	0	
Overtime schedule	0	1	1	0	0	0	
Satisfaction score	1(0)	+2(1)	+4(1)	+0	+0	+0	Total = 6

Parameter
$SF_{n,t}$ Satisfaction score of worker n on day t

Decision Variable
  MinScore Minimum satisfaction score

$$\text{Maximize } MinScore \qquad\qquad (13)$$

Subject to Eqs. (2–12) and,

$$MinScore \leq \sum\nolimits_{t=1}^{T} SF_{n,t} \cdot \sum\nolimits_{i=1}^{I} X_{i,j=OS,s,n,t} \qquad \forall s,n \qquad (14)$$

The objective (13) maximizes the minimum worker satisfaction score. Constraint (14) ensures that the value of *MinScore* is the minimum score among all workers.

3.3 Scenario 3: Multi-objective Optimization

A bi-objective noise-safe job rotation model is formulated to find a trade-off solution between cost and worker satisfaction. Additional parameters used in the model are defined below. Then, the model formulation is outlined.

Parameters

MinCost	Minimum total labor cost from Scenarios 1 and 2
MaxCost	Maximum total labor cost from Scenarios 1 and 2
MinScore	Minimum worker satisfaction Scenarios 1 and 2
MaxScore	Maximum worker satisfaction Scenarios 1 and 2

Decision Variables

Z	Overall performance of the solution
λ_{Cost}	Cost performance (in %) of the optimal solution from Scenario 3 relative to the upper and lower bound values, which are the costs of Scenarios 1 and 2, respectively
λ_{Score}	Satisfaction performance (in %) of the optimal solution from Scenario 3 relative to the upper and lower bound values, which are the minimum satisfaction levels of Scenarios 2 and 1, respectively

$$\text{Maximize } Z \qquad (15)$$

Subject to Eqs. (2–12, 14) and,

$$Z \leq (\lambda_{Cost}, \lambda_{Score}) \qquad (16)$$

$$\lambda_{Cost} = 1 - (Cost^* - MinCost)/(MaxCost - MinCost) \qquad (17)$$

$$\lambda_{Score} = (Score^* - MinScore)/(MaxScore - MinScore) \qquad (18)$$

The maximin technique is used in this scenario to find a compromise solution. Objective function (15) maximizes the overall performance (Z), also referred to as achievement or satisfaction level in the optimization literature. Due to Eq. (16) the objective function aims to maximize the minimum between λ_{Cost} and λ_{Score}, which are calculated based on Eqs. (17) and (18). In these equations, the cost and worker satisfaction performances are normalized to their corresponding optimal intervals, previously determined from the single-objective scenarios.

4 Numerical Example

A manufacturing system with 5 independent workstations is considered. There are 50 workers available for job rotation, with 15 skilled workers and 35 unskilled workers. The demand requirements over a 6-day planning period and noise levels of all workstations are given in Table 2. The regular and steady-state production rates of skilled and unskilled workers are also given in the table.

Table 2. Demand requirements and noise levels of workstations.

Workstation	Demand (Units)	Noise level (dBA)	Noise dose per shift	Regular production rate (Units/shift)		Steady-state production rate (Units/shift)	
				Skilled	Unskilled	Skilled	Unskilled
W1	16,000	91	0.57	130	110	195	165
W2	11,000	93	0.76	100	0	140	0
W3	12,000	82	0.16	110	0	132	0
W4	14,000	86	0.29	90	85	126	106
W5	17,600	84	0.22	130	120	170	150

Full daily wages are paid to workers, even when they are assigned to work only during a morning or afternoon shift. The daily wage, overtime wage, and overhead cost of skilled and unskilled workers are shown in Table 3. Overhead cost is incurred when a worker is selected to perform any task during the planning period. The overtime willingness values indicated by workers are used in the analysis, but are not shown in this paper.

Table 3. Direct and indirect labor costs.

Labor cost	Skilled workers	Unskilled workers
Daily wage (THB per workday)	400	300
Overtime (THB per OT shift)	150	112.5
Overhead (THB per worker)	1,500	1,500

5 Result and Discussion

In our analysis, Opensolver version 2.9.0 with the Gurobi optimizer is used as the optimization tool. Table 4 shows the optimal rotation schedule of Scenario 1, where workers are scheduled in response to the required demand at minimum cost. In Scenario 1, there are as many as 24 workers with a zero-satisfaction score.

The problem is then solved under Scenarios 2 and 3, to observe the changes in cost and satisfaction performances associated with the optimal result. A comparison of the

Table 4. Job rotation schedule of workers under Scenario 1.

	Worker no.	MON	TUE	WED	THU	FRI	SAT	Avg. DND	Satisfaction score
Skilled	1	W3, W2, -	W1, W4, -	W1, W4, -	W3, W2, -	W1, W4, -	W3, W2, -	0.882	0
	2	W2, W3, -	W2, W5, -	W3, W2, -	W2, W3, -	W1, W4, -	W2, W3, -	0.921	0
	3	W2, W3, -	W4, W1, -	W1, W4, -	W2, W3, -	W3, W1, -	W1, W4, -	0.862	0
	.	.	.	.	.	.	.	.	.
	.	.	.	.	.	.	.	.	.
	15	W4, W1, -	W2, W3, -	W1, W4, -	W2, W3, -	W2, W3, -	W1, W4, -	0.882	0
Unskilled	16	W5, W5, W5	W1, W4, -	W4, W1, -	W5, W5, W5	W1, W4, -	W4, W1, -	0.756	4
	17	W5, W5, W5	W1, W4, -	W5, W5, W5	W4, W1, -	W1, W4, -	W1, W4, -	0.756	6
	18	W1, W4, -	W5, W5, W5	W4, W1, -	W4, W1, -	W1, W4, -	W5, W5, W5	0.756	9
	.	.	.	.	.	.	.	.	.
	.	.	.	.	.	.	.	.	.
	50	W5, W5, W5	W4, W1, -	W1, W4, -	W5, W5, W5	W4, W1, -	W1, W4, -	0.756	7

Note: order of scheduling is morning shift, afternoon shift, overtime

cost and satisfaction performances of the optimal results obtained from the three scenarios is shown in Table 5. When the model is solved under Scenario 2, the average satisfaction score of workers is much higher, compared to that of Scenario 1. The lowest satisfaction score of 5 is obtained. The standard deviation of satisfaction scores also reduces significantly from 2.74 to 1.32. However, the total cost increases by about 11.5%. In Scenario 3, a compromise solution is obtained. In terms of job satisfaction, the schedule of Scenario 3 leads to better equality of satisfaction scores among the workers, compared to the other scenarios. The total cost of Scenario 3 is only 3.3% higher than that of Scenario 1. Hence, Scenario 3 is considered to be a promising solution.

Table 5. Cost and satisfaction performances of the optimal schedules.

		Scenario 1 (Min. cost)	Scenario 2 (Maximin satisfaction)	Scenario 3 (Multi-objective)
Total cost		168,038	187,388	173,623
Satisfaction score	(Min, Max)	(0, 9)	(6, 10)	(5, 8)
	Average	2.38	7.5	5.34
	Standard deviation	2.74	1.32	0.76

In Fig. 1, the horizontal axis represents workers who received the most the satisfaction score to the least, note that the numbers here do not indicate the type of workers, where the vertical axis represents the total satisfaction scores. The figure provides decision makers with an improved insight into the distribution of satisfaction scores among the worker. The figure also reveals the fact that cost and satisfaction are conflicting objectives in this study. In Scenario 1, under the cost minimization objective, there appears to be a large gap between the maximum and the minimum satisfaction scores. This means that some workers are assigned to work overtime during their preferred periods, while some are not. More proportionally distributed satisfaction scores are achieved in Scenarios 2 and 3. Decision makers can choose the model of Scenario 2, when job-satisfaction maximization is their priority. The model of Scenario 3 can be chosen when the scheduling objective is to promote the equality of worker satisfaction.

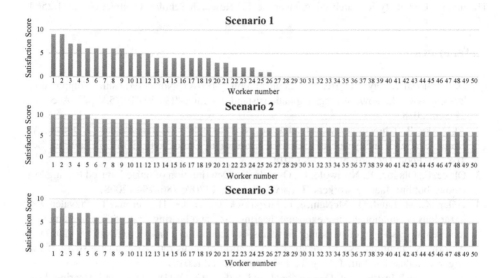

Fig. 1. Satisfaction score distribution

6 Conclusion

In labor-intensive industries, the consideration of worker safety and satisfaction is an integral component of the system design and planning. This research develops multi-period job rotation scheduling models for a heterogeneous workforce with different skill levels and preferences on preferred overtime periods. The willingness of workers to work overtime during a specific time period and its relation to job satisfaction are considered, to promote worker job satisfaction and satisfaction equality. A numerical example is used to demonstrate the usefulness of the proposed models. The general conditions of manufacturing processes are considered. This includes worker skills, task skill requirements, overtime shifts, and demand. The effect of job rotation on process continuity is taken into account.

In our analyses, the proposed model is first validated under the cost minimization scenario. The ability of the model to meet both demand and noise-exposure safety requirements is illustrated. Then, the Maximin modeling technique is used for worker job-satisfaction enhancement. After that, the proposed model is revised to account for the cost and job satisfaction objectives simultaneously, using multi-objective optimization techniques. Based on the outcomes, the proposed models can improve the job satisfaction and equality of job satisfaction among workers.

Our future studies aim to improve the practicality of the models by considering uncertainties related to demand and hazard intensity level. The need for problem solving approaches that are suitable for a more complex scheduling problem is anticipated.

Acknowledgement. The authors gratefully acknowledge the financial support provided by Thammasat University Research Fund under the TU Research Scholar, Contract No. 2/21/2560.

References

1. Occupational Safety and Health Administration (OSHA): Noise Exposure Computation. https://www.osha.gov/laws-regs/regulations/standardnumber/1910/1910.95AppA. Accessed 21 Sept 2018
2. Al-Dosky, B., Chowdhury, A., Mohammad, N., Haque, M., Manikandarajan, T., Eswar, A.: Noise level and annoyance of Industrial factories in Duhok city. IOSR J. Environ. Sci. Toxicol. Food Technol. **8**(5), 1–8 (2014)
3. Ologe, F., Olajide, T., Nwawolo, C., Oyejola, B.: Deterioration of noise-induced hearing loss among bottling factory workers. J. Laryngol. Otol. **122**(8), 786–794 (2008)
4. Feder, K., Michaud, D., McNamee, J., Fitzpatrick, E., Davies, H., Leroux, T.: Prevalence of hazardous occupational noise exposure, hearing loss, and hearing protection usage among a representative sample of working Canadians. J. Occup. Environ. Med. **59**(1), 92 (2017)
5. Anjorin, S., Jemiluyi, A., Akintayo, T.: Evaluation of industrial noise: a case study of two Nigerian industries. Eur. J. Eng. Technol. **3**(6), 59–68 (2015)
6. The National Institute of Occupational and Safety (NIOSH): Noise and Hearing Loss Prevention. https://www.cdc.gov/niosh/topics/noise/default.html. Accessed 21 Sept 2018
7. Rissén, D., Melin, B., Sandsjö, L., Dohns, I., Lundberg, U.: Psychophysiological stress reactions, trapezius muscle activity, and neck and shoulder pain among female cashiers before and after introduction of job rotation. Work Stress **16**(2), 127–137 (2002)
8. Kaymaz, K.: The effects of job rotation practices on motivation: a research on managers in the automotive organizations. Bus. Econ. Res. J. **1**(3), 69 (2010)
9. Tharmmaphornphilas, W., Green, B., Carnahan, B.J., Norman, B.A.: Applying mathematical modeling to create job rotation schedules for minimizing occupational noise exposure. AIHA J. **64**(3), 401–405 (2003)
10. Yaoyuenyong, S., Nanthavanij, S.: Hybrid procedure to determine optimal workforce without noise hazard exposure. Comput. Ind. Eng. **51**(4), 743–764 (2006)
11. Wongwien, T., Nanthavanij, S.: Ergonomic workforce scheduling under complex worker limitation and task requirements: mathematical model and approximation procedure. Songklanakarin J. Sci. Technol. **34**(5), 541–549 (2012)
12. Aryanezhad, M., Kheirkhah, A., Deljoo, V., Mirzapour Al-e-hashem, S.: Designing safe job rotation schedules based upon workers' skills. Int. J. Adv. Manuf. Technol. **41**(1–2), 193–199 (2009)
13. Deljoo, V., Mirzapour Al-e-hashem, S., Malekly, H., Bozorgi-Amiri, A., Aryanejad, M.: Applying multi objective modeling to create safe job rotation schedules based upon workers' skills and idleness. In: Computers & Industrial Engineering 2009, pp. 262–267. IEEE (2009)
14. Nanthavanij, S., Yaoyuenyong, S., Jeenanunta, C.: Heuristic approach to workforce scheduling with combined safety and productivity objective. Int. J. Ind. Eng. **17**(4), 319–333 (2010)

15. Moussavi, S., Mahdjoub, M., Grunder, O.: Reducing production cycle time by ergonomic workforce scheduling. IFAC-PapersOnLine **49**(12), 419–424 (2016)
16. Mossa, G., Boenzi, F., Digiesi, S., Mummolo, G., Romano, V.: Productivity and ergonomic risk in human based production systems: a job-rotation scheduling model. Int. J. Prod. Econ. **171**, 471–477 (2016)
17. Asawarungsaengkul, K., Nanthavanij, S.: Heuristic genetic algorithm for workforce scheduling with minimum total worker-location change over. Int. J. Ind. Eng. Theory Appl. Pract. **15**(4), 373–385 (2008)
18. Rerkjirattikarn, P., Satitanekchai, S., Olapiriyakul, S.: Safe job rotation scheduling with minimum setup time. Asia Pac. J. Sci. Technol. **22**(4), 9 (2017). APST-22-04-03
19. Rerkjirattikarn, P., Satitanekchai, S., Olapiriyakul, S.: Designing safe job rotation schedules with minimum productivity loss. J. Supply Chain. Oper. Manag. **14**(2), 48–60 (2016)
20. Niakan, F., Baboli, A., Moyaux, T., Botta-Genoulaz, V.: A bi-objective model in sustainable dynamic cell formation problem with skill-based worker assignment. J. Manuf. Syst. **38**, 46–62 (2016)
21. Rerkjirattikal, P., Kaorapapaong C., Olapiriyakul S.: Skill-based job rotation scheduling for occupational noise exposure control. In: International Conference on Artificial Life and Robotics 2018, pp. 161–165 (2018)
22. Adoly, A., Gheith, M., Nashat Fors, M.: A new formulation and solution for the nurse scheduling problem: a case study in Egypt. Alex. Eng. J. **57**, 2289–2298 (2018)
23. Rocha, M., Oliveira, J., Carravilla, M.: A constructive heuristic for staff scheduling in the glass industry. Ann. Oper. Res. **214**, 463–478 (2014)
24. Wongwien, T., Nanthavanij, S.: Priority-based ergonomic workforce scheduling for industrial workers performing hazardous jobs. J. Ind. Prod. Eng. **34**(1), 52–60 (2017)
25. Beckers, D.G., van der Linden, D., Smulders, P.G., Kompier, M.A., Taris, T.W., Geurts, S.A.: Voluntary or involuntary? Control over overtime and rewards for overtime in relation to fatigue and work satisfaction. Work Stress **22**(1), 33–50 (2008)
26. Yang, G., Tang, W., Zhao, R.: An uncertain workforce planning problem with job satisfaction. Int. J. Mach. Learn. Cybernet. **8**(5), 1681–1693 (2017)
27. Agrali, S., Taskin, Z.C., Unal, A.T.: Employee scheduling in service industries with flexible employee availability and demand. Omega **66**, 159–169 (2017)
28. Hollman, R.W.: Overtime working: employee willingness. Employee Relations **2**(5), 26–29 (1980)

Ambiguity Measures for Preference-Based Decision Viewpoints

Camilo Franco[1]($\boxtimes$), J. Tinguaro Rodríguez[2], Javier Montero[2], Daniel Gómez[3], and Ronald R. Yager[4]

[1] Department of Industrial Engineering, Andes University, Bogotá 111711, Colombia
c.franco31@uniandes.edu.co

[2] Department of Statistics and Operational Research,
Complutense University, 28040 Madrid, Spain
{jtrodrig,monty}@mat.ucm.es

[3] Statistics and Data Science, Complutense University, 28040 Madrid, Spain
dagomez@estad.ucm.es

[4] Machine Intelligence Institute, Iona College, New Rochelle, NY 10801, USA
yager@panix.com

Abstract. This paper examines the ambiguity of subjective judgments, which are represented by a system of pairwise preferences over a given set of alternatives. Such preferences are valued with respect to a set of reasons, in favor and against the alternatives, establishing a complete judgment, or *viewpoint*, on how to solve the decision problem. Hence, viewpoints entail particular decisions coming from the system of preferences, where the preference-based reasoning of a given viewpoint holds according to its *soundness* or *coherence*. Here we explore such a coherence under the frame of *ambiguity measures*, aiming at learning viewpoints with highest preference-score and minimum ambiguity. We extend existing measures of ambiguity into a multi-dimensional fuzzy setting, and suggest some future lines of research towards measuring the coherence or (ir)rationality of viewpoints, exploring the use of information measures in the context of preference learning.

Keywords: Ambiguity · Fuzziness · Rationality ·
Preference structures

1 Introduction

Any subjective decision process requires considering the formation of opinions, like preferences over a given set of possible choices. Examples of such a process can be political elections, an investment project competition, or much simpler

Supported by the Carolina Foundation (short postdoctoral research scholarship), the Government of Spain (grant TIN2015-66471-P), the Government of Madrid (grant S2013/ICE-2845, CASICAM-CM), and Complutense University research group (910149).

© Springer Nature Switzerland AG 2019
H. Seki et al. (Eds.): IUKM 2019, LNAI 11471, pp. 38–49, 2019.
https://doi.org/10.1007/978-3-030-14815-7_4

choices like deciding which product to buy. Then, based on a set of attributes, e.g. the nutritional attributes of different alternatives, referring to calories, minerals, and proteins, preferences are formed by considering the reasons in favor and against the alternatives, on how the amount of calories, minerals or proteins make us prefer and/or reject one alternative over the other. In general, the set of reasons act as *arguments* on which plan of action to take, like matching reasons one against the other, trying to balance the overall preference under a definite *viewpoint*. Hence, preferences have a key role in understanding the most suitable decision(s), and the less uncertain or *ambiguous* ones.

Traditionally, preference-based decision-making has been understood according to the theoretical (uni-polar) concept of *wants/desires*, as presented by classical decision theory (see e.g. [4]). Nonetheless, it has been suggested that decisions can be better understood according to the *pair* of explanatory concepts of *wants* and *needs* (see [10,14]), acting as drivers for decision-making. Following this line of research, the components of wants and needs can be inferred from global preferences under more general, opposite-paired semantics [11–13,17]. Besides, such a semantics also allows understanding different indecision states that can explain the choice of, actually, not making a decision [11]. In this same line, it is proposed that opposite-paired preferences allow representing the emotional meaning associated to judgments, stressing that emotion goes hand in hand with rational decision-making (see again [11]).

In this way, this study focuses on how to measure the ambiguity of preference arguments, addressing the question on how *irrational* can a decision be (see e.g. [3]). For doing so, the next section introduces the frame for subjective fuzzy preferences and the preference-aversion model. Then we present in Sect. 3 our proposal on decision viewpoints, obtaining a decision outcome and its overall ambiguity. In order to measure ambiguity, we explore in Sect. 4 the notion of ambiguity as it has been studied in decision theory literature, and extend classical ambiguity measures over multiple dimensions by fuzzy logic operators. Finally, some final comments are given for future research.

2 Decision Modeling by Fuzzy Preferences

Given a set of alternatives $\mathbb{A}$, standard preference modeling understands the preference predicate $R(a,b)$ as *"a is not worse than b"* or *"a is at least as wanted as b"* [8]. The pairwise relation representing such a predicate, namely the weak or global preference relation $R \in \{0,1\}$, can be decomposed into four distinct relations. These relations are *strict preference* P, the inverse strict preference P^{-1}, *indifference* I, and *incomparability* J, such that $P = R \cap \neg R^{-1}$, $I = R \cap R^{-1}$, and $J = \neg R \cap \neg R^{-1}$, where $\neg R = 1 - R$. Also consider that $R^d = \neg R^{-1} = (\neg R)^{-1}$. Under this classical/crisp setting, the relations I and J are assumed to be symmetrical, such that $I(a,b) = I(b,a)$ and $J(a,b) = J(b,a)$; I is assumed to be reflexive, such that $I(a,a) = 1$; J irreflexive, such that $J(a,a) = 0$; and P is assumed to be asymmetrical, such that $P(a,b)$ and $P(b,a)$ cannot hold simultaneously true.

2.1   The Standard Fuzzy Model

Following the classical/crisp preference setting, the standard preference structure $\langle P, I, J \rangle$ consists in the mutually exclusive relations P, I, J: $P \cap I = \emptyset$, $P \cap J = \emptyset$, and $J \cap I = \emptyset$, partitioning the valuation space in the following way:

$$P \cup I = R, \tag{1}$$

$$P \cup I \cup P^{-1} = R \cup R^{-1}, \tag{2}$$

$$P \cup J = R^d, \tag{3}$$

$$P \cup P^{-1} \cup I \cup J = \mathbb{A}^2. \tag{4}$$

Allowing the different relations to simultaneously co-exist with different intensities, the standard structure can be extended through fuzzy logic, affirming the existence of continuous functions $p, p^{-1}, i, j : [0,1]^2 \to [0,1]$ that maintain the classical properties Eqs. (1)–(4) as much as possible [8]. In this way, the preference relation R can be represented as a fuzzy relation, such that

$$R(a,b) = \{\langle a, b, \mu_R(a,b)\rangle | a, b \in \mathbb{A}\},$$

where $\mu_R(a,b) : \mathbb{A}^2 \to [0,1]$ is the membership function of R, measuring the degree or intensity in which the pair $(a,b) \in \mathbb{A}^2$ verifies the preference predicate represented by R.

Recalling some traditional principles of social choice theory (see e.g. [1,8]), the standard fuzzy model assumes *independence of irrelevant alternatives*, such that for every pair of alternatives $a, b \in \mathbb{A}$, the values of $P(a,b)$, $P^{-1}(a,b)$, $I(a,b)$ and $J(a,b)$ depend only on the pair (a,b), through the weak preference functions $x = \mu_R(a,b)$ and $y = \mu_R(b,a)$. So, it holds that $P(a,b) = p(x,y)$, $P^{-1}(a,b) = p(y,x)$, $I(a,b) = i(x,y)$ and $J(a,b) = j(x,y)$. Besides, other classical principles are assumed, like *monotonicity* or *positive association*, stating that functions $p(x,n(y))$, $p^{-1}(n(x),y)$, $i(x,y)$ and $j(n(x),n(y))$ are non-decreasing over both arguments, where n is a strict negation, and *symmetry*, affirming the symmetry of the functions $i(x,y)$ and $j(x,y)$ (see again [8], but also [18]).

2.2   Preference-Aversion Fuzzy Model

The standard fuzzy model can be generalized under a paired setting, aiming at clarifying the semantics for the preference predicate, and completely specify its valuation space [10,11]. Hence, two separate sources of information can be considered, representing the positive and the negative aspects of alternatives, with respect to the available criteria or attributes, serving as reasons or arguments for evaluating the verification of the preference predicate.

In this way, $R^+(a,b) = R(a,b) =$ *a is at least as wanted as b* and its inverse $R^+(b,a)$, evaluate the source of positive information, while the negative counterpart allows evaluating the negative preference predicate $R^-(a,b) =$ *a is at least as rejected as b* and its inverse $R^-(b,a)$. To simplify notation, from now on we say that $R^+(a,b) = Q(a,b)$, $R^+(b,a) = Q(b,a) = Q^{-1}(a,b)$, $R^-(a,b) = V(a,b)$ and $R^-(b,a) = V(b,a) = V^{-1}(a,b)$.

These positive and negative preference components can be aggregated into more complex structures, such as the Partial Comparability or the Preference-Aversion (P-A) structures. Following [11], the P-A structure in fact allows distinguishing between *wants* and *needs*, measuring positive aspects according to (Q, Q^{-1}), and measuring the negative ones through (V, V^{-1}).

Like a semi-dual or opposite of *preference*, the *aversion structure* can be seen as a separate, negative counterpart of the standard preference structure. In this sense, $\forall (a,b) \in \mathbb{A}^2$, the weak aversion predicates (V, V^{-1}) can be decomposed into the three relations Z, G, H, where $Z(a,b)$ holds if a is more rejected than b, $G(a,b)$ holds if a is as much as rejected as b, and $H(a,b)$ holds if a cannot be compared with b regarding their negative aspects (see e.g. [11]). Hence, the fuzzy P-A model can be defined by representing the predicates Q and V, as in

$$\mu_Q, \mu_V : \mathbb{A}^2 \to [0,1],$$

and defining fuzzy preference-aversion relations

$$p, i, j, z, g, h : [0,1]^2 \to [0,1]$$

such that [8,10]

$$P = p(\mu_Q, \mu_{Q^{-1}}) = T(\mu_Q, n(\mu_{Q^{-1}})),$$
$$I = i(\mu_Q, \mu_{Q^{-1}}) = T(\mu_Q, \mu_{Q^{-1}}),$$
$$J = j(\mu_Q, \mu_{Q^{-1}}) = T(n(\mu_Q), n(\mu_{Q^{-1}})),$$
$$Z = z(\mu_V, \mu_{V^{-1}}) = T(\mu_V, n(\mu_{V^{-1}})),$$
$$G = g(\mu_V, \mu_{V^{-1}}) = T(\mu_V, \mu_{V^{-1}}),$$
$$H = h(\mu_V, \mu_{V^{-1}}) = T(n(\mu_V), n(\mu_{V^{-1}})),$$

where T is a (conjunctive operator) t-norm, used for aggregating pairs of values of the same positive or negative nature, and i, j, g and h are symmetrical functions. Then, by means of a (disjunctive operator) t-conorm S, and following Eqs. (1)–(4), the classical properties can be formulated in fuzzy terms as

$$S(p, i) = \mu_Q, \tag{5}$$

$$S(p, i, p^{-1}) = S(\mu_Q, \mu_{Q^{-1}}), \tag{6}$$

$$S(p, j) = n(\mu_Q), \tag{7}$$

$$S(p, p^{-1}, i, j) = 1, \tag{8}$$

while the *aversion* ones are stated as

$$S(z, g) = \mu_V, \tag{9}$$

$$S(z, g, z^{-1}) = S(\mu_V, \mu_{V^{-1}}), \tag{10}$$

$$S(z, h) = n(\mu_V), \tag{11}$$

$$S(z, z^{-1}, g, h) = 1. \tag{12}$$

Some solutions for functions (p, i, j) and (z, g, h), fulfilling (5)–(8) and (9)–(12), are given by the De-Morgan triple (T^L, S^L, n) [8],

$$p(\mu_Q, \mu_{Q^{-1}}) = T^M(\mu_Q, n(\mu_{Q^{-1}})),$$

$$i(\mu_Q, \mu_{Q^{-1}}) = T^L(\mu_Q, \mu_{Q^{-1}}),$$

$$j(\mu_Q, \mu_{Q^{-1}}) = T^L(n(\mu_Q), n(\mu_{Q^{-1}})),$$

and

$$z(\mu_V, \mu_{V^{-1}}) = T^M(\mu_V, n(\mu_{V^{-1}})),$$

$$g(\mu_V, \mu_{V^{-1}}) = T^L(\mu_V, \mu_{V^{-1}}),$$

$$h(\mu_V, \mu_{V^{-1}}) = T^L(n(\mu_V), n(\mu_{V^{-1}})),$$

where n now denotes a strong negation, and $\forall x, y \in [0, 1], T^L(x, y) = \max(x + y - 1, 0), T^M = \min(x, y)$, and $S = S^L = \min(x + y, 1)$. Besides, the functions i and j are mutually exclusive, as well as g and h. Other solutions regarding t-(co)norms allow modeling p and z as strongly asymmetrical relations (by means of T^L), where the preference (5), (7)-(8), and the aversion conditions (9), (11)-(12), are satisfied, but not (6) nor (10). One last solution is given by the multiplicative De Morgan triple (T^P, S^P, N), such that $\forall x, y \in [0, 1], T^P(x, y) = x \cdot y$ and $S = S^P = x + y - x \cdot y$, where all basic relations in (p, i, j) and (z, g, h) can simultaneously co-exist, but without fulfilling (5)–(7) nor (9)–(11). This last solution only satisfies completeness (8) and (12) [19].

2.3   Modeling Decisions by Opposite Concepts

The decision problem can now be understood in terms of *opposites* [13, 17], where the positive and negative dimensions of the preference-aversion structure have to be aggregated into a unified outcome. One possibility for doing so is by inferring the behavioral components of preference, which guide the decision process. In this sense, wants and needs can be estimated from the preference statements, distinguishable under the P-A model (as shown in [11]). In this way, $W = wants$ and $D = needs$ are respectively defined with respect to the opposite poles of preference for $Q = wanting$ and for $V = rejecting$, such that

$$W = (Q \cap \neg Q^{-1}) \tag{13}$$

and

$$D = Q \cap \neg V, \tag{14}$$

where W and D are *paired concepts* in the sense of [17].

In this sense, the want-component W distinguishes a priority on only wanted alternatives, based on the positive reasons, while the need component D distinguishes a priority on wanted and non-rejected alternatives, based on both positive and negative reasons. As it has been shown in [11], the need component can be estimated from the specific preference semantics of the P-A structure, offering a decision-making criterion for a given system of preferences.

An operational example for estimating needs from preferences is given $\forall a, b \in \mathbb{A}$, by

$$\mu^D(a,b) = \phi_O[T(\mu_Q, n(\mu_{Q^{-1}})), n(T(\mu_V, n(\mu_{V^{-1}})))], \tag{15}$$

which can be simplified, in case n is an involutive strict negation, i.e. a strong negation, as

$$\mu^D(a,b) = \phi_O[T(\mu_Q, n(\mu_{Q^{-1}})), T(n(\mu_V), \mu_{V^{-1}})], \tag{16}$$

where $\phi_O : [0,1]^2 \to [0,1]$ is an overlapping (conjunctive, not necessarily associative) operator [2], modelling the intersection of positive (Q), and negative (V) preferences.

Let us note that the non-associativity of the aggregation operator ϕ_O is used here to stress that positive and negative reasons cannot be grouped arbitrarily without affecting the final outcome [11]. On the other hand, the want component W is given by $\mu^W = T(\mu_Q, n(\mu_{Q^{-1}}))$.

Hence, a decision can obey to the P-A overall preferential situations, making a choice because of wanting or needing, but also postponing it, due to the verification of some incomparable component, perhaps identifying *ambivalence* or lacking reasons for comparing alternatives on solid grounds [10,11]. Such reasoning for a preferential-argument should be explicit, by the following proposal on decision viewpoints.

3 Decision Viewpoints

We can understand the decision making process as an analytical search for the alternatives that allow satisfying, or even maximizing a given set of attributes or criteria. In general, decision making looks for a set of relevant reasons, or the one compelling reason, for choosing the most suitable alternatives. From a purely subjective perspective, by comparing pairs of alternatives regarding a set of attributes or criteria, the individual finds reasons, or arguments, assigning a positive or negative character to the specific property being measured by those criteria, having a paired-opposite meaning for each criterion/attribute. In this sense, a preference argument or *viewpoint*, refers to a whole system of preferences (entailing a global order or decision), which has an associated pair of values: one measuring the total value of preference (like an overall expected outcome), and another measuring the coherence of the preference argument.

It is worth noticing that this proposal can be related to a more standard multi-criteria decision approach on viewpoints [1], where a group of related criteria make up the more general concept of a *viewpoint*, allowing to independently compare the expected decision outcomes of different viewpoints. In this sense, we propose here a different approach, referring to *reasons* instead of groups of criteria, and considering besides the outcome of a viewpoint, an information measure on the (ir)rationality of that viewpoint.

In this way, a viewpoint V_h is defined on pairwise preference relations $\mu \in U$, where U is the set of all fuzzy preferences relations, and $\mu = (\mu_1, \mu_2, ..., \mu_m)$, being m the number of reasons, such that

$$V_h : U \to \Omega_h. \tag{17}$$

In consequence, every viewpoint V_h obtains an outcome that is here measured on a bi-variate space Ω_h, like e.g. $\Omega_h = [0, 1]^2$, where one variate ω_1^h takes on the total suitability-outcome of the system of preferences μ, and the other ω_2^h, measures the inconsistency or ambiguity of the resulting preference order or chain of preferences (what we call the preference argument or viewpoint).

The suitability-outcome ω_1^h expresses a global (positive and negative) total preference value for viewpoint V_h (estimating the overall intensity of preference-aversion). For instance, by denoting by m^+ and m^- the number of positive and negative reasons, respectively, the global suitability-outcome could be directly computed by

$$\omega_1^h = \sum_{i=1}^{m^+} \sum_{j=1}^{k} w_i^+ \mu_{ij}^{(+)} - \sum_{i=1}^{m^-} \sum_{j=1}^{k} w_i^- \mu_{ij}^{(-)}, \tag{18}$$

where w_i^+, w_i^- are weights assigned to each reason according to its relative importance (see e.g. [9] for a method on how to estimate robust weights); k is the number of comparisons among all the alternatives (excluding the diagonal/identity) on the preference and aversion dimensions of the P-A model, such that $k = |\mathbb{A}|(|\mathbb{A}| - 1)$; and for any pair $a, b \in \mathbb{A}$, indexed by $j = 1, ..., k$, under a specific reason $i = 1, ..., m$, it holds that $\mu_{ij}^{(+)} = \mu_{Q_i}(a, b)$, and $\mu_{ij}^{(-)} = \mu_{V_i}(a, b)$.

Notice that outcome ω_1^h can also be understood according to the want and need components of the P-A structure (as in Eq. (16)), defined for the need-viewpoint $V_{h=D}$. Then, by taking e.g. a *grouping* multidimensional operator over M dimensions [5], $\phi_G : [0, 1]^M \to [0, 1]$, we can compute the overall outcome $\omega_1^{h=D}$, by

$$\omega_1^D = \phi_G \left(\mu_{ij}^D \right). \tag{19}$$

In summary, the constitution of a decision viewpoint V_h requires the estimation of a pair of outcomes $\{\omega_1, \omega_2\}_h$. The first one, ω_1, the suitability score, expresses the overall intensity of preference which has to be computed from the system of pairwise relations U. The second one, ω_2, measures the soundness of the viewpoint. In order to address this second value, we explore next *ambiguity measures* (which have to be extended from their original formulations to operate on the P-A structure), referring to the incoherence or irrationality of a given viewpoint. In this sense, following Eq. (18), the decision problem can be solved by finding the viewpoint (or need-viewpoint according to Eq. (19)) with maximum outcome ω_1^h, while minimizing its *ambiguity* ω_2^h.

4   Ambiguity Measures for Decision Modeling

4.1   Fuzziness and Ambiguity Measures

Thinking about the notion of *consistency* for pairwise preference relations from a fuzzy perspective, it can be explored as how close a pairwise relation is from its crisp/extreme values. In this sense, it relates to *fuzziness*, and it serves as a first approximation to measuring how far away can a preference relation, and later a whole viewpoint be, from a completely *rational* strict judgment.

Examining the concept of fuzziness as the lack of distinction between a set Q and its negation $\neg Q$ [20], the intersection of both sets (both its affirmation and its negation), suggests the idea of blurry frontiers or fuzziness. Such fuzziness F can be defined as the intersection between the sets Q and $n(Q)$, such that

$$F(Q) = T(Q, n(Q)).$$

The relation between ambiguity and fuzziness becomes more clear when modeling *opposition* among concepts [13], identifying different types of opposition which may result not necessarily in a strict negation. Let's recall that an opposition operator [13] is defined for all relations $\mu \in [0, 1]$, by the function $O : [0, 1] \to [0, 1]$, such that O is involutive (i.e., $O^2 = \mathbf{I}$, where $\mathbf{I}$ is the Identity), and $\forall x, y, x', y' \in [0, 1]$, if $\mu(x, y) \leq \mu(x', y')$, then $O(\mu)(x', y') \leq O(\mu)(x, y)$ holds true. In this way, assuming a particular negation operator n^*, it can be said that the opposition will be of an antonym type, such that $O \leq n^*$, or on the contrary, O will be a sub-antonym operator such that $O > n^*$. This operator allows us to model a semantic relation of opposition regarding any chosen negation n^*, obtaining one of two main families of opposites, that of being antonym or sub-antonym.

Therefore, the antonym or the sub-antonym of a concept can be modeled by $O(Q)$, like e.g. $Q = high$, $O(Q) = low$ or $V = small$, $O(V) = big$, by means of the opposite operator O, and overlapping operator ϕ_O, such that

$$F(Q) = \phi_O(Q, O(Q)).$$

In this sense, fuzziness $F(Q)$ refers to an *imprecise* frontier for understanding the difference between both paired concepts Q and $O(Q)$, and the *middle state(s)* in between, capturing the *ambiguity* of the concept Q.

Ambiguity due to ignorance has been studied since the works of Knight [16] and Keynes [15], where the interest was explicitly focused on what they called, *measurable* and *unmeasurable* probabilistic-uncertainty. It was argued that ambiguity, understood as lack of knowledge [6,7], played an important role in rational learning for decision processes. Thereafter, Fishburn [7] suggested a function of ambiguity, which was later extended to the fuzzy setting [20].

In its original version, an ambiguity measure (α^*) is defined for a reference set X, and $\forall B, C \in X$, by

$$\alpha^* : X \to [0, 1]$$

if, and only if, it satisfies the following three axioms [7]:

1. A1. $\alpha^*(\emptyset) = 0$
2. A2. $\alpha^*(B) = \alpha^*(\neg B)$
3. A3. $\alpha^*(B \cup C) + \alpha^*(B \cap C) \leq \alpha^*(B) + \alpha^*(C)$

As mentioned above, these axioms have been extended into a fuzzy setting [20], using the min and max operators for representing conjunction and disjunction, respectively.

4.2   Fuzzy-Ambiguity Measures

Here we propose an extension of Fishburn's ambiguity axioms into a fuzzy setting by means of overlapping, grouping and opposition operators. In this way, ambiguity measures can be extended for measuring the overall ambiguity of fuzzy (P-A) relations.

Given a reference set of all fuzzy sets X (or all fuzzy preferences, where $X = U$), grouping and overlap operators ϕ_G, ϕ_O, and an opposition operator O, the function

$$\alpha : X \to [0, 1]$$

is a *fuzzy-ambiguity measure* if, and only if, it satisfies the following three axioms ($\forall \mu_B, \mu_C \in X$):

1. F1. $\alpha(\emptyset) = 0$
2. F2. $\alpha(\mu_B) = \alpha(O(\mu_B))$
3. F3. $\alpha(\phi_G(\mu_B, \mu_C)) + \alpha(\phi_O(\mu_B, \mu_C)) \leq \alpha(\mu_B) + \alpha(\mu_C)$

Notice that such a fuzzy formulation of ambiguity depends on the specific operator used to model opposition, and at the same time, on the overlap and grouping operators ϕ_O, ϕ_G, which allow the triangle inequality (F3) to hold.

In this way, for any pair of alternatives and their corresponding preference relations $\mu \in U$, some examples for fuzzy-ambiguity measures (modeling opposition by a strong negation) are Shannon's entropy (also a measure of fuzziness [20])

$$\alpha^{ent}(\mu) = - \left(\sum_{i=1}^{m} \mu_i \ln(\mu_i) + \sum_{i=1}^{m} n(\mu_i) \ln(n(\mu_i)) \right),$$

or the sum of specificity measures [20],

$$\alpha^{sp}(\mu) = \frac{m-1}{m} \left(Sp(\mu) + Sp(n(\mu)) \right)$$

where *specificity* measures the degree in which any set (of preferences) consists of one and only one element [20].

4.3   Ambiguity Measures for Decision Viewpoints

In this section we will examine ambiguity measures for estimating the soundness of decision viewpoints, denoted earlier by ω_2^h. Given a viewpoint V_h, its ambiguity is computed by extending the examples presented above for ambiguity measures for all the preference relations under V_h. That is, by extending the m-dimensional space for considering every pairwise comparison (between all pairs of alternatives) on $\mathbb{A}^2$.

Hence, the ambiguity of a given decision viewpoint can be computed by the entropy-ambiguity

$$\alpha_h^{ent}(V_h) = \frac{-1}{km} \left(\sum_{i=1}^{m}\sum_{j=1}^{k} \mu_{ij} \ln(\mu_{ij}) + \sum_{i=1}^{m}\sum_{j=1}^{k} n(\mu_{ij})) \ln(n(\mu_{ij}))) \right),$$

where the km constant stands for the number of k comparisons and m reasons.

Example. Consider a democratic voting example, where citizens vote for a public authority among three candidates, such that $\mathbb{A} = \{a, b, c\}$. Each candidate has their own proposals, and the subjective evaluations of those proposals constitute the reasons for valuing the preference-aversion intensities μ. That is, from each proposal there is an associated positive meaning, given by μ_Q, and a negative one, given by μ_V, acting in favor or against each candidate. The corresponding preference and aversion relations for two types of voters, an indecisive and a crisp voter, are given in Table 1.

Table 1. Preference and aversion relations for each pair of candidates (μ_Q, μ_V)

Indecisive	a	b	c	Strict	a	b	c
a		(0.3,0.8)	(0.5,0.3)	a		(1,0)	(1,0)
b	(0.9,0.2)		(0.3,0.6)	b	(0,1)		(0,1)
c	(0.9,0.4)	(0.4,0.5)		c	(0,1)	(1,0)	

Taking equal weights for the positive and negative reasons in Eq. (18), the score for the indecisive viewpoint is $\omega_1^{indecisive} = 0.5$, and if we take the need-outcome of Eq. (19), with a strong negation $n(x) = 1 - x$, and the overlap and grouping operators $\phi_O = $ min and $\phi_G = $ max, respectively, we obtain a score of $\omega_1^{indecisive-need} = 0.7$. On the other hand, computing $\omega_2^{indecisive}$ according to the entropy-ambiguity α^{ent}, we obtain that $\alpha^{ent} = 0.57$. Then it can be verified that there is presence of ambiguity in the preferences of this indecisive voter.

Under the same setting, if we take a second viewpoint $h = strict$, of a strictly crisp voter (with only strict transitive preferences), we obtain that $\omega_1^{strict} = 1 = \omega_1^{strict-need}$, and $\omega_2^{strict} = \alpha^{ent} = 0$. Then, for this type of strict voter with no ambiguous judgments, his preferences reflect that a clear decision follows from his complete argumentation.

As a result, the indecisive viewpoint reflects the greater ambiguity of the voter's preferences, having higher fuzziness than a strict voter, whose preferences are all strict judgments. As it could be expected, as preferences come closer to the higher fuzzy intensities of 0.5, their ambiguity becomes higher, and the decision problem appears to be more complex, or less clear.

Examining with greater detail the outcomes of the different viewpoints, it would be relevant to study the reasons that receive greater ambiguity, and understand the decision problem according to those dimensions that become more difficult to judge. Such dimensions may be of a complex nature, referring to a set of *simpler* reasons that could be evaluated separately, aiming at explaining the problem for the indecisive type of individuals.

5   Final Comments

We have explored the proposal on decision viewpoints, which guide and explain the decision making process according to its overall preference score, together with information on its levels of coherence or (ir)rationality. For this purpose we have examined ambiguity measures, presenting some examples for measuring the ambiguity of decision viewpoints. Focusing on the proposal for decision viewpoints, it is left for future research to examine viewpoints on specific strict relations (taking $\mu = p$), or only the indifference ($\mu = i$) or incomparability ($\mu = j$) relations, or even considering separate viewpoints for the preference and aversion structures. Then we could have strict, indifferent or incomparable viewpoints, as well as preference or aversion viewpoints.

Further studies over information measures and viewpoints could go beyond ambiguity, and consider also the coherence or rationality of arguments. Thinking about the ambiguity of a single viewpoint, we can also think of measuring the degree of concordance or coherence between groups of viewpoints. Then the question would be how *coherent* the different viewpoints are. This coherence could be understood as the consensus or dissention between two converging or diverging opinions.

Another approach for thinking on the soundness of decisions is to explore the *rationality* of judgments. In this sense, it would be desirable to think of fuzzy rationality measures [3] and their extension for the P-A framework, where the presence of cycles in preference and/or aversion chains should entail lower degrees of rationality. Besides, considering absolutely opposite opinions, or complementary opinions μ and $\neg\mu$, it is observed that complementary opinions should obtain the same value of rationality, irrespective if we look at μ or $\neg\mu$. In fact, this observation can be understood as a foundational basis for the notion of *rationality* (as suggested in [3]), affirming a *symmetric* condition between μ and $\neg\mu$. In this sense, rationality measures can be extended for both symmetric and asymmetric types of preference representation, modeling opposite opinions, or *arguments*, by μ and $O(\mu)$.

References

1. Bouyssou, D., Marchant, T., Pirlot, M., Tsoukiàs, A., Vincke, P.: Evaluation and Decision Models with Multiple Criteria. Springer, Heidelberg (2006). https://doi.org/10.1007/0-387-31099-1
2. Bustince, H., Pagola, M., Mesiar, R., Hullermeier, E., Herrera, F.: Grouping, overlap, and generalized bientropic functions for fuzzy modeling of pairwise comparisons. IEEE Trans. Fuzzy Syst. **20**, 405–415 (2012)
3. Cutello, V., Montero, J.: Fuzzy rationality measures. Fuzzy Sets Syst. **62**, 39–54 (1994)
4. Debreu, G.: Theory of Value. An Axiomatic Approach of Economic Equilibrium. Yale University Press, New York (1959)
5. De Miguel, L., et al.: General overlap functions. Fuzzy Sets and Systems. https://doi.org/10.1016/j.fss.2018.08.003
6. Ellsberg, D.: Risk ambiguity and the savage axioms. Q. J. Econ. **75**, 643–669 (1961)
7. Fishburn, P.: The axioms and algebra of ambiguity. Theor. Decis. **34**, 119–137 (1993)
8. Fodor, J., Roubens, M.: Fuzzy Preference Modelling and Multicriteria Decision Support. Kluwer Academic Publishers, Dordrecht (1994)
9. Franco, C.: Ranking fuzzy priorities. In: 2018 IEEE International Conference on Fuzzy Systems (FUZZ-IEEE). IEEE, Rio de Janeiro, pp. 1–7 (2018)
10. Franco, C., Montero, J., Rodríguez, J.T.: A fuzzy and bipolar approach to preference modelling with application to need and desire. Fuzzy Sets Syst. **214**, 20–34 (2013)
11. Franco, C., Rodríguez, J.T., Montero, J.: Building the meaning of preference from logical paired structures. Knowl.-Based Syst. **83**, 32–41 (2015)
12. Franco, C., Rodríguez, J.T., Montero, J.: Learning preferences from paired opposite-based semantics. Int. J. Approximate Reasoning **86**, 80–91 (2017)
13. Franco, C., Rodríguez, J.T., Montero, J., Gómez, D.: Modeling opposition with restricted paired structures. J. Multi-Valued Logic Soft Comput. **30**, 239–262 (2018)
14. Georgescu-Roegen, N.: The pure theory of consumer's behavior. Q. J. Econ. **50**, 545–593 (1936)
15. Keynes, J.M.: A Treatise on Probability. MacMillan, London (1963)
16. Knight, F.: Risk, Uncertainty, and Profit. University of Chicago Press, Chicago (1971)
17. Montero, J., et al.: Paired structures in knowledge representation. Knowl. Based Syst. **100**, 50–58 (2016)
18. Montero, J., Tejada, J., Cutello, C.: A general model for deriving preference structures from data. Eur. J. Oper. Res. **98**, 98–110 (1997)
19. Van der Walle, B., de Baets, B., Kerre, E.: Characterizable fuzzy preference structures. Ann. Oper. Res. **80**, 105–136 (1998)
20. Yager, R.R.: On a measure of ambiguity. Int. J. Intell. Syst. **10**, 1001–1019 (1995)

Logics of Dominance for Reasoning About Multi-criteria Decisions

Tuan-Fang Fan[1] and Churn-Jung Liau[2(✉)]

[1] Department of Computer Science and Information Engineering,
National Penghu University of Science and Technology, Penghu 880, Taiwan
dffan@npu.edu.tw
[2] Institute of Information Science, Academia Sinica, Taipei 115, Taiwan
liaucj@iis.sinica.edu.tw

Abstract. In this paper, we present modal logics of preference and dominance for reasoning about multi-criteria decisions. We first explore a basic logical framework based on multi-criteria preferences and its special case when the relations are total preorders. Then, we extend the basic formalism to a logic of modal dominances. The syntax, semantics, and complete axiomatization of each logic is presented. Finally, we also consider several extensions and variants of the proposed logics that can represent rules induced by dominance-based rough set analysis and beyond.

Keywords: Dominance-based rough set approach ·
Multi-criteria decision analysis · Modal logic · Logic of preference ·
Logic of dominance

1 Introduction

The rough set theory (RST) proposed in [12] is an effective tool for intelligent data analysis. The basic idea of RST is to partition objects into equivalence classes based on the indiscernibility of their attribute values. However, in many decision analysis applications, the domain of attribute values may be endowed with ordering. For example, in the multi-criteria decision analysis (MCDA) problem, each attribute corresponds to a criterion of decision-making that has a domain of ordered values. To deal with this kind of problem, RST is generalized to the dominance-based rough set approach (DRSA) in [6–8,13,14], where the indiscernibility between objects is replaced with the dominance relation.

It is well-known that RST can provide a set-theoretic semantics for the modal logic S5 system [4,9–11]. The implication of this fact is that modal logic can play a major role in the knowledge representation and reasoning for rules induced by rough set-based data analysis. Inspired by such result, the objective of the present paper is to develop a modal logic of dominance for reasoning about MCDA in the context of DRSA. The logic is based on a multi-modal version of the basic modal preference logic proposed in [1]. The language of our logic

© Springer Nature Switzerland AG 2019
H. Seki et al. (Eds.): IUKM 2019, LNAI 11471, pp. 50–62, 2019.
https://doi.org/10.1007/978-3-030-14815-7_5

contains two classes of modalities for reasoning about properties of better or worse options and each modality in these classes correspond to a criterion of decision-making. In addition, we use intersection modalities corresponding to a subset of criteria to represent dominating and dominated relations with respect to these criteria. We also provide complete Hilbert-style axiomatizations for some variants of the logic.

The remainder of the paper is organized as follows. In the next section, we review preliminaries on modal logic, RST, and DRSA. In Sect. 3, we first provide a logic of multi-criteria preference as the basic framework. Then, in Sect. 4, we extend the basic formalism to a modal logic of dominance. For these logics, we will present their syntax, semantics, and complete axiomatizations. Finally, we consider several possible extensions of the proposed logics and conclude the paper with some open questions for future research.

2 Preliminaries

2.1 Modal Logic

We start with a brief account of classical modal logic [3]. Let Φ denote the set of propositional symbols. Then, the formulas of classical modal logic are defined as follows:

$$\varphi ::= p \mid \bot \mid \varphi \to \varphi \mid \Box \varphi,$$

where $p \in \Phi$, $\bot$ is the logical constant representing *falsum* , $\to$ is the material implication, and $\Box$ is the necessity modality. Other logical connectives such as $\top, \neg, \wedge, \vee, \equiv$ and modality $\Diamond$ are defined as abbreviations as usual. We use $\mathcal{L}_\Box$ to denote the propositional modal language.

The standard semantics of modal logic is based on Kripke model, which is a triple $\mathfrak{M} = \langle W, R, \Vdash \rangle$, where W is a set of possible worlds (states), $R \subseteq W \times W$ is a binary *accessibility relation* on W, and $\Vdash \subseteq W \times \Phi$ is a *forcing relation* between possible worlds and propositional symbols such that $w \Vdash p$ means that p is satisfied in w. The forcing relation can be extended to a relation $\Vdash_\mathfrak{M}$ between W and $\mathcal{L}_\Box$ by the semantic rules. In addition to the standard rules for classical connectives, we have $w \Vdash_\mathfrak{M} \Box \varphi$ iff for any u such that $(w, u) \in R$, $u \Vdash_\mathfrak{M} \varphi$. We usually drop the subscript and simply write $\Vdash$ for the extended forcing relation when the model $\mathfrak{M}$ is clear from the context. For a set of formulas Σ, we also write $u \Vdash_\mathfrak{M} \Sigma$ if for all $\varphi \in \Sigma$, $u \Vdash_\mathfrak{M} \varphi$ holds.

A formula φ is said to be valid in a model $\mathfrak{M}$, denoted by $\models_\mathfrak{M} \varphi$, if for all worlds $w \in W$, $w \Vdash_\mathfrak{M} \varphi$. The pair $\mathfrak{F} = \langle W, R \rangle$ in a model $\mathfrak{M} = \langle W, R, \Vdash \rangle$ is called a Kripke frame. Let $\mathbf{C}$ be a class of Kripke frames and let $\Sigma \cup \{\varphi\}$ be a set of formulas. Then, $\Sigma \models_\mathbf{C} \varphi$ denotes that for any model $\mathfrak{M}$ based on a frame in $\mathbf{C}$ and any world w in $\mathfrak{M}$, $w \Vdash_\mathfrak{M} \Sigma$ implies $w \Vdash_\mathfrak{M} \varphi$. When $\Sigma = \emptyset$, we simply write $\Sigma \models_\mathbf{C} \varphi$ as $\models_\mathbf{C} \varphi$. We say that φ is a *logical consequence* (or semantic consequence) of Σ with respect to $\mathbf{C}$ if $\Sigma \models_\mathbf{C} \varphi$ holds and φ is *valid* for $\mathbf{C}$ if $\models_\mathbf{C} \varphi$ holds. We usually omit the subscript $\mathbf{C}$ if it is the class of all Kripke frames. In addition, we use **S5** to denote the class of all frames $\mathfrak{F} = \langle W, R \rangle$ where R is an equivalence relation.

2.2 Rough Set Theory

The basic construct of rough set theory is an *approximation space*, which is defined as a pair (U, R) such that U is a finite universe and $R \subseteq U \times U$ is an equivalence relation on U. We write an equivalence class of R as $[x]_R$ if it contains the element x. For any subset X of the universe, lower and upper approximations of X, denoted by $\underline{R}X$ and $\overline{R}X$ respectively, are defined as follows:

$$\underline{R}X = \{x \in U \mid [x]_R \subseteq X\}, \tag{1}$$

$$\overline{R}X = \{x \in U \mid [x]_R \cap X \neq \emptyset\}. \tag{2}$$

Although an approximation space is an abstract framework, it can easily be derived from a concrete data table. In [12], a data table[1] is defined as a tuple $T = (U, A, \{V_i \mid i \in A\}, \{f_i \mid i \in A\})$, where U is a nonempty finite set, called the universe; A is a nonempty finite set of primitive attributes; for each $i \in A$, V_i is the domain of values of i; and for each $i \in A$, $f_i : U \to V_i$ is a total function. An attribute in A is usually denoted by the lower-case letters i or j. In decision analysis (and throughout this paper), we assume the set of attributes is partitioned into $\{d\} \cup (A - \{d\})$, where d is called the *decision attribute*, and the remaining attributes in $C = A - \{d\}$ are called *condition attributes*. Given a subset of attributes B, we can define an equivalence relation, called the *indiscernibility relation*, as follows:

$$ind(B) = \{(x, y) \mid x, y \in U, f_i(x) = f_i(y) \forall i \in B\}. \tag{3}$$

Consequently, for each $B \subseteq A$, $(U, ind(B))$ is an approximation space.

Apparently, an approximation space corresponds to a Kripke frame in **S5**. Therefore, it is ready to see the connection between RST and the semantics of modal logic. For any approximation space (U, R), we can construct a Kripke model $\mathfrak{M} = \langle U, R, \Vdash \rangle$ for a modal language whose propositional symbols describe properties of objects in U. More precisely, for any propositional symbol p, $u \Vdash p$ iff the object u satisfies the property p. In addition, we define the truth set of a formula φ in $\mathfrak{M}$ as $|\varphi|_{\mathfrak{M}} = \{u \in U \mid u \Vdash \varphi\}$. Then, the following relationship between modal formulas and rough approximations holds (we omit the subscript $\mathfrak{M}$ for simplification):

$$|\Box\varphi| = \underline{R}|\varphi|, \quad |\Diamond\varphi| = \overline{R}|\varphi| \tag{4}$$

As a consequence of the above-mentioned relationship, we can easily represent and reason about knowledge induced from data tables with modal logic. For example, let φ and ψ describe properties of condition attributes and decision classes respectively. Then, $\varphi \to \Box\psi$ (resp. $\varphi \to \Diamond\psi$) can stand for a certain (resp. possible) rule derived from a data table by using RST [12].

---

[1] Also called knowledge representation systems, information systems, or attribute-value systems.

2.3   Dominance-Based Rough Set Approach

For MCDA problems, each object in a data table can be seen as a sample decision, and each condition attribute is a criterion for that decision. Since a criterion's domain of values is usually ordered according to the decision-maker's preferences, we define a preference-ordered data table (PODT) as a tuple $T = (U, C \cup \{d\}, \{(V_i, \succcurlyeq_i) \mid i \in C\} \cup \{V_d\}, \{f_i \mid i \in C \cup \{d\}\})$, where $(U, C \cup \{d\}, \{V_i \mid i \in C\} \cup \{V_d\}, \{f_i \mid i \in C \cup \{d\}\})$ is a classical data table; and for each $i \in C$, $\succcurlyeq_i \subseteq V_i \times V_i$ is a binary relation over V_i. The relation $\succcurlyeq_i$ is called a *weak preference relation* or *outranking* on V_i, and represents a preference over the set of objects with respect to the criterion i [13]. The weak preference relation $\succcurlyeq_i$ is supposed to be a preorder (aka quasiorder), i.e., a reflexive and transitive relation[2]. In addition, we assume that the domain of the decision attribute is a finite set $V_d = \{1, 2, \cdots, k\}$ such that r is strictly preferred to s if $r > s$ for any $r, s \in V_d$.

To deal with inconsistencies arising from violations of the dominance principle in MCDA, the indiscernibility is replaced by a dominance relation in DRSA. Let $P \subseteq C$ be a nonempty subset of criteria. Then, we can define the *P-dominance relation* $D_P \subseteq U \times U$ as follows:

$$(x, y) \in D_P \Leftrightarrow f_i(x) \succcurlyeq_i f_i(y) \forall i \in P. \tag{5}$$

When $(x, y) \in D_P$, we say that x *P-dominates* y, and that y is *P-dominated* by x. We usually use the infix notation $x D_P y$ to denote $(x, y) \in D_P$. Given the dominance relation D_P, the *P-dominating set* and *P-dominated set* of x are defined as $D_P^+(x) = \{y \in U \mid y D_P x\}$ and $D_P^-(x) = \{y \in U \mid x D_P y\}$ respectively. In addition, for each $t \in V_d$, we define the decision class Cl_t as $\{x \in U \mid f_d(x) = t\}$. Then, the *upward and downward unions of classes* are defined as $Cl_t^{\geq} = \bigcup_{s \geq t} Cl_s$ and $Cl_t^{\leq} = \bigcup_{s \leq t} Cl_s$ respectively. We can then define the *P*-lower and *P*-upper approximations of $Cl_t^{\geq}$ and $Cl_t^{\leq}$ by using the *P*-dominating sets and *P*-dominated sets instead of the equivalence classes. More specifically, we have the following definitions:

$$\underline{P}(Cl_t^{\geq}) = \{x \in U \mid D_P^+(x) \subseteq Cl_t^{\geq}\}, \tag{6}$$

$$\overline{P}(Cl_t^{\geq}) = \{x \in U \mid D_P^-(x) \cap Cl_t^{\geq} \neq \emptyset\}, \tag{7}$$

$$\underline{P}(Cl_t^{\leq}) = \{x \in U \mid D_P^-(x) \subseteq Cl_t^{\leq}\}, \tag{8}$$

$$\overline{P}(Cl_t^{\leq}) = \{x \in U \mid D_P^+(x) \cap Cl_t^{\leq} \neq \emptyset\} \tag{9}$$

Based on the approximations, it is shown that a generalized description of the preferential information contained in a PODT can be induced in terms of three types of decision rules [8][3]:

---

[2] In the original DRSA, it is also assumed that the relation is total, i.e., it satisfies that, for $x, y \in V_i$, $x \succcurlyeq_i y$ or $y \succcurlyeq_i x$. However, we will model the more general scenario in which there may exist incomparable values. Then, a PODT with all preference relations being total preorders is regarded as a special case.

[3] We omit the respective subscripts of $\preccurlyeq$ and $\succcurlyeq$ in the rules for brevity.

1. $D_{\geq}$ rule: if $f_{q_1}(x) \succcurlyeq r_{q_1}$ and $f_{q_2}(x) \succcurlyeq r_{q_2}$ and $\ldots f_{q_p}(x) \succcurlyeq r_{q_p}$, then $x \in Cl_t^{\geq}$,
2. $D_{\leq}$ rule: if $f_{q_1}(x) \preccurlyeq r_{q_1}$ and $f_{q_2}(x) \preccurlyeq r_{q_2}$ and $\ldots f_{q_p}(x) \preccurlyeq r_{q_p}$, then $x \in Cl_t^{\leq}$,
3. $D_{\geq\leq}$ rule: if $f_{q_1}(x) \succcurlyeq r_{q_1}$ and $\ldots f_{q_k}(x) \succcurlyeq r_{q_k}$ and $f_{q_{k+1}}(x) \preccurlyeq r_{q_{k+1}}$ and $\ldots$
   $f_{q_p}(x) \preccurlyeq r_{q_p}$, then $x \in Cl_t \cup Cl_{t+1} \cup \cdots \cup Cl_s$,

where $\{q_1, q_2, \cdots, q_p\} \subseteq C$, $r_{q_i} \in V_{q_i}$ for $1 \leq i \leq p$, and $t, s \in V_d$ such that $t < s$.
In the case of $D_{\geq\leq}$ rule, the two subsets $\{q_1, q_2, \cdots, q_k\}$ and $\{q_{k+1}, q_{k+2}, \cdots, q_p\}$
are not necessarily disjoint.

3   A Logic of Multi-criteria Preference

To prepare the logic of modal dominance, we first set up a basic logical framework
of multi-criteria preference in this section. Let $C = \{1, 2, ..., n\}$ be a set of criteria
and let Φ be a set of propositional symbols. Then, formulas for the logic of multi-
criteria preference (LMP) are defined by

$$\varphi ::= p \mid \bot \mid \varphi \rightarrow \varphi \mid [\leq_i]\varphi \mid [\geq_i]\varphi,$$

where $p \in \Phi$ and $i \in C$. We use $\mathcal{L}_{mp}$ to denote the language of LMP. As
in the case of $\mathcal{L}_\Box$, other Boolean connectives are defined as abbreviations. In
addition, the formulas $\neg[\leq_i]\neg\varphi$ and $\neg[\geq_i]\neg\varphi$ are abbreviated as $\langle\leq_i\rangle\varphi$ and $\langle\geq_i\rangle\varphi$
respectively.

For the semantics of $\mathcal{L}_{mp}$, an LMP model is defined as a tuple $\mathfrak{M} = \langle W,$
$(\succeq_i)_{i \in C}, \Vdash\rangle$, where W and $\Vdash$ are the same as in the standard Kripke models,
and $\succeq_i \subseteq W \times W$ is reflexive and transitive relation for every $i \in C$. Moreover,
we use $\preceq_i$ to denote the converse relation of $\succeq_i$ (i.e. for any $x, y \in W$, $x \preceq_i y$ iff
$y \succeq_i x$). Intuitively, W denotes the set of possible worlds (or options), and the
preorder $\succeq_i$ corresponds to the weak preference relation in DRSA. Hence, $x \succeq_i y$
means that the option x is *at least as good as* y with respect to the criterion i.
Then, the forcing relation can be extended to all formulas by using the following
rules:

1. $w \nVdash \bot$,
2. $w \Vdash \varphi \rightarrow \psi$ iff $w \nVdash \varphi$ or $w \Vdash \psi$,
3. $w \Vdash [\leq_i]\varphi$ iff for any $u \in W$ such that $w \preceq_i u$, $u \Vdash \varphi$,
4. $w \Vdash [\geq_i]\varphi$ iff for any $u \in W$ such that $w \succeq_i u$, $u \Vdash \varphi$.

According to the semantic rules, $[\leq_i]\varphi$ intuitively means that any world at least
as good as the current one satisfies φ. Or, conversely, it means that any world
not satisfying φ is worse than the current one if $\succeq_i$ is a total preorder. Loosely
speaking, this means that φ is desirable or $\neg\varphi$ is undesirable. An analogous
interpretation can be applied to $[\geq_i]\varphi$.

The notions of logical consequence and validity in a class of frames carry over
from those in modal logic and we use $\models_{mp}$ to denote the consequence relation
and validity in the class of all LMP frames. It is well-known that reflexive and
transitive Kripke frames can be characterized by S4 modal systems [3]. Therefore,

the validity of LMP can be axiomatized by mp system presented in Fig. 1. The system is essentially a multi-modal version of S4, where the axiom K and the Nec rule show that both $[\leq_i]$ and $[\geq_i]$ are normal modalities, and T and 4 correspond to the reflexivity and transitivity of preference relations respectively. In addition, the Cov axiom is a standard interaction axiom for bimodal systems when two modalities are converse to each other.

---

1. Axioms:
   (a) Prop: The standard set of axioms for classical propositional logic
   (b) $K_\leq$: $[\leq_i](\varphi \to \psi) \to ([\leq_i]\varphi \to [\leq_i]\psi)$;
   (c) $T_\leq$: $[\leq_i]\varphi \to \varphi$
   (d) $4_\leq$: $[\leq_i]\varphi \to [\leq_i][\leq_i]\varphi$
   (e) $K_\geq$: $[\geq_i](\varphi \to \psi) \to ([\geq_i]\varphi \to [\geq_i]\psi)$;
   (f) $T_\geq$: $[\geq_i]\varphi \to \varphi$
   (g) $4_\geq$: $[\geq_i]\varphi \to [\geq_i][\geq_i]\varphi$
   (h) Cov: $\varphi \to [\geq_i]\langle\leq_i\rangle\varphi$; $\varphi \to [\leq_i]\langle\geq_i\rangle\varphi$
2. Rules of inference:
   (a) Modus Ponens (MP): from φ and $\varphi \to \psi$, infer ψ
   (b) Necessitation (Nec): from φ infer $[\leq_i]\varphi$ and $[\geq_i]\varphi$.

Fig. 1. The axiomatic system mp

Let $\Sigma \cup \{\varphi\}$ be a subset of LMP formulas. Then, we use $\Sigma \vdash_{mp} \varphi$ and $\vdash_{mp} \varphi$ to denote that φ is derivable from Σ and that φ is a theorem in mp respectively. A set Σ is *inconsistent* if $\Sigma \vdash_{mp} \bot$, otherwise, Σ is consistent. Then, we have the following soundness and completeness theorem for the system mp.

Theorem 1. *For any $\Sigma \cup \{\varphi\} \subseteq \mathcal{L}_{mp}$, we have $\Sigma \vdash_{mp} \varphi$ iff $\Sigma \models_{mp} \varphi$.*

3.1 A Logic of Multi-criteria Total Preorder Preference

In the original DRSA, the weak preference relation is assumed to be a total preorder that demands all pairs of elements to be comparable. To model this special case, we can restrict our attention to the kind of frames $\mathfrak{F} = \langle W, (\succeq_i)_{i \in C}\rangle$ that satisfy the following condition:

$$x \succeq_i y \text{ or } y \succeq_i x, \text{ for all } i, x, \text{ and } y. \tag{10}$$

In modal logic, validity in totally pre-ordered Kripke frames is characterized by the S4.3 system, which is the addition of the following axiom to S4 [3]:

$$.3 : (\Diamond\varphi \wedge \Diamond\psi) \to (\Diamond(\varphi \wedge \Diamond\psi) \vee \Diamond(\varphi \wedge \psi) \vee \Diamond(\psi \wedge \Diamond\varphi)).$$

However, the axiom itself does not correspond to the totality condition like that in (10). Instead, it simply says that the binary relation for the modality has no branching to the right. That is, it is a relation R satisfying the following condition:

for any x, y, z, if $R(x, y)$ and $R(x, z)$ are both true, then $R(y, z)$ or $R(z, y)$ is true.

A preorder without branching to the right will become a proper total preorder if we consider its restriction to a subset of points generated from a single point (called the root) via the reachability of the preorder relation. This leads to the notion of *generated subframe or submodel* in modal logic [3]. It is well-known that generated submodels preserve the truth value of any modal formula. Hence, the validity in all frames determined by the system S4.3 is equivalent to the validity in totally pre-ordered ones. As a consequence, S4.3 is a system complete for the validity in totally pre-ordered Kripke frames.

To generalize the argument above to our setting, we must take the interaction of multiple modalities into account. On one hand, because we have a pair of bipolar modalities for each criterion, we can consider relations that have no branching not only to the right but also to the left. On the other hand, in the generated subframe, a point may be reachable from the root via the connection of different relations. Hence, we say that an LMP frame $\mathfrak{F} = \langle W, (\succeq_i)_{i \in C} \rangle$ has no branching if it satisfies the following two conditions:

1. for any $x, y, z \in W$ and $i, j \in C$, if $x \succeq_i y$ and $x \succeq_j z$, then $y \succeq_i z$ or $z \succeq_i y$;
2. for any $x, y, z \in W$ and $i, j \in C$, if $y \succeq_i x$ and $z \succeq_j x$, then $y \succeq_i z$ or $z \succeq_i y$.

These two conditions can be described by the following axioms in LMP:

$-\ .3^1_{\leq}:\ \langle \leq_i \rangle \varphi \wedge \langle \leq_j \rangle \psi \rightarrow$
$\quad \langle \leq_i \rangle (\varphi \wedge \langle \leq_i \rangle \psi) \vee (\langle \leq_i \rangle (\varphi \wedge \psi) \wedge \langle \leq_j \rangle (\varphi \wedge \psi)) \vee \langle \leq_j \rangle (\psi \wedge \langle \leq_i \rangle \varphi),$

$-\ .3^2_{\leq}:\ \langle \leq_i \rangle \varphi \wedge \langle \leq_j \rangle \psi \rightarrow$
$\quad \langle \leq_i \rangle (\varphi \wedge \langle \leq_j \rangle \psi) \vee (\langle \leq_i \rangle (\varphi \wedge \psi) \wedge \langle \leq_j \rangle (\varphi \wedge \psi)) \vee \langle \leq_j \rangle (\psi \wedge \langle \leq_j \rangle \varphi),$

$-\ .3^1_{\geq}:\ \langle \geq_i \rangle \varphi \wedge \langle \geq_j \rangle \psi \rightarrow$
$\quad \langle \geq_i \rangle (\varphi \wedge \langle \geq_i \rangle \psi) \vee (\langle \geq_i \rangle (\varphi \wedge \psi) \wedge \langle \geq_j \rangle (\varphi \wedge \psi)) \vee \langle \geq_j \rangle (\psi \wedge \langle \geq_i \rangle \varphi),$

$-\ .3^2_{\geq}:\ \langle \geq_i \rangle \varphi \wedge \langle \geq_j \rangle \psi \rightarrow$
$\quad (\langle \geq_i \rangle (\varphi \wedge \langle \geq_j \rangle \psi) \vee (\langle \geq_i \rangle (\varphi \wedge \psi) \wedge \langle \geq_j \rangle (\varphi \wedge \psi)) \vee \langle \geq_j \rangle (\psi \wedge \langle \geq_j \rangle \varphi).$

Then, the validity in the class of LMP frames without branching is characterized by the system mp^+, which is defined as $\mathsf{mp} \cup \{.3^1_{\leq}, .3^2_{\leq}, .3^1_{\geq}, .3^2_{\geq}\}$. That is, we have the following completeness theorem:

Theorem 2. *For any $\Sigma \cup \{\varphi\} \subseteq \mathcal{L}_{mp}$, we have $\Sigma \vdash_{mp^+} \varphi$ iff φ is a semantic consequence of Σ with respect to the class of all LMP frames without branching.*

To convert an LMP frame without branching into a totally pre-ordered frame, we define a *point-generated subframe* in a way similar to that in classical modal logic. Let $\mathfrak{F} = \langle W, (\succeq_i)_{i \in C} \rangle$ be an LMP frame and let $x \in W$ be a point in W. Then, the subframe of $\mathfrak{F}$ generated by x is $\mathfrak{F}_x = \langle W_x, (\succeq_i \cap (W_x \times W_x))_{i \in C} \rangle$, where W_x is the smallest set containing x and satisfying the following closure condition:

– for any $y \in W$ and $i \in C$, if $y \in W_x$, then $z \in W_x$ for any z such that $y \preceq_i z$ or $y \succeq_i z$.

It is obvious that if $\mathfrak{F}$ is an LMP frame without branching, then $\mathfrak{F}_x$ is a totally pre-ordered frame. In addition, the notion of point-generated submodel can be defined analogously. Then, following a standard argument in modal logic, we can see that a point-generated submodel preserves the truth of formulas in $\mathcal{L}_{mp}$. Consequently, the axiomatic system mp^+ also characterizes the class of totally pre-ordered frames.

Theorem 3. *Let $\models_{mp^+}$ denote the semantic consequence with respect to the class of totally pre-ordered LMP frames and let $\Sigma \cup \{\varphi\} \subseteq \mathcal{L}_{mp}$. Then, $\Sigma \vdash_{mp^+} \varphi$ iff $\Sigma \models_{mp^+} \varphi$.*

4 A Logic of Modal Dominance

Inspired by the relationship between RST and modal logic, the present paper is aimed at a logic of modal dominance (LMD) based on DRSA that is suitable for knowledge representation and reasoning in the context of MCDA. However, in the logic of multi-criteria preference, we simply consider the individual preference relation based on each criterion. To model the dominance relation, we have to aggregate preference relations based on multiple criteria. Hence, in this section, we will extend the basic framework of LMP to achieve the objective.

The alphabet of LMD is still comprised of a finite set of criteria $C = \{1, 2, ..., n\}$ and a set of propositional symbols Φ, but the formation rules of its formulas are now defined by

$$\varphi ::= p \mid \perp \mid \varphi \to \varphi \mid [\leq_P]\varphi \mid [\geq_P]\varphi,$$

where $p \in \Phi$ and $P \subseteq C$ is a nonempty subset of C. We use $\mathcal{L}_{md}$ to denote the modal dominance language. When $P = \{i\}$ is a singleton, we simply write $[\leq_i]\varphi$ and $[\geq_i]\varphi$ instead of the more cumbersome $[\leq_{\{i\}}]\varphi$ and $[\geq_{\{i\}}]\varphi$. As above, other Boolean connectives are defined as abbreviations, and $\neg[\leq_P]\neg\varphi$ and $\neg[\geq_P]\neg\varphi$ are abbreviated as $\langle\leq_P\rangle\varphi$ and $\langle\geq_P\rangle\varphi$ respectively.

For the semantics of $\mathcal{L}_{md}$, a model is defined as a tuple $\mathfrak{M} = \langle W, (\succeq_i)_{i \in C}, \Vdash \rangle$ as in the case of LMP. However, we further define

$$\succeq_P = \bigcap_{i \in P} \succeq_i, \tag{11}$$

for any nonempty $P \subseteq C$. Analogously, $\preceq_P = \bigcap_{i \in P} \preceq_i$ is the converse relation of $\succeq_P$. Note that $\succeq_P$ is also a preorder no matter whether each preference relation is total or not. Then, the forcing relation can be extended to all formulas by using the following rules:

1. $w \not\Vdash \bot$,
2. $w \Vdash \varphi \to \psi$ iff $w \not\Vdash \varphi$ or $w \Vdash \psi$,
3. $w \Vdash [\leq_P]\varphi$ iff for any $u \in W$ such that $w \preceq_P u$, $u \Vdash \varphi$,
4. $w \Vdash [\geq_P]\varphi$ iff for any $u \in W$ such that $w \succeq_P u$, $u \Vdash \varphi$.

Because we interpret $x \succeq_i y$ as the option x being *at least as good as* y with respect to the criterion i for each $i \in P$, $x \succeq_P y$ simply means that x P-dominates y. Therefore, we can define the P-dominating and P-dominated sets for a LMD model in the same way as in DRSA:

$$D_P^+(x) = \{y \in W \mid y \succeq_P x\}, \tag{12}$$
$$D_P^-(x) = \{y \in W \mid x \succeq_P y\}. \tag{13}$$

Now, for a given PODT $T = (U, C \cup \{d\}, \{(V_i, \succcurlyeq_i) \mid i \in C\} \cup \{V_d\}, \{f_i \mid i \in C \cup \{d\}\})$, we can associate a propositional symbol p_t with each $t \in V_d$ and define the following formulas in the language of LMD:

$$\varphi_t^+ = \bigvee_{s \geq t} p_t, \tag{14}$$

$$\varphi_t^- = \bigvee_{s \leq t} p_t. \tag{15}$$

Then, we can construct a model $\mathfrak{M} = \langle U, (\succeq_i)_{i \in C}, \Vdash \rangle$ from T such that the following two conditions are satisfied:

1. for any $x, y \in U$ and $i \in C$, $x \succeq_i y$ iff $f_i(x) \succcurlyeq_i f_i(y)$,
2. for any $x \in U$ and $1 \leq t \leq k$, $x \in Cl_t$ iff $x \Vdash p_t$.

As a consequence, we can prove a correspondence result similar to (4) in the context of DRSA as follows:

$$|[\leq_P]\varphi_t^+| = \underline{P}|\varphi_t^+|, \tag{16}$$
$$|\langle \geq_P \rangle \varphi_t^+| = \overline{P}|\varphi_t^+|, \tag{17}$$
$$|[\geq_P]\varphi_t^-| = \underline{P}|\varphi_t^-|, \tag{18}$$
$$|\langle \leq_P \rangle \varphi_t^-| = \overline{P}|\varphi_t^-|, \tag{19}$$

where, by definition, $|\varphi_t^+| = Cl_t^{\geq}$ and $|\varphi_t^-| = Cl_t^{\leq}$.

As for the axiomatization of validity in all LMD frames, it is almost the same as the system mp except that the subscript i is changed to P and two extra axiom regarding the monotonicity with respect to subsets of criteria are added as follows:

$$\text{Mon}_\leq : [\leq_{P1}]\varphi \to [\leq_{P2}]\varphi, \text{ if } P_1 \subseteq P_2,$$

$$\text{Mon}_\geq : [\geq_{P1}]\varphi \to [\geq_{P2}]\varphi, \text{ if } P_1 \subseteq P_2.$$

Let md be the resultant system and let $\models_{md}$ and $\vdash_{md}$ respectively denote the semantic consequence in the class of all LMD frames and the derivability relation in the md system. Then, we can prove the following theorem.

Theorem 4. *For any $\Sigma \cup \{\varphi\} \subseteq \mathcal{L}_{md}$, $\Sigma \vdash_{md} \varphi$ iff $\Sigma \models_{md} \varphi$.*

Note that, because we identify $\leq_{\{i\}}$ and $\geq_{\{i\}}$ with $\leq_i$ and $\geq_i$ respectively, $\mathcal{L}_{mp}$ can be regarded as a sub-language of $\mathcal{L}_{md}$. In fact, the system md is a conservative extension of mp in the following sense.

Proposition 1. *For any $\Sigma \cup \{\varphi\} \subseteq \mathcal{L}_{mp}$, $\Sigma \vdash_{mp} \varphi$ iff $\Sigma \vdash_{md} \varphi$.*

5   Extensions and Variants

5.1   Intersection of Arbitrary Modalities

In LMD, a modality $[\leq_P]$ corresponds to the intersection of a set of modalities $\{[\leq_i] \mid i \in P\}$. By using LMD formulas, we can represent the first two types of decision rules of DRSA in the following way:

$$D_\geq : \psi \to [\leq_P]\varphi_t^+ \tag{20}$$

$$D_\leq : \psi \to [\geq_P]\varphi_t^- \tag{21}$$

where $P = \{q_1, q_2, \cdots, q_p\}$ and ψ is a propositional formula representing $\bigwedge_{i \in P} f_i(x) = r_i$. However, we can not represent the third type of rule because it contains the intersection of $\leq_i$ and $\geq_j$ modalities. Hence, for the purpose of knowledge representation, we need further extension of LMD to a logic of general dominance (LGD).

Given a set of criteria C and a set of propositional symbols Φ, the formation rules for the LGD language are defined as follows:

$$\varphi :: = p \mid \bot \mid \varphi \to \varphi \mid [\mathfrak{m}]\varphi,$$

where $p \in \Phi$ and $\mathfrak{m} \subseteq \{\leq_i, \geq_i \mid i \in C\}$ is a nonempty subset of preference modalities. In general, there exist two subsets $P_1, P_2 \subseteq C$ such that $\mathfrak{m} = \{\leq_i \mid i \in P_1\} \cup \{\geq_i \mid i \in P_2\}$, Thus, we may also write $\mathfrak{m}$ as a pair $(\leq_{P_1}, \geq_{P_2})$. If $P_1 = \emptyset$ (resp. $P_2 = \emptyset$), we further abbreviate $\mathfrak{m}$ as $\geq_{P_2}$ (resp. $\leq_{P_1}$). In addition, if $P_1 = P_2 = P$, then we write it as $\sim_P$. Also, as above, we identify a singleton with its element. We use $\mathcal{L}_{gd}$ to denote the language.

For the semantics, the LGD model is the same as the LMP and LMD models. The only difference is the semantic condition for the modal formula:

- $w \Vdash [\mathfrak{m}]\varphi$ iff for any $u \in W$ such that $w \preceq_P u$ and $w \succeq_Q u$, $u \Vdash \varphi$, provided that $\mathfrak{m} = (\preceq_P, \succeq_Q)$.

In accordance with the semantics, the third type of decision rule for DRSA is now expressible in $\mathcal{L}_{gd}$ as follows:

$$D_{\geq\leq} : \psi \to [\mathfrak{m}](\bigvee_{i=t}^{s} p_i), \tag{22}$$

where $\mathfrak{m} = (\leq_P, \geq_Q)$ such that $P = \{q_1, q_2, \cdots, q_k\}$ and $Q = \{q_{k+1}, q_{k+2}, \cdots, q_p\}$.

Let $\models_{gd}$ denote the semantic consequence with respect to all LGD frames. Then, it can be axiomatized by the system **gd** presented in Fig. 2 and we have the completeness theorem of its derivability relation $\vdash_{gd}$.

---

1. Axioms:
    (a) Prop: The standard set of axioms for classical propositional logic
    (b) $K_{\mathfrak{m}}$: $[\mathfrak{m}](\varphi \to \psi) \to ([\mathfrak{m}]\varphi \to [\mathfrak{m}]\psi)$;
    (c) $T_{\mathfrak{m}}$: $[\mathfrak{m}]\varphi \to \varphi$
    (d) $4_{\mathfrak{m}}$: $[\mathfrak{m}]\varphi \to [\mathfrak{m}][\mathfrak{m}]\varphi$
    (e) $Mon_{\mathfrak{m}}$: $[\mathfrak{m}_1]\varphi \to [\mathfrak{m}_2]\varphi$, if $\mathfrak{m}_1 \subseteq \mathfrak{m}_2$
    (f) $5_\sim$: $\langle\sim_P\rangle\varphi \to [\sim_P]\langle\sim_P\rangle\varphi$
    (g) Cov: $\varphi \to [\geq_P]\langle\leq_P\rangle\varphi$; $\varphi \to [\leq_P]\langle\geq_P\rangle\varphi$
2. Rules of inference:
    (a) Modus Ponens (MP): from φ and $\varphi \to \psi$, infer ψ
    (b) Necessitation (Nec): from φ infer $[\mathfrak{m}]\varphi$.

Fig. 2. The axiomatic system **gd**

Theorem 5. *For any $\Sigma \cup \{\varphi\} \subseteq \mathcal{L}_{gd}$, $\Sigma \vdash_{gd} \varphi$ iff $\Sigma \models_{gd} \varphi$.*

5.2   Boolean Combination of Modalities

While LGD has been expressive enough to represent and reason with decision rules for MCDA in the context of DRSA, we can still extend it beyond the application to DRSA. On one hand, we have never considered strict dominance or preference in the logics presented so far. On the other hand, it is shown in [1] that global modality is needed for representing preference relations between propositions (i.e. subsets of options). To accommodate these features, a more general approach is to take Boolean combinations of modalities into account [5]. Therefore, we will briefly sketch a Boolean modal logic of dominance (BMLD) here.

Let the alphabet C and Φ be the same as above and let the set of *modal atoms* be $MA = \{\leq_i, \geq_i | i \in C\}$. Then, we can define *modal terms* (or simply modalities) by using the standard Boolean algebra:

$$\mathfrak{m} = \mathfrak{a} \mid 1 \mid \overline{\mathfrak{m}} \mid \mathfrak{m} \cap \mathfrak{m} \mid \mathfrak{m} \cup \mathfrak{m},$$

where $\mathfrak{a} \in MA$; and the BMLD language, denoted by $\mathcal{L}_{bmd}$, has the same formation rules as those of LGD. We can also define some abbreviations of modal terms in a natural way. For example, $\leq_P = \bigcap_{i \in P} \leq_i$, $<_P = \leq_P \cap \overline{\geq_P}$, etc.

The semantic model of BMLD remains the same as before. However, given a model $\mathfrak{M} = \langle W, (\succeq_i)_{i \in C}, \Vdash \rangle$, we now need to define a binary relation $R(\mathfrak{m}) \subseteq W \times W$ for each modality $\mathfrak{m}$. In essence, R is a homomorphism from the Boolean algebra of modal terms to the algebra of sets $2^{W \times W}$ such that $R(1) = W \times W$, $R(\geq_i) = \succeq_i$, and $R(\leq_i) = \preceq_i$ for every $i \in C$. Then, the truth condition of modal formulas is defined as:

- $w \Vdash [\mathfrak{m}]\varphi$ iff for any $u \in W$ such that $R(\mathfrak{m})(w, u)$, $u \Vdash \varphi$.

According to the semantics, $[<_P]\varphi$ and $[1]\varphi$ represent above-mentioned strict dominance and global modality respectively.

5.3   Hybrid Logic

In previous formalisms, we use propositional symbols to describe properties of options. However, we cannot talk about the options themselves although the preference relations are defined between them at the semantic level. The issue can be addressed by using hybrid logic [2]. By adding to the original alphabet C and Φ a set of nominals Ω disjoint from Φ, the formulas of the Boolean hybrid logic of dominance (BHLD) are defined as follows:

$$\varphi ::= a \mid p \mid \bot \mid \varphi \to \varphi \mid [\mathfrak{m}]\varphi \mid @_a\varphi,$$

where $p \in \Phi$, $a \in \Omega$, and $\mathfrak{m}$ is a modal term as before. A nominal is basically a special kind of propositional symbol denoting a possible world (an option) and $@_a\varphi$ means that φ is true in the world denoted by a. Hence, the semantic model is still a triple $\mathfrak{M} = \langle W, (\succeq_i)_{i \in C}, \Vdash \rangle$. However, $\Vdash$ is now a relation between W and $\Phi \cup \Omega$ satisfying that for each $a \in \Omega$, there is *exactly* one $w_a \in W$ such that $w_a \Vdash a$. In addition, the truth condition of $@_a\varphi$ is:

$$w \Vdash @_a\varphi \text{ iff } w_a \Vdash \varphi.$$

Then, in the logic, we can use $@_a\langle \geq_P \rangle b$ to express that w_a P-dominates w_b directly.

6   Conclusion

In this paper, we propose several modal logics of preference and dominance for knowledge representation and reasoning with multi-criteria decisions in the context of DRSA, which range from the most basic formalism to the more expressive one that can represent all three types of decision rules induced from DRSA. While we have presented complete axiomatizations for logics of multi-criteria preference (and its special case), modal dominance, and general dominance, the axiomatic systems for other extensions remain to be explored. In addition, for the special class of total pre-ordered frames in logic of modal dominance, logic of general dominance, Boolean modal logic of dominance, and Boolean hybrid logic of dominance, we do not have complete axiomatizations yet. Lastly, implementation aspects of these logics and their practical applications in real-world scenarios are also important open issues for the future work.

References

1. van Benthem, J., Girard, P., Roy, O.: Everything else being equal: a modal logic for ceteris paribus preferences. J. Philos. Logic **38**(1), 83–125 (2009)
2. Blackburn, P.: Representation, reasoning, and relational structures: a hybrid logic manifesto. Logic J. IGPL **8**(3), 339–365 (2000)
3. Blackburn, P., de Rijke, M., Venema, Y.: Modal Logic. Cambridge University Press, Cambridge (2001)
4. Fariñas del Cerro, L., Orlowska, E.: DAL - a logic for data analysis. Theoret. Comput. Sci. **36**, 251–264 (1985)
5. Gargov, G., Passy, S.: A Note on Boolean Modal Logic. In: Petkov, P.P. (eds) Mathematical Logic (1990). Springer, Boston. https://doi.org/10.1007/978-1-4613-0609-2_21
6. Greco, S., Matarazzo, B., Slowinski, R.: Rough set approach to multi-attribute choice and ranking problems. In: Proceedings of the 12th International Conference on Multiple Criteria Decision Making, pp. 318–329 (1997)
7. Greco, S., Matarazzo, B., Slowinski, R.: Rough approximation of a preference relation by dominance relations. Eur. J. Oper. Res. **117**(1), 63–83 (1999)
8. Greco, S., Matarazzo, B., Slowinski, R.: Rough set theory for multicriteria decision analysis. Eur. J. Oper. Res. **129**(1), 1–47 (2001)
9. Liau, C.: An overview of rough set semantics for modal and quantifier logics. Int. J. Uncertainty Fuzziness Knowl.-Based Syst. **8**(1), 93–118 (2000)
10. Orlowska, E.: Logic for reasoning about knowledge. Zeitschrift für Mathematische Logik und Grundlagen der Mathematik **35**(6), 559–572 (1989)
11. Pagliani, P.: A practical introduction to the modal-relational approach to approximation spaces. In: Polkowski, L., Skowron, A. (eds.) Rough Sets in Knowledge Discovery 1: Methodology and Applications, pp. 209–232. Physica-Verlag (1998)
12. Pawlak, Z.: Rough Sets-Theoretical Aspects of Reasoning about Data. Kluwer Academic Publishers, Dordrecht (1991)
13. Słowiński, R., Greco, S., Matarazzo, B.: Rough set analysis of preference-ordered data. In: Alpigini, J.J., Peters, J.F., Skowron, A., Zhong, N. (eds.) RSCTC 2002. LNCS (LNAI), vol. 2475, pp. 44–59. Springer, Heidelberg (2002). https://doi.org/10.1007/3-540-45813-1_6
14. Slowinski, R., Greco, S., Matarazzo, B.: Rough sets in decision making. In: Meyers, R.A. (ed.) Encyclopedia of Complexity and Systems Science, pp. 7753–7786. Springer, New York (2009). https://doi.org/10.1007/978-1-4614-1800-9

GY MEDIC: Analysis and Rehabilitation System for Patients with Facial Paralysis

Gissela M. Guanoluisa$^{(\boxtimes)}$, Jimmy A. Pilatasig$^{(\boxtimes)}$, and Víctor H. Andaluz$^{(\boxtimes)}$

Universidad de las Fuerzas Armadas ESPE, Sangolquí, Ecuador
{gmguanoluisa1, japilatasig2, vhandaluz1}@espe.edu.ec

Abstract. The diagnosis and treatment of facial paralysis is timely in the first seventy-two hours to avoid implications on the return of motor functions of facial muscles. The proposed solution is a software made up of two modules for the analysis and rehabilitation of facial paralysis: *(i) The analysis module* detects and quantifies the level of asymmetry in the face, this is achieved with the inclusion of techniques for calculating the divergence between the vital points of the face located in the eyes, nose and mouth, measurements optimized with the gamma correction method; *(ii) The rehabilitation module* stimulates the facial nerves with physical exercises and monitors the patient's progress during therapy, for this purpose, a 3D virtual environment was developed that tracks the patient's facial gestures and, through these, directs a series of activities supported in the physical rehabilitation processes.

Keywords: Facial paralysis · Diagnosis · Treatment ·
Kinect and virtual environment

1 Introduction

Facial paralysis is a condition characterized by degeneration of the motor and sensory function of the facial nerve [1]. This condition may be caused by causes such as facial nerve infections, head trauma, metabolic disorders [2, 3]. It is a mildly prevalent disorder in children under 10 years of age, the elderly; and high in pregnant women, diabetics and people with a previous history of the disease, 67% of patients exhibit excessive tearing of the eyes, 52% have postauricular pain, 34% have taste disorders and 14% may develop phonophobia [4].

The brief treatment of facial pathology is essential to prevent the clinical picture of the patient decline, the patient's recovery varies from 15 to 45 days or can even extend to 4 years, depending on the level of affection of the patient. Facial paralysis has a great impact on the quality of life of the sufferer, afflicts the capacity for facial-emotional expression, interferes with activities such as eating, ingesting liquids and speaking [2], loss of movement in the eyelid, dysfunctional tearing, nasal dysfunction and obstruction [5].

Among the methods for clinical evaluation of facial function are. *(i) The House-Brackmann scale*, which establishes 6 degrees of facial phase shift: Grade I: Normal functioning, Grade II: Mild dysfunction, Grade III: Moderate dysfunction,

© Springer Nature Switzerland AG 2019
H. Seki et al. (Eds.): IUKM 2019, LNAI 11471, pp. 63–75, 2019.
https://doi.org/10.1007/978-3-030-14815-7_6

Grade IV: Moderately severe dysfunction, Grade V: Severe dysfunction and Grade VI: Total paralysis [6, 7], while *(ii) The Burres-Fisch system* provides a linear measurement index on a continuous graduated scale. All methods quantify the distances between reference points, both at rest and during voluntary movement [5, 8]; *(iii) The scoring method* is performed according to the movement of the points extracted as reference from the affected side and relates them to the healthy side.

In recent years, the widespread use of cameras equipped with stereo vision (Kinect) has increased security in facial recognition systems, using stereo images in clinical applications with great success, aimed at functions for rehabilitation of the patient. This research proposes a facial recognition system that uses passive stereoscopic vision to capture facial information. So far, face recognition techniques assume the use of active measurement to capture facial features [9]. The main problem in research carried out when using stereo vision for facial detection is its low accuracy, and therefore a treatment of the images extracted from the sensor is carried out applying the methods studied for their correction. Several investigations have been conducted on the use of Kinect for virtual rehabilitation in recent years. *(i) The pecado and Lee* operate the Kinect device for rehabilitation in clinics where they demonstrate significant improvements in the motor function of stroke patients [10, 11]. *(ii) Chang et al.* employs Kinect-based rehabilitation for people with motor disabilities, where extracted results show greater motivation of patients towards treatment and correction of their movement [12]. *(iii) Gonzalez-Ortega et al.* developed a Kinect-based 3D computer vision system for evaluation and cognitive rehabilitation, which has been successfully tested as a percentage of surveillance [13, 14].

A virtual environment supports two important concepts: *(i) Interaction*, the user is directly involved with the virtual environment in real time, *(ii) Immersion*, through hardware devices the patient has the feeling of being physically in the virtual environment, while for a virtual rehabilitation are repetition, feedback and patient motivation [7]. Therefore, a virtual environment was used that simulates a recreation context directed directly at the patient and his rehabilitation.

It is imperative for facial nerve recovery that the patient follow a traditional exercise guide to overcome these limitations and regain sufficient control of their movements. However, physical rehabilitation is a complex, long-term process with repetitive instructions that can diminish the patient's motivation and interest in moving forward with their treatment [15].

The proposed virtual rehabilitation system can be used for rehabilitation under medical supervision or at home. The advantages include the recreation of different treatment exercises in a virtual way, the configuration of the characteristics of the rehabilitation exercise according to the obtaining of patient data, providing a more familiar, comfortable and attractive environment, with tasks based on games. With all these advantages, virtual clinical rehabilitation increases patient motivation, interest and adherence to treatment.

eHealth (medical informatics) involves the insertion of technology into medicine, aims at preventive diagnosis and, above all, maintains a certain prominence when making medical decisions. mHealth covers the use of information technologies in a safe and effective way, with the support of mobile devices in tasks such as: the evaluation and monitoring of the patient's condition, surveillance and control of health risks [16].

In this research work, we present a medical diagnosis and rehabilitation system directly dedicated to patients affected with facial paralysis using the Microsoft Kinect sensor, a device that allows the transmission of facial information to the virtual environment through the use of gestures and face mapping; the information extracted goes through an image correction to reduce error percentages. The first analysis quantifies the degree of paralysis of the patient shown in the interface dynamically. The second analysis presents a criterion for patient recovery, a proposal in the form of a virtual game.

2 System Structure

The system developed in Unity3D aims to operate autonomously and as a support to medical specialists in the diagnosis of facial paralysis, with the compilation of the patient's facial characteristics and the calculation of the symmetry index in the areas of the eyebrows, eyes and mouth. In addition, it contains an alternative module for the rehabilitation of the disease, based on the gestures that the patient can perform during his clinical picture. Figure 1 presents the general scheme of operation of the medical system segmented into three parts: *(i) The inputs*, which manage the acquisition of facial data; it is proposed to use the Kinect device for the detection of the patient's facial coordinates used in the diagnosis and recognition of gestures used in rehabilitation, *(ii) The Unity interfaces*, contain the virtual environments of the medical system, operated by several scripts that manage the facial data, the calculation of the symmetry index and the control of the game, *(iii) The outputs*, displayed by screen; they expose the results in real time of the percentage of affection of the pathology and the score obtained in the virtual rehabilitation process.

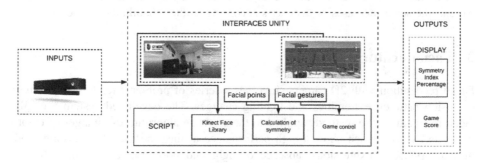

Fig. 1. Operation schema.

GY MEDIC system consists of the modules *(i) The analysis module*, described in Fig. 2 and *(ii) The rehabilitation module*, described in Fig. 3.

The module one subdivided in three sublayers *(a) Facial characteristics*, it processes the Kinect detection to capture the data, besides it maps the face in a model of visible points in screen for the final users, *(b) Error*, this section is in charge of filtering the facial coordinates used to estimate the damage of the disease, the software collects data only

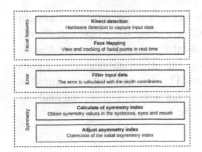

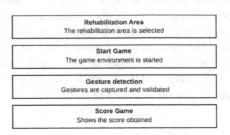

Fig. 2. Analysis module.   **Fig. 3.** Rehabilitation module.

when the user is in frontal position with respect to the Kinect with a margin of error less than 2%, *(c) Symmetry*, responsible for quantifying the degree of facial paralysis, applies correction methods to deliver accurate results of the analysis carried out.

Module two comprises the following structure: *(a) Rehabilitation area*, in sequence to the diagnostic process and immersed in the rehabilitation environment the user is assigned an exercise routine, this depends on the area affected by facial paralysis, *(b) Start Game*, the module begins with the registration of the patient in the Firebase Realtime Database of GY MEDIC housed in the cloud; This record is responsible for keeping a clinical picture of the progress of the user's treatment; below are presented series of physiotherapeutic exercises controlled by a time counter and distributed in series of times of 30, 60 and 90 s depends on the degree found in the user all this focused on three areas: eyebrows, eyes and mouth, *(c) Gesture detection*, the presence of the Kinect hardware is evident, the gestures detected in the patient are recognized and validated, *(d) Score Game*, in charge of weighing the results achieved by the patient, the score increases or decreases according to the rehabilitation process.

3   Facial Features Detection

Facial recognition with 2D cameras involves a series of problems related to gestures, occlusion between extremities and lighting changes. The developed system uses 3D facial recognition, the methods applied to the software algorithm use images of color, depth and space, which allow a high performance against the use of accessories and involuntary facial expressions, absence of light and noise tolerance. The software developed with this technology has a great impact on people with physical disabilities as well as on the diagnosis and rehabilitation of pathologies; their cost, performance and robustness make them very effective in medical support [13, 17]. The software solution incorporates an RGB panel provided by the *ColorFrameReader* attribute for the visualization of the patient's face; a 3D model superimposed on the real image is generated from the 2D reference points mentioned [9].

The tracking of the face in real time, the obtaining of coordinates and the decrease of errors when acquiring data implies the use of the following attributes: *(a) CameraSpacePoint*, represents a 3D point in the space of the camera (in meters);

(b) DepthSpacePoint, represents pixel coordinates of an image of depth inside [18–20]. The tracking of the face is a process that allows the visualization and tracking of the facial points superimposed on the image of the patient, offering a better interaction with the system in terms of movement, detection of features and diagnosis by a medical professional.

For the initial quantification of the pathology, sixteen reference points located in specific areas of the face detailed in Fig. 4 were captured, points extracted using the *Microsoft.Kinect.Face* library referring to the *HighDetailFacePoints* attribute that provides up to 1300 facial points [3].

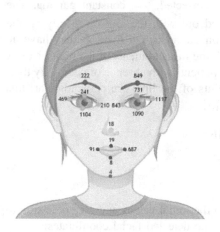

Coordinate	Detail
222	Middle of Eyebrows
849	Middle of Eyebrows
241	Eye Top
469	Eye Outer Corner
210	Eye Inner Corner
1104	Eye Bottom
731	Eye Top
843	Eye Inner Corner
1117	Eye Outer Corner
1090	Eye Bottom
18	Tip of the Nose
19	Mouth Top
91	Mouth Corner
687	Mouth Corner
8	Mouth Bottom
4	Chin Center

Fig. 4. Referential face points.

The process of estimating the symmetry index (IS) includes calculating the distances between the reference points of the face (see Table 1) and their differences expressed in percentages; these variations decrease or increase according to the patient's level of affection [3].

Table 1. IS distances.

Name	Distance	Difference
Right eyebrow - Nose	$d_1 = d_{222\_18}$	$\frac{d_1}{d_2}$
Eyebrow left - Nose	$d_2 = d_{849\_18}$	
Vertical right eye	$d_1 = d_{241\_1104}$	
Vertical left eye	$d_2 = d_{731\_1090}$	
Left lip - Nose	$d_1 = d_{91\_18}$	
Right lip - Nose	$d_2 = d_{687\_18}$	

(i) Gamma correction method

The gamma correction method consists of a non-linear mathematical operation that specifies the relationship between a pixel and its luminance; it is applied to images of modern visualization systems to improve contrast [21]. The proposed system adopts this method to reduce the error of quantification and magnify the measurements in the event that the patient presents the disease [3].

$$V_{output} = KV_{input}^{\gamma} \tag{1}$$

The gamma function with V_{output} represents the corrected output value, K represents the constant, V_{input} represents the value to be corrected, γ = constant gamma. The readjustment of the symmetry indices is carried out with the aim of specifying the results obtained and the correction is focused on the areas: *(a) Eyes*: as they have an elliptical shape, the areas can be calculated with the measurements of the major axis A and the least axis B by applying (2). *(b) Mouth*, patients with absence of pathology their vectors u (Nose - Chin) and v (left - right points of the lips) are perpendicular, this measure varies according to the condition (3) [3].

$$A_{elipse} = AB\pi \tag{2}$$

$$\theta = \frac{u.v}{|u||v|} \tag{3}$$

These modifications suppose a correction to the initial values and give precision to the system, being more sensitive to changes in the detected facial coordinates.

4   Virtual Environment

The presence of support systems in clinical decision making is insufficient in medical centers and rehabilitation programs, a system of this type considers some issues: cost, size, operation and automation. Rehabilitation software reduces the cost of personal therapy, increases patient motivation, and evaluates patient progress with satisfactory results [13].

One of the problems faced in the development of the application is the GUI, due to the patient-computer interaction. In particular, the concept of usability of software development maintains that the essential factor to be considered is the ease of use, parallel to the actions performed by the user that are described as in Fig. 5, this affects the precision requirements of the system in the mechanism of processing functions [22, 23].

The SketchUp software has the ability to design any 3D environment or virtual object and even allows assigning properties such as textures, colors and animations that can be used in Unity. The process for the development of the virtual environment in Unity, is guided by the steps: *(i) Import*, in Unity are used the models 3D * .fbx, generated in SketchUp or other modeling software, *(ii) Properties*, you have the option to scale, modify their textures and add new features of Unity, *(iii) Control*, the structure

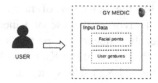

Fig. 5. Input data.

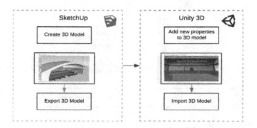

Fig. 6. Exporting a 3D model to unity.

of the system depends on the configuration and programming of the entries, the process of exporting a model to Unity is described in Fig. 6 [24, 25].

5 Experimental Results

5.1 Module I: Analysis

The following results allow to observe the performance that the user shows when using "GY MEDIC". The analysis module uses the coordinates of the facial points provided by the Kinect device, these coordinates are used in the calculation of the symmetry of the face. The main interface displays a window that allows to visualize the face of the user during his examination, superimposed we have sixteen points of green color that carry out the tracking of the face in coordination with the real image of the user, finally there are three text boxes with their respective indicators that correspond to the percentage of asymmetry of the analyzed zones, described in Fig. 7.

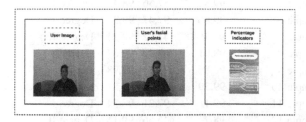

Fig. 7. Analysis and execution module.

The process of analysis for the detection of facial paralysis begins with the preparation of the scenario that involves the use of Kinect V2, a computer and the Kinect V2 adapter for computer. In sequence the patient's location with respect to the Kinect suggested in 1.4 m and the height of the Kinect with respect to the floor with a minimum of 0.6 and a maximum of 1.8 m. The system is switched on, which validates the presence of the sensor and initiates the tracking of the face. Finally, the analysis is carried out under medical assistance and the results of the examination are displayed on the screen in real time as described in Fig. 8.

Fig. 8. Analysis process.

In order to obtain experimental data in the analysis phase, samples from several participants have been used, the percentages are expected to vary according to the positioning with respect to the sensor, according to their facial characteristics and if they suffer any indication of facial paralysis. "GY MEDIC" uses a function to validate the input data; when the user is out of the recommended position the values are completely discarded to avoid erroneous results, this validation considers an error less than 2%. The values shown in Table 2 have been classified on the House-Brackmann scale which considers 6 degrees of dysfunction. In addition, the results showed some susceptibility of the system to changes in user gestures mainly by blinking.

Table 2. Results of the House-Brackmann gradation scale initial samples.

Samples	Percentage			House-Brackmann
	Eyebrows	Eyes	Mouth	
Sample 1	99,76	98,5	98,12	Degree I
Sample 2	99,72	57,31	97,2	Degree I
Sample 3	99,32	62,34	98,32	Degree I
Sample 4	99,21	94,4	97,98	Degree I
Sample 5	99,82	57,56	98,45	Degree I
Sample 6	99,86	97,47	99,2	Degree I
Sample 7	99,46	47,82	98,23	Degree II
Sample 8	98,91	63,47	97,87	Degree I
Sample 9	98,86	72,5	97,98	Degree I
Sample 10	99,39	57,68	98,21	Degree I
Sample 11	99,1	97,21	98,34	Degree I
Sample 12	98,7	45,56	99,11	Degree II
Average	99,34	70,99	98,25	Degree I
Standard deviation	0,40	20,35	0,53	

5.2 Module II: Rehabilitation

The rehabilitation module implements a total of twelve physiotherapeutic exercises, divided into three levels; that is, four games for each area: eyebrows, eyes and mouth. The game starts with the *(i) Initial Interface*, the data of the users are stored in the Firebase database of real time in the cloud; necessary record for the follow-up of the clinical picture of the patient (Figs. 9 and 10); *(ii) Main Menu*, through a menu of options of the three facial areas the user chooses with which option he wants to start the game, described in Fig. 11; *(iii) Indications Interface*, interface where the user in a period of three seconds receives indications of the exercise to be performed; it should be noted that the levels are being assigned according to their degree of affection (Fig. 12); *(iv) Game Interface*, dynamic interface contained of animations, sound effects where the user in a lapse of thirty seconds must collect thirty stars one for each second. The upper part of the game shows information about the muscular area that is being rehabilitated (Fig. 13). It should be pointed out that the user will only be able to go to the next level if the proposed exercise is carried out correctly in such a way that in all the interfaces of the virtual environment it contains a camera that shows in real time the tracking of the Kinect sensor.

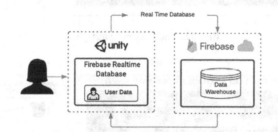

Fig. 9. Data warehouse schema.

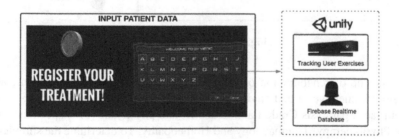

Fig. 10. Initial interface.

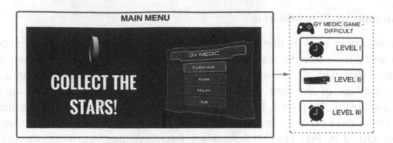

Fig. 11. Main menu interface.

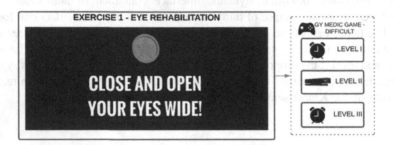

Fig. 12. Indications interface.

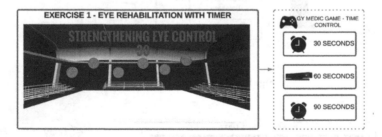

Fig. 13. Game interface.

The module of rehabilitation of patients with facial paralysis through interaction with a semi immersive virtual environment is put to validation with a group of users (FP) with facial paralysis Grade I and Grade II; and physicians (FM) specialists in the areas of Physiotherapy, Traumatology and Homeopathy. To determine the simplicity of the operation, a usability test is elaborated under several criteria of: learning, efficiency, user experience among others; through a bank of questions in Table 3, in a validation factor on ten points representing as a positive answer on the contrary a validation factor of zero as negative.

Table 3. Bank of questions under usability criteria.

FP1. It's a technological solution that reduces costs, time and space
FP2. "GY MEDIC" facilitates semi-immersive interaction with hardware and software
FP3. The management of virtual environments is no longer a myth, its use is relatively easy
FP4. The execution of the interface is simple and intuitive
FP5. The visual design, typographies, colors, animations are imperceptible
FP6. The system can be implemented in the work area of all specialists for pathology purposes
FP7. The hardware devices do not cause any discomfort, on the contrary, it generates in me the desire to surpass more levels in the game
FM1. Physiotherapist assures that the patients are enthusiastic while performing their rehabilitation, you can see that the patient's experience is positive
FM2. As a specialist in the area of Traumatology, "GY MEDIC" facilitates a precise clinical analysis in comparison to traditional techniques that are usually diagnosed
FM3. As a Homeopath the implementation of improvement levels seen in each game makes the patient's rehabilitation progress more rapid

6 Conclusion

This research presents a medical system for the analysis and rehabilitation of facial paralysis developed in Unity 3D, which can be used autonomously by the user or as a software that assists a medical specialist. The analysis module processes face depth coordinates in real time that optimizes the results when faced with problems such as involuntary movement of the patient and lack of light; the final quantification yields a percentage in the eyebrows, eyes and mouth area, with the highest values determining the presence of the pathology. The rehabilitation module presents a minimalist design according to the concepts of software usability, with an entertaining environment, intuitive and easy to use, which preserves the patient's attention for the treatment, the virtual rehabilitation exercises suggested by the system are based on current techniques used by physiotherapists, designed for each affected area, the final result of the game is an increase according to the recovery of the movement of the facial nerve of the patient.

Acknowledgements. In addition, the authors would like to thanks to the Corporación Ecuatoriana para el Desarrollo de la Investigación y Academia–CEDIA for the financing given to research, development, and innovation, through the CEPRA projects, especially the project CEPRA-XI-2017-06; Control Coordinado Multi-operador aplicado a un robot Manipulador Aéreo; also to Universidad de las Fuerzas Armadas ESPE, Universidad Técnica de Ambato, Escuela Superior Politécnica de Chimborazo, Universidad Nacional de Chimborazo, and Grupo de Investigación ARSI, for the support to develop this paper.

References

1. Pérez, E., et al.: Guía Clínica para rehabilitación del paciente con parálisis facial perférica. IMSS. **5**, 425–436 (2004)
2. Meléndez, A., Torres, A.: HOSPITAL GENERAL Perfil clínico y epidemiológico de la parálisis facial en el Centro de Rehabilitación y Educación Especial de Durango. México Hosp. Gen. México **69**, 70–77 (2006)
3. Gaber, A., Member, S., Taher, M.F., Wahed, M.A.: Quantifying facial paralysis using the Kinect v2. In: Proceedings of Annual International Conference IEEE Engineering in Medicine and Biology Society EMBS, pp. 2497–2501 (2015)
4. Park, J.M., et al.: Effect of age and severity of facial palsy on taste thresholds in Bell's Palsy patients. J. Audiol. Otol. **21**, 16–21 (2017)
5. Cid Carro, R., Bonilla Huerta, E., Ramirez Cruz, F., Morales Caporal, R., Perez Corona, C.: Facial expression analysis with kinect for the diagnosis of paralysis using Nottingham grading system. IEEE Lat. Am. Trans. **14**, 3418–3426 (2016)
6. Fonseca, K.M.D.O., Mourão, A.M., Motta, A.R., Vicente, L.C.C.: Scales of degree of facial paralysis: analysis of agreement. Braz. J. Otorhinolaryngol. **81**, 288–293 (2015)
7. Holtmann, L.C., Eckstein, A., Stähr, K., Xing, M., Lang, S., Mattheis, S.: Outcome of a graduated minimally invasive facial reanimation in patients with facial paralysis. Eur. Arch. Oto-Rhino-Laryngol. **278**, 3241–3249 (2017)
8. Brenner, M.J., Neely, J.G.: Approaches to grading facial nerve function. Semin. Plast. Surg. **18**, 13–22 (2004)
9. Uchida, N., Shibahara, T., Aoki, T., Nakajima, H., Kobayashi, K.: 3D face recognition using passive stereo vision. In: IEEE International Conference on Image Processing 2005, ICIP 2005, vol. 2, p. II-950-3 (2005)
10. Sin, H., Lee, G.: Additional virtual reality training using Xbox Kinect in stroke survivors with hemiplegia. Am. J. Phys. Med. Rehabil. **92**, 871–880 (2013)
11. Webster, D., Celik, O.: Systematic review of Kinect applications in elderly care and stroke rehabilitation. J. Neuroeng. Rehabil. **11**, 1–24 (2014)
12. Chang, Y.J., Chen, S.F., Huang, J.D.: A Kinect-based system for physical rehabilitation: a pilot study for young adults with motor disabilities. Res. Dev. Disabil. **32**, 2566–2570 (2011)
13. González-Ortega, D., Díaz-Pernas, F.J., Martínez-Zarzuela, M., Antón-Rodríguez, M.: A kinect-based system for cognitive rehabilitation exercises monitoring. Comput. Methods Programs Biomed. **113**, 620–631 (2014)
14. Zhao, L., Lu, X., Tao, X., Chen, X.: A kinect-based virtual rehabilitation system through gesture recognition. In: 2016 International Conference on Virtual Reality and Visualization, pp. 380–384 (2016)
15. Young, J., Forster, A.: Rehabilitation after stroke. N. Engl. J. Med. **334**, 86–90 (2007)
16. Silano, M.: La Salud 2.0 y la atención de la salud en la era digital. Rev. Médica Risaralda **19**, 1–14 (2013)
17. Arenas, Á.A., Cotacio, B.J., Isaza, E.S., Garcia, J.V., Morales, J.A., Marín, J.I.: Sistema de Reconocimiento de Rostros en 3D usando Kinect. In: XVII Symposium of Image, Signal Processing Artificial Vision (2012)
18. Microsoft: CameraSpacePoint Structure. https://msdn.microsoft.com/en-us/library/windows preview.kinect.cameraspacepoint.aspx
19. Microsoft: DepthSpacePoint Structure. https://msdn.microsoft.com/en-us/library/windows preview.kinect.depthspacepoint.aspx
20. Microsoft: ColorFrameReader Class. https://msdn.microsoft.com/en-us/library/windows preview.kinect.colorframereader.aspx

21. Cao, Y., Bermak, A.: An analog gamma correction method for high dynamic range applications, pp. 318–322 (2011)
22. Villaroman, N.H., Rowe, D.C.: Improving accuracy in face tracking user interfaces using consumer devices. In: Proceedings of the 1st Annual Conference on Research in Information Technology - RIIT 2012, p. 57 (2012)
23. Andaluz, V.H., et al.: Transparency of a bilateral tele-operation scheme of a mobile manipulator robot. In: De Paolis, L.T., Mongelli, A. (eds.) AVR 2016. LNCS, vol. 9768, pp. 228–245. Springer, Cham (2016). https://doi.org/10.1007/978-3-319-40621-3_18
24. Carvajal, C.P., Proaño, L., Pérez, J.A., Pérez, S., Ortiz, J.S., Andaluz, V.H.A.: Robotic applications in virtual environments for children with autism. In: Third International Conference, AVR 2016 Lecce, Italy, 15–18 June 2016 Proceedings, part II, vol. 1, pp. 402–409 (2016)
25. Andaluz, V.H., et al.: Unity3D virtual animation of robots with coupled and uncoupled mechanism. In: De Paolis, L.T., Mongelli, A. (eds.) AVR 2016. LNCS, vol. 9768, pp. 89–101. Springer, Cham (2016). https://doi.org/10.1007/978-3-319-40621-3_6

Revealed Preference for Network Design in Bilevel Linear Programming

Puchit Sariddichainunta[✉] and Masahiro Inuiguchi

Graduate School of Engineering Science, Osaka University, Toyonaka, Osaka, Japan
puchit@inulab.sys.es.osaka-u.ac.jp, inuiguti@sys.es.osaka-u.ac.jp

Abstract. We study the aggregation of user preferences in the network flow model. These users can be referred to the followers in the sequential decision model. We transform the preference data revealed by the followers into a bundle of linear constraints to represent the strategic norms of the followers in the centralized decision model. We also show that the revealed preference theory is a useful foundation to construct such a multiple users' rational norm without losing much the richness of preference by our simplification. Although the aggregated preference results in a set of ambiguous decision norms of the representative follower, in our case a convex polyhedron, we can apply this framework to the bilevel optimization and formulate this sequential decision problem as a maximin problem which is the centralized decision problem of the leader in our network flow model.

Keywords: Revealed preference · Network design ·
Bilevel linear programming · Data aggregation

1 Introduction

Bilevel optimization problems, a subclass of sequential decision problems, are found in many fields in the real world such as economic planning, management, transportation network, and known in non-cooperative game theory as sequential games. Stackelberg competition [22] is the original research of optimization problems for a duopoly quantity competition, which has a sequential strategic decision of the leader and the follower. First, the leader on the upper level decides a strategic choice. The follower on the lower level then makes a decision after considering the leader's decision. Following to the assumption that the leader can observe the follower's rational reaction, the leader takes the follower's reactions into his constraints and commit the optimal solution. Consequently, this optimal solution will be an equilibrium outcome of both decision makers.

In operations research, many scholars have incorporated linear programming problems into this kind of sequential decision model known as the bilevel linear programming (BLP). Bialas and Karwan [6] initially formulated bilevel linear optimization problem to model the role of central government to redistribute national budgets. Bard [2] proposed a model for decentralized organization with

© Springer Nature Switzerland AG 2019
H. Seki et al. (Eds.): IUKM 2019, LNAI 11471, pp. 76–85, 2019.
https://doi.org/10.1007/978-3-030-14815-7_7

superior and subordinate units controlling different decision variables. LeBlanc and Boyce [13] formulated a network design problem for the road traffic as a bilevel program which assumed linear investment functions. However, the solution method of [13] was conjectured by a parametrized single-level linear program, which did not fully reflect the rational response of the decision maker at the lower level. The legitimate solution method for such BLP formulation for a network design problem was later developed by Ben-Ayed et al. [5].

Some researchers applied the real dataset to analyze for an optimal solution in the sequential decision model. Ben-Ayed et al. [5] constructed a BLP model to optimize the investment in the inter-regional highway network of Tunisia. This study provided a network-based approach to find an optimal budget allocation to improve links in a road network system. The allocation by the central agency may conflict with those individual users. The utility maximization behavior of individual users could result in a different solution to what the centralized planner though optimal to the entire system. Hence, by considering such rational reactions of the users, BLP formulation is necessary to enhance a better decision which reflects the real world behavior.

The fact remains that the central agency's observation of the users' rational reaction is not crystal clear in the real world. The survey of user willingness to pay in each traffic route might be neither complete nor rational one. In our work, we apply the ideas of revealed preference theory which is studied in microeconomic analysis [8,14] to aggregate the users preference data in the network design (NWD) [5,13]. We propose a method to construct the preference data into a bundle of linear constraints to represent the strategic norms of the followers in the centralized decision model as a convex polyhedron. Although our proposed method to aggregate the preference data flow model results in an ambiguous decision norm, we can formulate it as a BLP with the follower's ambiguous objective function (BLPwAOF) [11,21]. In this model, we also derive a robust global optimal solution which will be the leader's optimal solution in such circumstance.

The organization of this paper is as follows. We formulate the NWD problem in BLP and clarify what should be the decision norms in Sect. 2. After introducing the aspect of revealed preference in Sect. 3, we propose the procedure to aggregate data to represent the rational behavior of the users. Finally, the conclusion and future works are discussed in Sect. 4.

2   Bilevel Linear Program for Network Design

Our NWD problems in this study refer to a challenge to improve the capacity of a given network flow [5,13,15]. It is common to formulate the NWD in the bilevel linear programming (BLP) [3,11,21]. NWD can have many settings especially the conformation of travelling cost of users in the network. We will model NWD as a minimal cost flow problem to represent the users rational behaviors and develop it as BLP to determine a set of links to be improved.

2.1 Problem Formulation

The network design (NWD) here is to find the optimal improvement of capacity in the linkage between nodes, corresponding to the choice of users in the network flow model. Let us consider a traffic network servicing several users who travel from an origin to a destination. We construct this situation like the conventional network flow problem [4] and can be generalized to the case of multiple ones [24].

Let $G = (N, A)$ be a directed network with $N = \{1, \dots, m\}$ the set of nodes and $A \subseteq N \times N$ the set of the directional link between nodes. The node i in the network G is associated with b_i, namely, the available supplies if $b_i > 0$ and the required demand if $b_i < 0$. When $b_i = 0$, the node i required none of unit and is called an intermediate. Corresponding to each directional link, $x_{ij} \geq 0$ is the amount of flow and $\tilde{c}_{ij}$ is the representative travelling cost on the link. We also assume that the total supply equals the total demand within the network, $\sum_{i=1}^{m} b_i = 0$. If not the case, we can add a dummy node $m+1$ with zero cost from each supply node to absorb the remaining ones. The constraints in the lower-level problem is the node-arc flow balancing equation [3,4]. Hence, we defined the lower-level problem of the network users. In BLP, we call them the followers.

Next, we consider the decision variables for the decision maker at the upper-level problem, so-called the leader in BLP. The decision maker such as the government has a mission to improve the network by increasing the capacity of the links. We define v_{ij} the decision variable to modify the capacity of the link with the cost coefficient τ_{ij}, while the current capacity of the link is u_{ij}. The leader has to observe the followers wisely to determine the most effective way to improve the link capacity, because the follower can affect to his objective function through the coefficients ψ_{ij}. It is straightforward to construct the BLP for the NWD as follows:

$$\begin{aligned}
\underset{v}{\text{Minimize}} \quad & \sum_{(i,j) \in A} \tau_{ij} v_{ij} + \sum_{(i,j) \in A} \psi_{ij} x_{ij}, \\
\text{Subject to} \quad & v_{ij} \geq 0, \quad \forall (i,j) \in A, \\
& \text{where } x \text{ solves,} \\
\underset{x}{\text{Minimize}} \quad & \sum_{(i,j) \in A} \tilde{c}_{ij} x_{ij}, \\
\text{Subject to} \quad & \sum_{j=1}^{m} x_{ij} - \sum_{k=1}^{m} x_{ki} = b_i, \forall i \in N, \\
& 0 \leq x_{ij} \leq u_{ij} + v_{ij}, \quad \forall (i,j) \in A.
\end{aligned} \tag{1}$$

2.2 The Decision Norms in the Network Flows

The decision norms of the leader and the followers are not yet explained in the above BLP formulation. It is not complicated for the leader's decision norm because he is a single entity to make a decision in the upper-level problem. So that, we can assume the leader's cost coefficients, τ_{ij} and ψ_{ij}, are known. However, there are multiple users in the traffic networks. They want to minimize their travelling cost, and their choices to use the network links might be varied by each individual preference. We review what could be the decision norms and some treatments for the coefficient vector in the lower-level problem of BLP, $\tilde{c}_{ij}$ in this subsection.

Observing the users' choices in the network, we can obtain the data such as travel costs that potentially demonstrate the preference of the network users. The travelling cost can be the summarized of all costs incurring in their trips such as opportunity costs of travelling time, road or transportation fee, special expenses on their commutes and so on. It also depends on the combination of choices to pass from nodes to nodes. The choices do not necessarily coincide and may conflict to each other because of the capacity limitation of flow in the network traffic. Hence, we expect the observed preferences in the dataset will represent the rationality which the users maximize each individual utility based on each preserved travel costs and budget constraints.

In the conventional BLP problem formulation [3,10], each decision maker is assumed to have the complete information about the program; however, the precise information of counterpart cannot be directly obtained in reality. Many researchers have noticed the difficulties in the treatment of the real data for the parameters in the follower's objective function. For example, the coefficients in the objective function of the lower-level problem in the model of Ben-Ayed et al. [5] was adjusted and estimated by the correlated data. As well, the fuel cost in the bilevel model of Islam [12] was the parameters estimated by the experts. On the other hand, there are many BLP models which assume the follower's coefficients are imprecise such as intervals [7,17], random variables [9,16] and fuzzy numbers [18,19].

The aforementioned BLP models assumed the follower's decision norms as the imprecise coefficients, and proposed the computation methods taking the strategic move of the lower-level into account. But, these models did not give any insights to the observable data of user choices to characterize the imprecise coefficients $\tilde{c}_{ij}$. They hypothesized rational reactions only in the strategic move. On the contrary, this research work utilizes the theory of revealed preference to fill the gap between the observable choices and the rational behavior achievable in the dataset.

3   Revealed Preference

Choices reveal preferences and underline the rational behaviours [8]. In microeconomics analysis, the consumption choices are based on the utility maximization hypothesis. It is common to assume the utility function for the construction of preference relations [14]. However, given the observed expenditure data with prices and chosen bundles, the principle of revealed preference can answer the questions about the consistency of observed behaviors to the utility maximization and form the utility function to represent the observations [23]. In this section, we introduce the principle of revealed preference and apply it to construct the lower-level decision norms of follower in BLP.

3.1   Mathematical Setting

We follows the mathematical setting in [8]. Let us suppose that we have the data of price and choice of consumption bundles, $D = \{(\boldsymbol{x}^k, \boldsymbol{p}^k) \in \mathbf{R}_+^L \times \mathbf{R}_{++}^L, k = 1, \ldots, K\}$ is the consumption dataset observed from K individuals. The quantity bundle of $\boldsymbol{x}^k$ is non-negative vector and the price vector $\boldsymbol{p}^k$ is strictly positive. We say,

it is revealed that $\boldsymbol{x}^k$ is preferred to $\boldsymbol{y}$ $\boldsymbol{x}^k \succeq \boldsymbol{y}$ if $\boldsymbol{p}^k \cdot \boldsymbol{y} \leq \boldsymbol{p}^k \cdot \boldsymbol{x}^k$,

it is revealed that $\boldsymbol{x}^k$ strictly preferred to $\boldsymbol{y}$ $\boldsymbol{x}^k \succ \boldsymbol{y}$ if $\boldsymbol{p}^k \cdot \boldsymbol{y} < \boldsymbol{p}^k \cdot \boldsymbol{x}^k$.

Also, we suppose that $\boldsymbol{x}^k$ is purchased at $\boldsymbol{p}^k$ and the total expenditure is $\boldsymbol{p}^k \cdot \boldsymbol{x}^k$.

Definition 1. *Weak rationalization*
A preference relation $\succeq$ weakly rationalizes a consumption dataset D if for all k and $\boldsymbol{y} \in X, \boldsymbol{p}^k \cdot \boldsymbol{x}^k \geq \boldsymbol{p}^k \cdot \boldsymbol{y}$ implies that $\boldsymbol{x}^k \succeq \boldsymbol{y}$.

Definition 2. *Weak axiom of revealed preference (WARP)*
D satisfies WARP, if for $(\boldsymbol{x}^k, \boldsymbol{p}^k), (\boldsymbol{x}^l, \boldsymbol{p}^l) \in D$, $\boldsymbol{x}^k$ is revealed preferred to $\boldsymbol{x}^l$ and $\boldsymbol{x}^k \neq \boldsymbol{x}^l$, then $\boldsymbol{x}^l$ is not revealed preferred to $\boldsymbol{x}^k$.

Definition 3. *Generalized axiom of revealed preference (GARP)*
D satisfies GARP if for $\{(\boldsymbol{x}^{k(n)}, \boldsymbol{p}^{k(n)}), n = 1, \ldots, N\} \subset D$, $\boldsymbol{x}^{k(n)}$ is revealed preferred to $\boldsymbol{x}^{k(n+1)}$ for $n = 1, \ldots, N + 1$ then $\boldsymbol{x}^{k(N+1)}$ is not strictly revealed preferred to $x^{k(1)}$.

Definition 4. *Local non-satiation*
A preference relation $\succeq$ on $X \in \mathbf{R}^n$ is locally nonsatiated if for all x and $\epsilon > 0$, there is some $y \in X$ such that $\|x - y\| < \epsilon$ and $y \succ x$.

Theorem 1. *Afriat's theorem [1, 8]*
Let X be a convex consumption space, and $D = \{(\boldsymbol{x}^k, \boldsymbol{p}^k)\}_{k=1}^K$ be a consumption dataset. The following statements are equivalent.

(I) D has a locally non-satiated weak rationalization;
(II) D satisfies GARP;
(III) There are strictly positive real number U^k and λ^k, for each k, such that

$$U^k \leq U^l + \lambda^l \boldsymbol{p}^l \cdot (\boldsymbol{x}^k - \boldsymbol{x}^l)$$

for each pair of observations $(\boldsymbol{x}^k, \boldsymbol{p}^k)$ and $(\boldsymbol{x}^l, \boldsymbol{p}^l) \in D$
(IV) D has a continuous, concave, and strictly monotonic rationalization
  $u : X \to \mathbf{R}$.

3.2   Data Aggregation Model

The principle of revealed preference is useful to check whether the rational behaviors coincide in the dataset. However, it is not straightforward for the dataset of choice in the network flows which are the combination of links between nodes to reflect the hypothesized rational reactions. In this subsection, we propose a method to handle this kind of reveal preferences in order to represent the rationality from the dataset.

Let D be a dataset surveyed from the network users that the amount of unit costs incurred when they use the link $(i, j) \in A$ of their concerned routes. We suppose that there are $|A| = L$ links in the network and N people who replied the survey. Then, we can define the nth observation: $\boldsymbol{x}^n = (x_1^n, \ldots, x_L^n)$ and $\boldsymbol{p}^n = (p_1^n, \ldots, p_L^n)$. When a network user revealed a willingness to pay costs for a certain link p_t^n, the corresponding quantity $x_t^n = 1$. On the other hand, $x_t^n = 0$ if users do not pass that link and no costs incur. Hence, $\boldsymbol{x}^n$ could contain many zero in the vector and create an answer pattern.

Let $S \in \mathbf{R}_+^{K \times L}$ be a matrix containing the unit cost of L links for K patterns. The L-element vector $\boldsymbol{\alpha} = (\alpha^1, \cdots, \alpha^L)^{\mathrm{T}}$ is the scope of reaction to the unit cost we can discover from the observed dataset D. In fact, the element $S^{i,j}$ in matrix S corresponds to the vector $\boldsymbol{x}^n$ and the vector $\boldsymbol{\alpha}^j$ corresponds to the vector $\boldsymbol{p}^n$ in the consumption dataset.

Moreover, we define a social welfare function to indicate the welfare value generating from the patterns domain maps to the number of answers, $\boldsymbol{W} = \boldsymbol{S}\boldsymbol{\alpha}$. The element α^j interacts with the column vector in the matrix S to indicate the contribution to the level of welfare. We also assume that the pattern with more answers is more meaningful in the dataset and also imply more social welfare. Intuitively we can consider that the pattern becomes important because many commuters rely on it. We use the following linear equations to represent the welfare of each answered pattern:

$$
\begin{aligned}
W^1 &= \alpha^1 S^{1,1} + \alpha^2 S^{1,2} + \alpha^3 S^{1,3} + \ldots + \alpha^L S^{1,L} \\
W^2 &= \alpha^1 S^{2,1} + \alpha^2 S^{2,2} + \alpha^3 S^{2,3} + \ldots + \alpha^L S^{2,L} \\
&\vdots \\
W^K &= \alpha^1 S^{K,1} + \alpha^2 S^{K,2} + \alpha^3 S^{K,3} + \ldots + \alpha^L S^{K,L}
\end{aligned}
\tag{2}
$$

Without loss of generality, we further assume that $W^1 \geq W^2 \geq \ldots \geq W^K$. After the subtraction between the equation of W^1 and the others, the system of equations in (2) becomes as follows,

$$
\boldsymbol{0} \leq
\begin{bmatrix}
(S^{1,1} - S^{2,1}) & (S^{1,2} - S^{2,2}) & (S^{1,3} - S^{2,3}) & \ldots & (S^{1,L} - S^{2,L}) \\
\vdots & \vdots & \vdots & \ddots & \vdots \\
(S^{1,1} - S^{K,1}) & (S^{1,2} - S^{K,2}) & (S^{1,3} - S^{K,3}) & \ldots & (S^{1,L} - S^{K,L})
\end{bmatrix}
\begin{bmatrix}
\alpha^1 \\
\alpha^2 \\
\alpha^3 \\
\vdots \\
\alpha^L
\end{bmatrix}
\tag{3}
$$

If we set $\alpha^1 = 1$, the system of equations in the matrix form is as follows:

$$\begin{bmatrix} S^{2,1} - S^{1,1} \\ \vdots \\ S^{K,1} - S^{1,1} \end{bmatrix} \leq \begin{bmatrix} (S^{1,2} - S^{2,2}) & (S^{1,3} - S^{2,3}) & \cdots & (S^{1,L} - S^{2,L}) \\ \vdots & \vdots & \ddots & \vdots \\ (S^{1,2} - S^{K,2}) & (S^{1,3} - S^{K,3}) & \cdots & (S^{1,L} - S^{K,L}) \end{bmatrix} \begin{bmatrix} \alpha^2 \\ \alpha^3 \\ \vdots \\ \alpha^L \end{bmatrix} \quad (4)$$

$$\beta \leq B\gamma. \quad (5)$$

Now, we derive the scope of the vector $\alpha = [1, \gamma]^T$ in the convex polyhedron set $\Gamma = \{\gamma \mid \beta \leq B\gamma\}$ for $\beta \in \mathbf{R}^{(K-1) \times 1}$, $\gamma \in \mathbf{R}^{(L-1) \times 1}$, and $B \in \mathbf{R}^{(K-1) \times (L-1)}$.

Proposition 1. *Given a network flow G, there exists a convex polyhedron set for the imprecise coefficient vector $\tilde{c}$ in the BLP problem (1) if Γ is not empty.*

The interpretation of the Γ set is necessary to be reviewed. We construct it by analogy of the revealed preference in the consumption dataset to be consistent to the utility maximization model. The statement (III) of Afriat's theorem is valid for an arbitrary index k and l, but in the case of a combination of choices we deploy just the observed patterns which are counted the most against the rest of patterns. When Γ contains all cases of pairs, we can say this data aggregation is equivalent to the Afriat's system of inequalities. We can prove, by using the fact of duality, the system of inequalities in (III) can be reduced to $p^l \cdot (x^k - x^l) \leq 0$.

Proposition 2. *The system of inequality in (3) is equivalent to the Afriat's system of inequalities.*

The set Γ not only represents the imprecise coefficient vector $\tilde{c}$, but it is also the evidence of rational behaviors from the observed choices in the network flows. This is the decision norms constructed by the observed data influenced by the Afriat's theorem. Therefore, we have appropriately constructed a representative follower's decision norms in the lower-level problem in our BLP for NWD with the application of real dataset.

3.3   BLP Problem Reformulation

From the construction of a convex polyhedron based on the observed dataset, we develop a three-step optimization for our NWD in BLP as follows.

When the leader knows exactly the coefficient vector of the follower's objective function, he can see the follower's rational reaction set $\text{Opt}(c, v) = \{x \in S(v) \mid c^T x = \max_{z \in S(v)} c^T z\}$. $S(v)$ is the feasible region which is determined by the additional capacity v in the links. In the case of imprecise coefficient vector $\tilde{c}$, the leader cannot know the follower's rational reaction exactly but he can explore the follower's rational reaction in a larger region. From (5), we can have the follower's cost coefficient vector $c = [1, \gamma]^T \in \Gamma'$ corresponding to the links in A and setting an arbitrary element as a numeraire price. Hence, we define the follower's possible rational reaction set as $\Pi S(v) = \bigcup_{c \in \Gamma'} \text{Opt}(c, v)$ and the inducible region set as $IR = \{(v, x) \mid (v, x) \in S, x \in \Pi S(v)\}$.

Furthermore, we assume that the leader will consider the worst effect of the follower's response to his strategy, and solve for the leader's optimal solution by adopting maximin criteria. Immediately, linear programming problem (1) is formulated as

$$\text{Maximize}_{v} \left(\sum_{(i,j)\in A} \tau_{ij} v_{ij} + \sum_{(i,j)\in A} \psi_{ij} x_{ij} + \left(\underset{x\in \Pi S(v)}{\text{Minimize}} \sum_{(i,j)\in A} c_{ij} x_{ij} \right) \right). \quad (6)$$

Problem (6) for NWD in BLP is rewritten with an element in the cost coefficient parameter c, derived from the survey data, is fixes to a numeraire price,

$$(OP) \begin{cases} \text{Maximize}_{v} \sum_{(i,j)\in A} \tau_{ij} v_{ij} + \sum_{(i,j)\in A} \psi_{ij} x_{ij}, \\ \text{Subject to } v_{ij} \geq 0, \quad \forall (i,j) \in A, \\ \qquad\quad \text{where } x, c \text{ solves,} \\ \\ (SP(v)) \begin{cases} \text{Minimize}_{x,c} \sum_{(i,j)\in A} \psi_{ij} x_{ij}, \\ \text{Subject to } \sum_{(i,j)\in A} c_{ij} x_{ij} = \underset{z\geq 0}{\max} \left\{ \sum_{(i,j)\in A} c_{ij} z_{ij} \;\middle|\; x_{ij} \leq u_{ij} + v_{ij} \right\}, \\ \qquad -B\gamma \leq -\beta, \\ \qquad \sum_{j=1}^{m} x_{ij} - \sum_{k=1}^{m} x_{ki} = b_i, \forall i \in N, \\ \qquad 0 \leq x_{ij} \leq u_{ij} + v_{ij}. \end{cases} \end{cases}$$

Proposing a method to aggregate revealed preference for the follower's decision norms, the formulation in (6) becomes three-step optimization. It starts at the lower-level problem $(SP(v))$ from the most internal part of our derived convex polyhedron, then the follower's optimal reaction. Consequently, the upper-level problem (OP) of the leader optimizes the most outside part of problem. We achieve our NWD in BLP in the structure of maximin problem. The solution algorithm can be found in our previous works [11,20,21] which modeled the follower's ambiguous objective coefficient vector is situated in a convex polyhedron.

4   Conclusion

We utilize the special unit cost structure in the network flow model to construct the aggregation of preference data. We transform the revealed preference data revealed by the network users into a bundle of linear constraints and show that the norm of the lower-level can be explained by the convex polyhedron. This comes from the celebrate Afriat's theorem to check the rationalization of the observed dataset. Although the aggregate preference results in a set of ambiguous decision norms of the representative follower in BLP for NWD, the solution method has been developed in our previous research [11,20,21]. Finally, the data aggregation should be fitting to the dataset to observe some problems might happen such as how to detect the data pattern that make the system of inequalities infeasible or how to remedy such failure from the dataset.

References

1. Afriat, S.N.: The construction of utility functions from expenditure data. Int. Econ. Rev. **8**(1), 67 (1967)
2. Bard, J.F.: Coordination of a multidivisional organization through two levels of management. Omega **11**(5), 457–468 (1983)
3. Bard, J.F.: Practical Bilevel Optimization: Algorithms and Applications. Kluwer Academic Publishers, Dordrecht (1998)
4. Bazaraa, M.S., Jarvis, J.J.: Linear Programming and Network Flows. Wiley, New York (1977)
5. Ben-Ayed, O., Blair, C.E., Boyce, D.E., LeBlanc, L.J.: Construction of a real-world bilevel linear programming model of the highway network design problem. Ann. Oper. Res. **34**(1), 219–254 (1992)
6. Bialas, W.F., Karwan, M.H.: On two-level optimization. IEEE Trans. Autom. Control **27**(1), 211–214 (1982)
7. Calvete, H.I., Galé, C.: Linear bilevel programming with interval coefficients. J. Comput. Appl. Math. **236**(15), 3751–3762 (2012)
8. Chambers, C.P., Echenique, F.: Revealed Preference Theory. Econometric Society Monographs, vol. 56. Cambridge University Press, Cambridge (2016)
9. Christiansen, S., Patriksson, M., Wynter, L.: Stochastic bilevel programming in structural optimization. Struct. Multi. Optim. **21**(5), 361–371 (2001)
10. Dempe, S.: Foundations of Bilevel Programming. Kluwer Academic, Dordrecht (2002)
11. Inuiguchi, M., Sariddichainunta, P.: Bilevel linear programming with ambiguous objective function of the follower. Fuzzy Optim. Decis. Making **15**(4), 415–434 (2016)
12. Islam, S.M.N.: Mathematical Economics of Multi-level Optimisation: Theory and Application. Physica, Heidelberg (1998)
13. LeBlanc, L.J., Boyce, D.E.: A bilevel programming algorithm for exact solution of the network design problem with user-optimal flows. Transp. Res. Part B Methodol. **20**(3), 259–265 (1986)
14. Mas-Colell, A., Whinston, M.D., Green, J.R.: Microeconomic Theory. Oxford University Press, New York (1995)
15. Migdalas, A.: Bilevel programming in traffic planning: models, methods and challenge. J. Global Optim. **7**(4), 381–405 (1995)
16. Patriksson, M.: On the applicability and solution of bilevel optimization models in transportation science: a study on the existence, stability and computation of optimal solutions to stochastic mathematical programs with equilibrium constraints. Transp. Res. Part B Methodol. **42**(10), 843–860 (2008)
17. Ren, A., Wang, Y.: A cutting plane method for bilevel linear programming with interval coefficients. Ann. Oper. Res. **223**(1), 355–378 (2014)
18. Ruzíyeva, A., Dempe, S.: Yager ranking index in fuzzy bilevel optimization. Artif. Intell. Res. **2**(1), 55 (2012)
19. Sariddichainunta, P., Inuiguchi, M.: Bilevel linear programming with lower-level fuzzy objective function. In: 2017 Joint 17th World Congress of International Fuzzy Systems Association and 9th International Conference on Soft Computing and Intelligent Systems (IFSA-SCIS), pp. 1–6 (2017)
20. Sariddichainunta, P., Inuiguchi, M.: The improvement of optimality test over possible reaction set in bilevel linear optimization with ambiguous objective function of the follower. J. Adv. Comput. Intell. Intell. Inform. **19**(5), 645–654 (2015)

21. Sariddichainunta, P., Inuiguchi, M.: Global optimality test for maximin solution of bilevel linear programming with ambiguous lower-level objective function. Ann. Oper. Res. **256**(2), 285–304 (2017)
22. von Stackelberg, H.: Market Structure and Equilibrium. Springer, Heidelberg (2011). https://doi.org/10.1007/978-3-642-12586-7. (Translated by Bazin, D., Hill, R., Urch, L.)
23. Varian, H.R.: Revealed preference and its applications. Econ. J. **122**(560), 332–338 (2012)
24. Williams, H.P.: Model Building in Mathematical Programming, 5th edn. Wiley, Chichester (2013)

Rule Induction Based on Rough Sets from Possibilistic Data Tables

Michinori Nakata[1](✉) and Hiroshi Sakai[2]

[1] Faculty of Management and Information Science, Josai International University,
1 Gumyo, Togane, Chiba 283-8555, Japan
nakatam@ieee.org
[2] Department of Mathematics and Computer Aided Sciences, Faculty of Engineering,
Kyushu Institute of Technology, Tobata, Kitakyushu 804-8550, Japan
sakai@mns.kyutech.ac.jp

Abstract. How to induce rules from data tables containing possibilistic information is described under using rough sets based on possible world semantics. A piece of possibilistic information is expressed in a normal and discrete possibility distribution. Under a degree of possibility, the incomplete data table is derived from a possibilistic data table. Rough sets and rules are derived from incomplete data tables. The rough sets and rules obtained under every degree of possibility are aggregated from the viewpoints of certainty and possibility. As a result, rough sets consist of objects with a degree expressed in an interval and an object also supports rules with a degree expressed in an interval value. Furthermore, a criterion is introduced to judge whether or not an object is regarded as validly supporting rules.

Keywords: Rough sets · Rule induction · Incomplete information · Non-deterministic information · Possibilistic information

1 Introduction

We live in a flood of possibilistic information. Possibilistic information is expressed in a possibility distribution. For example, "about 25" is expressed by the normal possibility distribution $\{(22, 0.1), (23, 0.5), (24, 1), (25, 1), (26, 1), (27, 0.5), (28, 0.1)\}_p$ in the sentence "His age is about 25." There ubiquitously exists such possibilistic information in the real world [13].

The framework of rough sets, proposed by Pawlak [10], is used as an effective tool for rule induction from data. The rough sets are based on indiscernibility of objects. When objects have the same value for an attribute, they are regarded as indistinguishable on the attribute. The fundamental framework is specified by lower and upper approximations under indiscernibility relations obtained from information tables containing only complete information. Rules are induced from the lower and upper approximations.

Some extensions are required to deal with possibilistic information. Słowiński and Stefanowski [12] proposed the concept of possible indiscernibility between

© Springer Nature Switzerland AG 2019
H. Seki et al. (Eds.): IUKM 2019, LNAI 11471, pp. 86–97, 2019.
https://doi.org/10.1007/978-3-030-14815-7_8

objects. Nakata and Sakai [5] expressed lower and upper approximations by using possible equivalence classes. The lower and upper approximations coincide with those obtained from the approach based on possible world semantics. Couso and Dubois expressed lower and upper approximations by using the degree to which objects possibly belong to the same equivalence class under indiscernibility relations [1]. These approaches focus on only the possibility that objects are indistinguishable, but not the possibility that objects are not indistinguishable. Considering this point, Nakata and Sakai [7,9] dealt with possibilistic information from viewpoints of certainty and possibility, as was done by Lipski in the field of databases [2,3]. As a result, the complementarity property linked with lower and upper approximations holds, as it is valid under complete information.

Rules are induced from rough sets. Nakata and Sakai [8] showed how to induce rules from possible tables with a degree of possibility that is derived from a possibilistic information table. The method has difficulty of computational complexity, because possible tables, whose number exponentially grows as the number of possibilistic values increases, are directly used. In this paper, we suggest a method in which the computational complexity can be avoided by using the approach based on non-deterministic information systems, called NIS [11]. The paper is organized as follows. In Sect. 2, an approach based on rough sets is briefly addressed under complete information. In Sect. 3, extended rough sets are described in a possibilistic information table. In Sect. 4, we show rule induction from the extended rough sets. In Sect. 5, conclusions are addressed.

2   Rough Sets in Complete Information Tables

A data set is represented as a table, called an information table, where each row and each column represent an object and an attribute, respectively. A mathematical model of an information table with complete information is called a complete information system. The complete information system is a triplet expressed by $(U, \mathcal{A}, \{D_{a_i} \mid a_i \in \mathcal{A}\})$. U is a non-empty finite set of objects called the universe, $\mathcal{A}$ is a non-empty finite set of attributes such that $a_i : U \to D_{a_i}$ for every $a_i \in \mathcal{A}$ where D_{a_i} is the domain of attribute a_i. Binary relation R_{a_i} for indiscernibility of objects on attribute $a_i \in \mathcal{A}$, which is called the indiscernibility relation for a_i, is:

$$R_{a_i} = \{(o, o') \in U \times U \mid a_i(o) = a_i(o')\}, \tag{1}$$

where $a_i(o)$ is the value for attribute a_i of object o. From the indiscernibility relation, equivalence class $[o]_{a_i}$ for object o is obtained:

$$[o]_{a_i} = \{o' \mid (o, o') \in R_{a_i}\}. \tag{2}$$

Finally, family $\mathcal{E}_{a_i}{}^1$ of equivalence classes on a_i is:

$$\mathcal{E}_{a_i} = \{[o]_{a_i} \mid o \in U\}. \tag{3}$$

---

[1] $\mathcal{E}_{a_i}$ is formally $\mathcal{E}_{a_i}(U)$. (U) is usually omitted.

Using the family of equivalence classes on a_i, lower approximation $\underline{apr}_{a_i}(\mathcal{O})$ and upper approximation $\overline{apr}_{a_i}(\mathcal{O})$ of set $\mathcal{O}$ of indiscernible objects are:

$$\underline{apr}_{a_i}(\mathcal{O}) = \{o \mid [o]_{a_i} \in \mathcal{E}_{a_i} \wedge [o]_{a_i} \subseteq \mathcal{O}\}, \tag{4}$$

$$\overline{apr}_{a_i}(\mathcal{O}) = \{o \mid [o]_{a_i} \in \mathcal{E}_{a_i} \wedge [o]_{a_i} \cap \mathcal{O} \neq \emptyset\}. \tag{5}$$

When $o \in \underline{apr}_{a_i}(\mathcal{O})$, $[o]_{a_i=u} \subseteq \mathcal{O}$, where $\mathcal{O}$ is a set of objects specified by value u' for a_j and $[o]_{a_i=u}$ shows that $[o]_{a_i}$ is characterized by value u that o has for a_i. Thus, o consistently supports the rule denoted by $a_i = u \rightarrow a_j = u'$ where o has u and u' for a_i and a_j, respectively. This is expressed by $(o, a_i = u \rightarrow a_j = u')$. When $o \in \overline{apr}_{a_i}(\mathcal{O})$, $[o]_{a_i=u} \cap \mathcal{O} \neq \emptyset$ if o has u for a_i. Thus, o inconsistently supports a rule denoted by $a_i = u \rightarrow a_j = u'$ where o has u for a_i, but does not always have u' for a_j, although this is also expressed by $(o, a_i = u \rightarrow a_j = u')$. All objects included in $[o]_{a_i=u}$ do not support $a_i = u \rightarrow a_j = u'$. The consistency degree, called accuracy, is evaluated by $|[o]_{a_i=u} \cap \mathcal{O}|/|[o]_{a_i=u}|$. Clearly, this degree is equal to 1, if $o \in \underline{apr}_{a_i}(\mathcal{O})$. To clarify what rule an object support, we use other expressions for rough approximations. Let lower and upper approximations be denoted by $\underline{r}_{a_i}(\mathcal{O})$ and $\overline{r}_{a_i}(\mathcal{O})$ that consist of pairs of an object and the rule that the object supports.

$$\underline{r}_{a_i}(\mathcal{O}) = \{(o, a_i = u \rightarrow a_j = u') \mid [o]_{a_i=u} \subseteq \mathcal{O}\}, \tag{6}$$

$$\overline{r}_{a_i}(\mathcal{O}) = \{(o, a_i = u \rightarrow a_j = u') \mid [o]_{a_i=u} \cap \mathcal{O} \neq \emptyset\}. \tag{7}$$

For formulae on sets A and A' of attributes whose individual attributes are denoted by a_i and a_j,

$$R_A = \cap_{a_i \in A} R_{a_i}, \tag{8}$$

$$[o]_A = \{o' \mid (o, o') \in R_A\} = \cap_{a_i \in A}[o]_{a_i}, \tag{9}$$

$$\mathcal{E}_A = \{[o]_A \mid o \in U\}, \tag{10}$$

$$\underline{apr}_A(\mathcal{O}) = \{o \mid [o]_A \in \mathcal{E}_A \wedge [o]_A \subseteq \mathcal{O}\}, \tag{11}$$

$$\overline{apr}_A(\mathcal{O}) = \{o \mid [o]_A \in \mathcal{E}_A \wedge [o]_A \cap \mathcal{O} \neq \emptyset\}, \tag{12}$$

$$\underline{r}_A(\mathcal{O}) = \{(o, A = \mathbf{u} \rightarrow A' = \mathbf{u}') \mid [o]_{A=\mathbf{u}} \subseteq \mathcal{O}\}, \tag{13}$$

$$\overline{r}_A(\mathcal{O}) = \{(o, A = \mathbf{u} \rightarrow A' = \mathbf{u}') \mid [o]_{A=\mathbf{u}} \cap \mathcal{O} \neq \emptyset\}, \tag{14}$$

where $\mathcal{O}$ is specified by $A' = \mathbf{u}'$ and $A = \mathbf{u}$ and $A' = \mathbf{u}'$ are equal to $\wedge_{a_i \in A} a_i = u_i$ and $\wedge_{a_j \in A'} a_j = u'_j$, respectively.

3   Rough Sets in Possibilistic Information Tables

In possibilistic information systems, $a_i : U \rightarrow \pi_{a_i}$ for every $a_i \in \mathcal{A}$ where π_{a_i} is the set of all normal possibility distributions over domain D_{a_i} of attribute a_i. When value $a_i(o)$ for attribute a_i of object o is expressed by a normal possibility distribution $\{(u, \pi_{a_i(o)}(u)) \mid u \in D_{a_i} \wedge \pi_{a_i(o)}(u) > 0\}_p$ in an information table, $\pi_{a_i(o)}(u)$ denotes the possibilistic degree that $a_i(o)$ has value $u \in D_{a_i}$ for attribute a_i. The information table is called a possibilistic information table.

When we focus on set B of attributes, set S_B^π of degrees of possibility in possibilistic information table T is:

$$S_B^\pi = \{\pi_{a_i(o)}(u) \mid a_i \in B \land o \in T \land u \in D_{a_i} \land \pi_{a_i(o)}(u) > 0\}. \tag{15}$$

α-cut $a_i(o)^\alpha$ of $a_i(o)$ is:

$$a_i(o)^\alpha = \{u \mid \pi_{a_i(o)}(u) \geq \alpha\}. \tag{16}$$

When we focus on B, by using attribute values with α-cut, set T_B of incomplete information table T^α with degree α of possibility is derived from T:

$$T_B = \{T^\alpha \mid \alpha \in S_B^\pi \land a_i \in B \land o' \in T^\alpha \land o \in T \land a_i(o') = a_i(o)^\alpha\}, \tag{17}$$

where T^α consists of set B of attributes.

In each incomplete information table lower and upper approximations are obtained [6].

Example 3.1. Let possibilistic information table T be obtained as follows:

$$T$$

U	a_1	a_2	a_3
1	$\{(x,1)\}_p$	$\{(c,1),(d,0.2)\}_p$	$\{(e,1)\}_p$
2	$\{(x,1),(y,0.2)\}_p$	$\{(a,0.9),(b,1)\}_p$	$\{(e,0.6),(f,1)\}_p$
3	$\{(y,1)\}_p$	$\{(b,1)\}_p$	$\{(g,1),(h,0.4)\}_p$
4	$\{(y,1),(z,1)\}_p$	$\{(b,1)\}_p$	$\{(h,1),(f,0.9)\}_p$
5	$\{(x,0.4),(w,1)\}_p$	$\{(c,1),(d,0.7)\}_p$	$\{(f,1)\}_p$

In information table T, $U = \{o_1, o_2, o_3, o_4, o_5\}$, where domains D_{a_1}, D_{a_2}, and D_{a_3} of attributes a_1, a_2, and a_3 are $\{w, x, y, z\}$, $\{a, b, c, d\}$, $\{e, f, g, h\}$, respectively. When we focus on attribute $\{a_1, a_2\}$, $S_{\{a_1, a_2\}}^\pi = \{0.2, 0.4, 0.7, 0.9, 1\}$. We have 5 incomplete information tables with $\{a_1, a_2\}$; namely, $T_{\{a_1, a_2\}} = \{T^{0.2}, T^{0.4}, T^{0.7}, T^{0.9}, T^1\}$.

	$T^{0.2}$			$T^{0.4}$			$T^{0.7}$			$T^{0.9}$			T^1	
U	a_1	a_2	U	a_1	a_2	U	a_1	a_2	U	a_1	a_2	U	a_1	a_2
1	$\{x\}$	$\{c,d\}$	1	$\{x\}$	$\{c\}$	1	$\{x\}$	$\{c\}$	1	$\{x\}$	$\{c\}$	1	$\{x\}$	$\{c\}$
2	$\{x,y\}$	$\{a,b\}$	2	$\{x\}$	$\{a,b\}$	2	$\{x\}$	$\{a,b\}$	2	$\{x\}$	$\{a,b\}$	2	$\{x\}$	$\{b\}$
3	$\{y\}$	$\{b\}$	3	$\{y\}$	$\{b\}$	3	$\{y\}$	$\{b\}$	3	$\{y\}$	$\{b\}$	3	$\{y\}$	$\{b\}$
4	$\{y,z\}$	$\{b\}$	4	$\{y,z\}$	$\{b\}$	4	$\{y,z\}$	$\{b\}$	4	$\{y,z\}$	$\{b\}$	4	$\{y,z\}$	$\{b\}$
5	$\{x,w\}$	$\{c,d\}$	5	$\{x,w\}$	$\{c,d\}$	5	$\{w\}$	$\{c,d\}$	5	$\{w\}$	$\{c\}$	5	$\{w\}$	$\{c\}$

Any α-cut table T^α is an incomplete information table with non-deterministic values on attributes to which are applied α-cut. We describe rough sets in an incomplete information table on the basis of Lipski's work using possible world semantics [2]. Suppose we focus on set B of attributes. The set of possible tables on B is obtained from an incomplete information table. In a possible table, the value that an object has for each attribute in B is one of possible values that the object has for the attribute in the incomplete information table. Set PT_{B,T^α} of possible tables on B, which are derived from incomplete information table T^α, is:

$$PT_{B,T^\alpha} = \{t : \forall o \in T^\alpha \; \forall a_i \in B \; a_i(o)^t = e \wedge e \in a_i(o)\}, \tag{18}$$

where $a_i(o)^t$ and $a_i(o)$ are values of attribute a_i for object o in possible table t and in incomplete information table T^α, respectively. Every possible table has complete information.

Example 3.2. Set $PT_{\{a_1,a_2\},T^{0.9}}$ of possible tables derived from $T^{0.9}$ in Example 3.1 is:

$$PT_{\{a_1,a_2\},T^{0.9}} = \{t_1, t_2, t_3, t_4\}.$$

t_1				t_2				t_3				t_4		
U	a_1	a_2		U	a_1	a_2		U	a_1	a_2		U	a_1	a_2
1	x	c		1	x	c		1	x	c		1	x	c
2	x	a		2	x	b		2	x	a		2	x	b
3	y	b		3	y	b		3	y	b		3	y	b
4	y	b		4	y	b		4	z	b		4	z	b
5	w	c		5	w	c		5	w	c		5	w	c

The formulae in complete information systems can be applied to every possible table on B. Therefore, lower and upper approximations of $\mathcal{O}$ by $A \subseteq B$, $\underline{apr}_A(\mathcal{O})^t$ and $\overline{apr}_A(\mathcal{O})^t$ in each possible table are derived by using the formulae described in the previous section. From aggregating lower and upper approximations in all the possible tables that are obtained from T^α, four approximations are obtained [6]: certain lower approximation $C\underline{apr}_A(\mathcal{O})^\alpha$, certain upper approximation $C\overline{apr}_A(\mathcal{O})^\alpha$, possible lower approximation $P\underline{apr}_A(\mathcal{O})^\alpha$, and possible upper approximation $P\overline{apr}_A(\mathcal{O})^\alpha$.

$$C\underline{apr}_A(\mathcal{O})^\alpha = \{o \mid o \in T^\alpha \wedge \forall t \in PT_{B,T^\alpha} o \in \underline{apr}_A(\mathcal{O})^t\}, \tag{19}$$

$$C\overline{apr}_A(\mathcal{O})^\alpha = \{o \mid o \in T^\alpha \wedge \forall t \in PT_{B,T^\alpha} o \in \overline{apr}_A(\mathcal{O})^t\}, \tag{20}$$

$$P\underline{apr}_A(\mathcal{O})^\alpha = \{o \mid o \in T^\alpha \wedge \exists t \in PT_{B,T^\alpha} o \in \underline{apr}_A(\mathcal{O})^t\}, \tag{21}$$

$$P\overline{apr}_A(\mathcal{O})^\alpha = \{o \mid o \in T^\alpha \wedge \exists t \in PT_{B,T^\alpha} o \in \overline{apr}_A(\mathcal{O})^t\}. \tag{22}$$

Proposition 3.1. If $\alpha \geq \alpha'$, then $C\underline{apr}_A(\mathcal{O})^\alpha \supseteq C\underline{apr}_A(\mathcal{O})^{\alpha'}$, $P\underline{apr}_A(\mathcal{O})^\alpha \subseteq P\underline{apr}_A(\mathcal{O})^{\alpha'}$, $C\overline{apr}_A(\mathcal{O})^\alpha \supseteq C\overline{apr}_A(\mathcal{O})^{\alpha'}$, and $P\overline{apr}_A(\mathcal{O})^\alpha \subseteq P\overline{apr}_A(\mathcal{O})^{\alpha'}$.

Using these approximations, membership degree $\mu_{P\underline{apr}_A(\mathcal{O})}(o)$, called a possible membership degree, with which object o possibly belongs to lower approximation $\underline{apr}_A(\mathcal{O})$ is:

$$\mu_{P\underline{apr}_A(\mathcal{O})}(o) = \max\{\alpha \mid o \in P\underline{apr}_A(\mathcal{O})^\alpha\}. \tag{23}$$

Membership degree $\mu_{C\underline{apr}_A(\mathcal{O})}(o)$, called a certain membership degree, with which object o certainly belongs to lower approximation $\underline{apr}_A(\mathcal{O})$ is:

$$\mu_{C\underline{apr}_A(\mathcal{O})}(o) = 1 - \max\{\alpha \mid o \notin C\underline{apr}_A(\mathcal{O})^\alpha\}. \tag{24}$$

Possible membership degree $\mu_{P\overline{apr}_A(\mathcal{O})}(o)$ with which object o possibly belongs to upper approximation $\overline{apr}_A(\mathcal{O})$ is:

$$\mu_{P\overline{apr}_A(\mathcal{O})}(o) = \max\{\alpha \mid o \in P\overline{apr}_A(\mathcal{O})^\alpha\}. \tag{25}$$

Certain membership degree $\mu_{C\overline{apr}_A(\mathcal{O})}(o)$ with which object o certainly belongs to upper approximation $\overline{apr}_A(\mathcal{O})$ is:

$$\mu_{C\overline{apr}_A(\mathcal{O})}(o) = 1 - \max\{\alpha \mid o \notin C\overline{apr}_A(\mathcal{O})^\alpha\}. \tag{26}$$

Proposition 3.2. $\forall o \in U$ $\mu_{C\underline{apr}_A(\mathcal{O})}(o) \leq \mu_{P\underline{apr}_A(\mathcal{O})}(o) \leq \mu_{C\overline{apr}_A(\mathcal{O})}(o) \leq \mu_{P\overline{apr}_A(\mathcal{O})}(o)$.

Four membership degrees are linked with each other.

Proposition 3.3. $\mu_{P\underline{apr}_A(\mathcal{O})}(o) = 1 - \mu_{C\overline{apr}_A(U-\mathcal{O})}(o)$ and $\mu_{C\underline{apr}_A(\mathcal{O})}(o) = 1 - \mu_{P\overline{apr}_A(U-\mathcal{O})}(o)$.

Each object has membership degrees for these four approximations denoted by formulae (23)–(26). Using these degrees, lower and upper approximations are expressed as follows:

$$\underline{apr}_A(\mathcal{O}) = \{(o, [\mu_{C\underline{apr}_A(\mathcal{O})}(o), \mu_{P\underline{apr}_A(\mathcal{O})}(o)] \mid \mu_{P\underline{apr}_A(\mathcal{O})}(o) > 0\}, \tag{27}$$

$$\overline{apr}_A(\mathcal{O}) = \{(o, [\mu_{C\overline{apr}_A(\mathcal{O})}(o), \mu_{P\overline{apr}_A(\mathcal{O})}(o)] \mid \mu_{P\overline{apr}_A(\mathcal{O})}(o) > 0\}. \tag{28}$$

These formulae show that each object has membership degrees expressed by not single, but interval values for lower and upper approximations, which is essential in possibilistic information systems. Degrees of imprecision for the membership degrees of o are evaluated by $\mu_{P\underline{apr}_A(\mathcal{O})}(o) - \mu_{C\underline{apr}_A(\mathcal{O})}(o)$ and $\mu_{P\overline{apr}_A(\mathcal{O})}(o) - \mu_{C\overline{apr}_A(\mathcal{O})}(o)$ in $\underline{apr}_A(\mathcal{O})$ and $\overline{apr}_A(\mathcal{O})$, respectively. Two approximations depend on each other.

Proposition 3.4.

$$\underline{apr}_A(\mathcal{O}) = U - \overline{apr}_A(U - \mathcal{O}),$$

where $1 - [\mu_{C\overline{apr}_A(U-\mathcal{O})}(o), \mu_{P\overline{apr}_A(U-\mathcal{O})}(o)] = [1 - \mu_{P\overline{apr}_A(U-\mathcal{O})}(o), 1 - \mu_{C\overline{apr}_A(U-\mathcal{O})}(o)]$.

Example 3.3. Let us go back to possibilistic information table T of Example 3.1. Let a set $\mathcal{O}$ of objects be $\{o_2, o_3, o_4\}$. Set T_{a_1} of incomplete information tables on a_1 is $\{T^{0.2}, T^{0.4}, T^1\}$. Using formulae (19)–(22), lower and upper approximations are obtained from $T^{0.2}$, $T^{0.4}$, and T^1, respectively. Using formulae (23)–(26), for object o_1,

$$\mu_{\underline{Capr}_{a_1}}(\mathcal{O})(o_1) = 0, \mu_{\underline{Papr}_{a_1}}(\mathcal{O})(o_1) = 0,$$
$$\mu_{\overline{Capr}_{a_1}}(\mathcal{O})(o_1) = 0.8, \mu_{\overline{Papr}_{a_1}}(\mathcal{O})(o_1) = 1.$$

Similarly, calculating membership degrees for the other objects,

$$\underline{apr}_{a_1}(\mathcal{O}) = \{(o_2, [0, 0.2]), (o_3, [1, 1]), (o_4, [1, 1])\},$$
$$\overline{apr}_{a_1}(\mathcal{O}) = \{(o_1, [0.8, 1]), (o_2, [1, 1]), (o_3, [1, 1]), (o_4, [1, 1]), (o_5, [0, 0.4])\}.$$

4   Rule Induction in Possibilistic Information Tables

When an object has multiple possible values for an attribute, the object supports multiple rules. Every object with a possible membership degree greater than zero for rough approximations possibly supports rules, whereas every object with a certain membership degree greater than zero does not certainly support a rule. In other words, an object does not always certainly support a rule, even if the object certainly belongs with a degree to approximations that are described in the previous section. From the approximations obtained in Sect. 3, we cannot induce concrete forms of rules that each object supports, although we can know that the object supports some rules with degrees. To clarify what rules objects support, we use other expressions $P\underline{r}_A(\mathcal{O})^\alpha$, $C\underline{r}_A(\mathcal{O})^\alpha$, $P\overline{r}_A(\mathcal{O})^\alpha$, and $C\overline{r}_A(\mathcal{O})^\alpha$ where rules that an object supports are individually described for approximations. When $\mathcal{O}$ is specified by $A' = \mathbf{u}'$,

$$P\underline{r}_A(\mathcal{O})^\alpha = \{(o, A = \mathbf{u} \to A' = \mathbf{u}') \mid \exists t \in PT_{A,T^\alpha}[o]^t_{A=\mathbf{u}} \subseteq \mathcal{O}\}, \tag{29}$$
$$C\underline{r}_A(\mathcal{O})^\alpha = \{(o, A = \mathbf{u} \to A' = \mathbf{u}') \mid \forall t \in PT_{A,T^\alpha}[o]^t_{A=\mathbf{u}} \subseteq \mathcal{O}\}, \tag{30}$$
$$P\overline{r}_A(\mathcal{O})^\alpha = \{(o, A = \mathbf{u} \to A' = \mathbf{u}') \mid \exists t \in PT_{A,T^\alpha}[o]^t_{A=\mathbf{u}} \cap \mathcal{O} \neq \emptyset\}, \tag{31}$$
$$C\overline{r}_A(\mathcal{O})^\alpha = \{(o, A = \mathbf{u} \to A' = \mathbf{u}') \mid \forall t \in PT_{A,T^\alpha}[o]^t_{A=\mathbf{u}} \cap \mathcal{O} \neq \emptyset\}, \tag{32}$$

where $[o]^t_{A=\mathbf{u}}$ is the equivalence class of o that has value $\mathbf{u}$ for A in possible table t.

We consider rules that objects support when $\mathcal{O}$ is specified by $A' = \mathbf{u}'$. Possible membership degree $\mu_{P\underline{r}_A(\mathcal{O})}((o, A = \mathbf{u} \to A' = \mathbf{u}'))$, with which $(o, A = \mathbf{u} \to A' = \mathbf{u}')$ belongs to possible lower approximation $P\underline{r}_A(\mathcal{O})$, is:

$$\mu_{P\underline{r}_A(\mathcal{O})}((o, A = \mathbf{u} \to A' = \mathbf{u}'))$$
$$= \max\{\alpha \mid (o, A = \mathbf{u} \to A' = \mathbf{u}') \in P\underline{r}_A(\mathcal{O})^\alpha\}. \tag{33}$$

Certain membership degree $\mu_{C\underline{r}_A(\mathcal{O})}((o, A = \mathbf{u} \to A' = \mathbf{u}'))$, with which $(o, A = \mathbf{u} \to A' = \mathbf{u}')$ belongs to certain lower approximation $C\underline{r}_A(\mathcal{O})$, is:

$$\mu_{C\underline{r}_A(\mathcal{O})}((o, A = \mathbf{u} \to A' = \mathbf{u}')) =$$
$$1 - \max\{\alpha \mid (o, A = \mathbf{u} \to A' = \mathbf{u}') \notin C\underline{r}_A(\mathcal{O})^\alpha\}. \tag{34}$$

Possible membership degree $\mu_{P\overline{r}_A(\mathcal{O})}((o, A = \mathbf{u} \to A' = \mathbf{u}'))$, with which $(o, A = \mathbf{u} \to A' = \mathbf{u}')$ belongs to possible upper approximation $P\overline{r}_A(\mathcal{O})$, is:

$$\mu_{P\overline{r}_A(\mathcal{O})}((o, A = \mathbf{u} \to A' = \mathbf{u}'))$$
$$= \max\{\alpha \mid (o, A = \mathbf{u} \to A' = \mathbf{u}') \in P\overline{r}_A(\mathcal{O})^\alpha\}. \tag{35}$$

Certain membership degree $\mu_{C\overline{r}_A(\mathcal{O})}((o, A = \mathbf{u} \to A' = \mathbf{u}'))$, with which object $(o, A = \mathbf{u} \to A' = \mathbf{u}')$ belongs to certain upper approximation $C\overline{r}_A(\mathcal{O})$, is:

$$\mu_{C\overline{r}_A(\mathcal{O})}((o, A = \mathbf{u} \to A' = \mathbf{u}')) =$$
$$1 - \max\{\alpha \mid (o, A = \mathbf{u} \to A' = \mathbf{u}') \notin C\overline{r}_A(\mathcal{O})^\alpha\}. \tag{36}$$

Using these membership degrees, we express a rule that object o supports and a degree with which o supports it without inconsistency and with inconsistency, respectively:

$$(o, A = \mathbf{u} \to A' = \mathbf{u}', [\mu_{C\underline{r}_A(\mathcal{O})}((o, A = \mathbf{u} \to A' = \mathbf{u}')),$$
$$\mu_{P\underline{r}_A(\mathcal{O})}((o, A = \mathbf{u} \to A' = \mathbf{u}'))])$$
$$(o, A = \mathbf{u} \to A' = \mathbf{u}', [\mu_{C\overline{r}_A(\mathcal{O})}((o, A = \mathbf{u} \to A' = \mathbf{u}')),$$
$$\mu_{P\overline{r}_A(\mathcal{O})}((o, A = \mathbf{u} \to A' = \mathbf{u}'))]).$$

Proposition 4.1. $[\mu_{C\underline{r}_A(\mathcal{O})}((o, A = \mathbf{u} \to A' = \mathbf{u}')), \mu_{P\underline{r}_A(\mathcal{O})}((o, A = \mathbf{u} \to A' = \mathbf{u}'))] \leq [\mu_{C\overline{r}_A(\mathcal{O})}((o, A = \mathbf{u} \to A' = \mathbf{u}')), \mu_{P\overline{r}_A(\mathcal{O})}((o, A = \mathbf{u} \to A' = \mathbf{u}'))].$

Example 4.1. Let us go back to Example 3.1. Using formulae (33) and (34), we obtain the degree with which each object supports rules without inconsistency. Rules that each object supports are as follows:

U	Rules without inconsistency
o_1	$(a_1 = x \to a_2 = c, [0, 0.2])$
	$(a_1 = x \to a_2 = d, [0, 0.2])$
o_2	$(a_1 = y \to a_2 = b, [0, 0.2])$
o_3	$(a_1 = y \to a_2 = b, [0.8, 1])$
o_4	$(a_1 = y \to a_2 = b, [0, 1])$
	$(a_1 = z \to a_2 = b, [0, 1])$
o_5	$(a_1 = w \to a_2 = c, [0.3, 1])$
	$(a_1 = w \to a_2 = d, [0, 0.7])$
	$(a_1 = x \to a_2 = c, [0, 0.2])$
	$(a_1 = x \to a_2 = d, [0, 0.2])$

Using formulae (35) and (36), we obtain the degree with which each object supports rules with inconsistency. Rules that each object supports are as follows:

U	Rules with inconsistency
o_1	$(a_1 = x \rightarrow a_2 = a, [0, 0.9])$
	$(a_1 = x \rightarrow a_2 = b, [0.1, 1])$
	$(a_1 = x \rightarrow a_2 = c, [0.8, 1])$
	$(a_1 = x \rightarrow a_2 = d, [0, 0.4])$
o_2	$(a_1 = x \rightarrow a_2 = a, [0, 0.9])$
	$(a_1 = x \rightarrow a_2 = b, [0.1, 1])$
	$(a_1 = x \rightarrow a_2 = c, [0.8, 1])$
	$(a_1 = x \rightarrow a_2 = d, [0, 0.7])$
	$(a_1 = y \rightarrow a_2 = a, [0, 0.2])$
	$(a_1 = y \rightarrow a_2 = b, [0, 0.2])$
o_3	$(a_1 = y \rightarrow a_2 = a, [0, 0.2])$
	$(a_1 = y \rightarrow a_2 = b, [1, 1])$
o_4	$(a_1 = y \rightarrow a_2 = a, [0, 0.2])$
	$(a_1 = y \rightarrow a_2 = b, [0, 1])$
	$(a_1 = z \rightarrow a_2 = b, [0, 1])$
o_5	$(a_1 = w \rightarrow a_2 = c, [0.3, 1])$
	$(a_1 = w \rightarrow a_2 = d, [0, 0.7])$
	$(a_1 = x \rightarrow a_2 = a, [0, 0.4])$
	$(a_1 = x \rightarrow a_2 = b, [0, 0.4])$
	$(a_1 = x \rightarrow a_2 = c, [0, 0.4])$
	$(a_1 = x \rightarrow a_2 = d, [0, 0.4])$

As is shown in Example 4.1, objects support lots of rules with low degrees. It does not seem that such a rule is true. To remove it, we introduce a criterion for valid support of an object to a rule, as is done for whether or not an object is acceptable in fuzzy relational databases [4]. From formulae (33)–(36), we can derive possible membership degrees $\mu_{\neg P \underline{r}_{a_i}(\mathcal{O})}((o, A = \mathbf{u} \rightarrow A' = \mathbf{u}'))$ and $\mu_{\neg P \overline{r}_A(\mathcal{O})}((o, A = \mathbf{u} \rightarrow A' = \mathbf{u}'))$ with which $(o, A = \mathbf{u} \rightarrow A' = \mathbf{u}')$ does not belong to $P \underline{r}_A(\mathcal{O})$ and $P \overline{r}_A(\mathcal{O})$, respectively:

$$\mu_{\neg P \underline{r}_A(\mathcal{O})}((o, A = \mathbf{u} \rightarrow A' = \mathbf{u}'))$$
$$= 1 - \mu_{C \underline{r}_A(\mathcal{O})}((o, A = \mathbf{u} \rightarrow A' = \mathbf{u}')), \tag{37}$$

$$\mu_{\neg P \overline{r}_A(\mathcal{O})}((o, A = \mathbf{u} \rightarrow A' = \mathbf{u}'))$$
$$= 1 - \mu_{C \overline{r}_A(\mathcal{O})}((o, A = \mathbf{u} \rightarrow A' = \mathbf{u}')). \tag{38}$$

Similarly, certain membership degrees $\mu_{\neg C \underline{r}_A(\mathcal{O})}((o, A = \mathbf{u} \rightarrow A' = \mathbf{u}'))$ and $\mu_{\neg C \overline{r}_A(\mathcal{O})}((o, A = \mathbf{u} \rightarrow A' = \mathbf{u}'))$ with which $(o, A = \mathbf{u} \rightarrow A' = \mathbf{u}')$ does not belong to $C \underline{r}_A(\mathcal{O})$ and $C \overline{r}_A(\mathcal{O})$, respectively are:

$$\mu_{\neg C \underline{r}_A(\mathcal{O})}((o, A = \mathbf{u} \rightarrow A' = \mathbf{u}'))$$
$$= 1 - \mu_{P \underline{r}_A(\mathcal{O})}((o, A = \mathbf{u} \rightarrow A' = \mathbf{u}')), \tag{39}$$

$$\mu_{\neg C\overline{T}_A(\mathcal{O})}((o, A = \mathbf{u} \rightarrow A' = \mathbf{u}'))$$
$$= 1 - \mu_{P\overline{T}_A(\mathcal{O})}((o, A = \mathbf{u} \rightarrow A' = \mathbf{u}')). \tag{40}$$

It is natural to use the following conditions to check whether or not object o validly supports rule $A = \mathbf{u} \rightarrow A' = \mathbf{u}'$, as is used for objects in fuzzy relational databases [4].

$$\mu_{P\underline{T}_A(\mathcal{O})}((o, A = \mathbf{u} \rightarrow A' = \mathbf{u}'))$$
$$\geq \mu_{\neg P\underline{T}_A(\mathcal{O})}((o, A = \mathbf{u} \rightarrow A' = \mathbf{u}')),$$
$$\mu_{P\overline{T}_A(\mathcal{O})}((o, A = \mathbf{u} \rightarrow A' = \mathbf{u}'))$$
$$\geq \mu_{\neg P\overline{T}_A(\mathcal{O})}((o, A = \mathbf{u} \rightarrow A' = \mathbf{u}')),$$
$$\mu_{C\underline{T}_A(\mathcal{O})}((o, A = \mathbf{u} \rightarrow A' = \mathbf{u}'))$$
$$\geq \mu_{\neg C\underline{T}_A(\mathcal{O})}((o, A = \mathbf{u} \rightarrow A' = \mathbf{u}')),$$
$$\mu_{C\overline{T}_A(\mathcal{O})}((o, A = \mathbf{u} \rightarrow A' = \mathbf{u}'))$$
$$\geq \mu_{\neg C\overline{T}_A(\mathcal{O})}((o, A = \mathbf{u} \rightarrow A' = \mathbf{u}')).$$

From them, we obtain the criteria for valid support of object o to rule $A = \mathbf{u} \rightarrow A' = \mathbf{u}'$ in lower and upper approximations:

$$\mu_{P\underline{T}_A(\mathcal{O})}((o, A = \mathbf{u} \rightarrow A' = \mathbf{u}'))$$
$$+ \mu_{C\underline{T}_A(\mathcal{O})}((o, A = \mathbf{u} \rightarrow A' = \mathbf{u}')) \geq 1, \tag{41}$$
$$\mu_{P\overline{T}_A(\mathcal{O})}((o, A = \mathbf{u} \rightarrow A' = \mathbf{u}'))$$
$$+ \mu_{C\overline{T}_A(\mathcal{O})}((o, A = \mathbf{u} \rightarrow A' = \mathbf{u}')) \geq 1. \tag{42}$$

Example 4.2. Let us go back to Example 4.1. We apply the criteria for valid support, denoted by (41) and (42), to the rule tables obtained in Example 4.1. Objects that validly support rules without inconsistency and the rules are as follows:

U	Rules without inconsistency
o_3	$(a_1 = y \rightarrow a_2 = b, [0.8, 1])$
o_4	$(a_1 = y \rightarrow a_2 = b, [0, 1])$
	$(a_1 = z \rightarrow a_2 = b, [0, 1])$
o_5	$(a_1 = w \rightarrow a_2 = c, [0.3, 1])$

Objects that validly support rules with inconsistency and the rules are as follows:

U	Rules with inconsistency
o_1	$(a_1 = x \rightarrow a_2 = b, [0.1, 1])$
	$(a_1 = x \rightarrow a_2 = c, [0.8, 1])$
o_2	$(a_1 = x \rightarrow a_2 = b, [0.1, 1])$
	$(a_1 = x \rightarrow a_2 = c, [0.8, 1])$
o_3	$(a_1 = y \rightarrow a_2 = b, [1, 1])$
o_4	$(a_1 = y \rightarrow a_2 = b, [0, 1])$
	$(a_1 = z \rightarrow a_2 = b, [0, 1])$
o_5	$(a_1 = w \rightarrow a_2 = c, [0.3, 1])$

From these tables, we find out by which object each rule is validly supported without inconsistency and with inconsistency, respectively:

Rule	Objects
$a_1 = y \rightarrow a_2 = b$	$(o_3, [0.8, 1]),(o_4, [0, 1])$
$a_1 = z \rightarrow a_2 = b$	$(o_4, [0, 1])$
$a_1 = w \rightarrow a_2 = c$	$(o_5, [0.3, 1])$

Rule	Objects
$a_1 = x \rightarrow a_2 = b$	$(o_1, [0.1, 1]),(o_2, [0.1, 1])$
$a_1 = x \rightarrow a_2 = c$	$(o_1, [0.8, 1]),(o_2, [0.8, 1])$
$a_1 = y \rightarrow a_2 = b$	$(o_3, [1, 1]),(o_4, [0, 1])$
$a_1 = z \rightarrow a_2 = b$	$(o_4, [0, 1])$
$a_1 = w \rightarrow a_2 = c$	$(o_5, [0.3, 1])$

As is shown in Example 4.2, we can select only rules that objects validly support among lots of rules.

5   Conclusions

We have shown rough sets and rule induction from them in a possibilistic information table. Incomplete information tables are obtained with a degree of possibility from the possibilistic information table. In each incomplete information table, lower and upper approximations and rules are obtained with a degree of possibility. Practically, this process can be done by using an approach based on NIS [11]. Therefore, the process has no difficulty of computational complexity, because the approach using NIS has no exponential computational complexity. The lower and upper approximations and the rules are aggregated. Similarly by using approximations considering rules that objects support in incomplete information tables, rules have been induced from a possibilistic information table. Criteria have been introduced for whether or not objects validly support rules, as is done for objects in fuzzy relational databases. This is because lots of rules appear with low degrees in possibilistic information systems. By using the criteria, we can select only rules that objects validly support from lots of rules.

References

1. Couso, I., Dubois, D.: Rough sets, coverings and incomplete information. Fundam. Inform. **108**(3–4), 223–347 (2011). https://doi.org/10.3233/FI-2011-421
2. Lipski, W.: On semantics issues connected with incomplete information databases. ACM Trans. Database Syst. **4**, 262–296 (1979)
3. Lipski, W.: On databases with incomplete information. J. ACM **28**, 41–70 (1981)
4. Nakata, M.: Unacceptable components in fuzzy relational databases. Int. J. Intell. Syst. **11**(9), 633–647 (1996). https://doi.org/10.1002/(SICI)1098-111X(199609)11:9
5. Nakata, M., Sakai, H.: Lower and upper approximations in data tables containing possibilistic information. In: Peters, J.F., Skowron, A., Marek, V.W., Orłowska, E., Słowiński, R., Ziarko, W. (eds.) Transactions on Rough Sets VII. LNCS, vol. 4400, pp. 170–189. Springer, Heidelberg (2007). https://doi.org/10.1007/978-3-540-71663-1_11

6. Nakata, M., Sakai, H.: Twofold rough approximations under incomplete information. Int. J. Gen. Syst. **42**, 546–571 (2013). https://doi.org/10.1080/17451000.2013.798898

7. Nakata, M., Sakai, H.: An approach based on rough sets to possibilistic information. Commun. Comput. Inf. Sci. **444**, 61–70 (2014)

8. Nakata, M., Sakai, H.: Rule induction based on rough sets from possibilistic information under Lipski's approach. In: Proceedings of the 2014 IEEE International Conference on Granular Computing (GrC), pp. 218–223. IEEE Computer Society (2014)

9. Nakata, M., Sakai, H.: Rough sets by indiscernibility relations in data sets containing possibilistic information. In: Flores, V., et al. (eds.) IJCRS 2016. LNCS (LNAI), vol. 9920, pp. 187–196. Springer, Cham (2016). https://doi.org/10.1007/978-3-319-47160-0_17

10. Pawlak, Z.: Rough Sets: Theoretical Aspects of Reasoning About Data. Kluwer Academic Publishers, Dordrecht (1991)

11. Sakai, H., Nakata, M., Watada, J.: NIS-apriori-based rule generation with three-way decisions and its application system in SQL. Inf. Sci. (2019, in press)

12. Słowiński, R., Stefanowski, J.: Rough classification in incomplete information systems. Math. Comput. Model. **12**, 1347–1357 (1989)

13. Zadeh, L.A.: Fuzzy sets as a basis for a theory of possibility. Fuzzy Sets Syst. **1**, 3–28 (1978)

Probability-Based Approach Explains (and Even Improves) Heuristic Formulas of Defuzzification

Christian Servin[1] , Olga Kosheleva[2] , and Vladik Kreinovich[2]([✉])

[1] Computer Science and Information Technology Systems Department,
El Paso Community College, 919 Hunter, El Paso, TX 79915, USA
cservin@gmail.com
[2] University of Texas at El Paso, El Paso, TX 79968, USA
{olgak,vladik}@utep.edu

Abstract. Fuzzy techniques have been successfully used in many applications. However, often, formulas for processing fuzzy information are heuristic: they lack a convincing justification, and thus, users are sometimes reluctant to use them. In this paper, we show that we can justify (and sometimes even improve) these methods if we use a probability-based approach.

Keywords: Defuzzification · Probabilistic approach ·
Heuristic algorithms · Optimality

1 Formulation of the Problem

Need for Fuzzy Knowledge. In many practical situations, ranging from medicine to driving, we rely on expert knowledge of how to cure diseases, how to drive in a complex city environment, etc. Some medical doctors are more qualified than others, some drivers are more skilled than others. It is therefore desirable to incorporate their skills and their knowledge in a computer-based system that will help other experts perform better – and ideally, make expert-quality decisions on its own, without the need for the experts.

One of the main obstacles to designing such a system is the fact that experts usually formulate their knowledge by using imprecise ("fuzzy") words from natural language like "close", "fast", "small", etc., and computers are not efficient in processing words, they are much more efficient in processing numbers. It is therefore desirable to represent the natural-language fuzzy knowledge in numerical terms.

A technique for such a representation was proposed in the 1960s by Lotfi Zadeh from Berkeley under the name of *fuzzy logic*. In fuzzy logic, to represent each word like "small" in numerical terms, we assign, to each possible value x

This work was supported in part by the US National Science Foundation grant HRD-1242122 (Cyber-ShARE Center of Excellence).

H. Seki et al. (Eds.): IUKM 2019, LNAI 11471, pp. 98–108, 2019.
https://doi.org/10.1007/978-3-030-14815-7_9

of the corresponding quantity, a degree $\mu(x) \in [0,1]$ to which, in the expert's option, the value x can be described by this word (e.g., to what extent x is small); see, e.g., [1,8,13,15,16,19].

Where Fuzzy Degrees Come from. There are many different ways to elicit the desired degrees.

If we are just starting the analysis and we do not have any records, then we can ask an expert to mark, on a scale, say, from 0 to 10, to what extent x is small. If the expert marks 7, we take 7/10 as the desired degree.

Usually, however, we already have a reasonably large database of records in which the experts used the corresponding word to describe different values of the corresponding quantity x. For example, when we describe the meaning of the word "small", then, for values x which are really small, we will have a large number of such records; on the other hand, for values x which are not too small, we will have a few such records – since few experts will consider these values to be small.

Based on the available records, we can estimate the probability density function (pdf) $\rho(x)$ that describes the frequency with which different values x appear in our records.

When x is really small, the value $\rho(x)$ is big; when x is not so small – so that fewer experts will consider this value to be small – the value $\rho(x)$ is much smaller. Thus, in principle, we could use the values $\rho(x)$ as the desired degrees. However, we want values of the membership function – and these values should be from the interval $[0,1]$. On the other hand, the pdf can take values larger than 1. For example, for a uniform distribution on the interval $[a,b]$, the value of the pdf $\rho(x)$ for value x inside this interval is $\rho(x) = 1/(b-a)$ – which can be very large when the width $b - a$ of the interval is small.

To make all the values smaller than or equal to 1, we can normalize these values, i.e., divide by the largest of them. As a result, we get

$$\mu(x) = \frac{\rho(x)}{\max\limits_{y} \rho(y)}.$$

This is a well-known way to get membership functions; see, e.g., [3–5,7,10,11].

Need for Defuzzification. By using expert knowledge transformed into the numerical form, we can determine, for each possible value u of the control, the degree $\mu(u)$ to which this value is reasonable.

These degrees can help an expert make better decisions. However, if we want to make an automatic system, we must select a single value u that the system will apply. Selecting such a value is known as *defuzzification*.

Centroid Defuzzification: Description, Successes, and Limitations. The most widely used defuzzification procedure is *centroid* defuzzification, in which we select the value

$$\overline{x} = \frac{\int x \cdot \mu(x)\,dx}{\int \mu(x)\,dx}.$$

It has led to many successful applications of fuzzy control; see, e.g., [1,8,13,15,16]. However, it has two related limitations:

- First, it is heuristic, it is not justified by a precise argument and therefore, we are not sure whether it will always work well.
- Second, it sometimes leads to disastrous results. For example, when a car encounters an obstacle on an empty road, it can go around it by veering to the left or by veering to the right. The situation is completely symmetric with respect to the direction to the obstacle. As a result, the centroid will lead exactly to the center – i.e., smack into the obstacle. The actual fuzzy control algorithms use some techniques to avoid such as a situation, but these techniques are also heuristic – and thus, not guaranteed to produce good results.

Optimization Under Fuzzy Constraints. Another class of situations in which fuzzy knowledge is important is optimization.

Traditional optimization techniques allows us to find the values of the parameters x for which the objective function attains its optimal value – largest or smallest depending on the problem. These techniques assume – explicitly or implicitly – that all possible combinations x are possible.

In practice, there are usually *constraints* restricting possible combinations. In some cases, constraints are formulated in precise terms – for example, there are regulations limiting noise level and pollution level from a plant. There are well-known techniques for dealing with such constraints – e.g., the Lagrange multiplier method that reduced the problem of optimizing an objective function $f(x)$ under constraint $g(x) = 0$ to the unconstrained optimization problem of optimizing the auxiliary objective function $f(x) + \lambda \cdot g(x)$, for an appropriate parameter λ.

Often, however, we also have imprecise (fuzzy) constraints. For example, a company that designs a plant in a city usually wants not just to satisfy all the legal requirements, but also to keep good relation with the city, and one way to do it is to make sure that the noise level is not high. This "not high" is clearly an example of an imprecise constraint.

Another case when fuzzy constraints are important is when one of the objectives is to make customers happy. For example, an elevator must be reasonable fast but also reasonably smooth.

We can describe the fuzzy constraint by a membership function $\mu(x)$, so that for each possible combination x of the corresponding parameters, $\mu(x)$ is a degree to which the alternative corresponding to these parameter values satisfies the constraint. How can we optimize an objective function $f(x)$ under such fuzzy constraints?

A well-known heuristic solution to this problem was proposed in a joint paper [2] by Lotfi Zadeh (the father of fuzzy logic) and Richard Bellman (the famous specialist in optimization): namely, to maximize an objective function $f(x)$ under fuzzy constraints, they proposed to maximize an auxiliary function

$$f_\& \left(\mu(x), \frac{f(x) - m}{M - m} \right),$$

where $f_\&(a, b)$ is usually either the minimum $\min(a, b)$ or the product $a \cdot b$, and m and M are, correspondingly, the minimum and the maximum of $f(x)$ over the set X of all theoretically possible combinations x:

$$m \stackrel{\text{def}}{=} \min_{x \in X} f(x), \quad M \stackrel{\text{def}}{=} \max_{x \in X} f(x).$$

The above formula is used when we want to maximize $f(x)$; when we want to minimize $f(x)$, then we maximize a slightly different auxiliary expression – which comes from the fact that minimizing $f(x)$ is equivalent to maximizing an auxiliary function $f'(x) \stackrel{\text{def}}{=} -f(x)$:

$$f_\& \left(\mu(x), \frac{M - f(x)}{M - m} \right).$$

Similarly to defuzzification:

– on the one hand, these heuristic formulas have led to many useful application, but
– on the other hand, the fact that these formulas are heuristic – and thus, lack a convincing justification – makes users often somewhat reluctant to use them.

What We Do in This Paper. In this paper, we show that if we take into account the widely spread probability-based origin of fuzzy techniques, then many heuristic techniques – including techniques related to defuzzification and optimization – become *justified*. Moreover, this use of probabilistic ideas sometimes enables us to *improve* the existing heuristic fuzzy techniques.

Comment. In our opinion, the above justification is a good example of the need for integrated uncertainty: in a situation where pure probabilistic methods are not natural, and where pure fuzzy techniques lack a convincing justification, a combination of probabilistic and fuzzy approaches helps.

2   Probability-Based Approach Explains Heuristic Formulas of Defuzzification

Let us start by showing that probability-based approach explains the main formula of centroid justification.

Crudely speaking, the membership function $\mu(x)$ describes the degree to which x corresponds to the optimal control. If the membership function comes from a probability distribution $\rho(x)$, this means that we do not know exactly which value x is optimal: different values x may turn out to be optimal, and the corresponding values $\rho(x)$ describes the probability of different values to be optimal.

Based on this information, we want to select a single value $\bar{x}$. Because of the probabilistic character of available information, no matter what value we select, there is a probability that this value will be not optimal. So, no matter what value we select, there will be a loss caused by this non-optimality. It is reasonable to select the value $\bar{x}$ for which the expected value of this loss is the smallest.

The loss happens if the optimal value x is different from the selected value x'. In other words, the loss $L(x, x')$ is caused by the fact that difference $x - x'$ is different from 0. The loss can thus be viewed as a function of this difference $L(x, x') = F(x - x')$ for some function $F(z)$.

It is reasonable to assume that the loss function $F(z)$ is continuous in z. Every continuous function on an interval can be approximated, with any given accuracy, by an analytical function – e.g., by a polynomial. Thus, it is safe to assume that the function $F(z)$ is analytical and can, therefore, be expanded in Taylor series (at least in a vicinity of $z = 0$):

$$F(z) = a_0 + a_1 \cdot z + a_2 \cdot z^2 + a_3 \cdot z^3 + \dots$$

The difference $z = x - x'$ is usually reasonable small. So, from the practical viewpoint, we can safely ignore higher order terms and keep only the first few terms in this expansion.

From the purely mathematical viewpoint, the simplest possible case is when we keep only the constant term $F(z) = a_0$. However, in this case, the loss does not depend on how far the selected value x' is from the unknown optimal value x – which does not make sense, since by definition, selecting the optimal value should lead to the best performance and thus, to the smallest possible loss.

Similarly, if we take into account linear terms, i.e., consider the loss function of the type $F(z) = a_0 + a_1 \cdot z$, this also does not make sense: the loss function should attains its smallest value $F(z) = 0$ when the selected value x' is optimal, so that $z = x' - x = 0$, but a linear function does not attains its minimum at 0.

The simplest possible approximate expression for the loss function $F(z)$ which makes sense is when we take quadratic terms into account, i.e., when we consider the expression

$$F(z) = a_0 + a_1 \cdot z + a_2 \cdot z^2.$$

When the selected value x' is exactly the optimal value x, i.e., when $z = 0$, then there is no loss: $F(0) = 0$. Substituting $z = 0$ into the above quadratic formula, we conclude that $a_0 = 0$. Also, when $z = 0$, the loss is the smallest; thus, for $z = 0$, the derivative $F'(0)$ is equal to 0 – which implies that $a_1 = 0$ – and the second derivative is non-negative – so $a_2 > 0$.

So, $F(z) = a_2 \cdot z^2$, so the loss is equal to $L(x, x') = a_2 \cdot (x - x')^2$, and the expected value of this loss is equal to

$$\int L(x, x') \cdot \rho(x) \, dx = \int a_2 \cdot (x - x')^2 \cdot \rho(x) \, dx.$$

We want to find the value x' that minimizes this loss. To find this value, we differentiate the above expression by x' and equate the resulting derivative to 0. As a result, we get

$$\int 2 \cdot a_2 \cdot (x - x') \cdot \rho(x) \, dx = 0.$$

Dividing both sides by $2a_2$ and representing the integral of the difference as the difference between the two integrals, we conclude that

$$\int x \cdot \rho(x) \, dx - x' \cdot \int \rho(x) \, dx = 0.$$

The second integral in this formula is simply the total probability, i.e., 1, so the optimal value $\bar{x}$ of the control x is equal to the mean

$$\bar{x} = \int x \cdot \rho(x) \, dx.$$

As we have mentioned earlier, the membership function $\mu(x)$ and the corresponding probability density function $\rho(x)$ differ only by a multiplicative constant:

$$\mu(x) = c \cdot \rho(x)$$

for an appropriate constant c – so that $\rho(x) = \dfrac{\mu(x)}{c}$. To find the constant c, we can integrate both sides of the equality $\mu(x) = c \cdot \rho(x)$. We thus get $\int \mu(x) \, dx = c \cdot \int \rho(x) \, dx$. If we take into account that, as we have recently mentioned, $\int \rho(x) \, dx = 1$, we conclude that $c = \int \mu(x) \, dx$. Thus, $\rho(x) = \dfrac{\mu(x)}{\int \mu(y) \, dy}$.

Substituting this expression for $\rho(x)$ into the above formula for $\bar{x}$, we get exactly the usual formula for centroid defuzzification.

3 Let Us Use the Probability-Based Justification to Improve the Heuristic Formulas for Defuzzification

Formulas for Defuzzification Need Improvement: Reminder. As we have mentioned earlier, the problem with the existing defuzzification formulas is not only that they are heuristic and thus need justification, but also that these formulas sometimes lead to disastrous (or at least suboptimal) results.

It is therefore desirable to improve these formulas.

Idea. A reasonable idea is to take into account that we are not just interested in finding the values x that minimize the total loss; ideally, the selected value x should also be optimal in relation to the original control problem. The corresponding degree of optimality is described by the membership function $\mu(x)$.

Thus, in effect, we have a problem of optimization under fuzzy constraint: minimize the expression

$$\int (x - \bar{x})^2 \cdot \rho(x)\, dx = \frac{\int (x - \bar{x})^2 \cdot \mu(x)\, dx}{\int \mu(x)\, dx}$$

under the fuzzy constraint described by the original membership function $\mu(x)$.

The denominator of the minimized expression does not depend on the selection of the control parameter $\bar{x}$; so, minimizing the above ratio is equivalent to minimizing the numerator $\int (x - \bar{x})^2 \cdot \mu(x)\, dx$.

To solve this problem, we can therefore use the Bellman-Zadeh approach. Let us see what we get.

Towards the Resulting Modification. Instead of the centroid defuzzification, we should select the value x' for which the following expression attains the smallest possible value

$$f_\& \left(\mu(x'), \frac{M - \int (x - x')^2 \cdot \mu(x)\, dx}{M - m} \right),$$

where

$$m \stackrel{\text{def}}{=} \min_{x'} \int (x - x')^2 \cdot \mu(x)\, dx$$

and

$$M \stackrel{\text{def}}{=} \max_{x'} \int (x - x')^2 \cdot \mu(x)\, dx.$$

Let us show how this formula can be simplified. To find m and M, we, correspondingly, minimize or maximize the expression $\int (x - x')^2 \cdot \mu(x)\, dx$. If we open parentheses, we can conclude that this expression is quadratic in terms of x':

$$\int (x - x')^2 \cdot \mu(x)\, dx = M_2 - 2M_1 \cdot x' + M_0 \cdot (x')^2,$$

where we denoted $M_2 \stackrel{\text{def}}{=} \int x^2 \cdot \mu(x)\, dx$, $M_1 \stackrel{\text{def}}{=} \int x \cdot \mu(x)\, dx$, and $M_0 \stackrel{\text{def}}{=} \int \mu(x)\, dx$. We know that the minimum of this expression is attained at the centroid value, $x_0 = \dfrac{M_1}{M_0}$. Substituting this value x_0 into the above expression for the integral $\int (x - x')^2 \cdot \mu(x)\, dx$, we conclude that

$$m = M_2 - 2M_1 \cdot \frac{M_1}{M_0} + M_0 \cdot \left(\frac{M_1}{M_0} \right)^2 = M_2 - \frac{M_1^2}{M_0}.$$

For the quadratic function which attains its minimum, its maximum on any interval is attained at one the interval's endpoints.

Thus, we arrive at the following modified version of the centroid defuzzification.

The Resulting Modification of Centroid Defuzzification. Once we know the membership function $\mu(x)$ on an interval $[x_-, x_+]$, then, to find the best value $\bar{x}$, we do the following:

- first, we compute the values $M_0 = \int \mu(x)\,dx$, $M_1 = \int x \cdot \mu(x)\,dx$, and $M_2 = \int x^2 \cdot \mu(x)\,dx$;

- then, we compute the values $m = M_2 - \dfrac{M_1^2}{M_0}$ and

$$M = \max(M_2 - 2M_1 \cdot x_- + M_0 \cdot x_-^2, M_2 - 2M_1 \cdot x_+ + M_0 \cdot x_+^2);$$

- finally, we find the value $\bar{x} = x'$ that maximizes the expression

$$f_\& \left(\mu(x'), \frac{M - (M_2 - 2M_1 \cdot x' + M_0 \cdot (x')^2)}{M - m} \right).$$

This is Indeed Better than Centroid. As we have mentioned earlier, the main problem of centroid defuzzification is that it sometimes leads to very bad decisions, i.e., decisions $\bar{x}$ for which the value $\mu(\bar{x})$ is 0 (or close to 0). This is possible for centroid defuzzification – since its algorithm does not take the value $\mu(\bar{x})$ into account at all.

However, for our new method, this is not possible. Indeed, for both $f_\&(a, b) = \min(a, b)$ and $f_\&(a, b) = a \cdot b$, we have $f_\&(0, a) = 0$ for all $a \in [0, 1]$. Thus, if $\mu(\bar{x}) = 0$, then the corresponding objective function is equal to 0 – i.e., to its smallest possible value – and thus, will never be selected under the new approach.

What if We Have Two Equally Possible Solutions? In the case of a symmetric obstacle, we will no longer go straight into this obstacle, so the corresponding angle $x = 0$ is not possible. Hence we select a value $\bar{x} \neq 0$.

Due to symmetry, if $\bar{x} \neq 0$ is a solution, then $-\bar{x}$ is a solution as well. Thus, we have at least two different solutions. Which one should we choose?

The situation is symmetric, so our decision should be symmetric as well. However, if we select one of the two possible solutions $\bar{x}$ or $-\bar{x}$, we violate $x \leftrightarrow -x$ symmetry. So what should we do?

The only way to preserve symmetry is to make a *probabilistic* decision, i.e., in this case, to select either $\bar{x}$ or $-\bar{x}$ with equal probability $1/2$.

Comment. Thus again, probabilistic ideas help: namely, they help to retain a natural symmetry of the situation.

Discussion. In fuzzy control, this may be a new idea, but in general, that symmetry sometimes naturally leads to randomness is a known fact.

The first such example is *game theory*; see, e.g., [12,14,18]. The fact that the optimal strategies are probabilistic has been known since the beginning of game theory. Indeed, suppose that we want to protect two equally valuable locations from a terrorist attack, but we only have resources for a single protection team. If we select a deterministic decision – i.e., send the team to one of the two locations – we will lose no matter which location we select, since the terrorists will successfully attack the remaining location. The best strategy is to each time send a team to one of the locations at random.

A more relaxed example is the rock-paper-scissors game, in which each of the two players selects either rock, or paper, or scissors. Paper beats rock, rock beats scissors, and scissors beat paper. If one side selects a deterministic strategy – i.e., selects the same choice every time – the opponent will always win by selecting the choice that beats this selection. The only way to avoid this defeat is to select each of the three choices with equal probability.

Another example is *physics*, for example, the radioactivity phenomenon, when some atoms spontaneously decay; see, e.g., [6,17]. Let us show that radioactivity cannot be deterministic, i.e., at the present moment t_0, we cannot predict the moment of time t at which the atom will decay (i.e., equivalently, the time period $t - t_0$ until the decay). Indeed, the laws of physics do not change if we simply change the starting point for measuring time. Thus, if the decay process was deterministic, then we will be able to conclude that when we observe the not-yet-decayed atom at a moment $t_0 + \varepsilon$ for some $\varepsilon > 0$, then we should also predict decay $t - t_0$ seconds in the future – but this cannot be, since there is only $t - t_0 - \varepsilon$ second to the deterministic decay. Thus, the decay cannot be deterministic, it has to be probabilistic.

Comment. This example shows, by the way, that the probabilistic character of quantum physics is not some counter-intuitive feature, it is a natural consequence of simple symmetries.

Remaining Problem. Our idea seems reasonable. However, there is still a problem: to come up with an improved defuzzification method, we used Bellman-Zadeh formulas – and, as we have mentioned earlier, these formulas are heuristic. It is thus desirable to come up with a justification for these formulas. Let us show that the probability-based approach provides exactly such a justification.

Comment. The main ideas behind this justification first appeared in [9].

4   Probability-Based Approach Explains Heuristic Formulas of Optimization Under Fuzzy Constraints

Let us consider the maximization case: we want to maximize the value objective function $f(x)$ under the fuzzy constraint described by the membership function $\mu(x)$. (The minimization case can be treated similarly.)

If we select a value x, and this value is possible, then we get the gain $f(x)$. On the other hand, if we select x, and this value x is *not* possible, then we will have to go back to the worst-case scenario m. Let us denote the probability of the value x to be possible by $p(x)$. Then:

– with probability $p(x)$, we get $f(x)$, and
– with the remaining probability $1 - p(x)$ we get m.

The expected gain is thus equal to $p(x) \cdot f(x) + (1 - p(x)) \cdot m$. This expression can be reformulated as

$$p(x) \cdot f(x) + m - p(x) \cdot m = m + p(x) \cdot (f(x) - m).$$

Adding a constant to all the values of an objective function (in this case, the constant m) does not change which values are larger and which values are smaller. Thus, maximizing the above objective function is equivalent to maximizing a simpler expression $p(x) \cdot (f(x) - m)$.

We consider the cases when the probabilities are proportional to the corresponding values of the membership function: $p(x) = c \cdot \mu(x)$. In this case, the above maximized expression takes the form $c \cdot \mu(x) \cdot (f(x) - m)$. Multiplying all the values of an objective function by the same constant does not change which values are larger and which are smaller – e.g., the richest person in Mexico remains the richest whether we count his net worth in US dollars or in Euros or in Mexican pesos. Thus, maximizing the above expression is equivalent to maximizing the product $\mu(x) \cdot (f(x) - m)$. Similarly, since the difference $M - m$ is also a constant not depending on x, the above maximization is equivalent to maximizing the expression

$$\mu(x) \cdot \frac{f(x) - m}{M - m}.$$

This is exactly Bellman-Zadeh formula for $f_\&(a, b) = a \cdot b$. Thus, the probability-based approach indeed explains this heuristic formula.

Acknowledgments. The authors are greatly thankful to the anonymous referees for valuable suggestions.

References

1. Belohlavek, R., Dauben, J.W., Klir, G.J.: Fuzzy Logic and Mathematics: A Historical Perspective. Oxford University Press, New York (2017)
2. Bellman, R.E., Zadeh, L.A.: Decision making in a fuzzy environment. Manage. Sci. **17**(4), B141–B164 (1970)
3. Coletti, G., Scozzafava, R.: Conditional probability and fuzzy information. Comput. Stat. Data Anal. **51**(1), 115–132 (2006)
4. Coletti, G., Scozzafava, R., Vantaggi, B.: Possibility measures in probabilistic inference. In: Dubois, D., Lubiano, M.A., Prade, H., Gil, M.Á., Grzegorzewski, P., Hryniewicz, O. (eds.) Soft Methods for Handling Variability and Imprecision. Advances in Soft Computing, vol. 48, pp. 51–58. Springer, Heidelberg (2008). https://doi.org/10.1007/978-3-540-85027-4_7
5. Coletti, G., Scozzafava, R., Vantaggi, B.: A bridge between probability and possibility in a comparative framework. In: Liu, W. (ed.) ECSQARU 2011. LNCS (LNAI), vol. 6717, pp. 557–568. Springer, Heidelberg (2011). https://doi.org/10.1007/978-3-642-22152-1_47
6. Feynman, R., Leighton, R., Sands, M.: The Feynman Lectures on Physics. Addison Wesley, Boston (2005)

7. Huynh, V.-N., Nakamori, Y., Lawry, J.: A probability-based approach to comparison of fuzzy numbers and applications to target-oriented decision making. IEEE Trans. Fuzzy Syst. **16**(2), 371–387 (2008)
8. Klir, G., Yuan, B.: Fuzzy Sets and Fuzzy Logic. Prentice Hall, Upper Saddle River (1995)
9. Kosheleva, O., Kreinovich, V.: Why Bellman-Zadeh approach to fuzzy optimization. Appl. Math. Sci. **12**(11), 517–522 (2018)
10. Lawry, J.: A voting mechanism for fuzzy logic. Int. J. Approximate Reasoning **19**(3–4), 315–333 (1998)
11. Lawry, J.: Borderlines and probabilities of borderlines: on the interconnection between vagueness and uncertainty. J. Appl. Logic **14**, 113–138 (2016)
12. Luce, R.D., Raiffa, H.: Games and Decisions: Introduction and Critical Survey. Dover, New York (1989)
13. Mendel, J.M.: Uncertain Rule-Based Fuzzy Systems: Introduction and New Directions. Springer, Cham (2017). https://doi.org/10.1007/978-3-319-51370-6
14. Myerson, R.B.: Game Theory: Analysis of Conflict. Harvard University Press, Harvard (1997)
15. Nguyen, H.T., Walker, E.A.: A First Course in Fuzzy Logic. Chapman and Hall/CRC, Boca Raton (2006)
16. Novák, V., Perfilieva, I., Močkoř, J.: Mathematical Principles of Fuzzy Logic. Kluwer, Boston (1999)
17. Thorne, K.S., Blandford, R.D.: Modern Classical Physics: Optics, Fluids, Plasmas, Elasticity, Relativity, and Statistical Physics. Princeton University Press, Princeton (2017)
18. von Neumann, J., Morgenstern, O.: Theory of Games and Economic Behavior. Princeton University Press, Princeton (1944)
19. Zadeh, L.A.: Fuzzy sets. Inf. Control **8**, 338–353 (1965)

Three Valued Representation of Opinions in Affective Design

Jonathan Lawry[1], Van-Nam Huynh[2(✉)], and Chris McMahon[3]

[1] Department of Engineering Mathematics, University of Bristol,
Bristol BS8 1UB, UK
j.lawry@bris.ac.uk
[2] School of Knowledge Science, Japan Advanced Institute of Science and Technology,
Nomi, Ishikawa 923-1292, Japan
huynh@jaist.ac.jp
[3] Department of Mechanical Engineering, Technical University of Denmark,
Kongens Lyngby, Denmark
chmcm@mek.dtu.dk

Abstract. Attributes based on natural language descriptions (Kansei words) are common in affective design questionnaire data. Such words are usually inherently vague and exhibit characteristics such as explicitly borderline cases and blurred boundaries between those cases to which the word does and those to which it does not apply. In this paper we propose an integrated treatment of vagueness and uncertainty which combines three value logic and probability by defining a probability distribution over valuations in Kleene's logic. Such an approach naturally results in lower and upper uncertainty measures on the sentences of the language, quantifying the uncertainty that a given sentence is true or that it is not false respectively. Within this framework we propose a representational model for opinions in the form of a graph of conjunctive clauses ordered by precision and weighted according to their respective lower and upper uncertainty measures. Furthermore, by extending the idea of scoring functions to a three valued setting we propose an approach for ranking different designs which takes into account both the level of belief in an opinion and also its relative strength. The potential of this approach is illustrated using a case study involving questionnaire data about Kutani traditional Japanese craft designs.

1 Introduction

On occasion the sales performance of a new product is disappointing despite it being functionally reliable and meeting high production standards. For whatever reason it does not connect emotionally with potential customers. Understanding the nature of this subjective connection so as to improve market performance is at the heart of Kansei Engineering and other affective design methodologies [1, 7]. For Kansei Engineering in particular, data collection and quantitative analysis play a central role. Data gathering techniques include questionnaires and other

© Springer Nature Switzerland AG 2019
H. Seki et al. (Eds.): IUKM 2019, LNAI 11471, pp. 109–121, 2019.
https://doi.org/10.1007/978-3-030-14815-7_10

direct elicitation methods, as well as psychological studies aimed at capturing sub-conscious reactions. The use of Kansei words in semantic scale questionnaires is commonly employed in conjunction with mathematical and statistical tools so as to analyse customer preference between different designs. Kansei words are natural language descriptions of subjective properties or feelings e.g. beautiful, colourful etc., and as such they tend to be inherently vague. This raises two important questions concerning the effective use of such questionnaire data. Assuming that opinions about different designs are expressed using Kansei words these are; (1) how do we model the inherent vagueness in combination with the uncertainty and natural variation between different individuals? And then using this model; (2) What is a good way of representing the different opinions present in the data so to inform the design process?

In this paper we will apply an integrated model of vagueness and uncertainty which defines probabilities over a three-valued truth model according to which a proposition can be true, false or borderline. We have argued [4–6] that such models capture several of the defining characteristics of vagueness in natural language. We will employ a graphical representation in which opinions correspond to conjunctive clauses ordered according to precision. Furthermore, each opinion is weighted by lower and upper measures respectively quantifying the probability that the clause is true and the probability that it is not false. These allow qualitative comparison between designs, taking account of the underlying vagueness and uncertainty, and provide measures of both the level of support for each opinion, and the strength with which it is asserted, with the latter represented by the probability of it being classified as a borderline case. In addition, we investigate scoring functions extended to a three valued context and show how these can be used for ranking competing designs. The potential of the proposed methodology is illustrated by its application to a Kansei data set relating to traditional Japanese porcelain designs [2].

An outline of the paper is as follows. Section 2 discusses the role of vagueness in affective design. Section 3 introduces Kleene's three valued logic as a way of formalizing explicit borderline cases in natural language propositions and then proposes an integrated treatment of vagueness and uncertainty in the form of probability distributions defined over Kleene truth models. This then results in lower and upper uncertainty measures on the sentences of the language. In Sect. 4 we propose a representational framework for opinions in the form of graphs comprised of conjunctive clauses as nodes weighted by their respective Kleene lower and upper measures, and with edges determined by a precision ordering. A case study about traditional Japanese porcelain designs is described in Sect. 5 as a means of illustrating our proposed approach to opinion representation. In Sect. 6 we extend the notion of scoring function to a three valued context. Finally Sect. 7 gives some discussions and conclusions.

2   Vagueness

Vagueness is ubiquitous in natural language although it is traditionally viewed in a negative light. Alternatively, from a positive perspective vagueness can be seen as characteristic of a more flexible and adaptive form of concept definitions than are captured by the classical Boolean model [13]. From this perspective simply replacing vague definitions in natural language descriptions with alternative crisp (i.e. non-vague) versions can significantly distort the intended meaning. For example, descriptions such as *beautiful* or *colourful* would seem to have a natural graded quality which the use of Likert or Semantic Distance (SD) scales in design questionnaires attempt to capture.

There has been significant discussion concerning linguistic vagueness in the Kansei engineering literature and a number of different methodologies have been applied. Fuzzy logic has been used to represent if-then rules linking design attributes to preferences of quality decisions [11,12,14]. In addition, [15] have proposed a target based approach in which fuzzy sets are defined over the elements of an SD scale. Here as in [16] membership functions are determined probabilistically either by using α–cuts or by means of a voting model. Rough set methods have also been applied in Kansei engineering as a model of vagueness [8]. In this context, for example, a *good design* is characterised by lower and upper sets of attribute values. To allow for natural variation between individuals rough sets can then be combined with probability theory by using an information theoretic threshold [9]. In this paper we propose a probabilistic model which also permits inherently borderline cases of vague predicates. It has been shown [4–6] that this approach generates a natural and expressive integrated calculus for representing both uncertainty and linguistic vagueness, and for which there is a clear operational semantics.

Following Keefe and Smith [3] we take a defining feature of vagueness as being the admittance of borderline cases about which a proposition is neither *absolutely true* nor *absolutely false*. For example, there are some height values which would neither be classified as being short nor not short. For propositions involving vague concepts this naturally results in truth-gaps. In other words, there are cases in which a proposition is neither absolutely true nor absolutely false. If Ethel's height lies in a certain intermediate range then the proposition 'Ethel is short' may be inherently borderline. Such truth-gaps suggest that a non-Tarskian notion of truth is required to capture this aspect of vagueness even in a simple propositional framework. Kleene's logic has been proposed as an appropriate formalism for truth gaps by a number authors including Shapiro [10] and Lawry [4]. The following section gives a brief introduction to Kleene's three valued logic.

3   Kleene's Three Valued Logic and Belief Pairs

In this paper we take beliefs as referring to sentences from a simple propositional logic language defined as follows. Let $\mathcal{L}$ be a language of propositional logic with connectives $\wedge$, $\vee$ and $\neg$ and propositional variables $\mathcal{P} = \{p_1, \ldots, p_n\}$. Let $S\mathcal{L}$ denote the sentences of $\mathcal{L}$ as generated recursively from the propositional variables by application of the three connectives, and let $L\mathcal{L} = \mathcal{P} \cup \{\neg p : p \in \mathcal{P}\}$ denote the literals of $\mathcal{L}$. In propositional logic a valuation is a function from the sentences of the language $S\mathcal{L}$ to truth values. For Boolean logic this function is defined recursively by initially allocating values **t** or **f** to the propositional variables and then extending the mapping to compound sentences by applying the standard truth tables for each of the connectives. Similarly for Kleene logic a valuation is a function $v : S\mathcal{L} \rightarrow \{\mathbf{t}, \mathbf{b}, \mathbf{f}\}$ defined recursively using the truth tables as shown in Table 1.

Table 1. Truth tables for the connectives $\wedge$, $\vee$ and $\neg$ in Kleene's three valued logic.

$\wedge$	t	b	f		$\vee$	t	b	f		$\neg$	
t	t	b	f		t	t	t	t		t	f
b	b	b	f		b	t	b	b		b	b
f	f	f	f		f	t	b	f		f	t

Let $\mathbb{V}$ denote the set of all Kleene valuations of $\mathcal{L}$ of which there are 3^n. Any Kleene valuation $v \in \mathbb{V}$ can be characterised by an *orthopair* (P, N) consisting of a pair of non overlapping sets of propositional variables; $P, N \subseteq \mathcal{P}$ and $P \cap N = \emptyset$. Here P denotes the set of true propositional variables, i.e. $P = \{p_j : v(p_j) = \mathbf{t}\}$, while N denotes the set of false propositional variables, i.e. $N = \{p_j : v(p_j) = \mathbf{f}\}$. Since this is a three valued logic then in general $P \cup N$ is not exhaustive since it excludes exactly those propositions with borderline truth value. In other words, the set of borderline propositional variables is given by $B = (P \cup N)^c$.

Example 1. Consider a new dress design described using the terms *colourful* (p_1), *patterned* (p_2), *modern* (p_3) and *formal* (p_4). Then the orthopair $(\{p_2, p_3\}, \{p_4\})$ represents the valuation according to which the dress design is classified as being absolutely patterned and modern, absolutely not formal and borderline colourful.

It is important to note that in this model truth gaps should not be understood as resulting from epistemic uncertainty about an underlying Boolean valuation. In other words, the borderline truth value should not be interpreted as meaning *uncertain*. Indeed in the context of modelling vagueness we may be certain that a particular proposition is a borderline case. For instance, having inspected a new dress design we may be completely sure that it is borderline colourful. Instead borderline cases are assumed to result from inherent vagueness or flexibility in the way many natural language predicates are actually defined. For instance, a three valued truth model would naturally arise if predicates were defined in terms of both positive and negative cases. None the less, it is often the case that such truth gaps occur in conjunction with epistemic uncertainty. In a propositional logic setting uncertainty concerns the truth value of propositions and consequently what is the correct valuation for the language $\mathcal{L}$. Hence, assuming an underlying three valued truth model, we now propose to model uncertainty concerning the sentences of $\mathcal{L}$ in terms of probability distributions defined over

Kleene valuations. Let w be a probability distribution on $\mathbb{V}$ then we naturally define the following three measures on $S\mathcal{L}$:

Lower Kleene Measure: $\underline{\mu} : S\mathcal{L} \to [0,1]$ such that $\forall \theta \in S\mathcal{L}$,

$$\underline{\mu}(\theta) = w(\{v \in \mathbb{V} : v(\theta) = \mathbf{t}\})$$

Upper Kleene Measure: $\overline{\mu} : S\mathcal{L} \to [0,1]$ such that $\forall \theta \in S\mathcal{L}$,

$$\overline{\mu}(\theta) = w(\{v \in \mathbb{V} : v(\theta) \neq \mathbf{f}\})$$

Truth Degree: $td : S\mathcal{L} \to [0,1]$ such that $\forall \theta \in S\mathcal{L}$,

$$td(\theta) = \frac{\underline{\mu}(\theta) + \overline{\mu}(\theta)}{2} = w(\{v \in \mathbb{V} : v(\theta) = \mathbf{t}\}) + \frac{w(\{v \in \mathbb{V} : v(\theta) = \mathbf{b}\})}{2}$$

Taken together we refer to lower and upper Kleene measures $\boldsymbol{\mu} = (\underline{\mu}, \overline{\mu})$ as a Kleene belief pair, and which satisfy the following properties: $\forall \theta, \varphi \in S\mathcal{L}$,

- $\underline{\mu}(\theta) \leq \overline{\mu}(\theta)$.
- $\underline{\mu}(\neg\theta) = 1 - \overline{\mu}(\theta)$ and $\overline{\mu}(\neg\theta) = 1 - \underline{\mu}(\theta)$.
- $\underline{\mu}(\theta \vee \varphi) = \underline{\mu}(\theta) + \underline{\mu}(\varphi) - \underline{\mu}(\theta \wedge \varphi), \overline{\mu}(\theta \vee \varphi) = \overline{\mu}(\theta) + \overline{\mu}(\varphi) - \overline{\mu}(\theta \wedge \varphi)$.

3.1 Conditional Kleene Measures

In this section we introduce conditioning for Kleene belief measures. In view of the inherently probabilistic nature of belief pairs, one obvious approach is based on conditional probabilities. In this context we assume that new knowledge takes the form of constraints on Kleene valuations, which it is then assumed that the *correct* valuation for $\mathcal{L}$ must satisfy. This knowledge then allows us to define conditional lower and upper belief measures by determining a posterior distribution on $\mathbb{V}$ from the prior w, according to the standard definition of conditional probability. More formally, let:

$$K = \{v(\varphi_j) \in \mathbf{z}_j : j = 1, \ldots t\} \text{ where } \mathbf{z}_j \subseteq \{\mathbf{t}, \mathbf{b}, \mathbf{f}\} \text{ and } \varphi_j \in S\mathcal{L} \text{ for } j = 1, \ldots, t$$

Then we define the lower and upper belief measures conditional on K as follows: $\forall \theta \in S\mathcal{L}$,

$$\underline{\mu}(\theta|K) = \frac{w(\{v \in \mathbb{V}(K) : v(\theta) = \mathbf{t}\})}{w(\mathbb{V}(K))} \text{ and } \overline{\mu}(\theta|K) = \frac{w(\{v \in \mathbb{V}(K) : v(\theta) \neq \mathbf{f}\})}{w(\mathbb{V}(K))}$$

Consider for example special cases where for sentence $\varphi \in S\mathcal{L}$, $K = \{v(\varphi) = \mathbf{t}\}$ and $K = \{v(\varphi) \neq \mathbf{f}\}$ and for which we have the following expressions:

$$\underline{\mu}(\theta|v(\varphi) = \mathbf{t}) = \frac{\underline{\mu}(\theta \wedge \varphi)}{\underline{\mu}(\varphi)} \text{ and } \overline{\mu}(\theta|v(\varphi) = \mathbf{t}) = \frac{\overline{\mu}(\theta \vee \neg\varphi) - \overline{\mu}(\neg\varphi)}{1 - \overline{\mu}(\neg\varphi)}$$

$$\underline{\mu}(\theta|v(\varphi) \neq \mathbf{f}) = \frac{\underline{\mu}(\theta \vee \neg\varphi) - \underline{\mu}(\neg\varphi)}{1 - \underline{\mu}(\neg\varphi)} \text{ and } \overline{\mu}(\theta|v(\varphi) \neq \mathbf{f}) = \frac{\overline{\mu}(\theta \wedge \varphi)}{\overline{\mu}(\varphi)}$$

4   Representing Opinions

For Kleene logic we take the *conjunctive clauses* of $\mathcal{L}$ as corresponding to the conjunctions of literals, so that:

$$CL = \{ \bigwedge_{p_j \in P} p_j \wedge \bigwedge_{p_j \in N} \neg p_j : P, N \subseteq \mathcal{P}, P \cap N = \emptyset, P \cup N \neq \emptyset \}$$

Notice that conjunctive clauses are also characterised by orthopairs in the same way as Kleene valuations. Indeed there is a natural correspondence between the set of Kleene valuations $\mathbb{V}$ and the set of conjunctive clauses CL, according to which for any valuation $v \in \mathbb{V}$ with orthopair (P, N) such that $P \cup N \neq \emptyset$, there is a unique clause $\alpha = \bigwedge_{p_j \in P} p_j \wedge \bigwedge_{p_j \in N} \neg p_j$, for which $v(\alpha) = \mathbf{t}$ and which is maximally precise.

Definition 1 (*Ordering Conjunctive Clauses*). *Let $\alpha_1, \alpha_2 \in CL$, characterised by orthopairs (P_1, N_1) and (P_2, N_2) respectively then $\alpha_1 \preceq \alpha_2$ if and only if $P_1 \supseteq P_2$ and $N_1 \supseteq N_2$. Note that $\preceq$ is a generality or precision ordering on conjunctive clauses so that $\alpha_1 \preceq \alpha_2$ means that α_1 is less general (more precise) than α_2 in the sense that for any Kleene valuation $v \in \mathbb{V}$, $v(\alpha_1) = \mathbf{t} \Rightarrow v(\alpha_2) = \mathbf{t}$ and $v(\alpha_2) = \mathbf{f} \Rightarrow v(\alpha_1) = \mathbf{f}$.*

Definition 2 (*Weighted Clausal Graph*). *A weighted clausal graph is a triple (A, E, μ) where $A \subseteq CL$, μ is a Kleene belief pair restricted to A, $E \subseteq A \times A$ is the set of edges such that:*

$$E = \{(\alpha_1, \alpha_2) \in A \times A : \alpha_1 \neq \alpha_2, \alpha_1 \preceq \alpha_2, \nexists \alpha_3 \neq \alpha_1, \alpha_2 \text{ such that } \alpha_1 \preceq \alpha_3, \alpha_3 \preceq \alpha_2\}$$

Note that (A, E) is a directed acyclic graph corresponding to the covering relation of the partially ordered set $(A, \preceq)$ for which the edges are between comparable clauses which are neighbours.

We refer to (CL, E, μ) as a *complete weighted clausal graph*. Notice that such a graph provides sufficient information to infer the underlying probability distribution w on $\mathbb{V}$, and consequently to determine the lower and upper Kleene belief values for any sentence of $\mathcal{L}$. The relevant inversion mapping is as follows: For $v \in \mathbb{V}$,

$$w(v) = \sum_{\alpha' \in CL : \alpha' \preceq \alpha} (-1)^{|B| - |B'|} \underline{\mu}(\alpha')$$

where if (P, N) is the orthopair representation of v then $\alpha = \bigwedge_{p_i \in P} p_i \wedge \bigwedge_{p_i \in N} \neg p_i$ and $\alpha' = \bigwedge_{p_i \in P'} p_i \wedge \bigwedge_{p_i \in N'} \neg p_j$. Furthermore, $B = (P \cup N)^c$ and $B' = (P' \cup N')^c$.

Note that $|CL| = 3^n - 1$ which means that complete clausal graphs can rapidly become unwieldy as a visualisation tool when the number of propositional variables becomes large. Consequently, in the sequel we will propose a thresholding method to generate a weighted clausal graph consisting of those clauses with both sufficient lower and upper support. In this context we now propose an approach in which weighted clausal graphs can be used as a means to represent the strong opinions about particular design as captured in questionnaire data.

Given an elicitation experiment in which s participants assess r designs in terms of the propositions $\mathcal{P}$, we assume a resulting data array of the form: $D = (v_{i,k})$ for $i = 1, \ldots, s$ and $k = 1, \ldots, r$ and where $\forall i, k, \ v_{i,k} \in \mathbb{V}$.

For each design k we define a probability distribution w_k on $\mathbb{V}$ according to:

$$\forall v \in \mathbb{V}, \ w_k(v) = \frac{|\{i : v_{i,k} = v\}|}{s}$$

The distribution w_k then defines a Kleene belief pair $\boldsymbol{\mu}_k$ as described in Sect. 3. Based on this we identify the following weighted clausal graph for design k where

$$A = \{\alpha \in C\mathcal{L} : \underline{\mu}_k(\alpha) \geq x^+, \overline{\mu}_k(\alpha) \geq x^-\}$$

In other words, this is the graph of conjunctive clauses α for which the proportion of participants who believe α to be absolutely true of design k is at least x^+ and for which the proportion of participants who believe that α is absolutely false of design k is less than $1 - x^-$.

Now from the definition of Kleene belief measures and Definition 1 it holds that if $\alpha_1 \preceq \alpha_2$ then both $\underline{\mu}(\alpha_2) \geq \underline{\mu}(\alpha_1)$ and $\overline{\mu}(\alpha_2) \geq \overline{\mu}(\alpha_1)$. Consequently, for weighted clausal graphs generated by applying a thresholding approach as above, it follows that if $\alpha_1 \in A$ then $\alpha_2 \in A$ for all $\alpha_2 \in C\mathcal{L}$ such that $\alpha_1 \preceq \alpha_2$. From this perspective then the minimal elements of the partially ordered set $(A, \preceq)$ have special status as the maximally precise opinions for which there is sufficiently strong support, and with each then characterising a set of more general opinions with at least as strong support.

Weighted clausal graphs where $A \subset C\mathcal{L}$ do not provide sufficient information to completely determine the underlying probability on w and hence $\boldsymbol{\mu}$ cannot be inferred for all sentences of the language. However, the graph does provide sufficient information to determine the belief values of certain sentences outside of the set A. In particular, for a clause $\alpha = \bigwedge_{p_j \in P} p_j \wedge \bigwedge_{p_j \in N} \neg p_j$ we refer to its dual as the clause $\alpha^- = \bigwedge_{p_j \in N} p_j \wedge \bigwedge_{p_j \in P} \neg p_j$. A weighted clausal graph provides sufficient information to calculate $\boldsymbol{\mu}(\alpha^-)$ for every clause $\alpha \in A$ by exploiting additivity. In general, for sentences $\theta, \varphi \in S\mathcal{L}$ then if the graph provides sufficient information to calculate $\boldsymbol{\mu}(\theta)$, $\boldsymbol{\mu}(\varphi)$ and $\boldsymbol{\mu}(\theta \wedge \varphi)$, additivity can be used to determine $\boldsymbol{\mu}(\theta \vee \varphi)$. Furthermore, the graph also provides sufficient information to evaluation the conditional measure values $\underline{\mu}(\theta|v(\varphi) = \mathbf{t})$ and $\overline{\mu}(\theta|v(\varphi) \neq \mathbf{f})$ as defined in Sect. 3.1.

5   Kansei Case Study

In this section we consider a case study in affective design to illustrate the above ideas. The study involves the evaluation of Kutani porcelain designs for coffee cups as described in [2]. Kutani porcelain is a traditional craft industry in

Japan originating from the Kutani Pottery Village in the Ishikawa prefecture, and with origins dating back to the seventeenth century. Here we will focus on the ten designs shown in Fig. 1. Data was collected from 60 subjects as part of a Kansei assessment process in which each participant was asked to place design on seven point SD scales defined by a number of different Kansei words. For the purpose of this case study we will focus on the three attributes *conventional*(*c*), *simple*(*s*) and *solemn*(*so*) these typically being viewed as positive characteristics for Kutani designs. We partitioned the seven point scale so as to correspond to the three truth values, where $\mathbf{t} = \{1,2\}$, $\mathbf{b} = \{3,4,5\}$ and $\mathbf{f} = \{6,7\}$[1].

Fig. 1. Coffee cup designs 1–10.

Weighted clausal graphs were generated from the data for each of the ten designs using the thresholding method with $x^+ = 0.2$ and $x^- = 0.6$. Figure 2 shows the graph for design #5 and Fig. 3 shows these graphs for the other nine designs. From Fig. 2 we see that for design #5 the clausal graph is characterised by the single most precise (maximal) clause $c \wedge s \wedge so$, which is believed to be true by 20% of the participants and not false by ≈82%. Although the opinions about this design are positive they are not especially strongly held in general. In particular, if we quantify the strength of opinion about a clause by the probability of it being classified as borderline, then we see that ≈62% of the subjects believe that for design #5 the clause $c \wedge s \wedge so$ is borderline. Indeed even the most strongly held opinions, in this case the propositions s and so, are both classified as borderline by 45% of participants. The graph does show, however, that there are some noticeable dependencies according to which a subject who has a strong opinion about one attribute is much more likely to also have a strong opinion

---

[1] Other partitions may also be reasonable including $\mathbf{t} = \{1,2,3\}$, $\mathbf{b} = \{4\}$ and $\mathbf{f} = \{5,6,7\}$.

about a different attribute. For example, applying the definition of conditional belief pairs given in Sect. 3.1 we have from Fig. 2 that:

$$\underline{\mu}(c|v(s) = \mathbf{t}) = \frac{0.26667}{0.31667} = 0.84211 \text{ and } \underline{\mu}(s|v(c \wedge so) = \mathbf{t}) = \frac{0.2}{0.21667} = 0.92306$$

Furthermore, by evaluating upper conditional measures we see that subjects who reject a completely negative evaluation of the design in terms of certain attributes are also highly likely to reject a negative evaluation in terms of other attributes. More formally, for the clausal graph in Fig. 2 it holds that for all clauses $\alpha_1, \alpha_2 \in A$, $\overline{\mu}(\alpha_1|v(\alpha_2) \neq \mathbf{f}) > 0.9$.

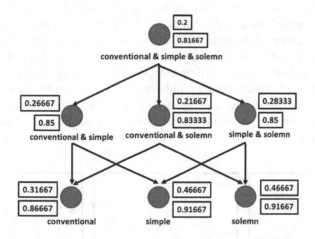

Fig. 2. Weighted clausal graph for design 5.

Table 2. Table showing most specific clauses for each of the designs

Cup #	Most precise opinions
1	Not conventional & not simple, not simple & not solemn
2	Not conventional & not simple & not solemn
3	Conventional, not conventional & not simple, not conventional & not solemn, not simple & not solemn
4	Not conventional & not simple, not simple & not solemn
5	Conventional & simple & solemn
6	Simple, not simple, not conventional & not solemn
7	Not conventional, simple & not solemn, conventional & simple
8	Conventional, simple, not solemn
9	Not conventional, conventional, simple, not solemn
10	Conventional, simple, not conventional & not simple, not conventional & not solemn

The maximally precise clauses from each graph (see Table 2) provides a summary of the different viewpoints held about each of the ten designs. More generally, we claim that clausal models of this kind are a useful qualitative representations of the most supported opinions together with their level and strength of support. However, in some cases a quantitative approach is required so as to obtain a more direct comparison between different designs. In the next section we explore how the notion of scoring function can be extended to the three valued context in order to allow for ranking of designs.

6   Ranking Designs

We now introduce a straightforward extension of scoring functions from Boolean logic to our proposed three valued models. Intuitively a scoring function associates a real number with each state of the world, quantifying the degree of utility received if that state holds. In the current context states of the world are represented by valuations on the language $\mathcal{L}$ so that we naturally define a score as a function from $\mathbb{V}$ into the real numbers $\mathbb{R}$. Given a probability distribution on the valuations $\mathbb{V}$ we can then determine expected score in the normal manner. As outlined in Sect. 4 a probability distribution w_k on $\mathbb{V}$, can be inferred from the data D for each design $k = 1, \ldots, r$. The expected score for each design as

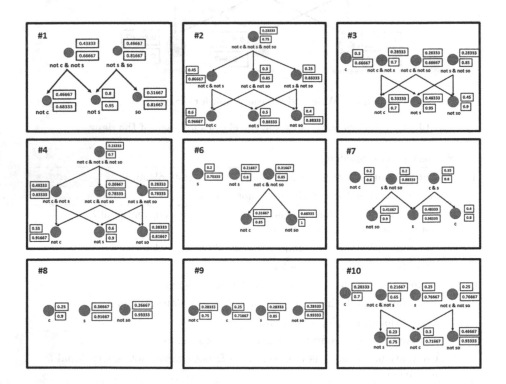

Fig. 3. Weighted clausal graphs for designs 1–4 and 6–10.

based on the marginal distributions $w_k : k = 1, \ldots, r$ and a common scoring function, can then provide a criterion for comparing different designs.

Definition 3 (*Expected Score*). *Let* $f : \mathbb{V} \to \mathbb{R}$ *be a real valued scoring function so that* $f(v)$ *is the score or utility of a design with features* v. *The expected score for design* k *is then given by:*

$$E_k(f) = \sum_{v \in \mathbb{V}} w_k(v) f(v)$$

The following example illustrates the use of some simple scoring functions for ranking designs. In general, the identification of useful scoring functions which capture aspects of design quality should be based on further data studies in which, for example, participants are asked to identify positive, negative and neutral attributes of a good design.

Example 2. Suppose that $\theta \in S\mathcal{L}$ is the description of a good design. Let scoring function f_1, f_2 and f_3 be defined such that $\forall v \in \mathbb{V}$,

$$f_1(v) = \begin{cases} 1 : v(\theta) = \mathbf{t} \\ 0 : \text{otherwise} \end{cases}, \quad f_2(v) = \begin{cases} 1 : v(\theta) \neq \mathbf{f} \\ 0 : \text{otherwise} \end{cases} \quad \text{and} \quad f_3(v) = \begin{cases} 1 : v(\theta) = \mathbf{t} \\ \frac{1}{2} : v(\theta) = \mathbf{b} \\ 0 : v(\theta) = \mathbf{f} \end{cases}$$

In this case $E_k(f_1) = \underline{\mu}_k(\theta)$, $E_k(f_2) = \overline{\mu}_k(\theta)$ and $E_k(f_3) = td_k(\theta)$.

In [2] it is proposed that given a set of positive requirements for a good design, a reasonable objective would be to identify designs which satisfy as many as possible of these requirements. In the context where we have three valued truth models this heuristic needs to be adapted so as to take account of borderline cases. The new objective should instead be to identify designs which satisfy as many of the requirements as possible whilst falsifying as few of them as possible. The following scoring function provides a quantitative version of this heuristic. Assuming that the sentence describing a good design is $\theta = \bigwedge_{j=1}^{n} p_j$ then define $f_4(v) = |P| - |N|$ where (P, N) is the orthopair characterising v.

We applied the above four scoring functions to the Kutani data. Table 3 gives the results for the ten coffee cup designs shown in Fig. 1, and where the description of a good design is taken to be $\theta = c \wedge s \wedge so$. Design #5 ranks top according to all four scoring functions, with designs #7 and #8 also tending to perform relatively well. This would seem to be consistent with the clausal graphs in Figs. 2 and 3 and characterised by the maximal clauses given in Table 2, which show that for design #5 all the opinions with lower and upper measures exceeding x^+ and x^- respectively, are positive. For designs #7 and #8 we find that there are a mixture of positive and negative viewpoints but with the weight of probability tending towards positive opinions. In contrast designs #1 to #4 and #6 are dominated by negative viewpoints and consequently have low expected scores.

Table 3. Designs ranked according to different scoring functions.

Cup k	$\underline{\mu}_k$	Rank	$\overline{\mu}_k$	Rank	td_k	Rank	$E_k(|P| - |N|)$	Rank
1	0.03333	6	0.15	10	0.09167	10	−0.56667	6
2	0	10	0.26667	8	0.13333	8	−1.23333	10
3	0	10	0.31667	6	0.15833	7	−0.83333	7
4	0.05	5	0.26667	8	0.15833	7	−1.16667	9
5	0.2	1	0.81667	1	0.50833	1	0.96667	1
6	0	10	0.23333	9	0.11667	9	−0.91667	8
7	0.06667	2	0.56667	4	0.31667	4	0.35	2
8	0.05	5	0.6833	2	0.36667	2	0.16667	3
9	0.01667	7	0.61667	3	0.31667	4	−0.06667	4
10	0.05	5	0.43333	5	0.24167	5	−0.41667	5

7   Conclusions

In this paper we have shown how by combining probability and three valued logic we can provide a representational model for opinions as based on Kansei words, which captures both uncertainty and inherent vagueness. Furthermore, by extending the notion of scoring function to a three valued setting we can use this framework in order to provide a ranking of different designs which takes account, not only of the level of belief in an opinion, but also the relative strength of that opinion.

Future work will focus on applying weighted clausal graphs to higher dimensional data involving many more Kansei words or attributes. In addition, we also plan to investigate statistical and machine learning approaches for identifying effective score functions of the type proposed in Sect. 7. This will require data linking Kansei attributes to design quality, which can be elicited in a number of ways, including by asking participants to identify positive design features, or by adding a direct quality assessment to questionnaires of the type described in Sect. 5. Any such learning method will, of course, need to be adapted to a three valued setting, which may involve modelling vagueness in the quality assessments themselves.

Acknowledgment. We would like to thank M. Ryoke, Y. Nakamori, and Y. Yamashita for providing us with the Kansei data used for the case study illustration in this paper.

References

1. Childs, T., et al.: Affective design (Kansei Engineering) in Japan. DTI International Technology Service Mission Report, University of Leeds (2003)
2. Huynh, V.-N., Nakamori, Y., Yan, H.: A comparative study of target-based evaluation of traditional craft patterns using Kansei data. In: Bi, Y., Williams, M.-A. (eds.) KSEM 2010. LNCS (LNAI), vol. 6291, pp. 160–173. Springer, Heidelberg (2010). https://doi.org/10.1007/978-3-642-15280-1_17
3. Keefe, R., Smith, P. (eds.): Vagueness: A Reader. MIT Press, Cambridge (2002)
4. Lawry, J., Gonzalez-Rodriguez, I.: A bipolar model of assertabity and belief. Int. J. Approximate Reasoning **52**, 76–91 (2011)
5. Lawry, J., Tang, Y.: On truth-gaps, bipolar belief and the assertability of vague propositions. Artif. Intell. **191–192**, 20–41 (2012)
6. Lawry, J.: Probability, fuzziness and borderline cases. Int. J. Approximate Reasoning **55**, 1164–1184 (2014)
7. Nagamachi, M.: Kansei Engineering. Int. J. Ind. Ergon. **15**, 3–11 (1995)
8. Nagamachi, M.: Kansei Engineering and rough sets model. In: Greco, S., et al. (eds.) RSCTC 2006. LNCS (LNAI), vol. 4259, pp. 27–37. Springer, Heidelberg (2006). https://doi.org/10.1007/11908029_4
9. Nishino, T., Nagamachi, M., Tanaka, H.: Variable precision Bayesian rough set model and its application to *Kansei* Engineering. In: Peters, J.F., Skowron, A. (eds.) Transactions on Rough Sets V. LNCS, vol. 4100, pp. 190–206. Springer, Heidelberg (2006). https://doi.org/10.1007/11847465_9
10. Shapiro, S.: Vagueness in Context. Oxford University Press, Oxford (2006)
11. Shimizu, Y., Jindo, T.: A fuzzy logic analysis method for evaluating human sensitivities. Int. J. Ind. Ergon. **15**, 35–47 (1995)
12. Tsuchiya, T., Maeda, T., Matsubara, Y., Nagamachi, M.: A fuzzy rule induction method using genetic algorithms. Int. J. Ind. Ergon. **18**, 135–145 (1996)
13. van Deemter, K.: Utility and language generation: the case of vagueness. J. Philos. Logic **38**, 607–632 (2009)
14. Xu, X., Hsiao, H.-H., Wang, W.: FuzEmotion as a backward Kansei Engineering tool. Int. J. Autom. Comput. **9**(1), 16–23 (2012)
15. Yan, H.-B., Huynh, V.-N., Murai, T., Nakamori, Y.: Kansei evaluation based on prioritized multi-attribute fuzzy target-oriented decision analysis. Inf. Sci. **178**(21), 4080–4093 (2008)
16. Yan, H-B., Nakamori, Y.: A probabilistic approach to Kansei profile generation in Kansei Engineering. In: Proceedings of the 2010 IEEE Conference on Systems, Man and Cybernetics, pp. 776–782 (2010)

Sampling Strategies for Fuzzy RANSAC Algorithm Based on Reinforcement Learning

Toshihiko Watanabe$^{(\boxtimes)}$

Osaka Electro-Communication University, Neyagawa, Osaka 5728530, Japan
t-wata@osakac.ac.jp

Abstract. The computer vision involves many modeling problems for preventing noise caused by sensing units such as cameras. In order to improve computer vision system performance, a robust modeling technique must be developed for essential models in the system. The RANSAC (Random Sample Consensus) and LMedS (Least Median of Squares) algorithms have been widely applied in such issues. However, the performance deteriorates as the noise ratio increases and the modeling time for algorithms tends to increase in industrial applications. As a promising technique, we proposed a new fuzzy RANSAC algorithm based on reinforcement learning concept for robust modeling. In this study, we investigated sampling strategies as an indispensable concept of reinforcement learning through modeling synthetic data of fuzzy modeling and camera homography experiments. Their results found the proposed ε-roulette strategy to be promising in improving calculation time, model optimality, and robustness in modeling performance.

Keywords: RANSAC algorithm · Computer vision · Reinforcement learning · Fuzzy set · Action strategy

1 Introduction

In computer vision systems, optical devices such as cameras, lasers, and projectors, are generally utilized to achieve non-contact measurement or reconstruction of the target shapes in the computer. Due to the characteristics of the computer vision systems, noise caused by the reflection of light, structural variation of installation, or optical characteristics of targets is practically unavoidable. Data used for constructing computer vision systems also often include significant non-negligible optical noise. Therefore, in almost all techniques for the computer vision system, it is one of the most important issues to prevent the effects of significant noises from processing or model estimation. In the computer vision system for 3-dimensional measurement [1, 2], the camera modeling process for internal and external parameters [3–5] should be performed so that it is invariant to such noise. Furthermore, in reconstructing 3-dimensional shapes from sensing data, dealing with the noise is also indispensable to obtain precise and appropriate shape of the target.

An important issue exists in developing robust modeling algorithms for computer vision system problems. Inherent outlier noise in the computer vision is generated unavoidably and its statistical distribution is almost not clear, but it can be defined as a

© Springer Nature Switzerland AG 2019
H. Seki et al. (Eds.): IUKM 2019, LNAI 11471, pp. 122–134, 2019.
https://doi.org/10.1007/978-3-030-14815-7_11

uniform distribution with a large range at best. The random sample consensus (RAN-SAC) algorithm [6, 7] and the least median of squares (LMedS) [8] have been widely and successfully applied to these problems as an essential and effective approach in computer vision. These algorithms are simple robust estimation algorithms and have good applicability to many problems. However, they need much computational time and the precision of the model is not always so high due to their algorithmic features. Moreover, the performance of these algorithms deteriorates as the noise ratio increases. Corresponding to such problems, we proposed a new fuzzy RANSAC algorithm [7, 14] based on reinforcement learning concept [9] for modeling of computer vision applications and various modeling problems. Through numerical experiments based on synthetic data and camera homography experiments, we evaluated the performance of the proposed algorithm and found to be quite effective compared to conventional algorithms. In this study, in order to improve the fuzzy RANSAC algorithm further, we investigate selecting action strategies, i.e., sampling methods, for the algorithm as a necessary concept of reinforcement learning scheme through experiments.

The remainder of the paper is constructed as follows. In Sect. 2, the computational robust estimation techniques such as RANSAC and LMedS are overviewed. The new fuzzy RANSAC algorithm based on reinforcement learning concept is introduced in Sect. 3. Selecting action strategies are discussed in Sect. 4. The experimental results of modeling problem and camera homography estimation are presented in Sect. 5. Finally, conclusions are drawn in Sect. 6.

2 Computational Robust Estimation Techniques

In the modeling and control areas such as mechanical control and plant control, although robust estimation techniques for modeling such as M-estimation have been important problems, we can consider that the computational technique such as the RANSAC algorithm or LMedS algorithm is more suitable methodology for various computer vision problems such as estimating homography containing various random optical outlier noises. The reason is that effects by outlier noises cannot be suppressed completely by the M-estimation. We review these computational robust estimation algorithms below. It is assumed that some data are measured and collected and that the model structure and method for estimating the model parameter are given. The number of collected dataset is assumed to be n. The objective of the robust estimation techniques is to estimate the parameters of the model precisely in short computational time period, preventing outlier noise effects.

2.1 RANSAC Algorithm

The RANSAC algorithm has been widely used for various modeling problems, especially in the computer vision area. The algorithm is simple and generally applicable to

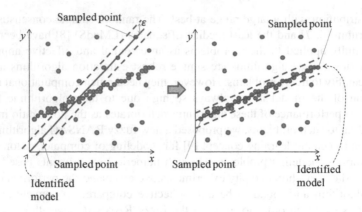

Fig. 1. Conceptual figure of RANSAC

robust estimation problems. Figure 1 shows the RANSAC algorithm concept as a simple 1-dimensional modeling problem. The RANSAC algorithm follows the following five steps:

Step1 Select samples randomly from collected data. The number of samples is set as the minimum number for model estimation in general.

Step2 Estimate the necessary model parameters by using the samples.

Step3 Count the number of data whose estimation error is within constant ε.

Step4 If the number of data within the error constant exceeds predefined value γ, the model is estimated using only data within the error constant and the algorithm is terminated.

Step5 If the number of data within the error constant is below predefined value γ, go to Step 1 and iterate the procedures.

Before the RANSAC algorithm is started, parameters ε and γ should be set properly. Based on the algorithm, effects from random noises are reduced by random sampling and a valid model is estimated by selecting data. The number of samplings is limited statistically. In order to improve the RANSAC algorithm, several techniques are proposed and shown of the effectiveness [10–12].

2.2 LMedS Algorithm

The LMedS algorithm [8] is based on a simple but effective idea using a median residual error. The LMedS finds parameters estimated by minimizing the median of squared residual errors corresponding to the whole data points as follows

$$\hat{\theta} = \underset{\theta}{\operatorname{argmin}} \ \underset{i}{med} \ e_i^2 \tag{1}$$

where θ is the parameter vector of the model, $\hat{\theta}$ is the estimated parameter vector, and e_i is the residual error of the model. Since there is no explicit solution algorithm for the formulation, estimation is performed based on random sampling as follows:

Step1 Select samples randomly from collected data.
Step2 Estimate the necessary model parameters using samples.
Step3 Evaluate the median value of residual errors of all measured data.
Step4 Iterate until a good median value is obtained.

Though the iteration could increase, good global robustness is expected to be attained by the LMedS algorithm. Considering comparison with RANSAC algorithm, there is no parameter in LMedS method.

3 Fuzzy RANSAC Algorithm Based on Reinforcement Learning

Though RANSAC algorithm is essential and effective for various problems such as computational estimation problems, there exist some problems in dealing with it, as described below. Although obvious noise data (outliers) may not effect to the modeling result by RANSAC sampling and procedures, estimated model precision is not always improved. This is because the number of data selection is fixed to the minimum value needed for estimation in general. Then we perform sampling varying the number of sampling data.

Assume that the number of samples is M in RANSAC. Because the model is estimated uniquely from M samples, RANSAC is considered to be a randomly selection of the best models from among up to ${}_nC_M$ candidate models. In terms of estimated model precision, we consider variation of the estimated models to be insufficient. We therefore perform sampling that varies the number of samples randomly to improve the lack of sufficient variation in candidate models. The number of samples h is varied randomly from minimum M to K, where $M < h < K < n$. K is decided in advance. Based on the proposed sampling method, the number of candidate models G becomes

$$G \leq \sum_{i=M}^{K} {}_nC_i \tag{2}$$

The number of candidate models clearly increases compared to conventional RANSAC. This leads to improvement of optimality, i.e., model precision.

Even though the number of samplings to obtain good candidate model is guaranteed statistically, the number of samplings may be huge in the worst case, depending on the objective model structure and the characteristics of collected data. This leads to much computational time in RANSAC method. In order to deal with the problem, we proposed a new RANSAC algorithm based on fuzzy set concept and reinforcement learning concept [9].

In RASAC algorithm, evaluation of modeling is performed based on the magnitude of each evaluation error by judging whether the absolute error is within predefined ε. However the performance of the modeling is affected by the setting ε. In the fuzzy RANSAC, a fuzzy set of the error is defined to evaluate the model. The fuzzy set of errors is defined as the membership function which center is 0 and width is b. Assuming the model error e_i of a measured data, the triangle type membership function is defined as:

$$m_i = \begin{cases} |e_i - b|/b & ; if\ |e_i| \leq b \\ 0 & ; if\ |e_i| > b \end{cases} \tag{3}$$

where m_i is the membership value of the ith data. Figure 2 shows the membership function. The modeling is performed using any estimation technique. The whole evaluation E of the model is calculated by:

$$E = \sum_{i=1}^{n} m_i \tag{4}$$

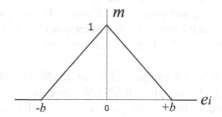

Fig. 2. Error membership function

From these definitions, we expect that the above described affection of the parameter ε in the conventional RANSAC will be relaxed. Moreover, the evaluation E can also be utilized to evaluate the estimation performance as an objective function.

We prepare evaluation value v_i for each measured data. Sampling is performed based on v_i applying the selecting action strategy described in the next section and model estimation is performed using the sampled data. The evaluation value is desirable to denote goodness of the data (inlier). In order to attain this, the evaluation value is learned based on reinforcement learning [9]. We define the learning algorithm with a necessary reward, the whole evaluation E in (3) and the membership value m of each data based on Monte Carlo method as follows:

$$r = E \tag{5}$$

$$v_i \leftarrow v_i + \alpha(m_i \cdot r - v_i), \ i = 1, \ldots, n \tag{6}$$

where r is a reward and α is the learning parameter $(0 < \alpha < 1)$ in the reinforcement learning.

We expect affection by random noise to be reduced due to the learning mechanism. This may also reduce calculation for modeling and improve model precision. The modeling process is so simple and applicable that it is applied to various problems in such areas as computer vision and control. Based on the method, the advantages of reinforcement learning that have a good balance of exploration and exploitation are utilized effectively. However the performance greatly depends on the selecting action (sampling) strategy. The strategy is an important characteristic in reinforcement learning framework. In this study, we focus on investigation of the sampling strategy for fuzzy RANSAC algorithm.

4   Sampling Strategies for Fuzzy RANSAC Algorithm

In reinforcement learning, the learning process is performed based only upon rewards from environment. It is important to make action (sampling) of exploration as well as exploitation to attain good performance of learning. In addition to the general strategies such as roulette strategy, ε-greedy and soft-max strategy based on Boltzmann distribution, we propose ε-roulette strategy.

4.1   Roulette Strategy and Soft-Max Strategy

In the roulette strategy, sampling is performed in proportion to evaluation values, i.e., the roulette strategy. Actual sampling is performed proportionally to normalized probability w as follows:

$$w_i = \frac{v_i}{\sum\limits_{j=1}^{n} v_j}, \ i = 1, \ldots, n \tag{7}$$

In this strategy, data of low evaluation value could be sampled in low probability. As for the soft-max strategy, the evaluation values are transferred using Boltzmann distribution with "temperature" parameter T and the sampling is performed in the same manner in (7) as:

$$w_i = \frac{exp(\frac{v_i}{T})}{\sum_{j=1}^{n} exp(\frac{v_j}{T})}, \ i = 1, \ldots, n \tag{8}$$

When T is large, all probabilities are equal and we have exploration. When T is small, better data are favored. Generally, the strategy is to start with a large T and decrease it gradually. Then, sampling action strategy moves from exploration to exploitation.

4.2 ε-Greedy Strategy

In ε-greedy strategy, with probability ε, we select h data (sampling) uniformly randomly among all data, namely explore, and with probability 1-ε, we select the best h data in terms of evaluation value v, namely exploit. By adjusting ε, we can decide the balance of exploration and exploitation easily for sampling.

4.3 ε-Roulette Strategy

We propose a new strategy named ε-roulette strategy to improve exploration performance of the roulette strategy applying ε-greedy concept. Namely, with probability ε, we select h data (sampling) uniformly randomly among all data, and with probability 1-ε, we select h data applying the roulette strategy based on the evaluation value v. Though the temperature parameter (cooling) is difficult to decide generally in the soft-max strategy, the setting for the ε-roulette strategy is simple that ε parameter can be set intuitively.

4.4 Modeling Procedures

The procedures of the Fuzzy RANSAC algorithm are as follows:

Step1 Decide the number of samples h randomly.

$$(M \leq h \leq K)$$

Step2 Perform sampling based on the selecting strategy.
Step3 Estimate the model based on samples.
Step4 Calculate the evaluation E and rewards for data.
Step5 Perform reinforcement learning and go to Step1.

The procedures are iterated until the predefined number of iterations or a good model is attained. The initial value of v is set to the same small number.

5 Experimental Results

We conduct the modeling experiments using synthetic data and camera calibration data to investigate essential modeling performance of the fuzzy RANSAC algorithm applying the selecting action strategies compared to the conventional RANSAC algorithm. It should be noted that LMedS algorithm cannot be compared because the evaluation function is quite different from the RANSAC evaluation.

5.1 Nonlinear Modeling Results Using Synthetic Data

We assume the true model shown in Fig. 3. The true model is assumed to be a simplified fuzzy model structure [13] using normal triangular membership functions as shown in Fig. 4:

$$\begin{cases} IF \ x \ is \ Small & THEN \ y = \delta_1 \\ IF \ x \ is \ Medium & THEN \ y = \delta_2 \\ IF \ x \ is \ Big & THEN \ y = \delta_3 \end{cases} \tag{9}$$

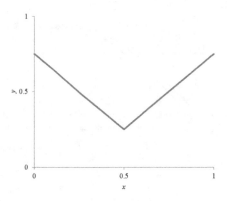

Fig. 3. True model for generation of measured data

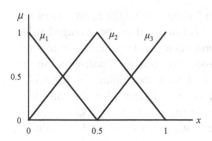

Fig. 4. Membership function for experiment model

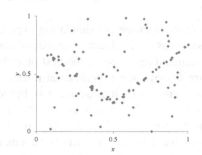

Fig. 5. Measured dataset for experiments

The output of the model $\hat{y}$ is calculated by:

$$\hat{y} = \sum\nolimits_{i=1}^{3} \mu_i \cdot \delta_i \qquad (10)$$

where μ is the value of the membership function.

The true values of δ are set as 0.75, 0.25, and 0.75, respectively. Base data are generated by sampling true data randomly and adding a small amount of white noise ($\sim N(0, 0.01^2)$). Random noise data are generated based on 2D uniform distribution. Then the measured dataset is prepared by mixing 60% of base data with 40% of the random noise data randomly. The number of data in the measured dataset is 100. We apply the fuzzy RANSAC algorithm with selecting action strategies and the conventional RANSAC algorithm to the measured dataset.

The problem in this experiment is to identify the fuzzy model estimating parameter δ by the least square errors method using the measured dataset. Measured dataset is shown in Fig. 5. Though at least 3 samples are required for the problem, estimation calculation of fuzzy model using 3 data falls into unstable situation according to the rank problem due to sparse dataset. For the reason, the number of samples is varied randomly from $M = 15$ to $K = 20$ in the modeling experiments.

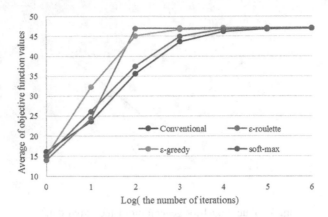

Fig. 6. Experimental results of fuzzy modeling

We conduct the 50 simulation experiments, including 1,000,000 iterations of the algorithm using the dataset. Estimation performance is evaluated by the averaged value and standard deviation of the best objective function values so far in each simulation. Figure 6 shows the experimental results. As for soft-max strategy ("soft-max" in the figure), amount of improvement is limited compared with the traditional RANSAC ("Conventional"). Also, ε-greedy ("ε-greedy") is not so effective for this problem. ε-roulette strategy ("ε-roulette"($\varepsilon = 0.1$)) outperforms the other strategies. The precision of the estimated models by ε-roulette strategy was also confirmed statistically compared with ε-greedy. The significance is investigated by using Welch's t-test for results (best values so far) at 100, 1000, 10000, 100000, and 1000000. All results are that null hypothesis, i.e., the means do not differ, is rejected with statistical significance level of 0.05. We investigate the effects of ε parameter in ε-roulette strategy as shown in Fig. 7. From the results, small value is suitable for the problem. Figure 8 shows the effects of ε parameter in ε-greedy strategy. In contrast to the results in Fig. 7, the appropriate value is large, i.e. almost random.

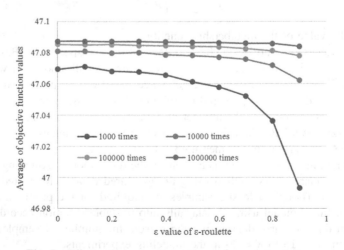

Fig. 7. Effects of ε parameter settings of ε-roulette strategy

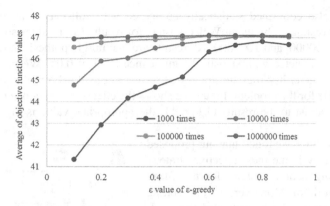

Fig. 8. Effects of ε parameter settings of ε-greedy strategy

5.2 Results of Camera Modeling

We conduct the experiments of camera model estimation using an industrial camera and a calibrator. We evaluate the proposed algorithm compared with the conventional RANSAC algorithm.

The camera model is generally expressed as the following essential formulation.

$$\omega \cdot U = \mathbf{P} \cdot W, U = [\,u \quad v \quad 1\,]^T, W = [\,X \quad Y \quad Z \quad 1\,]^T \tag{11}$$

where $\mathbf{P}$ is the 3×4 perspective view matrix, (u, v) are image coordinates of the target, (X, Y, Z) are corresponding world coordinates of the target, and ω is the scaling parameter. For application, we should estimate these 13 parameters. Firstly, data pairs of 2D image data and corresponding 3D target coordinates through the camera are collected using a calibrator. Then required parameters such as the perspective view matrix and the scaling parameter are estimated using collected datasets. The model estimation is performed by DLT algorithm based on the least square errors method.

The camera we used in this experiment is a high performance CMOS model with Gigabit Ethernet. Camera resolution is 2,592 $\times$ 1,944 pixels. The lens is aspheric with minimal distortion. The checkerboard calibrator shown in Fig. 9 is used for collecting data pairs (target 3D points in world coordinates system and corresponding image 2D pixels) for calibration. The camera direction is decided considering stereo vision installation as shown in Fig. 9. The number of collected data pairs is 105. We conducted estimation experiments for 50 simulations including 1,000,000 iterations of the algorithm by using collected dataset.

The results are summarized in Fig. 10. Though the strategies of the soft-max and ε-greedy are not so effective to improve the modeling performance, ε-roulette strategy is quite effective in contrast. The precision of the estimated models by ε-roulette strategy was also confirmed statistically compared with soft-max strategy. The significance is investigated by using Welch's t-test for results (best values so far) at 100, 1000, 10000, 100000, and 1000000. All results are that null hypothesis, i.e., the means do not differ, is rejected with statistical significance level of 0.05. We investigate the effects of ε parameter in ε-roulette strategy as shown in Fig. 11. From the results, small value is suitable for the problem. Figure 12 shows the effects of ε parameter in ε-greedy strategy. In contrast to the results in Fig. 11, the appropriate value is big, i.e. almost random.

From the results, we found that our proposed ε-roulette strategy is the most effective strategy as the selecting action strategy for the fuzzy RANSAC algorithm. Moreover, as for ε setting for ε-roulette strategy, we confirmed that variance of the objective function values tends to be lowered, i.e. stable, by setting of a small value for ε, compared with $\varepsilon = 0$ setting that is equivalent to the standard roulette strategy.

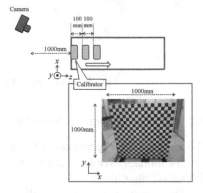

Fig. 9. Test field and calibrator

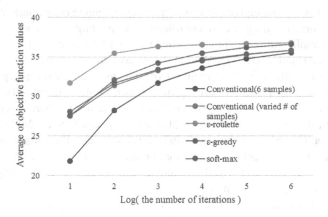

Fig. 10. Experimental results of camera modeling

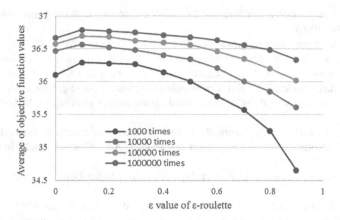

Fig. 11. Effects of ε parameter settings of ε-roulette strategy

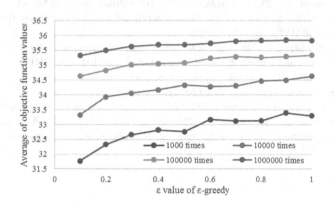

Fig. 12. Effects of ε parameter settings of ε-greedy strategy

6 Conclusions

In this study, we investigated selecting strategies as an indispensable concept of reinforcement learning for fuzzy RANSAC algorithm through modeling synthetic data of fuzzy modeling and camera homography experiments. Their results found the proposed ε-roulette strategy to be promising in improving calculation time, model optimality, and robustness in modeling performance. Our future plan includes application of the proposed method to the other modeling problem.

References

1. Hartly, R., Zisserman, A.: Multiple view geometry in computer vision. Cambridge University Press, Cambridge (2000)
2. Sato, J.: Computer Vision - Geometry of Vision. Corona Publishing, Tokyo (1999)

3. Zhang, Z.: A flexible new technique for camera calibration. IEEE Trans. PAMI **22**(11), 1330–1334 (2000)
4. Deguchi, K.: Foundation of Robot Vision. Corona Publishing, Tokyo (2000)
5. Watanabe, T., Saito, Y.: Camera modeling for 3D sensing using fuzzy modeling concept based on stereo vision. J. Adv. Comput. Intell. Intell. Inform. **19**(1), 158–164 (2015)
6. Fischler, M.A., Bolles, R.C.: Random sample consensus: a paradigm for model fitting with applications to image analysis and automated cartography. Commun. ACM **24**(6), 381–395 (1981)
7. Watanabe, T., Kamai, T., Ishimaru, T.: Robust estimation of camera homography by fuzzy RANSAC algorithm with reinforcement learning. J. Adv. Comput. Intell. Intell. Inform. **19**(6), 833–842 (2015)
8. Rousseeuw, P.J., Leroy, A.M.: Robust Regression and Outlier Detection. Wiley, New York (1987)
9. Sutton, R.S., Barto, A.G.: Reinforcement Learning. MIT Press, Boston (1998)
10. Lee, J., Kim, G.: Robust estimation of camera homography using fuzzy RANSAC. In: Gervasi, O., Gavrilova, Marina L. (eds.) ICCSA 2007. LNCS, vol. 4705, pp. 992–1002. Springer, Heidelberg (2007). https://doi.org/10.1007/978-3-540-74472-6_81
11. Botterill, T., Mills, S., Green, R.D.: New conditional sampling strategies for speeded-up RANSAC. In: BMVC, pp. 1–11 (2009)
12. Subbarao, R., Meer, P.: Beyond RANSAC: user independent robust regression. In: Computer Vision and Pattern Recognition Workshop, pp. 101–109 (2006)
13. Watanabe, T., Seki, H.: Modeling approach based on modular fuzzy model. J. Adv. Comput. Intell. Intell. Inform. **16**(5), 653–661 (2012)
14. Watanabe, T.: Chapter 1: a fuzzy RANSAC algorithm based on the reinforcement learning concept for modeling. In: Harvey, T., Mullis, D. (eds.) Fuzzy Modeling and Control – Methods, Applications and Research, pp. 1–22. Nova Science Publishers, New York (2018)

Interpretation of Variable Consistency Dominance-Based Rough Set Approach by Minimization of Asymmetric Loss Function

Yoshifumi Kusunoki[1]([⊠]) [iD], Jerzy Błaszczyński[3] [iD], Masahiro Inuiguchi[2],
and Roman Słowiński[3,4] [iD]

[1] Graduate School of Engineering, Osaka University,
2-1, Yamada-oka, Suita, Osaka 565-0871, Japan
kusunoki@eei.eng.osaka-u.ac.jp
[2] Graduate School of Engineering Science, Osaka University,
1-3, Machikaneyama, Toyonaka, Osaka 560-8531, Japan
inuiguti@sys.es.osaka-u.ac.jp
[3] Poznań University of Technology, Institute of Computing Science,
Piotrowo 3a, 60-965 Poznań, Poland
{jerzy.blaszczynski,roman.slowinski}@cs.put.poznan.pl
[4] Systems Research Institute, Polish Academy of Sciences, 01-447 Warsaw, Poland

Abstract. In this paper, we give a statistical interpretation of the variable consistency dominance-based rough set approach (VC-DRSA), which is a version of DRSA useful for practical reasoning about ordinal data. This study is building a bridge between theories of rough sets and statistics, and it is developing, moreover, a new direction of study about VC-DRSA. We consider a classification problem for each pair of complementary upward and downward unions of decision classes, and define an empirical risk function with an asymmetric loss function consisting of hinge and 0–1 loss functions. Then, we prove that approximations of two decision classes by VC-DRSA correspond to the minimum of the empirical risk function.

Keywords: Rough sets ·
Variable consistency dominance-based rough set approach ·
Ordinal data analysis · Empirical risk minimization

1 Introduction

The dominance-based rough set approach (DRSA) [5] is a methodology for multi-criteria classification problems. DRSA inductively builds a preference model from decision tables, which are data sets recoding past decisions or decision examples of decision makers. In multi-criteria classification problems, we commonly assume a monotonic relationship between evaluation of objects on criteria and their assignment to decision classes, where the decision classes are endowed with

© Springer Nature Switzerland AG 2019
H. Seki et al. (Eds.): IUKM 2019, LNAI 11471, pp. 135–145, 2019.
https://doi.org/10.1007/978-3-030-14815-7_12

a preference order. Consequently, if object a is evaluated at least as good as object b, then a should not be assigned to a worse decision class than b. This is called dominance principle. However, decision tables are often inconsistent with respect to the dominance principle. When the object set of each upward or downward unions of decision classes is inconsistent, the set is expressed by two consistent sets called lower and upper approximations, which stems from the rough set approach [9,10].

Since the strict application of the dominance principle may result in approximations of little information, Variable-Precision DRSA (VP-DRSA) [7] and Variable-Consistency DRSA (VC-DRSA) [1] were proposed. VP-DRSA and VC-DRSA provide parameterized approximations using consistency (or inconsistency) measures. Błaszczyński et al. [1] proposed several consistency measures and discussed their properties. In this paper, we especially study ϵ and ϵ' consistency measures.

In parallel to VP-DRSA and VC-DRSA, probabilistic extensions of approximations for the original rough set approach were also proposed. Furthermore, Yao [11,12] developed the decision-theoretic rough set model (DTRSM), which provides an interpretation of probabilistic approximations from conditional risk minimization of Bayesian decision theory. The authors [8] interpreted rough approximations in VP-DRSA successfully from the viewpoint of empirical risk minimization. More specifically, considering a classification problem for each complementary upward and downward unions of decision classes, we showed that classification by rough approximations in VP-DRSA coincides with the minimum risk solution. The risk is empirical, i.e., it is the average of losses for objects in a given decision table. In the case of VC-DRSA, the authors [3] also tried to explain the rough approximations by empirical risk minimization.

In this paper, we present another interpretation of VC-DRSA by the empirical risk minimization. We characterize approximations derived from the ϵ and ϵ' consistency measures by optimal solutions of certain empirical risk functions. Those functions are defined based on an asymmetric loss function consisting of the hinge and 0–1 loss functions.

This paper is organized as follows. In Sect. 2, we introduce definitions regarding DRSA and VC-DRSA. The ϵ and ϵ' consistency measures are also reminded. In Sect. 3, we define the classification problem and the asymmetric loss function. Then, we characterize the lower approximations with respect to the consistency measures by the empirical risk minimization. Finally, concluding remarks are given in Sect. 4.

2 Variable-Consistency Dominance-Based Rough Set Approach

2.1 Decision Table and DRSA

Dominance-based Rough Set Approach (DRSA) [5] is an extension of the classic rough set approach which involves dominance relation instead of the usual indiscernibility relation. Let us recall briefly the mathematical model of DRSA.

A decision table is defined by $(U, C \cup \{d\}, (V_a)_{a \in C \cup \{d\}})$, where U is a finite set of objects, $C \cup \{d\}$ is a finite set of attributes, and V_a is a set of attribute values with respect to $a \in C \cup \{d\}$. Moreover, each $a \in C$ is called a condition attribute, and d is called a decision attribute. Each attribute $a \in C \cup \{d\}$ is a map from U to V_a. For $A \subseteq C \cup \{d\}$, let $V_A = \prod_{a \in A} V_a$. Each attribute subset A is also regarded as a map from U to V_A.

An attribute $a \in C \cup \{d\}$ is called criterion, if there is a total order $\geq$ on its value set V_a. We assume that all criteria are of the gain-type, i.e., the greater the better. Let $\tilde{C}$ be the set of all criteria in C. The other attributes $a \in C \setminus \tilde{C}$ are called nominal attributes. Moreover, we assume that d is a criterion, and its value set is given by $V_d = \{1, 2, \ldots, p\}$. Hence, the decision attribute d assigns each object to one of totally ordered decision classes specified by V_d: $X_1, X_2, \ldots, X_p$, where $X_i = \{u \in U \mid d(u) = i\}$.

For $A \subseteq C$, a dominance relation D_A on U is defined by:

$$D_A = \left\{ (u, u') \in U \times U \;\middle|\; \begin{array}{l} a(u) \geq a(u'), \forall a \in A \cap \tilde{C} \\ \text{and } a(u) = a(u'), \forall a \in A \setminus \tilde{C} \end{array} \right\}. \tag{1}$$

D_A satisfies reflexivity and transitivity. From the dominance relation, we define dominating and dominated sets for each object u, which are denoted by $D_A^+(u)$ and $D_A^-(u)$, respectively:

$$D_A^+(u) = \{u' \in U \mid (u', u) \in D_A\}, \quad D_A^-(u) = \{u' \in U \mid (u, u') \in D_A\}. \tag{2}$$

Since the decision classes as well as the decision attribute values are ordered $X_1, X_2, \ldots, X_p$, one can define an upward union of decision classes $X_i^{\geq}$ with respect to i and a downward union of decision classes $X_{i-1}^{\leq}$ with respect to $i - 1$, $i = 2, \ldots, p$, as follows:

$$X_i^{\geq} = \bigcup_{j \geq i} X_j, \quad X_{i-1}^{\leq} = \bigcup_{j \leq i-1} X_j. \tag{3}$$

Moreover for the sake of convenience, we define $X_1^{\geq} = X_p^{\leq} = U$ and $X_0^{\leq} = X_{p+1}^{\geq} = \emptyset$. We have $X_i^{\geq} = \neg X_{i-1}^{\leq}$, where $\neg X$ is the complement set of $X \subseteq U$, i.e., $\neg X = U \setminus X$.

In DRSA, the dominance principle is supposed. That is, if an object u is better than or equal to another object u' with respect to all condition attributes, then the class of u should not be worse than that of u'. When the classification of an upward or downward union is inconsistent with the dominance principle, the union is approximated by two consistent sets, called lower and upper approximations. The vagueness of the upward or downward union is captured by the difference of the approximations. Given condition attribute set $A \subseteq C$, and $i \in \{2, \ldots, p\}$, the lower and upper approximations of $X_i^{\geq}$ with respect to A, denoted by $\underline{A}(X_i^{\geq})$ and $\overline{A}(X_i^{\geq})$, respectively, are defined as follows:

$$\begin{aligned} \underline{A}(X_i^{\geq}) &= \{u \in U \mid D_A^+(u) \subseteq X_i^{\geq}\}, \\ \overline{A}(X_i^{\geq}) &= \{u \in U \mid D_A^-(u) \cap X_i^{\geq} \neq \emptyset\}. \end{aligned} \tag{4}$$

Similarly, the lower and upper approximations of $X_{i-1}^{\leq}$ with respect to A are defined as follows:

$$\underline{A}(X_{i-1}^{\leq}) = \{u \in U \mid D_A^-(u) \subseteq X_{i-1}^{\leq}\},$$
$$\overline{A}(X_{i-1}^{\leq}) = \{u \in U \mid D_A^+(u) \cap X_{i-1}^{\leq} \neq \emptyset\}. \tag{5}$$

The differences of lower and upper approximations are called boundary sets.

2.2   VC-DRSA

When the number of objects is large and/or there are hidden attributes, e.g., multiple sources of objects or time dependence of evaluation, the sizes of lower approximations in DRSA risk to be small, because of strict application of the dominance principle. In such cases, it is reasonable to relax the conditions in (4) and (5), and accept a limited number of ambiguous (uncertain) objects. With this motivation, Variable-Consistency DRSA (VC-DRSA) was proposed [1,2,6].

Błaszczyński et al. [1] proposed several consistency measures, which quantify consistency of objects with respect to the dominance principle. In this paper, we study two cost-type consistency measures, namely inconsistency measures, ϵ and ϵ', reminded below.

First, we remind ϵ-consistency measure. For $A \subseteq C$ and $X_i^{\geq}$, $X_{i-1}^{\leq}$, $i = 2, 3, \ldots, p$, ϵ-consistency measures to $X_i^{\geq}$ and $X_{i-1}^{\leq}$ with respect to A, which are denoted by $\epsilon_{X_i^{\geq}}^A$ and $\epsilon_{X_{i-1}^{\leq}}^A$ respectively, are defined as follows. For $u \in U$,

$$\epsilon_{X_i^{\geq}}^A(u) = \frac{|D_A^+(u) \cap \neg X_i^{\geq}|}{|\neg X_i^{\geq}|} = \frac{|D_A^+(u) \cap X_{i-1}^{\leq}|}{|X_{i-1}^{\leq}|}, \tag{6}$$

$$\epsilon_{X_{i-1}^{\leq}}^A(u) = \frac{|D_A^-(u) \cap \neg X_{i-1}^{\leq}|}{|\neg X_{i-1}^{\leq}|} = \frac{|D_A^-(u) \cap X_i^{\geq}|}{|X_i^{\geq}|}. \tag{7}$$

The values of ϵ-consistency measures $\epsilon_{X_i^{\geq}}^A(u)$ and $\epsilon_{X_{i-1}^{\leq}}^A(u)$ can be interpreted as estimates of conditional probability $\Pr(u' \in D_A^+(u)|u' \in X_{i-1}^{\leq})$ and $\Pr(u' \in D_A^-(u)|u' \in X_i^{\geq})$, respectively. This type of conditional probability is called catch-all likelihood [4].

Next, we remind ϵ'-consistency measure. For $A \subseteq C$ and $X_i^{\geq}$, $X_{i-1}^{\leq}$, $i = 2, 3, \ldots, p$, ϵ'-consistency measures to $X_i^{\geq}$ and $X_{i-1}^{\leq}$ with respect to A, which are denoted by $\epsilon'_{X_i^{\geq}}^A$ and $\epsilon'_{X_{i-1}^{\leq}}^A$ respectively, are defined as follows. For $u \in U$,

$$\epsilon'_{X_i^{\geq}}^A(u) = \frac{|D_A^+(u) \cap \neg X_i^{\geq}|}{|X_i^{\geq}|} = \frac{|D_A^+(u) \cap X_{i-1}^{\leq}|}{|X_i^{\geq}|}, \tag{8}$$

$$\epsilon'_{X_{i-1}^{\leq}}^A(u) = \frac{|D_A^-(u) \cap \neg X_{i-1}^{\leq}|}{|X_{i-1}^{\leq}|} = \frac{|D_A^-(u) \cap X_i^{\geq}|}{|X_{i-1}^{\leq}|}. \tag{9}$$

The difference of $\epsilon'^A_{X_i^\geq}(u)$ (resp. $\epsilon'^A_{X_{i-1}^\leq}(u)$) from $\epsilon^A_{X_i^\geq}(u)$ (resp. $\epsilon^A_{X_{i-1}^\leq}(u)$) is only the denominator $|X_i^\geq|$ (resp. $|X_{i-1}^\leq|$).

Using the consistency measures, we can define parameterized lower and upper approximations of two complementary unions $X_i^\geq$ and $X_{i-1}^\leq$, $i = 2, 3, ..., p$, in the following way. Let $A \subseteq C$. Moreover, let $\varphi^A_{X_i^\geq}$ and $\varphi^A_{X_{i-1}^\leq}$ be ϵ- or ϵ'-consistency measure applied to $X_i^\geq$ and $X_{i-1}^\leq$ with respect to A, respectively. Additionally, let $\alpha_i^\geq$ and $\alpha_{i-1}^\leq$ be thresholds for $\varphi^A_{X_i^\geq}$ and $\varphi^A_{X_{i-1}^\leq}$, respectively. We assume that $\alpha_i^\geq, \alpha_{i-1}^\leq \in [0, 1]$ for ϵ-consistency measure, and $\alpha_i^\geq \in \left[0, \frac{|X_{i-1}^\leq|}{|X_i^\geq|}\right]$, $\alpha_{i-1}^\leq \in \left[0, \frac{|X_i^\geq|}{|X_{i-1}^\leq|}\right]$ for ϵ'-consistency measure. Then, lower and upper approximations of $X_i^\geq$ and $X_{i-1}^\leq$ with respect to A are defined as follows:

$$\underline{A}^\varphi(X_i^\geq \,|\, \alpha_i^\geq) = \{u \in X_i^\geq \,|\, \varphi^A_{X_i^\geq}(u) \leq \alpha_i^\geq\},$$

$$\overline{A}^\varphi(X_i^\geq \,|\, 1 - \alpha_i^\geq) = X_i^\geq \cup \{u \in X_{i-1}^\leq \,|\, 1 - \varphi^A_{X_{i-1}^\leq}(u) < 1 - \alpha_{i-1}^\leq\} \quad (10)$$

$$= \neg \underline{A}^\varphi(X_{i-1}^\leq \,|\, \alpha_i^\geq),$$

$$\underline{A}^\varphi(X_{i-1}^\leq \,|\, \alpha_i^\geq) = \{u \in X_{i-1}^\leq \,|\, \varphi^A_{X_{i-1}^\leq}(u) \leq \alpha_{i-1}^\leq\},$$

$$\overline{A}^\varphi(X_{i-1}^\leq \,|\, 1 - \alpha_i^\geq) = X_{i-1}^\leq \cup \{u \in X_i^\geq \,|\, 1 - \varphi^A_{X_i^\geq}(u) < 1 - \alpha_i^\geq\} \quad (11)$$

$$= \neg \underline{A}^\varphi(X_i^\geq \,|\, \alpha_i^\geq).$$

The lower approximation of $X_i^\geq$ (resp. $X_{i-1}^\leq$) is the set of positive objects u, whose consistency measures $\varphi^A_{X_i^\geq}(u)$ (resp. $\varphi^A_{X_{i-1}^\leq}(u)$) are at least $\alpha_i^\geq$ (resp. $\alpha_{i-1}^\leq$). The upper approximation of $X_i^\geq$ (resp. $X_{i-1}^\leq$) is given by the complement of the lower approximation of $X_{i-1}^\leq$ (resp. $X_i^\geq$). Interestingly, the lower approximations of an upward union (resp. downward union) of decision classes can be expressed as a union of the intersection of the class union and particular dominating sets (resp. dominated sets).

$$\underline{A}^\varphi(X_i^\geq \,|\, \alpha_i^\geq) = \bigcup_{u \in \underline{A}^\varphi(X_i^\geq \,|\, \alpha_i^\geq)} D_A^+(u) \cap X_i^\geq, \quad (12)$$

$$\underline{A}^\varphi(X_{i-1}^\leq \,|\, \alpha_{i-1}^\leq) = \bigcup_{u \in \underline{A}^\varphi(X_{i-1}^\leq \,|\, \alpha_{i-1}^\leq)} D_A^-(u) \cap X_{i-1}^\leq. \quad (13)$$

This fact indicates a connection between the lower approximations and definability [10] of rough sets.

3 Empirical Risk Minimization for VC-DRSA

In this section, we discuss only lower approximations of two classes X_1 and X_2 with a fixed subset A of criteria. Classes X_1 and X_2 correspond to any complementary downward and upward unions of decision classes. Let $X^\geq$ denote the upward union X_2, and $X^<$ the downward union X_1. Moreover, let α be a nonnegative value, and φ be ϵ or ϵ'. $\underline{X}^\geq_\varphi(\alpha)$ and $\underline{X}^<_\varphi(\alpha)$ denote the lower approximations $\underline{A}^\varphi(X^\geq|\alpha)$ and $\underline{A}^\varphi(X^<|\alpha)$, respectively. Similarly, we omit A in other symbols, for example in the dominance relation.

The aim of this study is to associate lower approximations $(\underline{X}^\geq_\varphi, \underline{X}^<_\varphi)$ with empirical risk minimization, which is widely used in statistical decision making. That is, we show that the pairs of lower approximations are minima of a risk function with different parameters. To specify the risk function, we provide a decision function, which gives a classification degree to $X^\geq$ or $X^<$ for each object, and a loss function, which measures misclassification by the decision function.

First of all, we define the following sets:

$$\mathcal{W}_P = \left\{ W \subseteq X^\geq \mid W = \bigcup_{u \in W} D^+(u) \cap X^\geq \right\}, \tag{14}$$

$$\mathcal{W}_N = \left\{ W \subseteq X^< \mid W = \bigcup_{u \in W} D^-(u) \cap X^< \right\}. \tag{15}$$

$\mathcal{W}_P$ (resp. $\mathcal{W}_N$) is the family of object sets which are upward (downward) definable in the sense that it is expressed by the union of the intersections of dominating sets and $X^\geq$ (resp. dominated sets and $X^<$). Then, we consider the Cartesian product $\mathcal{W} = \mathcal{W}_P \times \mathcal{W}_N$.

Associated with $W = (W_P, W_N) \in \mathcal{W}$, we define a decision function f_W as follows:

$$f_W(u) = \begin{bmatrix} f_{W_P}(u) \\ f_{W_N}(u) \end{bmatrix}, \tag{16}$$

where

$$f_{W_P}(u) = |D^-(u) \cap W_P|, \tag{17}$$

$$f_{W_N}(u) = |D^+(u) \cap W_N|. \tag{18}$$

Each object u is classified by the following rule:

$$u \text{ is classified to } \begin{cases} X^\geq & \text{if } f_{W_P}(u) > 0 \text{ and } f_{W_N}(u) \leq 0, \\ X^< & \text{if } f_{W_P}(u) \leq 0 \text{ and } f_{W_N}(u) > 0. \end{cases} \tag{19}$$

We can say that other objects which do not satisfy any of the conditions of the classification rule are not certainly classified to either $X^\geq$ or $X^<$. We remark that,

– for $u \in X^{\geq}$, $u \in W_P$ if and only if $f_{W_P}(u) \geq 1$,
– for $u \in X^{<}$, $u \in W_N$ if and only if $f_{W_N}(u) \geq 1$.

We could obtain the classification of W by f_W if we knew whether classified objects belong to $X^{\geq}$ or $X^{<}$.

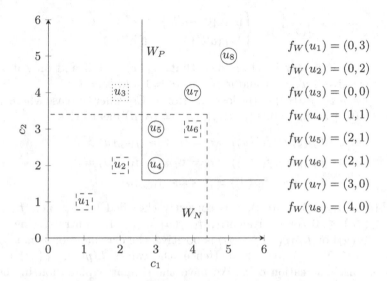

Fig. 1. Geometric interpretation of decision function f_W. Remark that $u_4, u_5 \notin W_N$ and $u_6 \notin W_P$.

We show a geometric interpretation of f_W in Fig. 1. This figure shows 8 objects $U = \{u_1, u_2, \ldots, u_8\}$ evaluated by two criteria $C = \{c_1, c_2\}$. Attribute values are given as follows: $C(u_1) = (1, 1)$, $C(u_2) = (2, 2)$, $C(u_3) = (2, 4)$, $C(u_4) = (3, 2)$, $C(u_5) = (3, 3)$, $C(u_6) = (4, 3)$, $C(u_7) = (4, 4)$, $C(u_8) = (5, 5)$. U is divided by $X^{\geq} = \{u_4, u_5, u_7, u_8\}$ and $X^{<} = \{u_1, u_2, u_3, u_6\}$. Let $W_P = \{u_4, u_5, u_7, u_8\}$ and $W_N = \{u_1, u_2, u_6\}$. Let us consider object u_8. We obtain $f_W(u_8) = (f_{W_P}(u_8), f_{W_N}(u_8)) = (4, 0)$, since $D^-(u_8) \cap W_P = \{u_4, u_5, u_7, u_8\}$ and $D^+(u_8) \cap W_N = \emptyset$. The objects in $D^-(u_8) \cap W_P$ are regarded as evidence for $u_8 \in X^{\geq}$ and those in $D^+(u_8) \cap W_N$ are regarded as evidence for $u_8 \in X^{<}$. For u_3 and u_4, we have $f_W(u_3) = (0, 0)$ and $f_W(u_4) = (1, 1)$. It means that f_W cannot classify u_3 and it has a contradiction at u_4.

We define the following function y that indicates the classification by $X^{\geq}$ and $X^{<}$.

$$y(u) = \begin{cases} 1 & u \in X^{\geq}, \\ -1 & u \in X^{<}. \end{cases} \tag{20}$$

For $a \in \{-1, 1\}$ and a pair of (nonnegative) real numbers $b = (b_P, b_N)$, we introduce a loss function L as follows:

$$L(a, b) = \begin{bmatrix} L_P(a, b) \\ L_N(a, b) \end{bmatrix} = \begin{bmatrix} l(a, b_P) \\ l(-a, b_N) \end{bmatrix}, \tag{21}$$

where l is a mixture of hinge and 0-1 loss functions defined by,

$$l(a', b') = \begin{cases} \max\{0, -a'b'\} & \text{if } a' = -1, \\ 1_{\leq 0}(a'b') & \text{if } a' = 1, \end{cases} \tag{22}$$

and $1_{\leq 0}$ is the indicator function such that $1_{\leq 0}(x) = 1$ if and only if $x \leq 0$. $l(a', b')$ is positive if $a' = -1$ and $b' > 0$ or $a' = 1$ and $b' = 0$.

We apply $(y(u), f_W(u))$ to the loss function L. Consider the case where $y(u) = 1$. Let $\alpha \in \{0, 1\}$ and $\beta \geq 0$. $L(y(u), f_W(u)) = (\alpha, \beta)$ means that

$$L_P(y(u), f_W(u)) = \alpha = 0 \iff f_{W_P}(u) > 0,$$
$$L_P(y(u), f_W(u)) = \alpha = 1 \iff f_{W_P}(u) = 0,$$
$$L_N(y(u), f_W(u)) = \beta \iff f_{W_N}(u) = \beta.$$

When $L(y(u), f_W(u)) = (0, 0)$, u is correctly classified by f_W, i.e., $f_{W_P}(u) > 0$ and $f_{W_N}(u) \leq 0$ (or equivalently, $f_{W_N}(u) = 0$). In contrast, when either $L_P(y(u), f_W(u))$ or $L_P(y(u), f_W(u))$ is positive, the classification is not correct, i.e., $f_{W_P}(u) \leq 0$ or $f_{W_N}(u) > 0$. Hence, the values $L(y(u), f_W(u))$ indicate degrees of misclassification of u. We have the similar explanation in the case where $y(u) = -1$.

An empirical risk of W is defined by the sum[1] of values of the loss vector function over U, which are scalarized by given weight vectors $\lambda^{\geq} = (\lambda_P^{\geq}, \lambda_N^{\geq})$ and $\lambda^{<} = (\lambda_P^{<}, \lambda_N^{<})$ for $X^{\geq}$ and $X^{<}$, respectively.

$$R(W \mid \lambda^{\geq}, \lambda^{<}) = \sum_{u \in X^{\geq}} (\lambda^{\geq})^{\top} L(y(u), f_W(u)) + \sum_{u \in X^{<}} (\lambda^{<})^{\top} L(y(u), f_W(u))$$

$$= \sum_{u \in \neg W_P \cap X^{\geq}} \lambda_P^{\geq} + \sum_{u \in X^{\geq}} \lambda_N^{\geq} |D^+(u) \cap W_N|$$

$$+ \sum_{u \in X^{<}} \lambda_P^{<} |D^-(u) \cap W_P| + \sum_{u \in \neg W_N \cap X^{<}} \lambda_N^{<}. \tag{23}$$

We show the following lemma, which provides another useful expression of the empirical risk function (23).

Lemma 1. *We have*

$$\sum_{u \in X^{\geq}} |D^+(u) \cap W_N| = \sum_{u \in W_N} |D^-(u) \cap X^{\geq}|,$$

$$\sum_{u \in X^{<}} |D^-(u) \cap W_P| = \sum_{u \in W_P} |D^+(u) \cap X^{<}|. \tag{24}$$

---

[1] We consider the sum instead of the average in order to simplify the formula by ignoring $\frac{1}{n}$.

Proof. We only prove the first equation. The others are proved in the same manner.

$$\sum_{u \in X^{\geq}} |D^+(u) \cap W_N| = |\{(u, u') \in U \times U \mid u \in X^{\geq}, u' \in D^+(u) \cap W_N\}|$$

$$= |\{(u, u') \in U \times U \mid u \in D^-(u') \cap X^{\geq}, u' \in W_N\}|$$

$$= \sum_{u' \in W_N} |D^-(u') \cap X^{\geq}|.$$

By applying Lemma 1 to the empirical risk function (23), we obtain the following expression:

$$R(W \mid \lambda^{\geq}, \lambda^{<}) = \sum_{u \in \neg W_P \cap X^{\geq}} \lambda_{\bar{P}}^{\geq} + \sum_{u \in W_N} \lambda_{\bar{N}}^{\geq} |D^-(u) \cap X^{\geq}|$$

$$+ \sum_{u \in W_P} \lambda_{\bar{P}}^{\leq} |D^+(u) \cap X^{<}| + \sum_{u \in \neg W_N \cap X^{<}} \lambda_{\bar{N}}^{\leq}. \qquad (25)$$

Using expression (25), we can easily obtain Lemma 2, which shows an optimality condition for the empirical risk minimization.

Lemma 2. *Given parameters* $\lambda^{\geq} = (\lambda_{\bar{P}}^{\geq}, \lambda_{\bar{N}}^{\geq}) \geq 0$ *and* $\lambda^{<} = (\lambda_{\bar{P}}^{\leq}, \lambda_{\bar{N}}^{\leq}) \geq 0$, $W^* = (W_P^*, W_N^*) \in U \times U$ *minimizes the empirical risk function* $R(W \mid \lambda^{\geq}, \lambda^{<})$ *under constraint* $W \in \mathcal{W}$ *if and only if* $W^* \in \mathcal{W}$ *and* W^* *satisfies the following implications: for all* $u \in U$

$$u \in X^{\geq} \text{ and } \lambda_{\bar{P}}^{\leq} |D^+(u) \cap X^{<}| < \lambda_{\bar{P}}^{\geq} \Longrightarrow u \in W_P^*,$$

$$u \in X^{\geq} \text{ and } \lambda_{\bar{P}}^{\leq} |D^+(u) \cap X^{<}| > \lambda_{\bar{P}}^{\geq} \Longrightarrow u \in \neg W_P^* \cap X^{\geq}, \qquad (26)$$

$$u \in X^{<} \text{ and } \lambda_{\bar{N}}^{\geq} |D^-(u) \cap X^{\geq}| < \lambda_{\bar{N}}^{\leq} \Longrightarrow u \in W_N^*,$$

$$u \in X^{<} \text{ and } \lambda_{\bar{N}}^{\geq} |D^-(u) \cap X^{\geq}| > \lambda_{\bar{N}}^{\leq} \Longrightarrow u \in \neg W_N^* \cap X^{<}.$$

Lemma 2 shows that an optimal solution can be attained by object-wise minimization. That is, we know that the optimal solution W_P^*, minimizing the empirical risk function value $R(W_P, W_N \mid \lambda^{\geq}, \lambda^{<})$, is obtained by the set of objects u such that $\lambda_{\bar{P}}^{\geq}$ is greater than $\lambda_{\bar{P}}^{\leq} |D^+(u) \cap X^{<}|$.

Using Fig. 2a, we discuss a geometric interpretation of the optimality condition for minimization of $R(W \mid \lambda^{\geq}, \lambda^{<})$. The setting is the same as in Fig. 1. The horizontal and vertical axses correspond to $s_{X^{\geq}} = |D^-(u) \cap X^{\geq}|$ and $s_{X^{<}} = |D^+(u) \cap X^{<}|$, respectively, and each object u has coordinates $(s_{X^{\geq}}(u), s_{X^{<}}(u))$. We call the space of $s_{X^{\geq}}$ and $s_{X^{<}}$ the support space.

In the left figure, the horizontal line is $S_P = \{(s_{X^{\geq}}, s_{X^{<}}) \mid \lambda_{\bar{P}}^{\leq} s_{X^{<}} = \lambda_{\bar{P}}^{\geq}\}$, and the vertical line is $S_N = \{(s_{X^{\geq}}, s_{X^{<}}) \mid \lambda_{\bar{N}}^{\geq} s_{X^{\geq}} = \lambda_{\bar{N}}^{\leq}\}$. The optimal solution is the pair of $W_P^* = \{u_4, u_5, u_7, u_8\}$ and $W_N^* = \{u_1, u_2, u_3\}$. The condition $\lambda_{\bar{P}}^{\leq} s_{X^{<}}(u) < \lambda_{\bar{P}}^{\geq} \Longrightarrow u \in W_P^*$ and $\lambda_{\bar{P}}^{\leq} s_{X^{<}}(u) > \lambda_{\bar{P}}^{\geq} \Longrightarrow u \in \neg W_P^*$ mean that objects under S_P should be included in W_P^* and those over S_P should be excluded from W_P^*, respectively. We have the similar explanation for W_N^*. The parameters

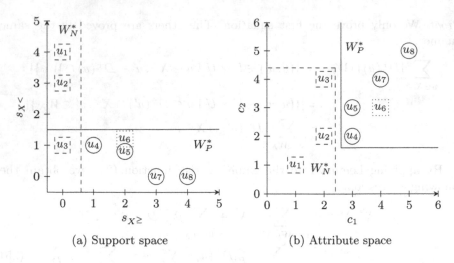

(a) Support space

(b) Attribute space

Fig. 2. Geometric interpretation of the optimality condition for minimization of $R(W \mid \lambda^{\geq}, \lambda^{<})$.

$\lambda^{\geq}$ and $\lambda^{<}$ control the positions of the lines. Additionally, Fig. 2b shows the optimal solution in the original attribute space.

From Lemma 2, we can characterize the lower approximations $\underline{X}_{\epsilon}^{\geq}$ and $\underline{X}_{\epsilon}^{<}$.

Theorem 1. *Given* $\lambda^{\geq} = (\lambda_P^{\geq}, \lambda_N^{\geq})$ *and* $\lambda^{<} = (\lambda_P^{<}, \lambda_N^{<})$, *where* $\lambda_P^{\geq}, \lambda_N^{<} \geq 0$, $\lambda_N^{\geq} > 0$ *and* $\lambda_P^{<} > 0$, *we define* $\alpha^{\geq} = \lambda_P^{\geq}/(\lambda_P^{<}|X^{<}|)$ *and* $\alpha^{<} = \lambda_N^{<}/(\lambda_N^{\geq}|X^{\geq}|)$. *Then,* $(\underline{X}_{\epsilon}^{\geq}(\alpha^{\geq}), \underline{X}_{\epsilon}^{<}(\alpha^{<}))$ *is an optimal solution of the problem of minimizing* $R(W \mid \lambda^{\geq}, \lambda^{<})$ *subject to* $W \in \mathcal{W}$.

Conversely, given $\alpha^{\geq}, \alpha^{<} \in [0, 1]$ *and* $s, t > 0$, *we define* $\lambda^{\geq} = (s\alpha^{\geq}|X^{<}|, t)$ *and* $\lambda^{<} = (s, t\alpha^{<}|X^{\geq}|)$. *Then,* $(\underline{X}_{\epsilon}^{\geq}(\alpha^{\geq}), \underline{X}_{\epsilon}^{<}(\alpha^{<}))$ *is an optimal solution of the problem of minimizing* $R(W \mid \lambda^{\geq}, \lambda^{<})$ *subject to* $W \in \mathcal{W}$.

Similarly, we can characterize the lower approximations $\underline{X}_{\epsilon'}^{\geq}$ and $\underline{X}_{\epsilon'}^{<}$.

Theorem 2. *There are given* $\lambda^{\geq} = (\lambda_P^{\geq}, \lambda_N^{\geq})$ *and* $\lambda^{<} = (\lambda_P^{<}, \lambda_N^{<})$, *where* $\lambda_P^{\geq}, \lambda_N^{<} \geq 0$, $\lambda_N^{\geq} > 0$ *and* $\lambda_P^{<} > 0$. *Moreover, we assume* $\lambda_P^{\geq} \leq \lambda_P^{<}|X^{<}|$ *and* $\lambda_N^{<} \leq \lambda_N^{\geq}|X^{\geq}|$, *and we define* $\alpha^{\geq} = \lambda_P^{\geq}/(\lambda_P^{<}|X^{\geq}|)$ *and* $\alpha^{<} = \lambda_N^{<}/(\lambda_N^{\geq}|X^{<}|)$. *Then,* $(\underline{X}_{\epsilon'}^{\geq}(\alpha^{\geq}), \underline{X}_{\epsilon'}^{<}(\alpha^{<}))$ *is an optimal solution of the problem of minimizing* $R(W \mid \lambda^{\geq}, \lambda^{<})$ *subject to* $W \in \mathcal{W}$.

Conversely, given $\alpha^{\geq} \in [0, |X^{<}|/|X^{\geq}|]$, $\alpha^{<} \in [0, |X^{\geq}|/|X^{<}|]$ *and* $s, t > 0$, *we define* $\lambda^{\geq} = (s\alpha^{\geq}|X^{\geq}|, t)$ *and* $\lambda^{<} = (s, t\alpha^{<}|X^{<}|)$. *Then,* $(\underline{X}_{\epsilon'}^{\geq}(\alpha^{\geq}), \underline{X}_{\epsilon'}^{<}(\alpha^{<}))$ *is an optimal solution of the problem of minimizing* $R(W \mid \lambda^{\geq}, \lambda^{<})$ *subject to* $W \in \mathcal{W}$.

4   Concluding Remarks

In this paper, we have shown the connection between VC-DRSA and the empirical risk minimization. Considering the classification problem for a pair of complementary upward and downward unions of decision classes and the empirical risk function based on the asymmetric loss function, we show that the pair of approximations corresponds to the optimal solution of the minimization problem. The obtained result makes a bridge between rough set reasoning about ordinal data and statistical machine learning from ordinal data. We believe that it will stimulate further development of VC-DRSA.

References

1. Błaszczyński, J., Greco, S., Słowiński, R., Szeląg, M.: Monotonic variable consistency rough set approaches. Int. J. Approximate Reasoning **50**, 979–999 (2009)
2. Błaszczyński, J., Greco, S., Słowiński, R., Szeląg, M.: On variable consistency dominance-based rough set approaches. In: Greco, S., et al. (eds.) RSCTC 2006. LNCS (LNAI), vol. 4259, pp. 191–202. Springer, Heidelberg (2006). https://doi.org/10.1007/11908029_22
3. Błaszczyński, J., Kusunoki, Y., Inuiguchi, M., Słowiński, R.: Empirical risk minimization for variable consistency dominance-based rough set approach. In: Yao, Y., Hu, Q., Yu, H., Grzymala-Busse, J.W. (eds.) Rough Sets, Fuzzy Sets, Data Mining, and Granular Computing, pp. 63–72. Springer International Publishing, Cham (2015)
4. Fitelson, B.: Likelihoodism, Bayesianism, and relational confirmation. Synthese **156**, 473–489 (2007)
5. Greco, S., Matarazzo, B., Słowiński, R.: Rough sets theory for multicriteria decision analysis. Eur. J. Oper. Res. **129**, 1–47 (2001)
6. Greco, S., Matarazzo, B., Słowiński, R., Stefanowski, J.: Variable consistency model of dominance-based rough sets approach. In: Ziarko, W., Yao, Y. (eds.) RSCTC 2000. LNCS (LNAI), vol. 2005, pp. 170–181. Springer, Heidelberg (2001). https://doi.org/10.1007/3-540-45554-X_20
7. Inuiguchi, M., Yoshioka, Y., Kusunoki, Y.: Variable-precision dominance-based rough set approach and attribute reduction. Int. J. Approximation Reasoning **50**, 1199–1214 (2009)
8. Kusunoki, Y., Błaszczyński, J., Inuiguchi, M., Słowiński, R.: Empirical risk minimization for variable precision dominance-based rough set approach. In: Lingras, P., Wolski, M., Cornelis, C., Mitra, S., Wasilewski, P. (eds.) RSKT 2013. LNCS (LNAI), vol. 8171, pp. 133–144. Springer, Heidelberg (2013). https://doi.org/10.1007/978-3-642-41299-8_13
9. Pawlak, Z.: Rough sets. Int. J. Inf. Comput. Sci. **11**, 341–356 (1982)
10. Słowiński, R., Yao, Y. (eds.): Rough Sets. Part C of the Handbook of Computational Intelligence, edited by J. Kacprzyk and W. Pedrycz. Springer (2015)
11. Yao, Y.: Probabilistic rough set approximations. Int. J. Approximation Reasoning **49**, 255–271 (2008)
12. Yao, Y., Zhou, B.: Two Bayesian approaches to rough sets. Eur. J. Oper. Res. **251**(3), 904–917 (2016)

Econometrics

An Econometric Study of Inbound Tourism Demand in Hong Kong, Macao and Taiwan: A Case Study of Mainland China

Bing Yang[2,3], Jianxu Liu[1,3(✉)], and Songsak Sriboonchitta[2,3]

[1] Faculty of Economics, Shandong University of Finance and Economics,
Jinan, China
jianxuliu1984@163.com
[2] Faculty of Economics, Chiang Mai University, Chiang Mai 50200, Thailand
[3] Puey Ungphakorn Center of Excellence in Economics, Chiang Mai University,
Chiang Mai 50200, Thailand

Abstract. The purpose of this paper is to estimate tourism demand using quarterly time series data (1998Q1–2017Q4) from Mainland China to Hong Kong, Macao, and Taiwan. Demand functions were used in the Autoregressive Distributive Lag (ARDL) model-Seemingly Unrelated Regression (SUR) model to estimate long-term and short-term tourism demand relationships in Hong Kong, Macao, and Taiwan. The results showed that word-of-mouth effect has a long-term relationship with tourism demand in Hong Kong, Macao, and Taiwan. Tourism price has a significant impact on Macao's tourism demand for long-term relationship. The seasonality is significant and negative effect in the first quarter. We suggest that launch products suitable for the Spring Festival family tour to attract tourists to travel with their families. We also found that the ARDL-SUR model is more robust than traditional ARDL model in this study.

Keywords: Inbound tourism · Tourism demand · ARDL · SUR · ECM

1   Introduction

The main tourist sources of tourist cities are from neighboring regions. The number of tourist arrival has always been a measure of the current state of tourism industry. Hong Kong, Macao and Taiwan are special regions of China. Due to historical issues, mainland residents to Hong Kong, Macao and Taiwan need to go through exit formalities. Many scholars did the inbound study of international tourists while studying the inbound tours of Hong Kong, Macao

Supported by Puey Ungphakorn Center of Excellence in Econometrics, Faculty of Economics, Chiang Mai University.

and Taiwan, and neglect the inbound study of mainland China. In this paper, it is of great significance to combines the four places to analysis the impact of Mainland China on the tourism demand of Hong Kong, Macao and Taiwan.

Tourism is one of the four major industries in Hong Kong. Over the past 10 years, Hong Kong's inbound tourism demand has increased significantly (Fig. 1). From the current situation of Hong Kong's tourism industry, Mainland China is the largest tourist market of Hong Kong's tourism industry. From 2009 to 2013, the number of Mainland Chinese tourists in Hong Kong increased by an average of 19.3% annually. In 2017, the number of Mainland Chinese tourists travelling to Hong Kong was 44.445 million, an increase of 3.9% year-on-year. In 2010, tourism accounts for about 4.3% of Hong Kong's GDP, and 8.9% of GDP in 2015.

Macao is one of the most developed and affluent regions in the world. In 2016, Macao's tourism revenue accounts for 43.9% of GDP, and the number of inbound tourists has increased year by year (Fig. 1). According to the latest study released by the World Tourism and Travel Council (WTTC), the Macao Special Administrative Region ranks third in the world's most dependent tourism economy, with tourism revenue accounting for 30% of Macao's GDP. More than 20 million Mainland tourists, an increase of 8.5% over the same period.

The tourism industry is one of the most important service industries in Taiwan. Before 2008, the number of Mainland tourists in Taiwan was almost zero (Fig. 1). Since 2008, the Taiwan authorities have opened up Mainland residents to visit Taiwan, the number of Mainland tourists has increased at an average annual rate of 19.52%. In 2012, the number of Mainland tourists from Taiwan's mainland reached 2.586 million, accounting for 35.57% of the total number of tourists in-bound in Taiwan. It shows that the Mainland China is gradually becoming the main source of tourists to Taiwan, further demonstrating the great potential of the Mainland China in Taiwan's tourism market.

Previous studies have found that income level and tourism price are important factors affecting tourism demand. Witt and Wit [1] found that income from the region of origin was considered one of the most influential factors that positively affected tourism demand. In cases where income is not included in a model, there is usually some form of aggregated expenditure variable in its place. Crouth [2] revealed that the income is the most important explanatory variable. Lim and McAleer [3] compared the explanatory power of real GDP and real private consumption expenditure in modelling tourist arrivals to Australia from Malaysia. Chang [4], Garin-Munoz and Amaral [5] found that price is very important for tourism demand. Some scholars used dynamic models to estimate the effects of long-term relationships between variables. Song and Li [6] used the ARDL model to find that the tourists' income and 'word-of-mouth' habit persistence effects are influenced mainly by tourist arrivals in Hong Kong. Habibi [7] got the results that Iranian tourist arrivals to Malaysia are positively influenced by Lag dependent variable (word of mouth), tourism price adjusted by exchange rate, tourism price substitute and trade value. Wang [8] revelated that income and foreign exchange rates are both significant explanatory variables. Song and Witt [9] used a vector

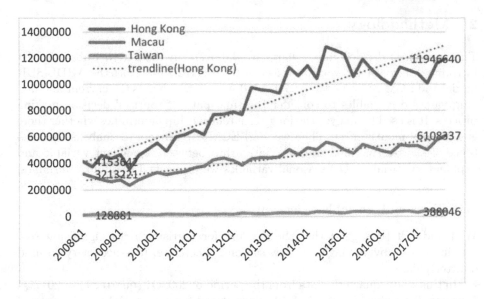

Fig. 1. Mainland Chinese Tourist Arrivals to Hong Kong, Macao and Taiwan (2008Q1–2017Q4).

autoregressive (VAR) model to predict international tourist flows to Macao, indicating that Macao will face the growing tourism demand from Mainland Chinese residents.

We found that many empirical researches on tourism demand have relied on the single equation model (Cho [10]; Law and Au [11]; Shan and Wilson [12]). The analysis of individual countries or regions ignored the competing interdependence of tourism destinations, which has a significant impact on the level of tourism demand for specific destinations and adjacent destinations. The SUR model estimates the information and relation between equations present in the error correlation of the cross equations. So that SUR model is more efficient than single equation estimation (Salman et al. [13]).

This study aims to extend the literature on tourism demand of Hong Kong, Macao and Taiwan from Mainland China in several aspects. First of all, this paper is the first study to examine the long-term and short-term relationship between inbound tourism demand, income, prices, and transportation cost from Mainland China to Hong Kong, Macao and Taiwan. Secondly, ARDL-SUR model was the first applied in the field of tourism.

The organization of this paper is as follows. Section 2 provides details of the methodology, including the theoretical basis for using and selecting data and models. Section 3 contains discussions on the empirical results. Finally, Sect. 4 contains conclusions and discussions.

2   Methodology

This section is divided into 2 parts to elaborate methodology. Subsection 2.1 introduces the data and variable selection; Subsect. 2.2 describes the ARDL-SUR model and ECM-SUR model. ARDL-SUR model, which uses the current and lagging values of variables to consider the time path of tourists' decision-making process. It is used to analyze the long-term relationship determines whether there is a word-of-mouth effect. Then we used ECM-SUR model to analysis the existence of short-term cointegration relationship between dependent variable and explanatory variable. The seasonal variables were specified as dummy variables.

2.1   Data and Variable Selection

Witt and Wit [1] suggest that the demand for tourism carries significant seasonality, with obvious high and low seasons. Therefore, it is necessary to used quarterly data.

In this study, quarterly data from the period of 2008Q1 (quarter 1) to 2017Q4 (quarter 4) making it a total of 40 observations for each variable. The data of tourist arrivals, GDP, CPI, exchange rate of Hong Kong obtained from Census and Statistics Department (The Government of the Hong Kong Special Administrative Region). The data of tourist arrivals, GDP, CPI, exchange rate of Macao obtained from Census and Statistics Department (The Government of the Macao Special Administrative Region). The data of tourist arrivals, GDP, CPI, exchange rate of Taiwan obtained from National Statistics, R.O.C.(Taiwan). The data of tourist arrivals, GDP, CPI of Mainland obtained from National Bureau of Statistics of the People Republic of China; the data of exchange rate of RMB against USD are obtained from the People's Bank of China. The data of international oil price obtained from a data-service provider Wind Info.

The tourism price index is an important factor in determining a potential visitor's decision. We divide it into two parts: (1) the cost of living for the visitor at the destination (relative price); and (2) the cost of travel or transportation to the destination (transportation cost). Relative prices are expressed with tourist prices of the destination country divided by the tourist prices of the source country (Lim and McAleer [3]). Transportation costs were assessed for all tourists. However, tourists will choose different transportation, such as airplanes, trains, boats, or coaches. And there will be difference price between high and low seasons, which is difficult to be measured. So this article uses international crude oil price as a proxy (Garin-Munoz [14]).

2.2   Model Specification

Consistent with the general literature, most international studies on tourism demand have adopted the number of tourist arrivals as the measure of demand. Several determinants of tourism demand have been identified and are widely used in econometric models; however, the most influential determinants are considered to be tourist income, tourism prices in a destination relative to those in the origin country (Song and Li [6]). The following demand function is proposed:

$$TA_{hk,t} = AY_t^{\beta_{01}} P_{hk,t}^{\beta_{02}} PT_t^{\beta_{03}} e_{1,t}, \tag{1}$$

$$TA_{mc,t} = BY_t^{\beta_{11}} P_{mc,t}^{\beta_{12}} PT_t^{\beta_{13}} e_{2,t}, \tag{2}$$

$$TA_{tw,t} = CY_t^{\beta_{21}} P_{tw,t}^{\beta_{22}} PT_t^{\beta_{23}} e_{3,t}, \tag{3}$$

where $TA_{hk,t}$, $TA_{mc,t}$, and $TA_{tw,t}$ are the number of tourist arrivals to Hong Kong, Macao and Taiwan from Mainland China at time t, respectively; Y_t is the income level of Mainland China at time t; $P_{hk,t}$ is the own price of tourism in Hong Kong relative to that in Mainland China at time t; $P_{mc,t}$ is the own price of tourism in Macao relative to that in Mainland China at time t; $P_{tw,t}$ is own price of tourism in Taiwan relative to that in Mainland China at time t; PT_t is the international crude oil price at time t; $e_{1,t}$, $e_{2,t}$ and $e_{3,t}$ are the error terms used to capture the influence of all other factors not include.

The definition of the tourism price is as below:

$$P_{i,t} = \frac{CPI_i \setminus EX_i}{CPI_{cn} \setminus EX_{cn}}, \tag{4}$$

where i = Hong Kong, Macao and Taiwan. CPI_{cn} and CPI_i are the consumer price indexes (CPI). EX_{cn} and EX_i are the exchange rate indexes.

By means of logarithmic transformation on both sides of the equation, the power functions (Eqs. (1)–(3)) are transformed into linear functions, which make it easier to estimate with ordinary least squares (OLS). The following functions are defined:

$$
\begin{aligned}
lnTA_{hk,t} =& \alpha_0 + \beta_{01} lnY_t + \beta_{02} lnP_{hk,t} + \beta_{03} lnPT_t \\
& + \beta_{04} DQ2 + \beta_{05} DQ3 + \beta_{06} DQ4 + u_{1,t},
\end{aligned} \tag{5}
$$

$$
\begin{aligned}
lnTA_{mc,t} =& \alpha_1 + \beta_{11} lnY_t + \beta_{12} lnP_{mc,t} + \beta_{13} lnPT_t \\
& + \beta_{14} DQ2 + \beta_{15} DQ3 + \beta_{16} DQ4 + u_{2,t},
\end{aligned} \tag{6}
$$

$$
\begin{aligned}
lnTA_{tw,t} =& \alpha_2 + \beta_{21} lnY_t + \beta_{22} lnP_{tw,t} + \beta_{23} lnPT_t \\
& + \beta_{24} DQ2 + \beta_{25} DQ3 + \beta_{26} DQ4 + u_{3,t},
\end{aligned} \tag{7}
$$

where $\alpha_0 = lnA$, $\alpha_1 = lnB$, $\alpha_2 = lnC$ are drift components; $u_{1,t} = lne_{1,t}$, $u_{2,t} = lne_{2,t}$ and $u_{3,t} = lne_{3,t}$. β_{01}, β_{11} and β_{11} are income elasticity; β_{02}, β_{12} and β_{22} are tourism price elasticity; β_{03}, β_{13} and β_{23} are the transportation cost elasticity. This paper drives the logarithms of all variables, such that the coefficient can be interpreted in a flexible manner. This is done to facilitate explanations of the model (Croes and Sr [15]; Song and Witt [16]; Vanegas and Croes [17]). DQ_2, DQ_3 and DQ_4 are the seasonal dummies for the second, third and fourth quarter of a year, which are used to capture the seasonality of tourism demand.

The above equation is a static model and does not take into account the dynamics of the tourist's decision-making process. The ARDL-SUR model initially takes the following form:

$$lnTA_{hk,t} = \alpha_0 + \sum_{j=1}^{p_1} \alpha_j lnTA_{hk,t-j} + \sum_{j=0}^{p_2} \beta_j lnY_{t-j}$$
$$+ \sum_{j=0}^{p_3} \gamma_j lnP_{hk,t-j} + \sum_{j=0}^{p_4} \delta_j lnPT_{t-j} + \phi_1 DQ_2 + \phi_2 DQ_3 + \phi_3 DQ_4 + \varepsilon_{1,t}, \tag{8}$$

$$lnTA_{mc,t} = \varphi_0 + \sum_{j=1}^{p_5} \varphi_j lnTA_{mc,t-j} + \sum_{j=0}^{p_6} \kappa_j lnY_{t-j}$$
$$+ \sum_{j=0}^{p_7} \eta_j lnP_{mc,t-j} + \sum_{j=0}^{p_8} \Theta_j lnPT_{t-j} + \phi_4 DQ_2 + \phi_5 DQ_3 + \phi_6 DQ_4 + \varepsilon_{2,t}, \tag{9}$$

$$lnTA_{tw,t} = \sigma_0 + \sum_{j=1}^{p_9} \sigma_j lnTA_{tw,t-j} + \sum_{j=0}^{p_{10}} \xi_j lnY_{t-j}$$
$$+ \sum_{j=0}^{p_{11}} \rho_j lnP_{tw,t-j} + \sum_{j=0}^{p_{12}} \psi_j lnPT_{t-j} + \phi_7 DQ_2 + \phi_8 DQ_3 + \phi_9 DQ_4 + \varepsilon_{3,t}, \tag{10}$$

where p_1 to p_{12} is the number of lags to be included as decide by the Akaike information criteria (AIC) and the Schwarz information criteria (SIC). The coefficients on the difference terms, namely α_j, β_j, γ_j, δ_j, φ_j, κ_j, η_j, Θ_j, σ_j, ξ_j, ρ_j and ψ_j represent the long-term dynamic of the model.

To estimate the equation for the tourism demand as specified above (Eqs. (8)–(10)), it is necessary to determine the statistical properties of the individual series.

When there is a long-term relationship between the variables, the ECM-SUR model is used to estimate the short-term effect adjustment speed of explained variable to explanatory variables. Therefore, the model can be written as:

$$\Delta lnTA_{hk,t} = \alpha_0 + \sum_{j=1}^{p_1} \alpha_j \Delta lnTA_{hk,t-j} + \sum_{j=0}^{p_2} \beta_j \Delta lnY_{t-j}$$
$$+ \sum_{j=0}^{p_3} \gamma_j \Delta lnP_{hk,t-j} + \sum_{j=0}^{p_4} \delta_j \Delta lnPT_{t-j} \tag{11}$$
$$+ \lambda_0 ECM_{hk,t-1} + \nu_{1,t}$$

$$\Delta lnTA_{mc,t} = \varphi_0 + \sum_{j=1}^{p_5} \varphi_j \Delta lnTA_{mc,t-j} + \sum_{j=0}^{p_6} \kappa_j \Delta lnY_{t-j}$$
$$+ \sum_{j=0}^{p_7} \eta_j \Delta lnP_{mc,t-j} + \sum_{j=0}^{p_8} \Theta_j \Delta lnPT_{t-j} \tag{12}$$
$$+ \lambda_1 ECM_{mc,t-1} + \nu_{2,t}$$

$$\Delta lnTA_{tw,t} = \sigma_0 + \sum_{j=1}^{p_9} \sigma_j \Delta lnTA_{tw,t-j} + \sum_{j=0}^{p_{10}} \xi_j \Delta lnY_{t-j}$$
$$+ \sum_{j=0}^{p_{11}} \rho_j \Delta lnP_{tw,t-j} + \sum_{j=0}^{p_{12}} \psi_j \Delta lnPT_{t-j} \qquad (13)$$
$$+ \lambda_2 ECM_{tw,t-1} + \nu_{3,t}$$

The speed of adjustment is measured by estimating the coefficient of λ_0, λ_1 and λ_2. Normally, the coefficient λ_0, λ_1 and λ_2 are statistically negative. Essentially, the speed of adjustment provides the information that the long-term equilibrium which converges to short-term.

3   Empirical Results

This section mainly describes the test results of the unit root test, ARDL-SUR and ECM-SUR models. We first performed the unit root test, and then estimated the single equation of the ARDL and ECM models, and finally estimated using the ARDL-SUR and ECM-SUR models. The unit root test is determined by the Phillips-Perron test. The estimated results are detailed below (Table 1).

Table 1. Unit root test results (PP-Test)

Variables	Intercept	Trend	Level	First difference	Order of integration
$lnTA_{hk}$	Yes	Yes	-2.358	-19.683^{***}	I(1)
$lnTA_{mc}$	Yes	Yes	-2.832	-7.915^{***}	I(1)
$lnTA_{tw}$	Yes	Yes	-6.496^{***}	–	I(0)
lnY	Yes	Yes	-10.071^{***}	–	I(0)
lnP_{hk}	Yes	Yes	-1.632	-5.762^{***}	I(1)
lnP_{mc}	Yes	Yes	-1.648	-4.969^{***}	I(1)
lnP_{tw}	Yes	Yes	-1.778	-5.479^{***}	I(1)
$lnPT$	Yes	Yes	-1.953	-5.192^{***}	I(1)

Note: (1) The null hypothesis: series has a unit root. (2) ***, ** and * is significant at 1%, 5% and 10% level, respectively.

The unit root test results show that tourist arrivals from Mainland China to Taiwan and the tourists income level are stationary at their zero order I(0). The rest of variables stationary in their first difference order I(1). Since the variables are not stationary at the same order, it is necessary to perform a cointegration test to check whether there is a cointegration relationship between the variables before estimating the demand function. Cointegration means that the tourism demand function can be specified using ECM model that estimates the short-term effect adjustment speed of explained variables to explanatory variables.

This study is based proceeds with the SUR-ARDL model (Pesaran [18]). There are several advantages to using this dynamic model. First, the ARDL-SUR model does not specify that all data series have the same order whether I(0) or I(1) or both (Pesaran and Pesaran [19]). Second, it is can be used in small sample datasets. Third, the ARDL-SUR model generally provides unbiased long-term estimates and valid t-statistics (Pesaran et al. [20–22]).

Table 2. Estimated coefficients of the tourism demand equations (long-term).

Variables	ARDL model			ARDL-SUR model		
	Hong Kong	Macao	Taiwan	Hong Kong	Macao	Taiwan
$lnTA(-1)$	0.573***	0.392***	0.336**	0.694***	0.511***	0.391***
	(4.617)	(2.830)	(2.151)	(7.880)	(5.035)	(3.063)
lnY	−0.654	0.138	−1.090	−0.797	0.080	−1.099
	(−1.091)	(0.225)	(−1.388)	(−1.648)	(0.160)	(−1.659)
$lnY(-1)$	1.141*	0.335	1.657**	1.114**	0.313	1.621**
	(1.191)	(0.565)	(2.229)	(2.328)	(0.649)	(2.584)
lnP	−0.623	−0.922	0.747	−0.693	−0.992*	0.769
	(−0.924)	(−1.327)	(1.039)	(−1.367)	(−1.845)	(1.308)
$lnP(-1)$	0.652	1.003	−0.558	0.885*	1.030*	−0.641
	(0.956)	(1.441)	(−0.735)	(1.722)	(1.903)	(−1.036)
$lnPT$	0.081	0.072	−0.002	0.100**	0.071	0.0006
	(1.375)	(1.249)	(−0.025)	(2.019)	(1.448)	(0.008)
$lnPT(-1)$	0.026	−0.003	−0.075	0.022*	−0.016	−0.068
	(0.481)	(−0.061)	(−0.838)	(0.499)	(−0.360)	(−0.900)
DQ_2	2.001	0.318	3.490**	2.062**	0.326	3.450**
	(1.557)	(0.245)	(2.124)	(1.995)	(0.309)	(2.489)
DQ_3	1.667	0.118	2.611*	1.824**	0.177	2.588**
	(1.516)	(0.105)	(1.830)	(2.055)	(0.193)	(2.150)
DQ_4	1.284	−0.104	2.210*	1.483*	−0.027	2.208*
	(1.219)	(−0.097)	(1.620)	(1.746)	(−0.031)	(1.919)
Constant	1.060	4.963***	2.124	0.431	3.846***	1.689
	(1.014)	(2.994)	(1.351)	(0.527)	(3.050)	(1.298)
R-squared	0.989	0.977	0.969	0.988	0.976	0.969
Adjusted R-squared	0.985	0.968	0.959	0.985	0.968	0.958
Durbin-Watson	2.247	1.987	2.202	2.450	2.110	2.292

Source: Authors' estimation; ***, ** and * is significant at 1%, 5% and 10% level, respectively. The number inside the parentheses is the t-statistics. AIC (Akaike Information Criterion) used in lag length selection criteria for the ARDL specification.

Table 2 indicates the results of the long-term dynamic relationships of Mainland Chinese tourist arrivals to Hong Kong, Macao and Taiwan. The results of short-term dynamic relationships shown in Table 3. Duo to the length of data and combined with the AIC and BIC, all the variables lag the first order is best. Given that the error terms of Eqs. (8), (9) and (10) are expected to be correlated, the ARDL-SUR model accounts for the entire matrix of correlations of all the equations, so that the SUR model produces more efficient estimators [23]. The estimators in the ARDL-SUR model minimize the determinant of covariance matrix of the disturbances [24]. Therefore, the estimation results of ARDL-SUR model and ECM-SUR model are better than ARDL model and ECM model. Select the ARDL-SUR model and ECM-SUR model results to explain.

For tourist arrivals to Hong Kong, Macao and Taiwan, in the long-term, the previous one quarter of tourist arrivals has significant positive effect. The

Table 3. The results of short-term dynamic effects of tourism demand.

Variables	ECM model			ECM-SUR model		
	Hong Kong	Macao	Taiwan	Hong Kong	Macao	Taiwan
$\Delta lnTA\,(-1)$	−0.333***	0.168	0.305**	−0.249***	0.145	0.297***
	(−3.009)	(1.308)	(2.448)	(−2.730)	(1.375)	(2.754)
ΔlnY	0.017	−0.002	0.105***	0.016	−0.002	0.106***
	(1.158)	(−0.294)	(5.236)	(1.276)	(−0.266)	(6.095)
$\Delta lnY\,(-1)$	0.086***	0.041***	−0.096***	0.085***	0.041***	−0.092***
	(5.520)	(3.870)	(−4.041)	(6.317)	(4.482)	(−4.484)
ΔlnP	−1.051*	−0.301	1.156	−1.126**	−0.599	1.341**
	(−1.747)	(−0.564)	(1.580)	(−2.238)	(−1.361)	(2.121)
$\Delta lnP\,(-1)$	−0.409	0.535	−0.471	−0.127	0.476	−0.492
	(−0.569)	(1.020)	(−0.635)	(−0.213)	(1.088)	(−0.770)
$\Delta lnPT$	−0.017	0.054	−0.085	−0.018	0.048	−0.093
	(−0.284)	(1.175)	(−1.082)	(−0.340)	(1.213)	(−1.364)
$\Delta lnPT\,(-1)$	0.030	−0.018	−0.023	0.032	−0.012	−0.026
	(0.516)	(−0.401)	(−0.290)	(0.640)	(−0.307)	(−0.368)
ECM(−1)	−0.215	−0.982***	−0.968***	−0.273**	−0.881***	−0.974***
	(−1.502)	(−4.928)	(−4.274)	(−2.378)	(−5.574)	(−4.994)
Constant	0.045***	0.008	0.014	0.042***	0.013	0.013
	(3.724)	(0.781)	(1.052)	(3.994)	(1.371)	(1.148)
R-squared	0.810	0.752	0.864	0.806	0.748	0.863
Adjusted R-squared	0.757	0.684	0.826	0.752	0.679	0.826
Durbin-Watson	1.509	1.405	2.282	1.626	1.581	2.237

Source: Authors' estimation; ***, ** and * is significant at 1%, 5% and 10% level, respectively. The number inside the parentheses is the t-statistics. AIC (Akaike Information Criterion) used in lag length selection criteria for the ARDL specification.

lagging dependent variables indicate that the "word-of-mouth" effect plays an important role in tourism demand. Tourism demand is stimulated by the "word of mouth" from friends and family, as well as tourism promotion and marketing strategies. The coefficient of current period income level are not significant, there is no statistical significance. The SUR model captures the current period tourism price and has a significant negative effect on tourist from Mainland China to Macao. Higher tourism price will lead to a decline in tourism demand. The model therefore predicts that, in the long-term, when the tourism price increase 1 percent, then the number of tourists from Mainland China arriving in Macao will decrease by 0.992 percent. Falling tourism price will increase their tourism demand, and more tourists will spend more money on accommodation, foods and drinks. The transportation cost of tourists to Hong Kong is statistically significant with a positive coefficient. It shows that increase in the transportation cost cannot directly affect the tourists' willingness to travel. There are many kinds of transportation from Mainland China to Hong Kong, such as by bus, by boat, by airplane. Many travel agencies adopt charter strategies and have great bargaining power. The rise in international crude oil price does not directly affect transportation costs, or the impact is not the expected to be negative effect.

Seasonality in Hong Kong and Taiwan was significant and positive in the second quarter, third and fourth quarters. Macao's seasonality was not significant in the second quarter, the third quarter and the fourth quarter. It indicates that the impact of the first quarter on Hong Kong and Taiwan is significant and negative. It also has a significant impact on Macao, but it cannot determine the positive coefficient or negative coefficient. We can explain this situation according to Chinese national conditions. New Year and the Spring Festival are two important festivals for Chinese people. In these two important festivals, Chinese people like to reunite with their families, most of them are at home, visit relatives and friends, rarely travel outbound. This is consistent with the decline in the number of tourists in the first quarter.

The outcomes prove that the estimated of lagged error correction term (ECM (-1)) is significant and negative. It suggests that the speed of adjustment from short-term towards long-term equilibrium path. The lagged error correction term of Hong Kong, Macao and Taiwan tourism demand are -0.273, -0.881, -0.974, respectively. It indicates that deviation from long-term demand to Hong Kong, Macao and Taiwan from Mainland China is adjusted by 27.3%, 88.1%, and 97.4% in the current quarter, respectively. The statistical significance and negative effects shown by the error correction coefficient further confirmation of the relationship between long-term tourists from Mainland China to Hong Kong, Macao and Taiwan, and their key independent variables.

4   Conclusion

This study used the ARDL-SUR model and the ECM-SUR model to analyze the long-term and short-term influence facotrs of Mainland Chinese tourist to Hong Kong, Macao and Taiwan. We found that word-of-mouth effect is an important factor in long-term relationship of Mainland Chinese tourist to Hong Kong,

Macao and Taiwan. The number and repeats of tourists should be given due attention by the Hong Kong, Macao and Taiwan tourism industry. The current tourism price has a negative impact on the number of tourists arriving to Macao in long-term. The speed of adjustment from short-term towards long-term equilibrium path showed that Hong Kong is the fastest, followed by Macao, and finally is Taiwan. Seasonal indicates that the first quarter was negatively significant. We also found that the ARDL-SUR model is more robust than traditional ARDL model.

The empirical results of long-term relationship in this study provide important policy implications for Hong Kong, Macao and Taiwan. First, the tourism price has significant and negative effect on Macao. We suggest that low-cost airplane tickets or travel packages be provided to reduce the cost of tourists and increase their travel intentions. Second, the seasonal impact of the first quarter is significant and negative, the New Year and the Spring Festival are two major Chinese festivals in this period. We recommend that products suitable for the Spring Festival family tourism be launched to attract tourists to travel with their families.

Hong Kong, Macao and Taiwan are very close to each other, and their culture is very similar. There may be a substitution effect between the three destinations. In the further study, it may be that 5–10 countries are selected as the source of tourists, Hong Kong, Macao and Taiwan as the destination, to study the substitution relationship between them.

References

1. Witt, S.F., Wit, C.A.: Forecasting tourism demand: a review of empirical research. Int. J. Forecast. **11**(3), 447–475 (1995)
2. Crouth, G.I.: The study of international tourism demand: a review of findings. J. Travel Res. **32**(1), 12–23 (1994)
3. Lim, C., McAleer, M.: Cointegration analysis of quarterly tourism demand by Hong Kong and Singapore for Australia. Appl. Econ. **33**(12), 1599–1619 (2001)
4. Chang, C., Khamkaew, T., McAleer, M.: Estimating price effects in an almost ideal demand model of outbound Thai tourism to East Asia. J. Tourism Res. Hospitality **1**(3) (2010)
5. Garin-Munoz, T., Amaral, P.A.: An econometric model for international tourism flows to Spain. Appl. Econ. Lett. **2000**(7), 525–529 (2000)
6. Song, H., Li, G.: Tourism demand modeling and forecasting: a review of recent research. Tourism Manage. **29**, 203–220 (2008)
7. Habibi, F.: Iranian Tourism demand for Malaysia: a bound test approach. Iran. Econmic Rev. **19**(1), 63–80 (2015)
8. Wang, Y.: The impact of crisis events and macroeconomic activity on Taiwan's international inbound tourism demand. Tourism Manage. **30**, 75–82 (2009)
9. Song, H., Witt, S.F.: Forecasting international tourist flows to Macao. Tourism Manage. **27**, 214–224 (2006)
10. Cho, V.: Tourism forecasting and its relationship with leading economic indicators. J. Hospitality Tourism Res. **25**, 399–420 (2001)
11. Law, R., Au, N.: Relationship modeling in tourism shopping: a decision rules induction approach. Tourism Manage. **21**, 241–249 (2000)

12. Shan, J., Wilson, K.: Causality between trade and tourism: empirical evidence from China. Appl. Econ. Lett. **8**, 279–283 (2001)

13. Salman, A., Arnesson, L., Sorensson, A., Shukur, G.: Estimating and Swedish and Norwegian international tourism demand using ISUR technique. CESIS Electronic Working Paper Series, vol. 198 (2009)

14. Garin-Munoz, T.: Inbound international tourism to Canary Islands: a dynamic panel data approach. Tourism Manage. **27**, 281–291 (2006)

15. Croes, R.R., Sr, M.V.: An econometric study of tourist arrivals in Aruba and its implications. Tourism Manage. **26**, 879–890 (2005)

16. Song, H., Witt, S.F.: Tourism demand modelling and forecasting: modern econometric approaches. Pergamon, Oxford (2000)

17. Vanegas, S., Croes, R.R.: Evaluation of demand: US tourists to Aruba. Ann. Tourism Res. **27**(4), 946–963 (2000)

18. Hashem, P.M.: An Autoregressive Distributed Lag Modelling approach to cointegration analysis. Cambridge (1999). University:134–150

19. Pesaran, M., Pesaran, B.: Wording with Microfit 4.0. In Interactive Econometric Analysis. Oxford University Press, Oxford (1997)

20. Pesaran, M., Shin, Y., Smith, R.: Bounds testing approaches to analysis of level relationship. J. Appl. (2001). Econometrics:289–326

21. Alhassan, A., Fiador, V.: Insurance-growth nexus in Ghana: an autoregressive distributed lag bounds cointegration approach. Rev. Dev. Finance **4**, 88–96 (2014)

22. Odhiambo, N.: Energy consumption and economic growth nexus in Tanzania: an ARDL bounds testing approach. Energy Policy **37**(2), 617–622 (2009)

23. Haden, K.: The demand for cigarettes in Japan. Am. J. Agric. Econ. **72**(2), 446–450 (1990)

24. Zellner, A.: An efficient method of estimating Seeming Unrelated Regression equations and test of aggregation bias. J. Am. Stat. Assoc. **57**, 500–509 (1962)

The Dependence Structure and Portfolio Optimization in Economic Cycles: An Application in ASEAN Stock Market

Jittima Singvejsakul[1]([✉]), Chukiat Chaiboonsri[1], and Songsak Sriboonchitta[1,2]

[1] Faculty of Economics, Chiang Mai University, Chiang Mai 50200, Thailand
Jittimasvsk@gmail.com
[2] Puey Ungphakorn Center of Excellence in Economics, Chiang Mai University, Chiang Mai 50200, Thailand

Abstract. Investigation is made on the dependence structure of ASEAN stock markets, including Thailand, Indonesia, Malaysia, the Philippines, Vietnam and Singapore. Technically, data is divided into boom and recession periods during 2008–2017. The econometric tools employed for the analysis are the Markov-Switching model (MS-model), D-vine trees and Markowitz Portfolio selection model. Empirically, the results of MS-model prescribe 649 and 1,600 stock trading days in the ASEAN stock markets for the bull and the recession period, respectively. Second, the findings of the relationship among ASEAN stock markets show that there are strongly positive dependent structures in bull period. On the other hand, in the bear periods, Vietnam stock market has the negative relation among stock markets in ASEAN countries. Third, the empirical results from the Markowitz portfolio selection indicated that the choice to minimize risk value in the bull period is more efficient than to maximize the return. The proportion to invest in this period is to invest Singapore that is the sensible choice to make the investment. Conversely, the best alternative way in recession period is to maximize return, and the diversification investment is more suitable to get lower risks since their dependent structure has the negative connection.

Keywords: MS-model · D-vine trees · Bull markets · Bear markets · ASEAN stock market · Markowitz portfolio selection

1 Introduction

The emerging financial markets offer alternative choices for investors from advanced economies. The Association of Southeast Asian Nations (ASEAN) stock exchanges is one of the destinations for portfolio investments. There are six countries included in the ASEAN's stock exchanges: Indonesia, Malaysia, the

Supported by Puay Ungpakoyn Centre of Excellence in Econometrics, Faculty of Economics, Chiangmai University.

H. Seki et al. (Eds.): IUKM 2019, LNAI 11471, pp. 161–171, 2019.
https://doi.org/10.1007/978-3-030-14815-7_14

Philippines, Singapore, Thailand, and Vietnam. ASEAN remains a major destination of global foreign direct investment (FDI), receiving around 16% of the world FDI among developing economies with total FDI flows of $120 billion in 2016 and $130 billion in 2017. The increasing of investment has a great impact on the business and economics in ASEAN countries that they rapidly grow regionally and internationally. Particularly from 2016 to 2017, the GDPs of Indonesia, Malaysia, the Philippines, Singapore, Thailand, and Vietnam grew 5.01%, 6.2%, 1.9%, 11.2%, 3.9% and 6.28%, respectively. Parallel to the economic growth, the ASEAN stock markets were also growing in 2017, with increase of the market capitalization of stock market 14.08% in Indonesia, 7.5% in the Philippines, 15% in Singapore, 13.61% in Thailand SET and 24.64% in Vietnam. Therefore, the numbers of investments in ASEAN stock markets have been assumed that they have the relationship. Consequently, the purpose of this paper is to clarify the relationship among the ASEAN stock markets and the choices to invest, which is crucial information for the investors whether risk lovers or risk averters to plan their decisions.

Historically, there are several studies addressing the relationship between stock markets and economic systems. The stock market integration in ASEAN was studied by Click and Plummer [10] that considers the relation of the five stock markets in ASEAN, which are correlated by using time series and cointegration. The main objective is to find the long-term dependence. There is only one cointegrating vector, the stock markets of Indonesia, Malaysia, the Philippines, Singapore, and Thailand during 1998–2002. Thus they are not completely correlated. From the perspective of the international portfolio investor, benefits of international portfolio diversification across the five markets are reduced but not eliminated. Abd [8] examine the long-run market integration among the five founding members in the ASEAN emerging markets, namely Malaysia, Indonesia, Thailand, Singapore and the Philippines. The study found that they are cointegrated, especially after the Asian financial crisis. Jiang, Nie and Monginsidi [7] analyze the co-movement and the volatility fluctuations of stock markets in ASEAN countries by using the Variation Modes Decomposition (VMD) based on copula models. The results show that ASEAN stock markets are stronger in the short term, especially following external shocks, and Vietnam (Indonesia) has the lowest (highest) interdependence with the rest of ASEAN trading participants. Sriboonchitta, Liu, Kreinovich, and Nguyen [12] used the copula approach to analyze the financial risks and co-movement trends of stock markets in Indonesia, the Philippines and Thailand and capture the pairwise and conditional dependences between the variables by using the vine copula model. The empirical evidence shows that all the leverage affects the capacity for the explanation of the three stock returns. The D-vine is more reliable than the C-vine for describing the dependence of three stock markets.

Economic cycle represent a structural change of stock markets into 2 regimes by applying the Markov-switching model. Regarding to the study of regimes switching vine copula models for global equity and volatility indices by Fink et al. [5]. They explore the relationships between stock and volatility indices in

Asia, Europe and the USA and identify times of normal and abnormal states. They found that there are strong predominance of normal regime while the abnormal seems to be mainly driven by SPX-BBC as the general dependence structure and the relationship in this regime are lower dependency than the normal states. Chokethaworn, Chaitip, Sriwichailamphan and Chaiboonsri [4] study the dependence structure and co-movement between Thai currency and Malaysian currency using Markov switching model in dynamic copula approach. The study found that the first regimes represented appreciation against the US dollar. And the second regimes represented depreciation against the US dollar. In addition, both first regimes and second regimes found dependent structure and co-movement between two currencies explained by Elliptical copula. In terms of recommendation, an estimation result of the experiment can significantly develop the prediction accuracy of Markov Switching Model in Dynamic Copula approach (MSDC) algorithm. Furthermore, the applicable solutions from the computations indicated that the Markov Switching Model in dynamic Copula (MSDC) algorithms represented the evidence used to support the existence of identically distributed densities.

This paper is proposed for analyzing the stock market relationship, which is divided into the 2 regimes ASEAN financial markets. Since the financial crisis in 2008, boom and bear periods have predominantly existed. Additionally, this study also finds the suitable proportion to investment for both of two regimes. The main contributions of the paper are as follows: the first objective is to use Markov switching model (MS-model) to find regimes in the stock markets. The second purpose in this paper is to describe the relationship and co-movement among the stock markets by using D-vine trees copula in the two regimes. Third, the study is conducted to find the mean and variance of stock markets in each period by using the Markowitz portfolio selection model.

2   Methodology

In this study, multi-step procedures is introduced to find the relationship of structure and the proportion of investment in economic cycles. Firstly, the Markov Switching model is used to classified individual return series to be two regimes, namely bull and bear markets. Second, AR-GARCH is employed to find the estimation of regression and volatility of six stock markets. Third, the best-fit AR-GARCH will give the standardized residuals which are transformed into a uniform distribution in [0, 1] and plug into the D-vine copula model that standard deviation obtained from Maximum entropy bootstrapping. Finally, the Markowitz portfolio selection model is used to indicate the mean, variance and proportion to investment in ASEAN stock market.

2.1   The Markovian Switching Model

A Markovian switching is constructed by combining two or more dynamic models via a Markovian switching mechanism [6] and can be shown in Eq. (1)

$$r_{st} = \alpha_{st} + \beta r_{st-1} + \epsilon_{st}, \tag{1}$$

where $\epsilon_t =$ i.i.d. random variables with mean zero and variance σ_t^2 and $|\beta| < 1$. This model admits two dynamic structures at different levels, depending on the unobserved value of the state variable s_t.

The evaluation of the latent variable driving the regime changes, s_t, is governed by first-order Markov chain with constant transition probabilities collected in the (SxS) transition probability matrix P: in Eq. 3.

$$pr(s_t = j|s_{t-1} = i) = p_{ij},$$ (2)

$$p = \begin{pmatrix} p_{11} & p_{12} & \cdots & p_{1s} \\ p_{21} & p_{22} & \cdots & p_{2s} \\ \cdot & \cdot & \cdots & \cdot \\ p_{1s} & p_{2s} & \cdots & p_{ss} \end{pmatrix}.$$ (3)

2.2 AR-GARCH Model

Generalized Auto-Regressive Conditional Heteroskadasticity (GARCH) model is the extension of the ARCH model. This is named as Generalized ARCH model [1]. This is used to estimate the marginal distributions for fitting the copula model. The autoregressive process of order p or AR (p) is defined by the Eq. (4).

$$r_t = c + \Sigma_{i=1}^p \phi_j r_{t-i} + \epsilon_t,$$ (4)

$$\epsilon_t = h_t \eta_t,$$ (5)

$$h_t^2 = \omega + \sum_{i=1}^k \alpha_i \epsilon_{t-i}^2 + \sum_{i=1}^l \beta_i h_{t-i}^2,$$ (6)

where $\phi = (\phi_1, \phi_2, ..., \phi_p)$ is the vector of model coefficient and p is a Non-negative integer $\Sigma_{j=1}^p \phi_j < 1, \omega > 0, \alpha_i > 0$ and $\beta_i > 0$. The Eqs. (4) and (6) are the mean equation and variance equation; respectively. The Eq. 5 demonstrates the residuals ϵ_t is consist of the standard variance h_t and the standardized residuals η_t, which assumed to be normal distribution.

2.3 The Vine Copulas Construction

Copula, the function that joins multivariate marginal distribution functions to form joint distribution function, was first proposed as Sklar theorem and it can be expressed by Eq. (7).

$$C(u_1, u_2, ..., u_n) = H(F^{-1}(u_1)), F^{-1}(u_2), ..., F^{-1}(u_n)),$$ (7)

Where $X = (X_1, X_2, ..., X_n)$ is a random vector with joint distribution function H and the marginal distributions $F_1, F_2, ..., F_n$.

Vine structures are developed from pair-copula constructions, in which $d(d-1)/2$ pair-copulas are arranged in $d-1$ trees. The so-called pair copula constructions (PCCs) was initially introduced by Joe and developed in more detail in the work of Bedford and Cook. The D-vine copula is a very flexible model since multivariate copulas can easily accommodate complex dependence structures such as asymmetric dependences or strong joint tail behaviors [9]. The D-Vine tree has a path structure which leads to the construction of the D-vine density, which can be constructed as follows:

$$f(x) = \prod_{k=1}^{d} f_k(x_k) \prod_{j=1}^{d-i} c_{j,j+1|(j+i):(j+i-1)}(F(x_j|x_{j+1}, ..., x_{j+i-1}),$$

$$F(x_{j+i}|x_{j+1}, ..., x_{j-i-1})|\theta_{j,j+1|(j+1):(j+i+1)}), \tag{8}$$

The marginal conditional distribution in the D-vine can be given by

$$h(x|v, \theta) = f(x|\nu) = \frac{\partial c_{x\nu_j|\nu_{-j}}(F(x|\nu_{-j}), F(\nu_j|\nu_{-j}|\theta)}{\partial F(\nu_j|\nu_{-j}}, \tag{9}$$

Where ν_j is an arbitrary component of ν_j and ν_{-j} denotes the (m-1)-dimensional vector ν excluding ν_j and θ is the parameters set of $c_{x\nu_j|\nu_{-j}}$.

2.4   The Maximum Entropy Bootstrapping

As stated by Vinod [13], Maximum entropy is a powerful tool for avoiding all unnecessary distributional assumptions. This computational estimation follows a given finite state space and constraints, which are finite sets of n states with probabilities p and they were defined as a prior estimation formed with new information I; $\sum_i p_1 a_{ki} = 0$ or $\sum_i p_1 b_{ki} > 0$ where known numbers are a_{ki} and b_{ki}. Accordingly, the entropy maximization is expressed as [11].

$$H(q) = -\sum_{i>1} p_i log(p_i), \tag{10}$$

The bootstrap is the computationally statistical approach regarding the relation between samples and (unknown) populations by a comparable linkage between the samples at hand and appropriately designed (observable) resamples [13]. Technically, let us compute the bootstrap percentile of the function $y = f(q_i)$, where q_i is the a motif statistic. The samples, $q_1, ..., q_n$, are assumed to be independent and identically distributed variables (i.i.d) and the function, $P(q|I)$, stands for the parametric density of choices. Hence, we can set $y = f(q_i)$ for each I, the bootstrap percentile is given by

$$P_{BS}(f, q) = \frac{|y_i|y < y|}{n}. \tag{11}$$

Interestingly, combining the Maximum Entropy and Bootstrapping approach results in a powerful technique of modern time series inferences while algorithms for estimation by the Maximum Entropy Bootstrapping approach (MEboot) can generate an ensemble of worldwide changing time [2]. The overview of the steps in Vinod's ME bootstrap algorithm due to seven steps were found and described in the study of Chaitip and Chaiboonsri [3].

2.5   The Markowitz Portfolio Selection Model

Markowitz's portfolio selection theory is one of the pillars of theoretical finance which was developed by Harry Markowitz in 1950. The Markowitz identified the trade-off between the investment risk and expected return. This illustrate geometrically relations between beliefs and choice of portfolio according to the expected returns and the variance of returns. There are expected outcome, μ, and the variances, σ_i^2, for each variable, i = 1,...,n. The proportion of the total amount in variable i and x_i, which can compute the expected return and the variance of the resulting portfolio $x_i = (x_i, ..., x_n)$. This can be written as the mean and variance in Eqs. 12 and 13, respectively.

$$E(x) = x_1 u_i + ... + x_n u_i = u^T x, \tag{12}$$

$$Var(x) = \sum_{ij} \rho_{ij} \sigma_i \sigma_j x_i x_j = x^T x, \tag{13}$$

Where $\rho_{ij} = 1, Q_{ij} = \rho_{ij}\sigma_i\sigma_j$, the correlation coefficient ρ_{ij} between two variables i and j is technically assumed to be known. In addition, the portfolio vector x must satisfy $\sum_i = 1$ and the feasible portfolio x is explained as efficient.

3   Empirical Results of Research

3.1   The Data Description

Generally, Table 1 presents descriptive information of daily stock returns in ASEAN financial markets. There are six stock markets operating to provide the returns such as Thai stock market, Indonesia stock market, Malaysia stock market, the Philippines stock market, Vietnam stock market and Singapore stock market. The data collected covering the period 2008–2017 are the daily time-series and are transformed into log-return for estimation.

3.2   The Estimation of AR-GARCH in Bull and Bear Markets

Empirically, the estimated results of Markov switching model indicate that there are 649 and 1,600 stock trading days of bull markets and bear markets, respectively. Table 2 shows the results of the marginal assumption of the AR-GARCH and use the Maximum Entropy bootstrapping that is a powerful tool for avoiding all unnecessary distributional assumptions in order to obtain the standard deviation. As displayed in Table 2, most of the estimated parameters in the conditional mean and variance equations are statistically significant. This can be concluded

Table 1. Data description of ASEAN stock markets' daily returns.

	SET (Thailand)	IDX (Indonesia)	KLCI (Malaysia)	PSEI (Philippine)	HOSE (Vietnam)	STI (Singapore)
Mean	0.000327	0.000365	8.76E−05	0.000368	5.58E−06	−5.34E−06
Median	0.000701	0.000966	0.000232	0.000773	0.000638	0.000219
Maximum	0.075487	0.103196	0.057165	0.070560	0.091664	0.093453
Minimum	−0.085892	−0.113412	−0.102366	−0.130887	−0.099324	−0.111228
Std. Dev.	0.012596	0.014244	0.007629	0.013125	0.015876	0.011945
Skewness	−0.450556	−0.668514	−1.012388	−0.903582	−0.556300	−0.310472
Kurtosis	8.967669	14.30417	22.68379	12.43032	8.309878	15.78956
Jarque-Bera	3398.156	12087.97	36528.46	8601.189	2745.824	15295.96
Probability	0.000000	0.000000	0.000000	0.000000	0.000000	0.000000
ADF testing	0.000000	0.000000	0.000000	0.000000	0.000000	0.000000
(t-statistics)	(−42.2469)	(−29.4313)	(−43.5319)	(−28.9956)	(−39.2787)	(−44.9584)
Observations	2239	2239	2239	2239	2239	2239

that all series show significant autoregressive components and the parameters that are statistically significant at 1% are found close to one indicating the high volatility. The optimal models that seem to fit the ASEAN stock market are AR (1, 1) and GARCH (1, 1). The estimated parameters, their standard errors and BIC of bull markets and bear markets are reported in Table 2.

Table 2. The estimation of AR-GARCH in Bull and Bear markets.

		SET (Thailand)	IDX (Indonesia)	KLCI (Malaysia)	PSEI (Philippine)	HOSE (Vietnam)	STI (Singapore)
Bull markets	c	−0.00094*	−0.001098*	−0.00024*	−0.00007*	−0.0012*	−0.00164*
		(0.00075)	(0.000871)	(0.000449)	(0.00073)	(0.00088)	(0.000752)
	Ar1	0.101937**	0.13705***	0.09783*	0.10500*	0.33882***	0.076765*
		(0.04127)	(0.04119)	(0.04129)	(0.04127)	(0.03905)	(0.04136)
	ω	0.000016*	0.000025**	0.000002*	0.000029*	0.00002*	0.0000047*
		(0.000008)	(0.0000083)	(0.000001)	(0.00016)	(0.00001)	(0.000002)
	α	0.09955**	0.11500***	0.046150**	0.10950**	0.11800**	0.12060***
		(0.03163)	(0.02678)	(0.01595)	(0.03426)	(0.04126)	(0.02601)
	β	0.84970***	0.82800***	0.93270***	0.79870***	0.85010***	0.86760***
		(0.0450)	(0.03322)	(0.02301)	(0.07424)	(0.05190)	(0.02367)
	BIC	−5.239105	−4.993973	−6.206327	−5.256580	−4.82029	−5.456244
Bear markets	c	−0.000944*	0.0010687***	0.000195*	0.0004205*	0.000728*	0.000607**
		(0.000752)	(0.000266)	(0.000153)	(0.000271)	(0.000321)	(0.000220)
	Ar1	0.101937**	−0.07296**	0.033046*	0.14058***	−0.03980**	0.048731*
		(0.04127)	(0.024948)	(0.025009)	(0.024759)	(0.024999)	(0.024978)
	ω	0.0000048*	0.0000014***	0.000001*	0.000007*	0.0000012*	0.000006**
		(0.000021)	(0.0000004)	(0.000002)	(0.000003)	(0.0000005)	(0.000002)
	α	0.09219***	0.068870***	0.08092***	0.1239***	0.072290***	0.06819***
		(0.01358)	(0.010420)	(0.01318)	(0.03136)	(0.01205)	(0.009464)
	β	0.90850***	0.91790***	0.89540***	0.8134***	0.9229***	0.92390***
		(0.01189)	(0.010740)	(0.01575)	(0.05605)	(0.01238)	(0.00925)
	BIC	−6.68420	−6.53134	−7.701825	−6.421015	−6.171113	−6.983395

Source: Authors estimation; ***, ** and * indicate statistical significance at the 0.01, 0.05, and 0.1 level, respectively. The number in the parentheses is the standard deviation obtained from Maximum entropy bootstrapping.

3.3 The Results of Co-movement Structure of D-Vine Trees in Bull and Bear Markets

The estimated residuals from AR-GARCH were transformed into cumulative distributions uniform (cdf) [0,1] in order to graphically estimate D-vine trees. Technically, D-vine trees based on Gaussian copula were employed to clarify the 6 stock markets. Considering Fig. 1, the D-vine trees copulas are five dependence structure in bull periods which explain the relation among ASEAN stock markets. According to the empirical Kendall's tau, the sequence for the D-vine copula is SET (Thailand), PSE (Philippines), IDX (Indonesia), KLCI (Malaysia), STI (Singapore) and HOSE (Vietnam), respectively. Thailand is the first node since it is at the center of most ASEAN countries and also the gateway to Asia. Obviously, the movement of six stock markets exhibits the strongly structural dependence in terms of capital flows. There are empirically the positive correlations, which are 0.99 and 0.98, respectively. This can imply that when the financial market system in one of the stock markets in ASEAN is stimulated, a cross-country fund will be induced in the other countries. In the bear market, there are dependence structures, and they are very compilcated. As a restult, there are three major stock markets that consist of Thai stock market the Philippines stock market and Indonesia stock market show the correlation between STI index, Singapore stock market and Malaysia stock market, which is the positive relation, reported as 0.32. On the other hand, the relation in Vietnamstock market is the negative relationship among the six stock markets, which shows the correlation as –0.21. The details are graphically illustrated in Fig. 1.

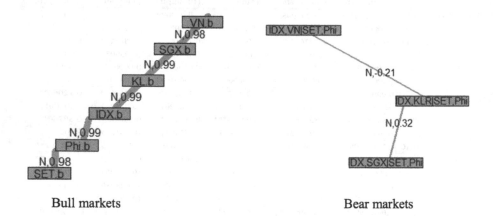

Bull markets   Bear markets

Fig. 1. The dependence structure of D-vine trees in boom and bear periods.

3.4 The Markovian Optimization for Portfolio Selection

The Markovian optimization for portfolio selection is employed to find the appropriate choice for investing in ASEAN stock market. Table 3 shows the computation of optimal portfolio in six stock markets in ASEAN. There are two conditions that are the minimum risk value (variance) and the maximum return (mean). Empirically, the Sharpe ratio is the estimator to select the optimal portfolio as provided in Table 3. In the bull period, the value of Sharpe ratio in the minimum risk value is more than the maximum return which is 0.0835 and 0.0351, respectively. Conversely, the Sharp ratio of maximum return which equals 0.0764, is the highest value in the bear periods. Accordingly, in order to get the high-return portfolio, the minimum variance is suitable plan to make decision in the bull periods. On the other hand, the bear period should select the maximum return choice.

Table 3. The mean-variance portfolio optimization.

		The optimal portfolio (Minimizing variances)	The optimal portfolio (Maximizing, means)
Bull periods	The average return (μ)	0.0005	0.0002
	The variance value (σ)	0.0069	0.0083
	The Sharp ratio (μ/σ)	0.0835	0.0351
Bear periods	The average return (μ)	0.0003	0.0006
	The variance value (σ)	0.0085	0.0085
	The Sharp ratio (μ/σ)	0.0418	0.0764

Considering the proportion of investment from Markowitz portfolio selection, Table 4 shows the proportion of investment from Markowitz Portfolio Selection, relying on the decision making from the Sharpe ratio in Table 3. As a results, in the bull periods, the optimal portfolio of minimizing risk value represents 100% for investing in the STI index (Singapore) which implied that Singapore contains the lowest risk when ASEAN stock market is in the boom period. In contrast, the target to maximizing return was selected to invest in the bear period. The results show the diversification investment is better in this period which should invest in IDX index (Indonesia stock market), KLCI index (Malaysia stock market), PSE index (Philippine stock market) and HOSE index (Vietnam stock market) that is 20%, 10%, 23% and 47%, respectively.

Table 4. The proportion of investment from Markowitz Portfolio Selection.

ASEAN stock market	The proportion investment in ASEAN stock market	
	Bull markets	Bear markets
	Minimizing variances (%)	Maximizing means(%)
SET (Thailand)	0.00	0.00
IDX (Indonesia)	0.00	20.00
KLCI (Malaysia)	0.00	10.00
PSE (Philippine)	0.00	23.00
HOSE (Vietnam)	0.00	47.00
STI (Singapore)	100.00	0.00

4   Conclusion

In this study, ASEAN stock markets have been inevitably the crucial emerging market in the world since they are continuously growing in terms of the investment during 2008–2017. The Markov switching is proposed to clarify bull markets and bear markets which are presented as 649 and 1,600 periods, respectively. The optimal model used to estimate is the marginal assumption (AR-GARCH) estimating by the maximum entropy bootstrapping and the estimation of residuals for the financial markets are successfully clarified. Since the maximum entropy bootstrapping is a powerful tool for avoiding all unnecessary distributional assumptions. The main estimated results can be summarized all series, which show significances for both autoregressive (AR) and generalized autoregressive conditional heteroscedasticity (GARCH) components.

Furthermore, the results of the D-vine trees show there are five dependence structures which explain the relation among ASEAN stock markets. In the bull markets, the co-movement of 6 stock markets was the strongly structural dependence in terms of capital flows, which are the positive correlations. Conversely, in the bear market, there are the dependence structures that are very complicated. There are three major stock markets consisted of SET index (Thailand), PSE index (Philippine) and IDX index (Indonesia) show the correlation between STI index (Singapore), and KLCI (Malaysia). The correlation is positive. In contrast, the relation in HOSE index (Vietnam) is the negative relationship among the six stock markets. The Markowitz portfolio selection model is employed to investigate the mean and variance of portfolio in order to find the best alternative for investing in the ASEAN stock markets. To express the results of the optimal portfolio by using the Sharpe ratio, the best choice to invest in the bull period is the minimizing risk value (Variance) while the maximum return (mean) is more efficient to invest during the recession period or bear market. Additionally, the computational proportion of investments from Markowitz Portfolio Selection related to Table 3. This depicts that Singapore stock market contains the lowest risk for investing in the bull markets. In contrast, the diversification investment is better to get the lower risk when ASEAN stock market in the recession period, which should divide the investment to stock markets in Indonesia, Malaysia, the Philippines and Vietnam.

References

1. Brechmann, E., Schepsmeier, U.: Modeling Dependence with C- and D-Vine Copulas. the R Package CDVine. J. Stat. Softw. **52**(3), 1–27 (2013)
2. Chaitip, P., Chaiboonsri, C.: AEC demand for ICT. maximum entropy bootstrap approach in panel data. Procedia Econ. Finan. **5**, 125–132 (2013)
3. Chaiboonsri, C., Chaitip, P.: A boundary analysis of ICT firms on Thailand stock market. a maximum entropy bootstrap approach and Highest Density Regions (HDR) approach. Int. J. Comput. Econ. Econometrics **3**, 14–26 (2013)

4. Chokethaworn, K., Chiatip, P., Sriwichailamphan, T., Chaiboonsri, C.: The dependence structure and co-movement toward between Thai's currency and Malaysian's currency. markov switching model in dynamic copula approach. Procedia Econ. Finan. **5**, 152–161 (2013)
5. Fink, H., Klimova, Y., Czado, C., Stober, J.: Regime switching vine copula models for global equity and volatility indices. Econometrics **5**, 3 (2017)
6. Hamilton, J.: Time Series Analysis. Princeton University Press, New Jersey (1994)
7. Jian, Y., Nie, H., Yohance Monginsidi, J.: Co-movement of ASEAN stock markets: new evidence from wavelet and VMD-based copula tests. J. Econ. Model. **64**, 384–398 (2017)
8. Shabri, M., Majid, A., Mohd, O., Hassanuddeen, A.: Dynamic linkages among ASEAN 5 emerging stock markets. Int. J. Emerg. Markets **4**(2), 160–184 (2009)
9. Nikoloulopoulos, A., Joe, H.: Vine copulas with asymmetric tail dependence and applications to financial return data. Comput. Stat. Data Anal. **56**(11), 3659–3673 (2012)
10. Click, R., Plummer, M.: Stock market integration in ASEAN after the Asian financial crisis. J. Asian Econ. **16**, 5–28 (2004)
11. Shore, J., Johnson, R.W.: Axiomatic Derivation of the principle of maximum entropy and the principle of minimum cross-entropy. IEEE Trans. Inf. Theor. **26**(1), 26–37 (1980)
12. Sriboonchitta, S., Liu, J., Kreinovich, V., Nguyen T.H.: Modeling Dependence in Econometrics, vol. 7, pp. 245–257 (2016)
13. Vinod, H.D.: Maximum entropy bootstrap for time series. J. Stat. Softw. **29**(5), 1–18 (2009)

Hedging Benefit of Safe-Haven Gold in Terms of Co-skewness and Covariance in Stock Market

Sukrit Thongkairat[1,2], Woraphon Yamaka[2,3(✉)], and Songsak Sriboonchitta[2,3]

[1] International College of Digital Innovation, Chiang Mai University,
Chiang Mai 50200, Thailand
[2] Center of Excellence in Econometrics, Chiang Mai University,
Chiang Mai 50200, Thailand
woraphon.econ@gmail.com
[3] Faculty of Economics, Chiang Mai University, Chiang Mai 50200, Thailand

Abstract. This study revisits the question of whether or not gold offers a hedging benefit for stock returns. Thus, we examine this benefit in terms of conditional co-skewness in which relate to the selected stock markets, conditional beta risk, and correlation. We use two-step approach to assess the impact of these factors, including Shanghai, GDAXI, FTSE100, S&P500, and Nikkei225 stock index returns. We firstly estimate the Markov-Switching Dynamic Conditional Correlation GARCH (MS-DCC-GARCH) to obtain the correlation, volatility, and covariance which we further use in co-skewness and beta risk computation. In the second step, the linear regression is employed to investigate the effect of the co-skewness and beta risk on stock returns and examine hedging benefit of gold on stocks. We find some evidences that gold can be acted as a safe haven asset for some major stock markets.

Keywords: MS-DCC-GARCH · Co-skewness · Hedging benefit · Risk

1 Introduction

This paper was inspired by the perspective of Chan, Yang and Zhou [5]. They studied the demand for safe-haven currencies after the global financial crisis in 2008 and examined hedging benefit of safe-haven currencies which refer to MSCI world stock index. However, in this study, we examine the hedging benefit of gold on global stock market. Gold is a popular market and it has attracted investors for many years. Gold has played an important role as a safe haven investment during times of turbulent economic period and equity crashes [17]. Pastpipatkul, Yamaka, and Sriboonchitta [17] examined the relationship between volatility of gold and stock returns and also tested whether or not gold serves as a hedge in the stock markets. They found that gold plays a vital role as it can lower the risk of the stock portfolios.

© Springer Nature Switzerland AG 2019
H. Seki et al. (Eds.): IUKM 2019, LNAI 11471, pp. 172–183, 2019.
https://doi.org/10.1007/978-3-030-14815-7_15

In the traditional hedging analysis, many studies employed correlation or covariance between two assets to construct the hedging portfolio weight or an optimal hedge ratio. Baur and Lucey [1] tried to examine whether gold acts as a hedge or a safe haven for investors by using a constant and time-varying correlations between U.S., U.K. and German stock returns and bond return and gold return. They found that gold was a good hedge against investors in turmoil market conditions. Coudert and Raymond-Feingold [8] revealed a similar result when they analyzed the relationship for some developed countries. The recent literature has provided supportive empirical evidence that gold is a safe haven against stocks (see e.g. [4, 7, 11, 14, 15, 18]).

However, the hedging benefit cannot be completely explained by the correlation alone. Many evidences illustrate that the normal distribution of returns is too restrictive an assumption in the real practice (see, [9, 13, 16]). They suggested that many financial asset returns exhibit a high kurtosis and skewness. Positive skewness means that a distribution has a long right tail, and hence it is more likely for extremely high returns to occur. That is, investors are facing an investment opportunity in which they have greater chance to capture upside potentials than to be trapped in a deep loss in return. This preference for a right-skewed portfolio is generally considered an inherent result of risk aversion [19]. On the other hand, the negative skewness refers to extremely loss return. That is, investors are facing a higher chance of loss than gain. So, in the portfolio context, the presence of co-skewness of the asset returns in the portfolio is possible whereas high and low co-skewness present a different return of the portfolio. Portfolio with higher co-skewness should have more desirable hedging value for the overall portfolio and thereby lowering expected returns and risks. In contrast, lower co-skewness indicates a higher expected returns and risks. Therefore, the co-skewness should be considered in the hedging strategy analysis.

As we mentioned above, there might exhibit a co-skewness in the portfolio and it will contribute some effect to the expected return. Unlike the conventional one, in this study, we take the co-skewness (measured by the covariance between gold returns and global equity market volatility) into account and follow procedures of, Chan, Yang, and Zhou [5] to examine whether gold acts as a safe haven. We investigate the effect of co-skewness on Shanghai Stock Exchange Composite Index (Shanghai), the 30 major German companies trading (GDAXI), Financial Times-Stock Exchange 100 Index (FTSE100), the Standard & Poor's 500 from American stock market index (S&P500), and the Tokyo Stock Exchange Nikkei 225 (Nikkei225). This is due to the fact that these markets are the most important and largest stock markets in the world. In the view of investors, this will generate a potential benefit to the investors who invest their money in these five stock markets and intend to diversify their portfolio's risk.

From the estimation point of views, we differ from Chan, Yang, and Zhou [5] by incorporating dynamic correlation for the time varying correlation between gold and stock returns. Thus, this study uses the bivariate Markov Switching-Dynamic Correlation GARCH model of Chodchuangnirun et al. [6] to quantify the conditional co-skewness, the conditional correlation, and the conditional variance of gold and stock pair. Then, the obtained conditional co-skewness, conditional correlation, and the conditional variance are further used to investigate their effects on the stock returns.

The remainder of this paper is divided into four sections. Section 2 briefly explains the methodology for use in this paper. Section 3 presents descriptive statistics of the data. Section 4 provides and discusses the empirical results from the regime-switching model and regression model. The concluding remarks and suggestions for future works are presented in Sect. 5.

2   Methodology

To examine the bivariate regime-switching model for gold and stock returns of each market, we firstly explain the Markov Switching-Dynamic Correlation GARCH model and the derivation of conditional co-skewness, the conditional correlation, and the conditional variance. Then, we examine the effect of estimated co-skewness series by linear regression context.

2.1   Markov Switching Dynamic Conditional Correlation GARCH

The regime-switching model is the nonlinearity modeling in time series to assume different behavior. In this study, we employ bivariate Markov Switching Dynamic Conditional Correlation (MS-DCC) GARCH to analyze dynamic correlation between assets, following Chodchuangnirun et al. [6]. This model has extended MS-DCC-GARCH of Billio and Caporin [2] by allowing the parameters in mean, variance and correlation equations to switch across different regimes or to be state dependent according to the first order Markov process. Let $r_t = (r_t^G, r_t^S)'$ be the vector of gold return (r_t^G) and stock return (r_t^S), the model can be written as

$$
\begin{aligned}
r_t &= \mu(s_t) + \varepsilon_{i,t} \\
\varepsilon_t(s_t) &= e_t(s_t)\sqrt{H_t}(s_t)
\end{aligned}
\tag{1}
$$

where $\mu(s_t)$ is 2×1 vector of regime dependent intercept parameter of gold and stock mean equations. $\varepsilon_t(s_t)$ is 2×1 vector of regime dependent error which is assumed to follow an i.i.d. bivariate normal distribution. $H_t(s_t)$ is the regime dependent variance-covariance matrix, which is defined as follows

$$
H_t(s_t) = D_t(s_t)R_t(s_t)D_t(s_t), \; D_t(s_t) = \begin{pmatrix} \sqrt{h_t^G(s_t)} & 0 \\ 0 & \sqrt{h_t^S(s_t)} \end{pmatrix},
$$
$$
R_t(s_t) = \begin{pmatrix} 1 & \rho_t(s_t) \\ \rho_t(s_t) & 1 \end{pmatrix},
\tag{2}
$$

where ρ_t is the dynamic conditional correlation (DCC) of gold and stock pair. $R_t(s_t)$ is given by:

$$
R_t(s_t) = Q_t^{*-1/2}(s_t)Q_t(s_t)Q_t^{*-1/2}(s_t),
\tag{3}
$$

then, the regime dependent DCC equation can be specified by

$$Q_t(s_t) = (1 - \theta_1(s_t) - \theta_2)(s_t)\bar{Q}(s_t) + \theta_2(s_t)Q_{t-1}(s_t)$$
$$+ \theta_2(s_t)\varepsilon_{t-1}(s_t)\varepsilon'_{t-1}(s_t), \tag{4}$$

where, $\bar{Q}(s_t) = \frac{1}{T}\sum_{t=1}^{T}\varepsilon_{t-1}(s_t)\varepsilon'_{t-1}(s_t)$ is the regime dependent unconditional covariance matrix of the standardized residuals and $Q_t^*(s_t)$ is a diagonal matrix with the square root of the diagonal elements of Q_t. The coefficients $\theta_1(s_t)$ and $\theta_2(s_t)$ must satisfy: $0 \leq \theta_1(s_t) + \theta_2(s_t) < 1$. $D_t(s_t) = diag(\sqrt{h_t^S(s_t)}, \sqrt{h_t^G(s_t)})$ represents a diagonal matrix of conditional volatility of stock and gold return obtained from the estimation of the GARCH process [3], which is expressed by

$$h_t(s_t) = a_0(s_t) + a_1(s_t)\varepsilon_{t-1}^2 + a_2(s_t)h_{t-1}(s_t), \tag{5}$$

where $a_0(s_t)$ is regime dependent constant term, $a_1(s_t)$ and $a_2(s_t)$ are regime dependent coefficients of error term and volatility term respectively. The restriction $a_1(s_t), a_2(s_t) > 0$ and $a_1(s_t) + a_2(s_t) \leq 1$ are introduced to satisfy the stationary GARCH process. The latent regime is usually parameterized as a first-order Markov chain. Given the information available at time t − 1, the conditional probability of observing the state is given as follows:

$$P(s_t = s_t | F_{t-1}) = P(s_t = s_t | s_{t-1}) \times P(s_{t-1} = s_{t-1} | F_{t-1}) \tag{6}$$

where F_{t-1} contains past information of stock and gold joint distribution, including returns, volatilities and estimated parameters. As described in Hamilton [12], for two regimes, the state transition can be constructed by

$$p_{ij} = P(s_t = j | s_{t-1} = i, F_{t-1}), \quad i, j \in 1, 2, \tag{7}$$

where p_{ij} is the transition probability of state i to state j.

2.2 Generating Co-skewness and Covariance Processes

The important issue in this study is how co-skewness variable series affects stock return. Additionally, we also take into account the effect of risk on stock return. Thus, the linear regression is constructed for this purpose.

$$r_t^S = \alpha_0 + \alpha_1\beta_t^S + \alpha_2 std_t^S + \alpha_3\beta_{SKD,t}^S + \alpha_4 skew_t^S + \varepsilon_t^S, \tag{8}$$

where r_t^S is the stock excess return, β_t^S is the beta risk, std_t^S is stock idiosyncratic volatility, proxied by the residual from the auxiliary regression of conditional standard deviation orthogonal to β_t^S. $\beta_{SKD,t}^S$, is the residual of stock standardized conditional co-skewness orthogonal to β_t^S and std_t^S. $skew_t^G$ is the estimated stock conditional idiosyncratic skewness, which is the residual of stock conditional skewness orthogonal to β_t^S, std_t^S and cos_t^S. cos_t^S is estimated stock conditional co-skewness with gold return.

Following Chan, Yang, and Zhou [5], we can approximate the conditional moments of the bivariate regime-switching model in general, and the conditional covariance and correlation, conditional beta and co-skewness as well as volatility and skewness as follows

$$\beta_t^S = \text{cov}_t / E(h_t^S), \tag{9}$$

where $E(h_t^S) = p_{1t}h_t^S + p_{2t}h_t^S$ is the expected volatility, p_{1t} and p_{2t} are filtered probabilities of regime 1 and 2, respectively. cov_t is the estimated conditional covariance between stock and gold returns and it is given by

$$\text{cov}_t = p_{1t}\text{cov}_1 + p_{2t}\text{cov}_2 + p_{1t}p_{2t}[(\mu^S(s_t = 1) - \mu^S(s_t = 2))(\mu^G(s_t = 1) - \mu^G(s_t = 2))], \tag{10}$$

where $\text{cov}_i = \rho(s_t = i)\sqrt{h_i^S(s_t = i)h_i^G(s_t = i)}$ is the covariance in regime i, $i = 1, 2$ and conditional correlation is computed from Eq. (2). The co-skewness can be calculated by

$$\begin{aligned}
\cos_t^S &= p_{1t}p_{2t}[(p_{2t} - p_{1t})(\mu^S(s_t = 1) - \mu^S(s_t = 2))^2(\mu^G(s_t = 1) - \mu^G(s_t = 2)) \\
&+ (\mu^G(s_t = 1) - \mu^G(s_t = 2))(h_t^S(s_t = 1) - h_t^S(s_t = 2)) \\
&+ 2(\mu^S(s_t = 1) - \mu^S(s_t = 2))(\text{cov}_1 - \text{cov}_2)]
\end{aligned} \tag{11}$$

The standardized conditional co-skewness of stock with the gold spot price is $\beta_{SKD,t}^S$ which can be calculated by

$$\beta_{SKD,t}^S = A/B$$

$$\begin{aligned}
A &= p_{1t}p_{2t}[(p_{2t} - p_{1t})(\mu^S(s_t = 1) - \mu^S(s_t = 2))^2(\mu^G(s_t = 1) - \mu^G(s_t = 2)) + \\
&(\mu^G(s_t = 1) - \mu_{2,t}^G(s_t = 2))(h^S(s_t = 1) + \\
&h^S(s_t = 2)) + 2(\mu^G(s_t = 1) - \mu^G(s_t = 2))(\text{cov}_1 - \text{cov}_2)] \\
B &= p_{1t}h^S(s_t = 1) + p_{2t}h^S(s_t = 2) + \\
&\frac{p_{1t}p_{2t}(\mu^S(s_t = 1) - \mu^S(s_t = 2))^2}{\sqrt{p_{1t}h^G(s_t = 1) + p_{2t}h_2^G(s_t = 2) + p_{1t}p_{2t}(\mu^G(s_t = 1) - \mu^G(s_t = 2))^2}}
\end{aligned} \tag{12}$$

Then the stock skewness is derived by

$$skew_t^S = \frac{p_{1t}p_{2t}(\mu^S(s_t = 1) - \mu^S(s_t = 2)) \cdot C}{[E(h_t^S) + p_{1t}p_{2t}(\mu^S(s_t = 1) - \mu^S(s_t = 2))^2]^{3/2}}, \tag{13}$$

where $C = [(p_{2t} - p_{1t})(\mu^S(s_t = 1) - \mu^S(s_t = 2))^2 + 3(h^S(s_t = 1) - h^S(s_t = 2))]$. Finally, the standard deviation (std_t^S) is given by

$$std_t^S = \sqrt{p_{1t}h_1^S(s_t = 1) + p_{2t}h_2^S(s_t = 2) + p_{1t}p_{2t}(\mu^S(s_t = 1) - \mu^S(s_t = 2))^2}. \tag{14}$$

3   Data Description

This study conducts a monthly time series (Jan, 2008 to Aug, 2018) of gold spot price and stock index in five countries (Shanghai, GDAXI, FTSE100, S& 500, Nikkei225). This data set is retrieved from Thomson Reuters database and then transformed into a log return. Table 1 presents the descriptive statistics of the log return data series. In the sample, the mean value of all data series is very close to zero. However, the FTSE index has less variation than the others. All log return data series show non-normality with negative skewness and high kurtosis values (>3). In addition, the Jarque-Bera (J-B) test is considered to investigate the normal distribution property.

With the ban of p-value in 2016, in this study, the statistical inference is based on the Minimum Bayes factor (MBF). Following Goodman [10] for H1, a MBF10 between 1 and 1/3 is considered weak evidence, 1/3–1/10 moderate evidence, 1/10–1/30 substantial evidence, 1/30–1/100 strong evidence, 1/100–1/300 very strong evidence, and <1/300 decisive evidence. As a result, MBF values of Jarque-Bera test are close to zero. This indicates that there is decisive evidence for normal distribution. Also, the Augmented Dickey Fuller (ADF) test is conducted to show the stationarity of the data series. The results show the substantial and decisive evidence for stationarity as presented in Table 1.

Table 1. Descriptive statistics

	Gold	Shanghai	GDAXI	FTSE100	S&P500	Nikkei225
Mean	0.0021	−0.0038	0.0047	0.0020	0.0060	0.0042
Median	0.0004	0.0045	0.0130	0.0045	0.0102	0.0127
Maximum	0.1278	0.1655	0.1786	0.1202	0.1274	0.1209
Minimum	−0.1762	−0.2663	−0.2378	−0.1395	−0.2009	−0.2207
Std. Dev.	0.0529	0.0761	0.0568	0.0421	0.0444	0.0569
Skewness	−0.3158	−0.8051	−0.6554	−0.3884	−0.9766	−0.6956
Kurtosis	3.5520	4.7436	5.8022	4.2179	6.3264	4.1599
Jarque-Bera	3.7229	29.8076	50.6427	11.0422	78.7406	17.3620
MBF	0.0062	0.0000	0.0000	0.0000	0.0000	0.0000
ADF-test	−13.5284[a]	−9.7921[a]	−9.7135[a]	−11.4297[a]	−9.7680[a]	−9.5191[a]

Note: "a"   denotes   the   interpretation   of   MBF   as   Substantial evidence   for   stationarity,   MBF   is   calculated   by $MBF_{01}(p)$ $=$
$\begin{cases} -\exp(1)p\log p & \text{for } p < 1/\exp(1) \\ \quad 1 & \text{for } p \geq 1/\exp(1) \end{cases}$ [20, 21].

4   Empirical Summarization

4.1   Results on Regime-Switching Model Estimation

The estimation proceeds with the two-regime-switching DCC-GARCH model. The estimated parameters of each stock and gold pairs are provided in Table 2. In this table, the parameter estimates for the conditional variance make divide the higher volatility regime as a bear market for the gold spot price and stock pairs, while the lower volatility regime as a bull market. Table 1 presents the parameter estimation results of MS-DCC models. The maximization the log-likelihood function with respect to the unknown parameters is used to find the parameter estimates. We can observe that the conditional means $\mu^S(s_t)$ show both negative and positive values, indicating the presence of positive stock returns of these five stock markets, except for Shanghai stock return. This is consistent with unconditional average returns of raw data in Table 1. In the volatility equation of stock returns, we find that Nikkei225 and FTSE 100 stock index exhibit the largest volatility persistence $(a_1^S + a_2^S)$ that is equal to 0.7906 and 0.7292 for first regime and second regime, respectively. While, Shanghai stock index has the smallest $(a_1^S + a_2^S)$ that is equal to 0.2206 and 0.6037 for first regime and second regime, respectively. In the correlation equation, $\theta_1(s_t)$ and $\theta_2(s_t)$ reflect the memory in correlation in state-independent DCC for gold and stock returns. For example, in the case of gold and GDAXI, $\theta_1(s_t)$ is equal to 0.0601 and 0.1465 for the first regime and the second regime, respectively of gold and $\theta_2(s_t)$ is equal to 0.7845 and 0.7641 for the first regime and the second

Table 2. MS-DCC-GARCH model estimation for the gold and stock pairs.

	Shanghai		GDAXI		FTSE100		S&P500		Nikkei225	
	R1	R2	R1	R2	R1	R2	R1	R2	R1	R2
Volatility equation										
a_0^G	5.7533	0.9727	0.7621	0.0416	0.1119	0.0002	0.8224	0.0347	1.2126	0.0002
a_1^G	0.244	0.157	0.2572	0.2079	0.1804	0.1429	0.2138	0.1714	0.1916	0.1438
a_2^G	0.3804	0.0635	0.5203	0.2171	0.5605	0.5863	0.4762	0.3664	0.5986	0.5751
a_0^S	1.6424	0.6706	0.0114	0.1203	0.6466	0	1.6972	0.0202	0.1477	0.0003
a_1^S	0.241	0.195	0.2452	0.168	0.0786	0.0548	0.2971	0.2318	0.0984	0.0605
a_2^S	0.6307	0.4087	0.6241	0.131	0.2975	0.7251	0.6698	0.5341	0.644	0.655
Correlation equation										
$\theta_1(s_t)$	0.0297	0.0001	0.0601	0.1465	0	0.1	0.1038	0.191	0.0099	0.1
$\theta_2(s_t)$	0.773	0.7322	0.7845	0.7641	0.783	0.75	0.8056	0.806	0.8793	0.75
Mean equation										
μ^G	0.4303	0.3588	0.1884	0.0097	0.1827	0.1	0.0544	0.0013	0.147	0.1
μ^S	−0.5543	−0.3995	0.1266	0.0389	0.2162	0.05	0.1242	0.0044	0.1168	0.0500.

Note: R denotes regime.

Table 3. Transition probability matrix of regime-switching model.

	Shanghai		GDAXI		FTSE100		S&P500		Nikkei225	
	R1	R2	R1	R2	R1	R2	R1	R2	R1	R2
p_1	0.9288	0.0712	0.9188	0.0812	0.9999	0.0001	0.8792	0.1208	0.9999	0.0001
p_2	0.0001	0.9999	0.0001	0.9999	0.0514	0.9486	0.0086	0.9914	0.0526	0.9474

Note: R denotes regime.

regime respectively of stock. These results indicate a high correlation persistence of these two returns. We compare the correlation persistence of these five pairs and find that there are highest correlation persistence for S&P500 and gold as the summation of $\theta_1(s_t)$ and $\theta_2(s_t)$ are, respectively, 0.9094 and 0.9970.

Table 3 provides the probability of transition regime. The estimated probability means the probability based on the information available throughout the whole sample period at future date t. For example, for the gold and S&P500 pair, the probability for switching from Regime 1 to Regime 2 is 0. 1208, while remaining in Regime 1 is 0.8792. On the other hand, the probability of switching from Regime 2 to Regime 1 is 0.0086, while remaining in Regime 2 is 0. 9914. This suggests a higher probability for staying in regime 2 when compared to regime 1. In the case of gold and GDAXI pair, we can observe that the result of the probability for switching from Regime 1 to Regime 2 is 0.0812, while remaining in Regime 1 is 0.9188. On the other hand, the probability of switching from Regime 2 to Regime 1 is 0.0001, while remaining in Regime 2 is 0.9999. This suggests a higher probability for staying in regime 2 when compared to regime 1. For other pairs, the similar interpretation is obtained. We can conclude that gold and stocks pairs are likely to remain in their own regime in high probability and only the extreme event can switch the series to change from one to another regime.

Table 4. Average of conditional moment estimates derived from the gold - stock pairs MS-DCC-GARCH

	Shanghai	GDAXI	FTSE100	S&P500	Nikkei225
Conditional beta	0.9426	−0.131	0.7327	0.486	5.4001
Conditional standard deviation	2.078	3.9385	0.0025	2.2209	−0.0161
Conditional skewness	13.1427	4.2788	22.2289	2.1456	24.5199
Conditional correlation	0.2522	−0.0052	3.3299	0.0221	2.1519
Conditional covariance	1.0582	−0.0167	0.0004	0.0018	0.0003
Conditional standardized co-skewness	−0.0001	0.0035	0.0065	−0.0121	−0.0091
Conditional co-skewness	2.911	0.4298	0.0374	0.7837	0.1083

Note: Conditional moment estimates are derived as in Eqs. (9)–(14).

4.2   The Effects of Co-skewness and Correlation on Stock Returns

The summary statistics of conditional beta risk, correlation, skewness and co-skewness and standardized co-skewness are presented in Table 4. The results show that all stock index returns have a positive conditional beta and correlation; except for GDAXI that has the negative beta risk and correlation, and negative standardized conditional co-skewness with the gold return on average. This results indicate that gold could be a safe asset only for GDAXI and prevent the risk of a decline in GDAXI market. In other words, gold is not suitable as a hedge in the Shanghai, FTSE100, S&P500 and Nikkei225 stock markets. Consider the standardized conditional co-skewness of various stocks with the gold market, the Shanghai, S&P500, and Nikkei225 co-skewness are negative, indicating the decrease of excess returns for these three markets when the world stock and gold markets which both of them are continuing to be more volatile.

Table 5. Regression results.

Stock index	Intercept	Conditional beta	Idiosyncratic volatility	Std co-skewness	Skewness
Shanghai	13.5569	−3.9693	−8.6932	8.5389	−1.1402
MBF	(0.1427)	(0.1501)	(0.1649)	(0.2638)	(0.1676)
GDAXI	0.6399	4.9781	−1.4744	−4.4758	−0.1841
MBF	(1.0000)	(0.8264)	(1.0000)	(0.8990)	(1.0000)
FTSE100	0.06775	−0.01896	−0.07708	−0.04165	0.01366
MBF	(1.0000)	(1.0000)	(1.0000)	(1.0000)	(1.0000)
S&P500	0.2643	−0.1497	−0.1747	0.0912	−0.2776
MBF	(0.9466)	(0.0131)	(0.6669)	(0.9361)	(0.9941)
Nikkei225	−0.3534	0.1772	−0.5155	−0.2491	−0.06
MBF	(0.9951)	(0.7514)	(0.7977)	(0.8233)	(0.5876)

Note: () denotes MBF, I denotes Idiosyncratic and Std denotes standardized.

We then investigate the effect of the risks and co-skewness on the stock price returns using the regression model. The results are reported in Table 5. According to the results, we find a strong evidence for the effect of conditional beta, idiosyncratic volatility, standardized co-skewness and skewness on Shanghai stock returns as the MBF values lie in 1/3–1/10 interval. All factors contribute a negative effect to Shanghai stock returns, except for standardized co-skewness. In contrast, there are many weak evidences supporting the effect of risks and co-skewness on other stock market returns.

Firstly, if we are to interpret the results of the coefficients; the conditional beta of Shanghai, FTSE100, and S&P500 is negative, indicating a hedging pattern on these three stock returns. As we expect, the higher volatility induces a decrease of five stock returns. We also observe a negative effect of standardized co-skewness on GDAXI, FTSE100, and Nikkei225. One standard deviation increase in the gold co-skewness induces a decrease of three stock returns. While, one standard deviation increase in co-skewness induces an increase of 8.5389 and 0.0912 in Shanghai and S&P500, respectively.

There is an evidence to confirm that gold's co-skewness contributes both negatively and positively price effect in each stock index across the markets. Such a desirable co-skewness property is negatively priced in the stock returns and it leads to lower expected stock returns. Specifically, when global equity volatility increases, investors flock to the gold market of safe-haven asset economies, driving up their prices. Thus, investors are willing to accept a lower expected return for the safe-haven assets because of the good hedging property. In contrast, the co-skewness of other assets is lower, and their averages are negative. Such an undesirable co-skewness property is priced in asset returns so that expected returns on other asset are higher. The higher expected return is needed to compensate for the co-skewness risk, as investors will move their money from stock markets which will drop during the turmoil market. For the positive co-skewness, investors can view gold as another speculative asset and there is bad hedging property here.

5   Conclusion

In this study, we aim to investigate whether gold acts as a safe haven asset for the stock markets. We firstly use a bivariate Markov Switching dynamic conditional correlation (MS-DCC) GARCH model to construct the conditional correlation and volatility of stock and gold pair. In this study, five stock markets are considered, that are Shanghai Stock Exchange Composite Index (Shanghai), the 30 major German companies trading (GDAXI), Financial Times-Stock Exchange 100 Index (FTSE100), The Standard & Poor's 500 (S&P500), and Tokyo Stock Exchange (Nikkei225). It is common in the literature to focus on co-skewness rather than skewness or kurtosis, to reflect the argument that investors are not compensated for diversifiable skewness in equilibrium. We thus compute the conditional skewness, co-skewness, standardized co-skewness and consider two more traditional factors, that are beta risk and correlation. To achieve our purpose, MS-DCC-GARCH model is employed to obtain the parameter estimates and use the obtained parameters to compute those factors further. These obtained variables can be viewed as the factor affecting the return of our selected stock markets. In the last step, we investigate the effect of these factors using the linear regression model. This study allows us to examine the hedging benefit of gold in terms of co-skewness with the stock market (that is, the covariance between gold return and stock market volatility).

Overall, we find some evidence that gold show effect of hedging benefit on five stock indexes. We find that gold can be acted as a safe-haven asset for Shanghai and FTSE100 stock indexes, both exhibit negative effect of conditional beta risk. Furthermore, we consider the effect of the co-skewness that provide the information on extreme volatility event, there is an evidence help confirm that co-skewness characteristic of gold contributes both negative and positive price effects in each stock index across the markets.

In this study, we made up of only gold spot price and five major stock indexes, that might have some argument that this study doesn't cover on the emerging stocks. It would be interesting to extend the analysis of characteristic and compare the result of developed and emerging stocks. Emerging stock is a popular target of carrying trades as there is room to gain a higher gain. How does gold play a role in hedging emerging stock? It is also interesting to explore changes in hedging with other types of assets. For investors in five stock markets, the optimal hedging analysis with co-skewness on various assets in portfolio investment is a crucial issue that can help investors understand more about the characteristic on higher moment preference, which could help the investors put their money in the right way and the right time.

Acknowledgement. The authors are grateful to Centre of Excellence in Econometrics, Faculty of Economics, Chiang Mai University for the financial support.

References

1. Baur, D.G., Lucey, B.M.: Is gold a hedge or a safe haven? An analysis of stocks, bonds and gold. Financ. Rev. **45**(2), 217–229 (2010)
2. Billio, M., Caporin, M.: Multivariate Markov switching dynamic conditional correlation GARCH representations for contagion analysis. Stat. Methods Appl. **14**(2), 145–161 (2005)
3. Bollerslev, T.: Generalized autoregressive conditional heteroskedasticity. J. Econometrics **31**(3), 307–327 (1986)
4. Capie, F., Mills, T.C., Wood, G.: Gold as a hedge against the dollar. J. Int. Financ. Markets Inst. Money **15**(4), 343–352 (2005)
5. Chan, K., Yang, J., Zhou, Y.: Conditional co-skewness and safe-haven currencies: a regime switching approach. J. Empirical Finance **48**, 58–80 (2018)
6. Chodchuangnirun, B., Yamaka, W., Khiewngamdee, C.: A regime switching for dynamic conditional correlation and GARCH: application to agricultural commodity prices and market risks. In: Huynh, V.-N., Inuiguchi, M., Tran, D.H., Denoeux, T. (eds.) IUKM 2018. LNCS (LNAI), vol. 10758, pp. 289–301. Springer, Cham (2018). https://doi.org/10.1007/978-3-319-75429-1_24
7. Ciner, C.: Commodity prices and inflation: testing in the frequency domain. Res. Int. Bus. Finance **25**(3), 229–237 (2011)
8. Coudert, V., Raymond, H.: Gold and financial assets: are there any safe havens in bear markets. Econ. Bull. **31**(2), 1613–1622 (2011)
9. Dittmar, R.F.: Nonlinear pricing kernels, kurtosis preference, and evidence from the cross section of equity returns. J. Finance **57**(1), 369–403 (2002)
10. Goodman, S.N.: Toward evidence-based medical statistics. 2: the Bayes factor. Ann. Intern. Med. **130**(12), 1005–1013 (1999)
11. Gurgun, G., Unalmis, I.: Is gold a safe haven against equity market investment in emerging and developing countries? Finance Res. Lett. **11**(4), 341–348 (2014)
12. Hamilton, J.D.: Time Series Analysis, vol. 2, pp. 690–696. Princeton University Press, Princeton (1994)
13. Harvey, C.R., Siddique, A.: Conditional skewness in asset pricing tests. J. Finance **55**(3), 1263–1295 (2000)

14. Joy, M.: Gold and the US dollar: Hedge or haven? Finance Res. Lett. **8**(3), 120–131 (2011)
15. Kumar, D.: Return and volatility transmission between gold and stock sectors: application of portfolio management and hedging effectiveness. IIMB Manage. Rev. **26**(1), 5–16 (2014)
16. Lai, J.Y.: An empirical study of the impact of skewness and kurtosis on hedging decisions. Quant. Finance **12**(12), 1827–1837 (2012)
17. Pastpipatkul, P., Yamaka, W., Sriboonchitta, S.: Analyzing financial risk and co-movement of gold market, and Indonesian, Philippine, and Thailand stock markets: dynamic copula with Markov-switching. In: Huynh, V.-N., Kreinovich, V., Sriboonchitta, S. (eds.) Causal Inference in Econometrics. SCI, vol. 622, pp. 565–586. Springer, Cham (2016). https://doi.org/10.1007/978-3-319-27284-9_37
18. Reboredo, J.C.: Is gold a safe haven or a hedge for the US dollar? Implications for risk management. J. Bank. Finance **37**(8), 2665–2676 (2013)
19. Scott, R.C., Horvath, P.A.: On the direction of preference for moments of higher order than the variance. J. Finance **35**(4), 915–919 (1980)
20. Sellke, T., Bayarri, M.J., Berger, J.O.: Calibration of p values for testing precise null hypotheses. Am. Stat. **55**(1), 62–71 (2001)
21. Vovk, V.G.: A logic of probability, with application to the foundations of statistics. J. Roy. Stat. Soc.: Ser. B (Methodol.), 317–351 (1993)

Markov Switching Beta-skewed-t EGARCH

Woraphon Yamaka, Paravee Maneejuk[✉], and Songsak Sriboonchitta

Center of Excellence in Econometrics, Chiang Mai University,
Chiang Mai 50200, Thailand
mparavee@gmail.com

Abstract. This study extends the work of Harvey and Sucarrat [15] and present Markov regime-switching (MS) Beta-skewed-t-EGARCH (exponential generalized autoregressive conditional heteroscedasticity) model to predict the volatility. To examine the performance of our model, in-sample point forecast precision and AIC and BIC weights are conducted. We study the volatility of five Exchange Traded Fund returns for period from January 2012 to October 2018. Our proposed model is not found to outperform all the other models. However, the dominance of MS-Beta-skewed-t-EGARCH for SPY, VGT, and AGG may support the application of the MS-Beta-skewed-t-EGARCH model for some financial data series.

Keywords: Markov Switching · Beta-skewed-t-EGARCH ·
Volatility forecasting · AIC and BIC weights

1 Introduction

The study on financial market is to a large extent related to volatility as it really serves as an important indicator of the financial returns. In some cases, it poses a challenge of how difficult it might be for the present to predict the future. Nonetheless, volatility connotes risk and affects the values of asset return and might even lead to a more intense and wildly fluctuation. Hence, the understanding of volatility behavior is important for risk management, asset allocation, and option pricing as the information about risk from volatility assessment may offer useful insights to profit maximizing investors regarding the extent and severity of the uncertainty. In the past literature, several theoretical models assumed that volatility in stock market is constant overtime, see Merton [19] and Black and Scholes [3]. However, it is currently well-known that volatility of asset returns is persistent and varies overtime while the constant volatility cannot present and explain the high fluctuation of the asset return over long period of time. Therefore, the conditional variance was introduced and firstly quantified by Autoregressive Conditional Heteroskedastic (ARCH) in Engle [11] and later by Generalized Autoregressive Conditional Heteroskedastic (GARCH) in Bollerslev [7]. The later extension has been intensively adopted for estimating

© Springer Nature Switzerland AG 2019
H. Seki et al. (Eds.): IUKM 2019, LNAI 11471, pp. 184–196, 2019.
https://doi.org/10.1007/978-3-030-14815-7_16

the conditional volatility of financial returns as it can capture complex patterns like heavy tails and volatility clustering.

Nowadays, GARCH model has been extended into different specifications such as Exponential GARCH ([20]), GJR-GARCH ([12]), APARCH ([10]). FIGARCH ([1]), HYGARCH ([9]), etc. The survey of various GARCH specifications is referred to Bauwens, Laurent and Rombouts [2]. Those models are usually used for both volatility modeling and forecasts of financial time series; however, from the search for the best volatility specification still found no decisive evidence on what type of specifications is the best volatility model. Wei, Wang, and Huang [25] mentioned that the model comparison depends on not only the given data sample but also the corresponding forecasting purpose with the selected loss function. Thus, it is not easy to find the best fit volatility model. Recently, the new GARCH specification, namely Beta-t-EGARCH (1,1) is introduced by Harvey and Chakravarty [14]. Many studies have investigated the performance of this model and confirmed the superiority of this model over the others (see, [5,22]). This model has been extended by Harvey and Sucarrat [15] for the skewed case and it is consequently referred to as Beta-skewed-t-EGARCH model.

Harvey and Sucarrat [15] highlights a number of advantages of the model as follows: First, the model is robust to the outlier and performs well on the financial data. Second, the model accounts for prominent characteristics of leverage, fat-tailedness and skewness. Third, the positive parameter assumption of ARCH, GARCH and leverage are not hold in this model and fourth, the unconditional moments of return exist in long term forecasting. Although this model has many advantages, the nonlinear behavior or structural break of financial variable are not considered. Many studies have argued that nonlinear GARCH model is better capturing this nonlinear behavior than the linear ones. In particular, Markov Switching (MS) model has been shown to properly capture the dynamics of several financial assets data series which occasionally exhibit periodical breaks in their behavior associated with events such as financial crises or abrupt changes in some severe events or shocks (see, [4,6,8,18,26]). More specifically, Markov Switching model for asset returns is motivated by the observation that in financial markets, low- and high-volatility periods with random duration follow each other.

Based on the aforementioned considerations, this paper extends the work of Harvey and Sucarrat [15] and related researches in three ways. First, we extend the Markov switching approach to Beta-skewed-t-EGARCH(1,1) model and we thus propose MS Beta-skewed-t-EGARCH(1,1) to depict the important stylized facts about fat tails, asymmetries, autocorrelation, volatility clustering or mean reversion in different financial regimes. Second, we adopt various model selections and loss functions as the forecasting criteria to compare the performance of our proposed model with that of conventional one-regime Beta-skewed-t-EGARCH(1,1). Lastly, the AIC weight and BIC weight are employed to get more robust results. To the best of our knowledge, this paper represents one of the first studies focused on volatility forecasting of financial assets using MS Beta-skewed-t-EGARCH(1,1) processes (under a skewed Student-t distribution).

This paper is organized as follows: The MS Beta-skewed-t-EGARCH(1,1) is presented in Sect. 2. In Sect. 3, we present the criteria for model comparison and test. The empirical results are shown in Sect. 4. Finally, Sect. 5 presents the conclusions.

2   Methodology

2.1   Beta-skewed-t GARCH Model

Prior to explaining our proposed model, let us briefly touch on the single-regime Beta-skewed-t-GARCH model. This model was proposed by Harvey and Sucarrat [15]. The volatility specification for is the following model:

$$y_t = exp(\lambda_t)\varepsilon_t, \tag{1}$$

$$\lambda_t = \omega + \alpha e_{t-1} + \beta_1 \lambda_{t-1} + \beta_2 \left(sgn(-y_{t-1})(e_{t-1} + 1) \right), \tag{2}$$

where $exp(\lambda_t) = \sigma_t$ is the conditional volatility and $\varepsilon_t \sim st(0, \sigma_\varepsilon^2, v, \gamma)$ is a skewed t with zero mean, scale $\sigma_\varepsilon^2 > 0$, degrees of freedom parameter $v > 2$, and skewness parameter $\gamma > 0$. If $\gamma = 1$, Beta -t-EGARCH (1,1) is obtained. λ_t is the conditional logarithm of the scale which is defined as Eq. (2). ω is the log-scale intercept and can be interpreted as the long-term log-volatility, β_1 is the scale persistence parameter and is restricted to be $|\beta_1| < 1$, β_2 is the leverage effects parameter and sgn denotes the signum function. In Eq. (2), e_t is unobserved conditional score, which can be derived by the derivative of the log-likelihood of y_t at time t with respect to λ_t. Therefore, e_t is given by

$$e_t = \frac{(v+1)(y_t^2 - y_t \mu_{\varepsilon*} exp(\lambda_t))}{v\gamma^{2sgn(y_t - \mu_{\varepsilon*} exp(\lambda_t))} + (y_t - \mu_{\varepsilon*} exp(\lambda_t))^2} - 1, \tag{3}$$

where μ_ε^* is the expected simulated skewed t distribution with mean zero, variance one, degree of freedom v and skewness γ.

2.2   Markov Switching Beta-skewed-t GARCH Model

Roughly speaking, Markov regime Switching model of Hamilton [13] is extended into Beta-skewed-t GARCH model of Harvey and Sucarrat [15], thus, all parameters in this model can vary across regimes or states, i.e. high and low volatility. The regime switches are governed by a hidden Markov chain to gain more ability to capture some stylized facts of non-linear behavior of financial time series. The general form of the model can be written as

$$y_t = exp(\lambda_t(s_t))\varepsilon_t(s_t), \tag{4}$$

$$\lambda_t(s_t) = \omega(s_t) + \alpha(s_t)e_{t-1} + \beta_1(s_t)\lambda_{t-1}(s_t) + \beta_2(s_t)\left(sgn(-y_{t-1})(e_{t-1}(s_t) + 1)\right), \tag{5}$$

$$e_t(s_t) = \frac{(v(s_t) + 1)(y_t^2 - y_t\mu_{\varepsilon^*}(s_t)exp(\lambda_t(s_t)))}{v(s_t)\gamma(s_t)^{2sgn(y_t - \mu_{\varepsilon^*}(s_t)exp(\lambda_t(s_t)))} + (y_t - \mu_{\varepsilon^*}(s_t)exp(\lambda_t(s_t)))^2} - 1.$$

(6)

The estimated parameters $\omega(s_t)$, $\alpha(s_t)$, $\beta_1(s_t)$, $\beta_2(s_t)$, $v(s_t)$ and $\gamma(s_t)$; and the conditional score $e_t(s_t)$ and the conditional logarithm of the scale $\lambda_t(s_t)$ are regime dependent. $s_t \in (0, 1)$ is the unobserved regime or state variable which is the probabilistic structure of the switching regime indicator and is defined by first-order Markov process with constant transition probabilities P.

$$P = \begin{bmatrix} Pr(s_t = 0 \,|s_{t-1} = 0) & Pr(s_t = 1 \,|s_{t-1} = 0) \\ Pr(s_t = 0 \,|s_{t-1} = 1) & Pr(s_t = 1 \,|s_{t-1} = 1) \end{bmatrix},$$

(7)

$$P = \begin{bmatrix} p_{00} & 1 - p_{00} \\ 1 - p_{11} & p_{11} \end{bmatrix},$$

(8)

where p_{00} and p_{11} are transition probability parameters, taking values between 0 and 1. To estimate the parameter set in this model, the maximum likelihood method is used and the general form of the likelihood can be defined as

$$L(\Theta(s_t = j) \,|y) = f(\Theta(s_t = j) \,|y)Pr(s_t = j \,|y_{t-1}),$$

(9)

$$L(\Theta(s_t = j) \,|y) = f(\Theta(s_t = 0) \,|y)Pr(s_t = 0 \,|y_{t-1}) \\ + f(\Theta(s_t = 1) \,|y)Pr(s_t = 1 \,|y_{t-1}),$$

(10)

where $f(\Theta(s_t = j) \,|y_t)$ is the skewed-t density function, $\Theta(s_t) = \{\omega(s_t), \alpha(s_t), \beta_1(s_t), \beta_2(s_t), v(s_t), \gamma(s_t)\}$ is state dependent parameter set of the model and $Pr(s_t = j \,|y_{t-1})$ is the filtered probabilities of regime j. In the MS Beta-skewed-t-EGARCH, once the parameters are estimated, we can make inferences on s_t by approximating the filtered probability $Pr(s_t = j \,|y_{t-1})$, and we employ a Hamilton's filter of Hamilton [13] which can be written as

$$Pr(s_t = j \,|y_{t-1}) = \frac{f(\Theta(s_t = j) \,|y)Pr(s_t = j \,|y_{t-1})p_{jj}}{\sum_{j=1}^{2} f(\Theta(s_t = j) \,|y)Pr(s_t = j \,|y_{t-1})p_{jj}}.$$

(11)

3   Forecasting Methodology and AIC and BIC Weights

3.1   Forecasting Evaluation Method

The forecasting performance of our proposed model is investigated in this study. We compare one-step ahead predictive performance of MS Beta-skewed-t-EGARCH and other EGARCH-class models: single regime Beta-skewed-t-EGARCH of Harvey and Sucarrat [15], single regime Beta-t-EGARCH of Harvey and Chakravarty [14] and MS Beta-t-EGARCH of Blazsek, and Ho [6]. Several criteria for measuring the predictive ability of the models were developed namely MSE (Mean Squared Error), MAE (Mean Absolute Error) and heteroskedasticity adjusted MSE and MAE. The following four loss functions will be used in the models out-of-sample forecast performance evaluation

$$MSE = T^{-1}\sum_{t=1}^{T}\left(\sigma_t^2 - \widehat{\sigma}_t^2\right)^2,$$ (12)

$$MAE = T^{-1}\sum_{t=1}^{T}\left(\sigma_t^2 - \widehat{\sigma}_t^2\right),$$ (13)

$$HMSE = T^{-1}\sum_{t=1}^{T}\left(1 - \sigma_t^2/\widehat{\sigma}_t^2\right)^2,$$ (14)

$$HMAE = T^{-1}\sum_{t=1}^{T}\left(1 - \sigma_t^2/\widehat{\sigma}_t^2\right).$$ (15)

Since volatility itself is unobservable, the comparison of volatility forecasts relies on an observable proxy for the latent volatility process. We follow the approach suggested by Kang, Kang, Yoon, [16], who approximate true volatility by

$$\widehat{\sigma}_t^2 = y^2.$$ (16)

We note that lower loss function evidences more precise one-step-ahead in-sample predictive performance.

3.2   Model Selection: Akaike Weights and Bayesian Weights

To make a robust result, we consider two more approaches, namely AIC weights and BIC weights of Wagenmakers and Farrell, [24], which is the transformation of the raw AIC and BIC. In contrast to forecasting evaluation techniques in Sect. 2.2, these two statistics can be interpreted as conditional probabilities for each model. It is vital to assess the weight of evidence in favor of the best model as we may make a wrong decision in the comparison. As discussed by Lopez [17], it is not obvious which loss function or criterion is more appropriate for the evaluation of volatility models. The other candidate models are simply discarded when a higher loss functions or AIC and BIC are observed. However, it may not make sense to confirm the higher performance of one model compared to the other when the difference between the loss functions or AIC and BIC are very small. The acceptance of a single model may lead to a false sense of confidence. In addition, those values cannot tell us what the weight of evidence is in favor of each model. Thus, in this study, we also consider AIC weight and BIC weight.

Following Wagenmakers and Farrell, [24], the AIC weight and BIC weight of each candidate model can be computed by

$$w_i\left(AIC\right) = \frac{exp\left\{-\frac{1}{2}\Delta_i\left(AIC\right)\right\}}{\sum_{k=1}^{K} exp\left\{-\frac{1}{2}\Delta_k\left(AIC\right)\right\}},\tag{17}$$

$$w_i\left(BIC\right) = \frac{exp\left\{-\frac{1}{2}\Delta_i\left(BIC\right)\right\}}{\sum_{k=1}^{K} exp\left\{-\frac{1}{2}\Delta_k\left(BIC\right)\right\}},\tag{18}$$

where $\Delta_i\left(AIC\right) = AIC_i - \min AIC$ and $\Delta_i\left(BIC\right) = BIC_i - \min BIC$. min AIC and min BIC are the minimum AIC and BIC of the best model. We note that these two weight statistics can be viewed as the probability that model i. (M_i) is the best. We use the restrictions $\sum w_i\left(AIC\right) = 1$ and $\sum w_i\left(BIC\right) = 1$.

4   Estimate Results

4.1   Data Description

We are going to investigate the performance of our proposed MS Beta-skewed-t-EGARCH model and see whether the model outperforms the conventional single regime Beta-skewed-t-EGARCH, single regime Beta-t-EGARCH and MS Beta-t-EGARCH [23]. The data employed in this study consists of daily closing prices of five Exchange Traded Funds (ETFs) which are AGG (a US bond fund), DBC (a commodities fund), EFA (a non-US equity fund), SPY (an S&P500 ETF), VGT (a technology fund). The data are obtained from the web http://finance.yahoo. com. The daily data last seven years, i.e., 2012 to 2018, are used to evaluate the in-scale volatility forecasts. All daily sample prices are converted into a daily nominal percentage return series $r_t = \log(y_t) - \log(y_{t-1})$. Table 1 provides the descriptive statistics of the five-return series and their stationarity test. The standard deviation of returns is the highest for VGT (0.0096) and the lowest for AGG (0.002). We observe the negative skewness, thus the distribution is more pronounced to the right tail for our sample returns. We evidence that the kurtosis of all returns are higher than three, the distribution is higher peak than found in the normal distribution. Finally, we perform augmented Dickey-Fuller (ADF) unit root tests for all returns. The t-test statistics and the corresponding Minimum Bayes factor (MBF) values show a decisive evidence for stationary returns as the MBF values are 0.0000. The interpretation and derivation of MBF can be found in Page and Satake [21].

4.2   Forecast Evaluation

As we mentioned before, the study uses MAE, MSE, HMAE and HMSE as a loss functions for making a comparison of one-step ahead predictive performance of MS Beta-skewed-t-EGARCH and other competing EGARCH models. Table 2

Table 1. Data description

	SPY	VGT	EFA	DBC	AGG
Mean	0.0005	0.0007	0.0002	−0.0003	0.0000
Median	0.0005	0.0011	0.0005	0.0000	0.0001
Maximum	0.039	0.0487	0.0354	0.0424	0.0084
Minimum	−0.0427	−0.0444	−0.0898	−0.0469	−0.011
Std. Dev.	0.0078	0.0096	0.0094	0.009	0.002
Skewness	−0.4931	−0.3853	−0.8658	−0.1138	−0.4034
Kurtosis	5.9914	5.1752	9.4174	4.5227	4.4419
ADF-test	−41.8273	−40.6543	−42.1729	−43.2276	−42.5108
	[0.0000]	[0.0000]	[0.0000]	[0.0000]	[0.0000]

Note: [] denote Minimum Bayes factor which is the p-value calibration.

reports the in-sample statistical performance results and shows that our MS-Beta-EGARCH is superior to other Beta-EGARCH specifications for SPY, VGT, ad AGG as the lowest MAE, MSE, HMAE and HMSE are obtained. However, Beta-t-EGARCH is superior for EFA and DBC. Therefore, a loss function-based model comparison does not evidence statistical dominance of the MS- Beta-EGARCH model.

In short, our model is not found to be superior to the other ones. It dominates only with respect to 3 out of five estimations. This result reminds us that a model which performs very well in a particular market may not be reliable in other markets. Therefore, nonlinear GARCH models do not always outperform the linear ones. Thus, we should be careful when using a Beta-EGARCH-class model for volatility forecast. However, as noted above, it may not make sense to confirm the higher performance of one model over the other when the difference between loss functions is small. We can observe that value of loss function in some case is close to another case. Thus, we further investigate the forecasting performance using AIC and BIC weights to make a robustness result.

4.3   AIC and BIC Weights

Table 3 presents the AIC and BIC weights results for the various volatility models. The bold number represents the best model (lowest AIC and BIC). Thus, the remaining three models are treated as competitive ones. Note that the AIC and BIC weights are viewed as the probability that the model is the best among the whole set of candidate models, therefore, the higher weight presents the higher probability that the model is the best and outperforms all the other models. In other words, [the better is the forecasting performance of the best model compared to that of the alternative ones]. According to the results, we can observe that all of the best models show the high probability that the model is the best. For instance, AIC weight of MS-Beta-skewed-t-EGARCH is 1.0000 for SPY, VGT, and AGG, indicating that it has a 100% chance of being the best one

Table 2. Forecasting comparison

	Beta-EGARCH		MS- Beta-EGARCH	
SPY	*std*	*sstd*	*std*	*sstd*
MAE	0.0404	0.0401	0.0069	**0.0061**
MSE	0.0813	0.0739	0.0386	**0.0169**
HMAE	1.6809	1.6443	0.7046	**0.6503**
HMSE	3.5365	3.0110	0.9680	**0.709**
VGT	*std*	*sstd*	*std*	*sstd*
MAE	0.0505	0.0730	0.0265	**0.0096**
MSE	0.0113	0.0132	0.0533	**0.0263**
HMAE	0.7972	0.7911	0.6483	**0.4505**
HMSE	0.8275	0.8267	0.8078	**0.8047**
EFA	*std*	*sstd*	*std*	*sstd*
MAE	**0.0051**	0.0070	0.0076	0.0085
MSE	**0.0113**	0.0130	0.0320	0.0331
HMAE	**0.5709**	0.7990	0.8004	0.8861
HMSE	**0.6825**	0.8807	0.8662	0.9276
DBC	*std*	*sstd*	*std*	*sstd*
MAE	**0.0049**	0.0070	0.0400	0.0487
MSE	**0.0113**	0.0127	0.016	0.0810
HMAE	**0.6627**	0.6987	0.7436	0.7390
HMSE	**0.6618**	0.6911	0.7419	0.7211
AGG	*std*	*sstd*	*std*	*sstd*
MAE	0.0263	0.0210	0.0010	**0.0009**
MSE	0.0213	0.0203	0.0014	**0.0013**
HMAE	0.8103	0.6097	0.5667	**0.5554**
HMSE	0.9110	0.6691	0.4110	**0.4014**

among those considered in the set of candidate models. This result provides a similar statistical characteristic, as shown in Table 2, our proposed model is not found to outperform all the other models. However, the dominance of MS-Beta-skewed-t-EGARCH for SPY, VGT, and AGG may support the application of the MS-Beta-skewed-t-EGARCH model for some financial data series.

4.4   Estimation Parameter Results

Table 4 presents estimation results of the best EGARCH (1,1) specification for all returns. For the single-regime Beta-t-EGARCH for EFA and DBC returns. $\alpha(s_t = 0)$ is close to 1, indicating that there is high volatility persistent in EFA and DBC returns. Regarding the degrees of freedom, $v(s_t = 0)$, estimates are

Table 3. Results of AIC and BIC analysis for four competing models

	Beta-EGARCH		MS- Beta-EGARCH	
SPY	*std*	*sstd*	*std*	*sstd*
AIC	5732.58	5312.26	3282.63	**3207.87**
BIC	5754.56	5344.88	3336.99	**3283.98**
$\Delta_i\,(AIC)$	2524.71	2104.39	74.76	**0.0000**
$\Delta_i\,(BIC)$	2470.58	2060.9	53.01	**0.0000**
$w_i\,(AIC)$	0.0000	0.0000	0.0000	**1.0000**
$w_i\,(BIC)$	0.0000	0.0000	0.0000	**1.0000**
VGT	*std*	*sstd*	*std*	*sstd*
AIC	3355.15	3322.45	3249.08	**2989.48**
BIC	3376.9	3311.48	3303.44	**3045.59**
$\Delta_i\,(AIC)$	365.67	332.97	259.6	**0.0000**
$\Delta_i\,(BIC)$	331.31	265.89	257.85	**0.0000**
$w_i\,(AIC)$	0.0000	0.0000	0.0000	**1.0000**
$w_i\,(BIC)$	0.0000	0.0000	0.0000	**1.0000**
EFA	*std*	*sstd*	*std*	*sstd*
AIC	**3135.362**	3139.228	3266.818	3298.844
BIC	**3151.109**	3157.981	3321.184	3334.956
$\Delta_i\,(AIC)$	**0.0000**	3.866	131.456	163.482
$\Delta_i\,(BIC)$	**0.0000**	6.872	170.075	183.847
$w_i\,(AIC)$	**0.8736**	0.1264	0.0000	0.0000
$w_i\,(BIC)$	**0.9688**	0.0312	0.0000	0.0000
DBC	*std*	*Sstd*	*std*	*sstd*
AIC	**3195.98**	3205.55	4119.554	3949.827
BIC	**3224.54**	3239.24	4173.92	4025.939
$\Delta_i\,(AIC)$	**0.0000**	9.57	923.574	753.847
$\Delta_i\,(BIC)$	**0.0000**	14.7	949.3800	801.399
$w_i\,(AIC)$	**0.9917**	0.0083	0.0000	0.0000
$w_i\,(BIC)$	**0.9994**	0.0006	0.0000	0.0000
AGG	*std*	*sstd*	*std*	*sstd*
AIC	3815.22	3716.314	3145.549	**3137.621**
BIC	3847.58	3748.933	3239.915	**3213.733**
$\Delta_i\,(AIC)$	677.6	578.69	7.93	**0.0000**
$\Delta_i\,(BIC)$	633.85	535.2	26.18	**0.0000**
$w_i\,(AIC)$	0.0000	0.0000	0.0186	**0.9814**
$w_i\,(BIC)$	0.0000	0.0000	0.0000	**1.0000**

between 6 and 8.5, confirming that the t distribution is appropriate choice for this model. For the results of MS- Beta-skewed-t-EGARCH for SPY, VGT, and AGG, the coefficients across the two regimes are all different. We find that $\omega(s_t = 0)$ is higher than $\omega(s_t = 1)$, thus we can interpret regime 0 as high volatility regime while regime 1 as low volatility regime. The degrees of freedom $v(s_t = 0)$ and $v(s_t = 1)$ are both high and hover around 6.0. There is also evidence of negative skewness which is estimated to be 0.85 and 0.7 on average for regime 0 and regime 1, respectively. This corresponds to the descriptive statistics that provide an

Table 4. Forecasting comparison

Parameter	SPY	VGT	EFA	DBC	AGG
	MS- Beta-EGARCH	MS- Beta-EGARCH	Beta-EGARCH	Beta-EGARCH	MS- Beta-EGARCH
Distribution	sstd	Sstd	std	std	sstd
$\omega(s_t = 0)$	−0.5026 (0.2025)	−0.1609 (0.0001)	−4.9368 (0.0682)	−4.9125 (0.0984)	−5.6144 (0.0117)
$\alpha(s_t = 0)$	0.9236 (0.0035)	0.9311 (0.0002)	0.9501 (0.0152)	0.9898 (0.0054)	0.9944 (0.0009)
$\beta_1(s_t = 0)$	0.0710 (0.0115)	0.0452 (0.0004)	0.0815 (0.0122)	0.0296 (0.0064)	0.0001 (0.0001)
$\beta_2(s_t = 0)$	0.0902 (0.0928)	−1.0925 (0.0000)			0.1987 (0.0125)
$v(s_t = 0)$	5.5814 (0.5345)	5.2103 (0.0001)	6.5206 (0.9054)	8.2624 (1.4644)	10.9876 (0.0016)
$\gamma(s_t = 0)$	0.8527 (0.1661)	0.8168 (0.0007)			1.0394 (0.0005)
$\omega(s_t = 1)$	−4.0212 (0.0003)	−0.1213 (0.0009)			0.1779 (0.0097)
$\alpha(s_t = 1)$	0.7389 (0.0005)	0.7448 (0.0337)			0.7768 (0.0014)
$\beta_1(s_t = 1)$	0.0568 (0.0001)	0.0361 (0.0244)			0.0003 (0.0001)
$\beta_2(s_t = 1)$	0.0721 (0.0002)	0.0603 (0.0142)			0.7412 (0.0012)
$v(s_t = 1)$	4.4651 (0.1578)	4.1683 (1.0036)			9.5432 (1.0206)
$\gamma(s_t = 1)$	0.6822 (0.0003)	0.6535 (0.0314)			0.9708 (0.0233)
p_{00}	0.9500 (0.0376)	0.9500 (0.0086)			0.9944 (0.0234)
p_{11}	0.9501 (0.0230)	0.0501 (0.0110)			0.9900 (0.0124)

Note: () is standard error

evidence of negative skewness in the SPY, VGT, and AGG returns. The volatility in these three returns also tend to persist as the $\alpha(s_t = 0)$ and $\alpha(s_t = 1)$ are close to one. Regarding the leverage effect, $\beta_2(s_t = 1)$ is positive, indicating that positive innovations are less destabilizing than negative innovations.

In addition, the performance of our best EGARCH(1,1) specification is also illustrated in Fig. 1. We fitted the conditional volatility of each return and compare it with the actual volatility. Figure 1 reveals that all the models can track the actual volatilities.

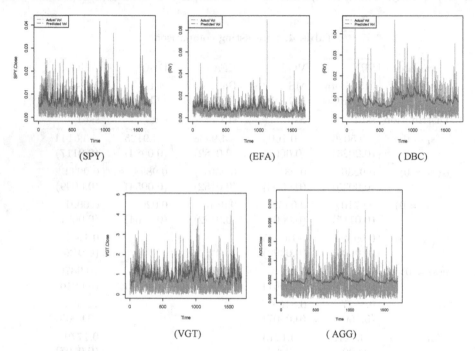

(SPY)   (EFA)   (DBC)

(VGT)   (AGG)

Fig. 1. One-day-ahead volatility forecasts of WTI crude oil based on the best EGARCH (1,1) specification models and the actual volatility measures (as red). (Color figure online)

5   Conclusion

In this study, we extend the work of Harvey and Sucarrat [15] and propose a MS-Beta-skewed-t-EGARCH to quantify the volatility under structural change in the financial behavior. The proposed model is then examined and compared its volatility forecasting performance using various loss functions. In addition, we are also concerned that the loss functions may not be enough to clarify the performance of the forecasting model as there are many studies mentioned that the comparison of volatility forecasting models is influenced by the criterion or the loss function. The acceptance of a single model may lead to a false sense of confidence. In addition, those values cannot tell us what the weight of evidence is in favor of each model. Hence, we confirm the reliability and robustness of the forecasts by using the AIC and BIC weights. The use of these two approaches gives us the greater insight into the relative merits of the competing forecasting models. We note that various EGARCH models are considered as the competing models (i.e., single regime Beta-skewed-t-EGARCH, single regime Beta-t-EGARCH and MS Beta-t-EGARCH).

We have used data on returns of five Exchange Traded Funds (ETFs) obtained from yahoo finance from January 2012 to October 2018. The results suggest that four loss functions, namely MAE, MSE, HMAE and HMSE, have not shown that our MS- Beta-skewed-t-EGARCH is completely superior to the competing models. Nevertheless, the dominance of MS-Beta-skewed-t-EGARCH for SPY, VGT, and AGG may support the application of the MS-Beta-skewed-t-EGARCH model for some financial data series. Moreover, we have presented AIC and BIC weights and the results show a similar statistical characteristic, our model perform high probability that the model is the best forecasting model.

In summary, our proposed model is not found to be absolutely superior to the other ones. However, there are some evidences supporting the application of the MS-Beta-skewed-t-EGARCH model for some financial data series.

References

1. Andersen, T.G., Bollerslev, T.: Intraday periodicity and volatility persistence in financial markets. J. Empir. Financ. 4(2–3), 115–158 (1997)
2. Bauwens, L., Laurent, S., Rombouts, J.V.: Multivariate GARCH models: a survey. J. Appl. Econ. 21(1), 79–109 (2006)
3. Black, F., Scholes, M.: The pricing of options and corporate liabilities. J. Polit. Econ. 81(3), 637–654 (1973)
4. Blazsek, S., Downarowicz, A.: Regime switching models of hedge fund returns. Working Papers (Universidad de Navarra. Facultad de Ciencias Económicas y Empresariales) (12), 1 (2008)
5. Blazsek, S., Villatoro, M.: Is Beta-t-EGARCH (1, 1) superior to GARCH (1, 1)? Appl. Econ. 47(17), 1764–1774 (2015)
6. Blazsek, S., Ho, H.C.: Markov regime-switching Beta-t-EGARCH. Appl. Econ. 49(47), 4793–4805 (2017)
7. Bollerslev, T.: Generalized autoregressive conditional heteroskedasticity. J. Econ. 31(3), 307–327 (1986)
8. Chodchuangnirun, B., Zhu, K., Yamaka, W.: Pairs trading via nonlinear autoregressive GARCH models. In: Huynh, V.-N., Inuiguchi, M., Tran, D.H., Denoeux, T. (eds.) IUKM 2018. LNCS, vol. 10758, pp. 276–288. Springer, Cham (2018). https://doi.org/10.1007/978-3-319-75429-1_23
9. Davidson, J.: Moment and memory properties of linear conditional heteroscedasticity models, and a new model. J. Bus. Econ. Stat. 22(1), 16–29 (2004)
10. Ding, Z., Granger, C.W., Engle, R.F.: A long memory property of stock market returns and a new model. J. Empir. Financ. 1(1), 83–106 (1993)
11. Engle, R.F.: Autoregressive conditional heteroscedasticity with estimates of the variance of United Kingdom inflation. Econ. J. Econ. Soc. 50, 987–1007 (1982)

12. Glosten, L.R., Jagannathan, R., Runkle, D.E.: On the relation between the expected value and the volatility of the nominal excess return on stocks. J. Financ. **48**(5), 1779–1801 (1993)
13. Hamilton, J.D.: A new approach to the economic analysis of nonstationary time series and the business cycle. Econ. J. Econ. Soc. **57**, 357–384 (1989)
14. Harvey, A.C., Chakravarty, T.: Beta-t-(E) GARCH. University of Cambridge, Faculty of Economics, Working paper CWPE 08340 (2008)
15. Harvey, A., Sucarrat, G.: EGARCH models with fat tails, skewness and leverage. Comput. Stat. Data Anal. **76**, 320–338 (2014)
16. Kang, S.H., Kang, S.M., Yoon, S.M.: Forecasting volatility of crude oil markets. Energy Econ. **31**(1), 119–125 (2009)
17. Lopez, J.A.: Evaluating the predictive accuracy of volatility models. J. Forecast. **20**(2), 87–109 (2001)
18. Maneejuk, P., Yamaka, W., Sriboonchitta, S.: A Markov-switching model with mixture distribution regimes. In: Huynh, V.-N., Inuiguchi, M., Tran, D.H., Denoeux, T. (eds.) IUKM 2018. LNCS, vol. 10758, pp. 312–323. Springer, Cham (2018). https://doi.org/10.1007/978-3-319-75429-1_26
19. Merton, R.C.: Lifetime portfolio selection under uncertainty: the continuous-time case. Rev. Econ. Stat. **51**, 247–257 (1969)
20. Nelson, D.B.: Conditional heteroskedasticity in asset returns: a new approach. Econ. J. Econ. Soc. **59**, 347–370 (1991)
21. Page, R., Satake, E.: Beyond P values and hypothesis testing: using the minimum Bayes factor to teach statistical inference in undergraduate introductory statistics courses. J. Educ. Learn. **6**(4), 254 (2017)
22. Salisu, A.A.: Modelling oil price volatility with the Beta-Skew-t-EGARCH framework. Econ. Bull. **36**(3), 1315–1324 (2016)
23. Sucarrat, G.: betategarch: simulation, estimation and forecasting of Beta-Skew-t-EGARCH models. R J. **5**(2), 137–147 (2013)
24. Wagenmakers, E.J., Farrell, S.: AIC model selection using Akaike weights. Psychon. Bull. Rev. **11**(1), 192–196 (2004)
25. Wei, Y., Wang, Y., Huang, D.: Forecasting crude oil market volatility: further evidence using GARCH-class models. Energy Econ. **32**(6), 1477–1484 (2010)
26. Zhu, K., Yamaka, W., Sriboonchitta, S.: Pair trading rule with switching regression GARCH Model. In: Huynh, V.-N., Inuiguchi, M., Le, B., Le, B.N., Denoeux, T. (eds.) IUKM 2016. LNCS, vol. 9978, pp. 586–598. Springer, Cham (2016). https://doi.org/10.1007/978-3-319-49046-5_50

Building Fuzzy Levy-GJR-GARCH American Option Pricing Model

Huiming Zhang[1]([⊠]) and Junzo Watada[2]

[1] Graduate School of Information, Production and Systems,
Waseda University, Kitakyushu, Japan
h.zhang@kurenai.waseda.jp
[2] Computer and Information Sciences Department,
University of Technology PETRONAS, Seri Iskandar, Malaysia
junzo.watada@gmail.com

Abstract. Taking into account the time-varying, jump and leverage effect characteristics of asset price fluctuations, we first obtain the asset return rate model through the GJR-GARCH model (Glosten, Jagannathan and Rundle-generalized autoregressive conditional heteroskedasticity model) and introduce the infinite pure-jump Levy process into the asset return rate model to improve the model's accuracy. Then, to be more consistent with reality and include more uncertainty factors, we integrate the more generalized parabolic fuzzy variable (which can cover the triangle and trapezoid fuzzy variable) to represent asset price volatility. Next, considering more general situations with fuzzy variables with mixed distributions, we apply fuzzy simulation technology to the least squares Monte Carlo algorithm to create fuzzy pricing numerical algorithms, that is the fuzzy least squares Monte Carlo algorithm. Finally, by using American options data from the Standard & Poor's 100 index, we empirically test our fuzzy pricing model with different widely used infinite pure-jump Levy processes (the VG (variance gamma process), NIG (normal inverse Gaussian process) and CGMY (Carr-Geman-Madan-Yor process) under fuzzy and crisp environments. The results indicate that the fuzzy option pricing model is more reasonable; the fuzzy interval can cover the market prices of options and the prices that obtained by the crisp option pricing model, the fuzzy option pricing model is feasible one.

Keywords: American option · Fuzzy set theory ·
Fuzzy simulation technology · Levy process · GJR-GARCH model ·
Least squares Monte Carlo approach

1 Introduction

The pricing problem of American options is usually solved with either analytical or numerical methods. Earlier studies mainly used analytical methods to determine the price of American options: Johnson [1] used approximate analysis

© Springer Nature Switzerland AG 2019
H. Seki et al. (Eds.): IUKM 2019, LNAI 11471, pp. 197–209, 2019.
https://doi.org/10.1007/978-3-030-14815-7_17

to determine the value of an American option under the assumption of no dividend; Geske et al. [2] constructed a model to analyse an American option with a dividend pay-out, but no closed-form solution was obtained. Therefore, numerical methods began discussed to solve American option pricing; these include commonly used binomial tree, finite difference, and least square Monte Carlo methods, among others. Cox et al. [3] proposed the binomial tree method, which offers simple and effective solutions, and it provides an accurate numerical solution by continuously shrinking the time step; therefore, it is often used as a reference to evaluate the accuracy of other numerical approaches. However, when the model includes multiple random influencing factors, the number of values increases exponentially to calculate in the binomial tree method, which often leads to the curse of dimensionality. The finite difference method mainly converts the asset pricing differential equation into a difference equation, and by obtaining solutions through an iterative method, it mitigates the difficulty in directly solving the differential equation. In 1978, Brennan et al. [4] applied this calculation method in the pricing of American options, but the curse of dimensionality persists when this method is used to solve high-dimensional problems. The Monte Carlo method has the characteristic of forward simulation, so it cannot be applied directly for the pricing of American options, which have a backward iterative search characteristic. Longstaff et al. [5] modified the Monte Carlo method by using the least squares approach and proposed the least squares Monte Carlo algorithm, which solves the application difficulty of the said method in the pricing of American options; they also provided empirical evidence of the method effectiveness. This method uses the least squares approach to estimate the expected value of continuous holding for each path. By comparing the values to the value associated with immediate exercise, the exercise point of each path is determined. Finally, the value of the American option is obtained by computing the discounted average value of each path's exercise point.

The above calculation methods are effective in pricing American options; however, these studies use the Black-Scholes (B-S) model as their theoretical basis, in which the asset price random process is treated as a geometric Brownian motion, which is unfit for real-life financial markets. Empirical studies have demonstrated that fluctuations in asset price and rate of return are often characterized by non-continuity, clustering and leverage effects; consequently, we need to construct a more flexible asset pricing model to accurately reflect how asset prices change in reality. Asset price usually jumps in movements, and by adding a Levy process in the pricing model, we can construct a jump model with random jumps of different strengths. Moreover, generalized autoregressive conditional heteroskedasticity (GARCH) models are most frequently used to express the volatility in asset price fluctuations and leverage effects, and such models are highly expandable and more capable of providing accurate descriptions of volatility; therefore, by combining the two models to form the Levy-GARCH model, we can better capture the characteristics of the volatility of the underlying asset. The Levy-GARCH model is widely used in the pricing of European options, but due to the complexity of American options, the model is less frequently applied as a theoretical model for American options. Based on the background described above, jump measure, time-varying volatility and leverage effects are incorporated in this study to construct the Levy-GARCH pricing model for American options proposed by Glosten, Jagannathan and Rundle (Levy-GJR-GARCH) on the basis of an infinite pure jump Levy process and an asymmetric GARCH model. In addition, in real-life financial markets, many subjective and objective uncertainty factors lead to randomness and fuzziness in the price of the option. Therefore, it is necessary to incorporate fuzzy set theory in the pricing model to improve the classic pricing theory. Hence, this study analysed the American option pricing model under a fuzzy environment, incorporated fuzzy simulation technology, and used the least squares Monte Carlo algorithm to analyse the model. Lastly, through empirical analysis, we compared the option pricing simulation results of the three infinite pure jump Levy processes (variance gamma (VG), normal inverse Gaussian (NIG), Carr-Geman-Madan-Yor (CGMY)) combined with the GJR-GARCH model.

Through the review of the existing literature, we found abundant studies regarding European option, but studies about American option pricing are still limited. Furthermore, the existing studies mainly focus on numerical algorithm improvements, and insufficient research was pursued to improve the theoretical model. Therefore, we constructed the fuzzy Levy-GJR-GARCH American option pricing model, which is more consistent with reality, and verify the model's efficiency by using empirical analysis. The rest of this chapter is structured as follows: Sect. 2 introduces the parabolic fuzzy variable; Sect. 3 deduces the Levy-GJR-GARCH American option pricing model under a fuzzy environment; Sect. 4 provides a brief introduction of fuzzy simulation technology, then based on it design the algorithms for fuzzy American option pricing model, that is fuzzy least squares Monte Carlo algorithm. Section 5 combines the Standard & Poor's 100 index (S&P 100 Index) American put option prices to perform empirical testing. Section 6 summaries the findings of this study.

2   Parabolic Fuzzy Variable

The concept of fuzzy sets was first proposed by Zadeh in 1965 [6]. It gradually developed into a more complete fuzzy theory, which revealed a new direction for asset pricing theories.

Let $\tilde{A}$ be a mapping of the domain X to $[0,1]$, that is, $\tilde{A}\colon X \to [0,1]$, $x \to \tilde{A}(x)$ is called a fuzzy set on X. $\tilde{A}(x)$ is called the membership function of the fuzzy set $\tilde{A}$, and the set of all fuzzy sets on X is denoted as $\tilde{F}(X)$. If $\alpha \in [0,1]$, $\tilde{A}_{\alpha} = \{x \in X \mid \tilde{A}(x) \geq \alpha\}$, then $\tilde{A}_{\alpha}$ is called the α-level set of fuzzy set $\tilde{A}$. If $\tilde{a}$ is a regular convex fuzzy set with a upper semi-continuous membership function $\tilde{a}(x)$ and the level set $\tilde{a}_{\alpha}$ is bounded, i.e., $\alpha \in [0,1]$, then $\tilde{a}$ is called a fuzzy number.

If the membership function form of fuzzy number $\tilde{A}$ is:

$$\mu_{\tilde{A}}(x) = \begin{cases} \left(\dfrac{x-a_1}{a_2-a_1}\right)^n, & a_1 \leq x \leq a_2 \\ 1, & a_2 \leq x \leq a_3 \\ \left(\dfrac{x-a_4}{a_3-a_4}\right)^n, & a_3 \leq x \leq a_4 \\ 0, & \text{others} \end{cases} \tag{1}$$

where $\tilde{A}$ is a parabolic fuzzy number called $\tilde{A} = (a_1, a_2, a_3, a_4)_n$. If $n = 1$, the above is a trapezoidal fuzzy number; if $n = 1$ and $a_2 = a_3$, the above is a triangular fuzzy number. Therefore, triangular fuzzy numbers and trapezoidal fuzzy numbers are special cases of parabolic fuzzy numbers (See Fig. 1). At this point, the α level set of $\tilde{A}$ can be expressed as $\tilde{A}_\alpha = [\tilde{A}_\alpha^L, \tilde{A}_\alpha^U] = [a_1 + \sqrt[n]{\alpha}(a_2 - a_1), a_4 - \sqrt[n]{\alpha}(a_4 - a_3)]$, where $\tilde{A}_\alpha^L$ is the α pessimistic value of fuzzy variable $\tilde{A}$ and $\tilde{A}_\alpha^U$ is the α optimistic value of $\tilde{A}$.

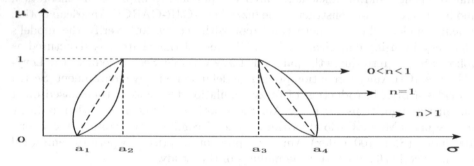

Fig. 1. Plot of membership function of a parabolic fuzzy number

3   Fuzzy Levy-GJR-GARCH American Option Pricing Model

As the notations used in the remainder of this paper are listed as follows:

Variable	Description
S_0	The underlying asset price at initial time
S_t	The underlying asset price at time t
T	Time to expiration
K	Exercise price
r	Risk-free interest rate
$V(S_t, t)$	Option price at time t
σ_t	Volatility at time t
X_t	Levy process
θ	Drift rate
v	Jump rate
g	Gamma function

3.1   The Process of the Underlying Asset Price

We assumed the fluctuation of the underlying asset price has the characteristics of time-varying, jump and leverage effect, thus the sequence of the rate of return of the underlying asset is described using an asymmetric conditional heteroskedasticity model. GJR-GARCH model can express the conditional heteroskedasticity "leverage effect". Therefore, we chose to use GJR-GARCH model proposed by Glosen et al. [7] as the specific form of the asset return rate model, specifically as follows:

$$
\begin{cases}
R_t = \ln(\dfrac{S_t}{S_{t-1}}) = u_t - \gamma_t + \sigma_t z_t \\
\sigma_t^2 = w + \alpha \sigma_{t-1}^2 + \beta \sigma_{t-1}^2 z_{t-1}^2 + \delta I_{t-1} \sigma_{t-1}^2 z_{t-1}^2 \\
I_t = \begin{cases} 1, z_t < 0 \\ 0, z_t \geq 0 \end{cases} \\
z_t | F_{t-1} \sim D(0, 1; \theta_D)
\end{cases}
\tag{2}
$$

In the asset return rate model (2), R_t is asset's logarithmic return rate, u_t is the expected rate of return under the condition of information set F_{t-1}, γ_t is the mean correction factor, and σ_t^2 is the time-varying variance sequence, I_t represents the indicator function. w represents intercept, α is the influence coefficient of the variance of previous period to the variance of current period, β is the influence coefficient of the residual of previous period to the residual of current period, δ represents asymmetric effect coefficient. z_t represents the innovation of the mean equation, and it follows distribution $D(\bullet)$ with mean value of 0, variance of 1, and parameter θ_D, for which this study will establish several different infinite pure jump Levy processes, such as VG, NIG and CGMY process.

(1) VG process:

$$E(e^{iuX_t}) = \varphi(u; \sigma, v, \theta)$$
$$= (1 - iuv\theta + \frac{1}{2}\sigma^2 vu^2)^{-\frac{1}{v}} \tag{3}$$

of which,

$$C = \frac{1}{v} > 0$$

$$M = (\sqrt{\frac{1}{4}\theta^2 v^2 + \frac{1}{2}\sigma^2 v} + \frac{1}{2}v\theta)^{-1} > 0$$

$$G = (\sqrt{\frac{1}{4}\theta^2 v^2 + \frac{1}{2}\sigma^2 v} - \frac{1}{2}v\theta)^{-1} > 0$$

(2) NIG process:

$$E(e^{iuX_t}) = \varphi(u; \lambda, \eta, \kappa)$$
$$= \exp(\eta\sqrt{\lambda^2 - \eta^2} - \kappa\sqrt{\lambda^2 - (\eta + iu)^2}) \tag{4}$$

of which, $\lambda > 0$, $\kappa > 0$, $-\lambda < \eta < \lambda$.

(3) CGMY process:

$$E(e^{iuX_t}) = \varphi(u; C, G, M, Y)$$
$$= \exp(Cg(-Y)(M - iu)^Y$$
$$+ (G + iu)^Y - M^Y - G^Y) \tag{5}$$

of which, $C > 0$, $G > 0$, $M > 0$, $Y < 2$, g represents gamma function.

3.2   The Risk-Neutral Conversion of the Underlying Asset Pricing

In theory, there should be no arbitrage in the option value; therefore, the asset return rate model (see Eq. (2)) requires risk-neutral conversion to ensure the validity of the no-arbitrage assumption. Under risk-neutral measure Q, $E^Q(S_t|S_{t-1}) = S_{t-1}e^{r_t}$, where r_t represents the risk-free rate of return. Here, the risk-neutral model is,

$$S_t = S_{t-1}e^{r_t - \varphi^Q(\sigma_t) + \sigma_t \varsigma_t^Q} \tag{6}$$

Above, $\varphi^Q(\sigma_t) = E^Q(e^{\sigma_t \varsigma_t^Q})$ is the mean correction factor, where ς_t^Q is white noise with mean of 0 and variance of 1. Using the Christofersen et al. [8] method to construct the pricing kernel $\{\varsigma_t\}$, we establish a Radon-Nikodym derivative sequence that can materialise real measurement of risk-neutral measure conversion:

$$\frac{dQ}{dP}|F_{t-1} = \exp(-\sum_{i=1}^{t}(\varsigma_i \sigma_i z_i + \psi(\varsigma_i))) \tag{7}$$

Under a non-normal environment, the kernel sequence $\{\varsigma_t\}$ is not the only one that fulfils the following formula:

$$\psi_t(\varsigma_t - 1) - \psi_t(\varsigma_t) + u_t - r_t - \gamma_t = 0 \tag{8}$$

Here, $\psi(\bullet)$ represents the exponential part of the moment-generating function. Based on the characteristics of the moment-generating function $\psi_t'(0) = E_{t-1}[\sigma_t z_t]$, $\psi_t''(0) = Var_{t-1}[\sigma_t z_t] = \sigma_t^2$, we obtain the following analytical expression for the kernel sequence $\{\varsigma_t\}$:

$$\varsigma_t \approx \frac{1}{2} + \frac{u_t - r_t - \gamma_t - \psi_t'(0)}{\psi_t''(0)} = \frac{1}{2} + \frac{u_t - r_t - \gamma_t}{\sigma_t^2} \tag{9}$$

After obtaining the kernel sequence $\{\varsigma_t\}$, we can perform risk-neutral adjustment on the stochastic item $\varepsilon_t = \sigma_t z_t$ and obtain the following formula:

$$\varepsilon_t^Q = \varepsilon_t - E_{t-1}^Q[\varepsilon_t] = \varepsilon_t - \psi_t'(\varsigma_t) \tag{10}$$

Therefore, under the risk-neutral measure, the mean equation can be expressed as follows:

$$R_t^Q = r_t - \psi_{\varepsilon_t^Q}^Q(1) + \varepsilon_t^Q = r_t - \psi_{z_t^Q}^Q(\sigma_t^Q) + \sigma_t^Q z_t^Q \tag{11}$$

The conditional variance formula for the risk-neutral asset return rate model can be expressed as

$$(\sigma_t^2)^Q = w^Q + \alpha^Q(\sigma_{t-1}^2)^Q + \beta^Q(\varepsilon_{t-1}^Q + \psi'(\varsigma_{t-1}))^2 + \delta^Q I_{t-1}(\varepsilon_{t-1}^Q + \psi'(\varsigma_{t-1}))^2 \tag{12}$$

At this point, we can see that there is some discrepancy between the risk-neutral measure and the real measure of sequence ε_t^Q and $(\sigma_t^2)^Q$; therefore, it is necessary to perform parameter adjustment using kernel sequence $\{\varsigma_t\}$.

4   The Algorithm Design for Fuzzy American Option Pricing Model

4.1   Fuzzy Simulation Technology

Fuzzy simulation technology only provides a statistical estimate of the model, not the precise result, but it is the effective method for complex problems for which analytical results are unattainable.

If ξ is a fuzzy variable with probability space $(\Theta, P(\Theta), Pos)$, the function $f(\xi)$ is also a fuzzy variable; at the same time, the membership function of $f(\xi)$ can be obtained using the following simulation method:

Step 1. Randomly and evenly extract a number ξ_k $(k = 1, 2 \ldots N)$ from the level set of fuzzy variable ξ, calculate ξ_k membership from the membership function of ξ, and denote it as v_k.

Step 2. Based on the formula for function $f(\xi_k)$, calculate the function value $f(\xi_k)$.

Step 3. Repeat Steps 1 through 2 N times.

Step 4. Calculate the expected value $E(f(\xi)) = \frac{1}{N} f(\xi_k)$ of function $f(\xi)$ and draw the membership function of $f(\xi)$ based on $(f(\xi_k), v_k)$.

4.2 Fuzzy Least Squares Monte Carlo Algorithm

Presuming that the number of Monte Carlo algorithm-simulated paths is N and that the time to expiration T is divided into M periods, at time t_i, the exercise value of path-j is $I_{i,j}(S_{i,j}) = \max(K - S_{i,j}, 0)$, where K is the exercise price and $S_{i,j}$ is the asset price on path j during t_i. The conditional expected value of continuous option holding can only obtained using backward inference $E_{i,j}(S_{i,j}) = E[\exp(-r\Delta t)V_{i+1,j}(S_{i+1,j})|S_{i,j}]$. Therefore, the conventional Monte Carlo method is not suitable for numerical simulation of the American option pricing model. The least squares Monte Carlo approach regards the discounted value of the option value at time t_{i+1}, $\exp(-r\Delta t)V_{i+1,j}(S_{i+1,j})$, as the Y variable and $S_{i,j}$ and $S_{i,j}^2$ as X variable, constructing a least squares regression model for Y as a function of X and obtaining regression coefficients a_1, a_2 and a_3. The following formula can yield an approximation for $E_{i,j}(S_{i,j})$:

$$E_{i,j}(S_{i,j}) \approx a_1 + a_2 S_{i,j} + a_3 S_{i,j}^2 \tag{13}$$

Based on the above method, compare the value of continuation holding and the value of exercise at each node of N paths, thereby obtaining the optimal exercise strategy for each path. Discount the option value of each path to the present period and obtain the average of each path's discounted option value; this said average value is the acquired option price.

If the asset price volatility σ is a fuzzy number, the asset price $S_{i,j}$ is also a fuzzy number, whereas the exercise value $I_{i,j}(S_{i,j})$ and value of continuous holding $E_{i,j}(S_{i,j})$ are both functions of $S_{i,j}$; therefore, $I_{i,j}(S_{i,j})$ and $E_{i,j}(S_{i,j})$ are also fuzzy numbers. Their α level set can be expressed as follows:

$$\tilde{I}_{i,j}^{\;\alpha}(S_{i,j}) = [\max\{K - \tilde{S}_{i,j}^{\;U}(\alpha)\}^+, \max\{K - \tilde{S}_{i,j}^{\;L}(\alpha)\}^+] \tag{14}$$

$$\tilde{E}_{i,j}^{\;\alpha}(S_{i,j}) = [a_1 + a_2 S_{i,j}^L + a_3 S_{i,j}^{2\;L}, a_1 + a_2 S_{i,j}^U + a_3 S_{i,j}^{2\;U}] \tag{15}$$

Because the asset price and option value are fuzzy variables, when comparing and solving the least squares regression equation, the expected value of fuzzy variable is used in the calculation. The calculation of the option price using the least squares Monte Carlo algorithm is as follows:

Step 1. Randomly and evenly extract a number $\sigma_k (k = 1, 2 \ldots N)$ from the α level set of the fuzzy variable $\tilde{\sigma}$, calculate the membership degree of σ_k from the membership function of $\tilde{\sigma}$, and denote it as v_k.

Step 2. Based on the asset price formula, calculate the asset price $S_{k,j}(j = 1, 2 \ldots M)$ at each node of path j.

Step 3. Find the option exercise value $I_{k,j}(S_{k,j})$ at each node of path j, and calculate the value of continuous option holding $E_{k,j}(S_{k,j})$ at each node using the least squares method.

Step 4. Repeat Steps 1 through 3 N times.

Step 5. Calculate the expected option value $E(V_0) = V_{k,0}/N$, and based on $(V_{k,0}, v_k)$, draw the fuzzy option value membership function diagram.

5 Empirical Analysis

5.1 Source of Data

This study used the S&P 100 Index and American put options acquired from S&P 100 Index as the data for empirical analysis. The S&P 100 Index prices were selected from the closing prices of data of 1526 days dating from March 22, 2011 to March 23, 2017 (data source: Yahoo!Finance), and the American S&P 100 Index put option prices were selected to be the average prices of the final transacted prices for different expiry dates and different exercise prices for put options on March 23, 2017 (data source: Chicago Board of Options Exchange). The data used in this research excluded options with same month expiry. We only took into account American options with exercise prices within the range of 95%–105% of the index prices and eliminated contracts with option values close to 0. Consequently, we obtained 70 data points, of which 22 expire in April, 22 expire in May, 11 expire in June, 5 expire in July, 5 expire in September and 5 expire in December.

Table 1. Estimated results for the Levy-GJR-GARCH model parameters

GJR-GARCH model parameters				VG process parameters			
ω	α	β	δ	θ	σ	υ	-
0.0001***	0.0050	0.7364***	0.4746***	−0.2131	1.0676***	0.3740**	-
(2.9711)	(0.4902)	(40.3428)	(9.9941)	(−1.2262)	(6.1442)	(2.1527)	
NIG process parameters				CGMY process parameters			
λ	η	κ	-	C	G	M	Y
1.7806***	−0.2396	0.8883***	-	6.6594***	3.1636***	2.7937***	1.6961***
(5.1771)	(−0.6968)	(2.5827)		(76.5830)	(36.3812)	(32.1268)	(19.5048)

Remark: The numerical values in parentheses correspond to the t-statistics of the parameter values, * indicates significant at the 10% significance level, ** indicates significant at the 5% significance level, and *** indicates significant at the 1% significance level.

Table 2. Descriptive statistics regarding volatility and innovations

Indicator	Sample size	Maximum value	Minimum value	Average value	Standard deviation	Skewness	Kurtosis
Time-varying volatility	1525	0.0005	0.0000	0.0000	0.0000	7.7168	88.3883
Innovation	1525	3.4999	−6.6548	0.0753	1.0338	−0.4951	4.9872

5.2 Parameter Estimation

To reduce the complexity of parameter estimation, this study used a two-step method to estimate the parameters of GJR-GARCH model and Levy process: in step 1, set innovations as Gaussian distribution and use maximum likelihood estimation to estimate the parameters of the GJR-GARCH model; in step 2, based on the innovations data obtained in step 1, use the generalised method of moments to estimate the parameters of the VG, NIG and CGMY models. The results of the parameters estimation are presented in Table 1, from which we can see that the "leverage effect" parameter δ of GJR-GARCH model is greater than 0. At the 1% significance level, the significance is not 0, indicating that the changes in volatility are clearly asymmetric, with downward fluctuations stronger than upward fluctuations.

Figures 2 and 3 show the time-varying volatility sequence and the innovations sequence, and Table 2 presents the descriptive statistics of innovations. From the characteristics of the innovation data, we can see that volatility in innovations is not white noise; the skewness and kurtosis are $-0.4951 < 0$ and $4.9872 > 3$, respectively, indicating a leptokurtic, fat-tailed distribution. Therefore, a Levy process can provide higher accuracy than a Gaussian distribution.

5.3 Empirical Result Analysis

The multiplier of S&P 100 Index options is 100 USD (i.e. each point represents 100 USD). The closing price of S&P 100 Index on March 23, 2017 was 1,040, i.e. $S_0 = 1,040$. We used the 10-year T-bond yield as of March 23, 2017 as the risk-free interest rate, $r = 2.4\%$ (data source: official website of US Treasury Department). To examine the pricing effect of the Levy-GJR-GARCH model under a fuzzy environment, we compared the pricing result with that of the Levy-GJR-GARCH model under a crisp environment. Under fuzzy theory, the volatility σ of an asset price is set as a fuzzy variable, whereas in the GARCH

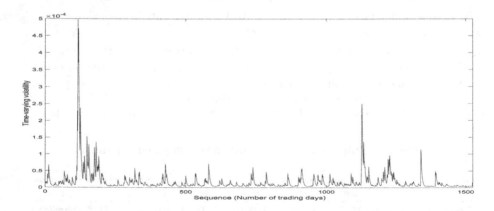

Fig. 2. Time-varying volatility

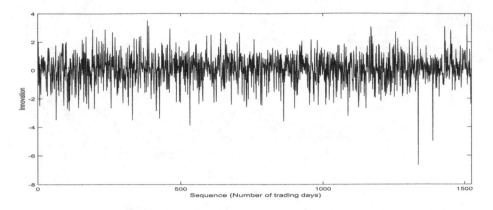

Fig. 3. Innovations sequence diagram

model, the volatility σ is set as a time-varying variable. To reduce the complexity of the fuzzy calculation, the membership function of the time-varying volatility $\{\widetilde{\sigma}_t\}$ was set as an equal form of parabolic membership function. Because estimation of four parameter values was required for parameter interval of parabolic fuzzy numbers, the historical rates of return of 1525, 1200, 800 and 400 trading days before March 23, 2017 were selected as observation samples based on different market information reflected by different sampling intervals.

The expected values under a fuzzy environment were obtained from the upper and lower weights of the $\alpha = 0.95$ level set of fuzzy number $\widetilde{V}_t$, and the exact formula is as follows:

$$
M(\widetilde{V}) = \frac{M(\widetilde{V})^L + M(\widetilde{V})^U}{2}
$$
$$
= \frac{\int_0^1 f(\alpha)\widetilde{V}_\alpha^L \, d\alpha + \int_0^1 f(\alpha)\widetilde{V}_\alpha^U \, d\alpha}{2}
$$
$$
= \int_0^1 \frac{f(\alpha)}{2}(\widetilde{V}_\alpha^L + \widetilde{V}_\alpha^U)d\alpha \tag{16}
$$

Furthermore, we selected 22 short-term option pricing results with expiry in April 2017 to further analyse the differences in option pricing under fuzzy and crisp environments. The result is shown in Fig. 4. From the simulation result, we can see that all market prices fall within the fuzzy interval of the VG, NIG and CGMY models under a fuzzy environment, which shows that the market prices of options are better covered when a fuzzy price interval is used. In contrast with the smaller fuzzy interval of the VG model and the greater fuzzy interval of the NIG model, the fuzzy interval of the CGMY model offers better simulation results. Simultaneously, we observe that under a crisp environment, the simulation results of the VG and CGMY models are greater than the market price when the exercise price is lower and less than the market price when the exercise

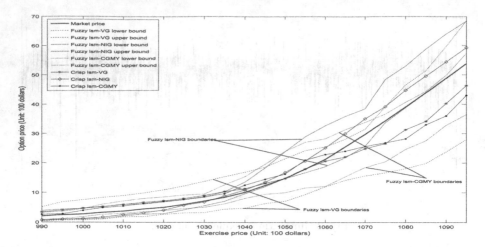

Fig. 4. Option pricing results for April 2017 expiry (Remark: "Fuzzy" denotes simulation under a fuzzy environment, "Crisp" denotes simulation under a crisp environment, "lsm" denotes the least squares Monte Carlo algorithm. VG, NIG and CGMY are Levy processes.)

price is higher, whereas the simulation result of the NIG model is less than the market price when the exercise price is lower and greater than the market price when the exercise price is higher.

6   Conclusions

The decision of the optimal stopping time makes American option pricing problems more complicated than the European option pricing problem, and the traditional BS (Black Schloes) model is not capable of deciding American option pricing. Taking into account the time-varying, jump and leverage effect characteristics of the asset price fluctuation, this study built a Levy-GJR-GARCH American option pricing model based on an infinite pure jump process. Meanwhile we incorporated fuzzy set theory and set the underlying asset price volatility as the more generalized parabolic fuzzy variable and considering more general situations with the fuzzy variables with mixed distributions, based on fuzzy simulation technology established fuzzy least squares Monte Carlo numerical algorithms for the proposed model. Lastly, using the S&P 100 Index and data for the corresponding American put options, we empirically tested our fuzzy pricing model with different widely used infinite pure-jump Levy processes (the VG, NIG and CGMY processes) under fuzzy and crisp environments. The results indicate that the fuzzy option pricing model is more reasonable; the fuzzy interval can cover the market prices of options and the prices that obtained by the crisp option pricing model, the fuzzy option pricing model is feasible one.

References

1. Johnson, H.E.: An analytic approximation for the American put price. J. Financ. Quant. Anal. **18**(01), 141–148 (1983)
2. Geske, R., Johnson, H.E.: The American put option valued analytically. J. Finance **39**(5), 1511–1524 (1984)
3. Cox, J.C., Ross, S.A., Rubinstein, M.: Option pricing: a simplified approach. J. Financ. Econ. **7**(3), 229–263 (1979)
4. Brennan, M.J., Schwartz, E.S.: Finite difference methods and jump processes arising in the pricing of contingent claims: a synthesis. J. Financ. Quant. Anal. **13**(03), 461–474 (1978)
5. Longstaff, F.A., Schwartz, E.S.: Valuing American options by simulation: a simple least-squares approach. Rev. Financ. stud. **14**(1), 113–147 (2001)
6. Zadeh, L.A.: The concept of a linguistic variable and its application to approximate reasoning-I. Inf. Sci. **8**(3), 199–249 (1975)
7. Glosten, L.R., Jagannathan, R., Runkle, D.E.: On the relation between the expected value and the volatility of the nominal excess return on stocks. J. Finance **48**(5), 1779–1801 (1993)
8. Christoffersen, P., Elkamhi, R., Feunou, B., Jacobs, K.: Option valuation with conditional heteroskedasticity and nonnormality. Rev. Financ. Stud. **23**(5), 2139–2183 (2009)

Nonlinear Dependence Structure in Emerging and Advanced Stock Markets

Roengchai Tansuchat[✉] and Woraphon Yamaka

Puey Ungphakorn Center of Excellence in Econometrics, Faculty of Economics,
Chiang Mai University, Chiang Mai, Thailand
roengchaitan@gmail.com

Abstract. The aim of this paper is to propose smooth transition (ST) copula as new model to capture nonlinear or two regimes dependence structure between emerging and advanced stock markets, and compare the performance of ST copula with Markov-Switching (MS) copula and traditional copula. The data consists of two sets of stock markets, namely five emerging stock markets: China, India, Brazil, Indonesia, and Turkey, and two advanced stock markets: United Kingdom and United States of America. The results show that ST student-t copula for two-regime dependent structure outperforms MS copula, and one regime copula. Thus, ST copula is more appropriate model for the dependence structure between emerging and advanced stock markets.

Keywords: Nonlinear copula · Markov switching · Smooth transition

1 Introduction

Emerging markets are the markets of developing countries that are expanding their economies, trade and investment. Their importance lies in their role in driving economic growth in the global economy. They include notably China, India, Indonesia, Brazil, Turkey, and others. Emerging markets have five characteristics, namely lower-than-average per capita income, rapid growth, high volatility, small capital markets compared to developed markets, and higher-than-average return for investors. Because of lower-cost consumer goods, most of them focus on an export-driven strategy to satisfy market needs of advanced economies. Meanwhile, investors from developed countries taking the advantage of emerging markets' features make the investment in the latter. These have resulted in an extraordinary relationship between emerging economies and advanced economies in terms of both international trade and investment; and this is to say not only in real sector but also in financial sector.

During 2007–2010, the United States subprime mortgage crisis was a nationwide financial crisis that contributed to the U.S. and global recession (Manda 2010). The report of the World Economic Outlook of IMF explained the interdependence between emerging equity market and advanced market which in line with the Global Risks Report of World Economic Forum that describes the dependence in volatility in the stock markets between emerging and advanced economies.

From Fig. 1, it is clear that the world's major global indices (United Kingdom: FTSE 100 and United States of America: DJI) have fallen sharply during the crisis.

© Springer Nature Switzerland AG 2019
H. Seki et al. (Eds.): IUKM 2019, LNAI 11471, pp. 210–221, 2019.
https://doi.org/10.1007/978-3-030-14815-7_18

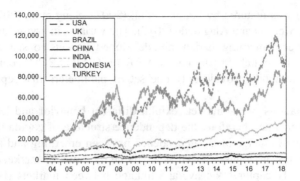

Fig. 1. Stock market indices

In the subprime crisis, the capital was transferred from advanced economies to emerging economics. It represents the dependence of each other in the global stock market. Furthermore, there is abundant evidence in the literature that there is a dependence in global stock markets. For example, Cha and Oh (2000) studied the relationship between the United States and Japan and the four emerging Asian equity markets including Hong Kong, Korea, Singapore and Taiwan. Wong's et al. (2004) studied the relationship between stock markets of major developed countries and Asian emerging markets. Samitas et al. (2007) examined the financial crises and stock market dependence of BRIC and developed markets. Tripathi and Sethi (2010) studied the relationship of Indian stock market with major global Stock Markets. Mensah and Alagidede (2017) examined the dependency structure of emerging African markets and two developed markets. Recently, Song et al. (2018) studied the dynamic conditional relationships between US and Korean markets.

The dependency is very important to the movement of the stock price level, and it is necessary to understand how many markets around the world influence on another particularly between emerging and advanced stock markets. The impacts and results of linked and interdependent financial markets become evidently pronounced during financial crisis. This effect was a result of increased stock market movements during the financial crisis and after the financial crisis. In addition, evidence from the past financial crisis suggests that financial market volatility has resulted in higher levels of outbreaks and relationships, reducing the chance of risk diversification thus causing a higher risk. Dependency between international stock markets strongly affects the risk management in each country. The interdependence of stock markets and diversification of international portfolios contribute to the mutual risk and return. Consequently, the level of interdependence between markets has a significant impact in terms of forecasting, portfolio diversification and asset allocation. Therefore, understanding the structure of interdependence between emerging and advanced stock markets is critical for global investors in risk management and risk diversification and strategic risk diversification.

In the past, the relationship between emerging markets and developed markets has been studied using linear models and nonlinear models. Linear model was adopted by Cha and Oh (2000), Wong et al. (2004), Tripathi and Sethi (2012) and Wei (2014).

Nonlinear model was featured by Samitas et al. (2007), Song et al. (2018), Seifoddini et al. (2017) and Mensah and Alagidede (2017). The variants in nonlinear model for the relationship between emerging markets and developed markets consist of time-varying copula with Markov switching parameters, VAR-DCC-MGARCH, copula and others. However, these methods provide only one set of parameters that represents all the data set.

In many situations, stock markets exhibit different behavior and lead to different dependencies over time; therefore, the dependence structure of copula model may not reflect the structural change in the dependence between emerging and advanced markets. In addition, it is no longer viewed as a reliable measure for market dependence as it is static and only captures the average linkage between markets (BenSaïda et al. 2018). This is to say, such market returns tend to be more dependent during crisis periods and periods of high volatility in international equity markets than otherwise. Prior studies have mainly measured cross market dependence with structural change through Markov Switching (MS) copula model, pioneered by Rodriguez (2007). This model has emerged as a promising method for dependence modeling owing to its ability to simultaneously deal with asymmetric and structural change dependence as documented in many recent studies (Pastpipatkul et al. 2016a; Pastpipatkul et al. (2016b) and Zhu et al. 2016) The model becomes more flexible to the dependence structural change as it allows the dependence copula parameter to be governed by hidden Markov process.

Nevertheless, in many situations, a specific event can lead to changes in the marginal cumulative distributions. The question then becomes whether the Markov Switching (MS) copula model has an ability to capture specific event changes in the copula dependence or not. Therefore, in this study, we introduce another non-linear copula model, called "Smooth Transition (ST) copula." To the best of our knowledge, this model has not been proposed in the literature yet. Extending copula approaches used in the dependence measure literature, we apply Smooth Transition of Silvennoinen and Teräsvirta (2009) to copula model that allows dynamic structure of the correlations to be controlled by exogenous transition variables. The appealing feature of this model is that it can capture both gradual and sudden changes in dependence patterns. The comparison between Smooth Transition and Markov Switching approaches under the Autoregressive model context has been investigated by Deschamps (2008). He mentioned that Markov Switching model incorporates less prior information than Smooth Transition model as the filtered or smoothed regime probability in an MS model can be interpreted as a transition function which is estimated flexibly from the data alone. However, this transition function of ST model needs some prior information on the factors determining the transitions between regimes, in the form of a particular transition function and transition variable. It is therefore expected that *an appropriate* ST model should make better use of available information and perform better than MS model.

The aim of this paper is to propose a new nonlinear ST copula to model the dependence between the emerging and advanced stock markets. Also, it is of interest to investigate and compare the performance of MS copula and ST copula. Finally, the best fit model is used to investigate the nonlinear dependence between the emerging and advanced stock markets. These are three contributions of the paper. The remainder of this study is structured as follows. Section 2 describes the methodology used in this study. Section 3 describes the two-step estimation procedure for the model. Section 4, presents the marginal modeling. In Sect. 5, the data and empirical results are presented and the last section concludes the study.

2 The Model

In this section, we first provide a brief account of copula theory, before we turn to introducing two nonlinear copula models, namely smooth transition copula and Markov Switching copula.

2.1 Bivariate Copula

Copula is a joint distribution function which can be decomposed into marginal distributions of n random variables (Trivedi and Zimmer 2007). Suppose that $x_1, \ldots, x_n$ are continuous random variables with marginals $F_1(x_1), \ldots, F_n(x_n)$, and let $H(x_1, \ldots, x_n)$ be a joint distribution function. Then, the copula $C(u_1, \ldots, u_n)$ exists such that $\forall x_i \in \mathbb{R}^n$, where $i = 1, \ldots, n$.

$$H(x_1, \ldots, x_n) = C(F_1(x_1), \ldots, F_n(x_n)), \tag{1}$$

$$C(u_1, \ldots, u_n) = H(F_1^{-1}(u_1), \ldots, F_n^{-1}(u_n)), \tag{2}$$

where $u_1, \ldots, u_n$ are cumulative distribution function of standardized residuals which are all uniform on the $[0, 1]$ interval. Another advantage of copula is that it distinguishes the marginal behavior and dependence structure from a joint distribution function (Mahfoud and Michael 2012). In this study, we have concentrated our attention on bivariate copulas, thus we model the dependence structure of the univariate marginal distributions of two random variables. The two bivariate copulas considered are Gaussian and Student-t copulas. Interested readers can refer to Nelsen (2007); Joe (2014) for more information about copulas and their properties.

2.2 Smoot Transition Copula

This section presents the first nonlinear copula called "Smoot Transition (ST) copula model". This model combines the copula theory with Smooth Transition model of Silvennoinen and Teräsvirta (2009). Basically, three basic functions, namely logistic function, exponential function, and second order logistic function, are used as a smooth transition function in order to allow for higher degree of flexibility in copula parameters. However, in this study, we adopt only logistic function as a transition function in the bivariate copula model. The model can be described as in the following

$$\begin{aligned} C(u_1, u_2 | \theta_{S_t}) = & G(v, \delta, c) \cdot \left(F_1^{-1}(u_1), F_2^{-1}(u_2) | \theta_{S_t=1} \right) \\ & + (1 - G(v, \delta, c)) \cdot \left(F_1^{-1}(u_1), F_2^{-1}(u_2) | \theta_{S_t=2} \right)' \end{aligned} \tag{3}$$

where θ_{S_t} is the regime dependent parameters with value assumed to be $\theta_{S_t=1}$ for high dependence regime, and $\theta_{S_t=2}$ for low dependence regime. S_t denotes the unobserved regime at time t. As we can see in Eq. (3), the dependence structure is driven by logistic transition function

$$G(v, \delta, c) = (1 + \exp[-\delta(v - c)])^{-1}, \tag{4}$$

where c is the threshold parameter, δ determines the speed and smoothness of the transition. v is the transition variable, which can be either u_1 or u_2; or both. Note that when $\delta \to \infty$, our model becomes the sudden switch Threshold copula. When $\delta \to 0$, our model becomes the one regime bivariate copula.

2.3 Markov Switching Copula

Another type of nonlinear copula considered in this study is Markov Switching (MS) copula. Different from STC model, this model combines the bivariate copula with Markov Switching approach of Hamilton (1989). This model is based on the idea that the process is evolving according to an unobserved Markov chain. The specification used here has the following form:

$$\begin{aligned} C(u_1, u_2 | \theta_{S_t}) = {} & Pr(S_t = 1 | u_1, u_2) \cdot \left(F_1^{-1}(u_1), F_2^{-1}(u_2) | \theta_{S_t=1}\right) \\ & + (1 - Pr(S_t = 1 | u_1, u_2)) \cdot \left(F_1^{-1}(u_1), F_2^{-1}(u_2) | \theta_{S_t=2}\right), \end{aligned} \tag{5}$$

where $\theta_{S_t=1}$ and $\theta_{S_t=2}$ are governed by the first order Markov Chain. A Markov Chain is defined as a process for which the present realization of the process only depends on the most recent past realization. As is standard in the literature we assume that S_t can be completely characterized by its transition matrix P

$$P = \begin{bmatrix} p_{11} & p_{11} & \cdots & p_{1K} \\ p_{21} & p_{22} & \cdots & p_{2K} \\ \vdots & \vdots & \ddots & \vdots \\ p_{K1} & p_{K2} & \cdots & p_{KK} \end{bmatrix}, \tag{6}$$

where p_{ij} is the probability of switching from regime i to regime j, which can be defined as

$$\begin{aligned} p_{ij} &= Pr\{S_t = j | S_{t-1} = i, S_{t-2} = k, \ldots\} \\ &= Pr\{S_t = j | S_{t-1} = i\} \text{ and } \sum\nolimits_{ij=1}^{K} p_{ij} = 1; i, j = 1, \ldots, K. \end{aligned} \tag{7}$$

According to the specification of this MS-copula model, we have K regimes dependence parameters and transition probability parameters to be estimated. As seen in Eq. 5, $Pr(S_t = 1 | u_1, u_2)$ is not observed filtered probabilities. In order to compute this, we use the Hamilton Filter methodology, as explained in Hamilton (1989). For two regimes model, $Pr(S_t = j | u_1, u_2)$ can be obtained by

$$Pr(S_t = j | u_1, u_2) = \frac{f(\theta(S_t = j) | u_1, u_2) Pr(S_t = j | u_1, u_2) p_{jj}}{\sum\limits_{j=1}^{2} f(\theta(S_t = j) | u_1, u_2) Pr(S_t = j | u_1, u_2) p_{jj}} \tag{8}$$

where $f(\Theta(S_t = j) | u_1, u_2)$ is the copula density.

3 Estimation

The estimation of our two nonlinear copula models, namely MS copula and ST copula, is performed by Maximum Likelihood estimator (MLE). We note that two-step estimation or the Inference Function of Margins (IFM) method (Joe and Xu 1996) is used in this study as the one step estimation normally could be computationally intensive in the case of large parameter estimates, as it requires us to jointly estimate the parameters of the margins and the parameters of the dependence and probability. Generally speaking, we estimate the parameters in a two-step procedure that separates the marginals from the dependence structure.

In this study, we consider only two regimes model where the marginal densities are not regime-dependent and copula density is assumed to be regime dependent. Thus, the full likelihood of our two models are given as

$$l(\Theta_{S_t} | Y) = l(\phi_1 | y_1) + l(\phi_2 | y_2) + \sum_{S_t = j}^{2} c(u_1, u_2 | \theta_{S_t}, p_{11}, p_{22}) \mathbf{P}, \tag{9}$$

where Y is the vector of the random variables y_1 and $y_2 \cdot \Theta_{S_t} = \{\phi_1, \phi_2, \theta_{S_t}, p_{11}, p_{22}\}$. $c(\cdot)$ is the density function of the bivariate copula (see, Schmidt 2007). $\mathbf{P}$ is logistic function, Eq. 4, for ST copula is a filtered probability, Eq. 8, for MS copula. $l(\phi_1 | y_1)$ and $l(\phi_2 | y_2)$ are the likelihood of the marginal model. Note that, we consider the Glosten-Jagannathan-Runkle GARCH (GJR-GARCH) as our marginal model. The brief description of this model is provided in the next section. As we already mentioned, two-step estimation method is employed, thus we firstly estimate the univariate marginal distributions for y_1 and y_2. This estimation is straightforward, as it does not depend on the regime switching, thus we can estimate each GJR-GARCH model separately:

$$\hat{\phi}_m = \arg\max_{\theta_m} \log l(\phi_m | y_m), \quad m = 1, 2. \tag{10}$$

We then collect the coefficients in vector: $\hat{\phi} = \{\hat{\phi}_1, \hat{\phi}_2\}$ and use these marginal parameters as the fixed parameters in our full likelihood Eq. 9. Thus, in the second step, we estimate only the dependence and probability parameters.

$$\hat{\Theta}_{S_t} = \arg\max_{\hat{\Phi}} \log l(\theta_{S_t}, p_{11}, p_{22} | Y) + \log l(\hat{\phi}_1 | y_1) + \log l(\hat{\phi}_2 | y_2). \tag{11}$$

Under certain regularity conditions, the IFM estimator verifies the property of asymptotic normality and can be seen as a highly efficient estimator compared to the one step estimation (Joe 1997).

4 Marginal Distribution Modeling

In order to take into account the dynamics of the volatility clustering and the leverage effects (asymmetry) on return volatility arising from a negative shock versus a positive shock, we model the marginal distributions of each one of our returns using the univariate Glosten-Jagannathan-Runkle GARCH (GJR-GARCH) of Glosten, Jagannathan and Runkle (1993). In this study, we consider ARMA (1,1)- (GJR-GARCH(1,1).

Specifically, the model is expressed as

$$y_t = a_0 + a_1 y_{t-1} - a_2 \varepsilon_{t-1} + \varepsilon_t, \tag{12}$$

$$\varepsilon_t = \sigma_t z_t \tag{13}$$

$$\sigma_t^2 = \omega + \alpha \varepsilon_{t-1} + \beta \sigma_{t-1}^2 + \gamma I_{t-1} \varepsilon_{t-1}, \tag{14}$$

where $I_{t-i} = 1$ if $\varepsilon_{t-i} < 0$ else $I_{t-i} = 0$ if $\varepsilon_{t-i} \geq 0$. σ^2 is conditional variance obtained from the GJR-GARCH (1,1) process in Eq. 14. $\phi = \{a_0, a_2, a_3, \omega, \alpha, \beta, \gamma\}$ is the vector of model parameters. z_t is a standardized residual which is assumed to have skewed-t distribution. Then, the standardized residuals are to be transformed into a uniform distribution in (0, 1) using cumulative skewed student-t distribution.

5 Empirical Study

5.1 Data and Summary Statistics

In this study, we aim to examine the nonlinear dependence between five emerging stock markets and two advanced stock markets. Therefore, our data consists of two sets of stock markets. In the first set are five emerging stock markets of China (SSE Composite), India (S&P BSE 30), Brazil (BOVESPA), Indonesia (IDX Composite), and Turkey (BIST 100). The second set contains two advanced stock markets, namely United Kingdom (FTSE 100) and United States of America (DJI). The daily closing price indexes for these seven stock markets start from 2004, January 2nd to 2018, October, 15th, with a total of 3,858 observations. All data are transformed to be log-returns.

Table 1 presents the descriptive statistics of seven returns composing of mean, median, maximum, minimum, standard deviation, skewness, kurtosis, normality Jarque-Bera (JB) test and Augmented Dickey Fuller (ADF) unit root test. All market returns have a positive mean and median. In addition, all returns have slightly left-tailed (negative skewness) with their kurtosis values higher than the kurtosis values of the normal distribution (3). This means that both distributions are leptokurtic. As a result, the Jarque-Bera (JB) test confirms a strong evidence supporting a non-normal

Table 1. Data description

	China	India	Brazil	Indonesia	Turkey	UK	USA
Mean	0.0001	0.0002	0.0001	0.0002	0.0002	0.0001	0.0001
Median	0.0000	0.0001	0.0000	0.0002	0.0002	0.0001	0.0001
Maximum	0.0392	0.0694	0.0594	0.0331	0.0527	0.0408	0.0456
Minimum	−0.0402	−0.0513	−0.0525	−0.0476	−0.0481	−0.0402	−0.0356
Std. Dev.	0.0069	0.0060	0.0073	0.0056	0.0071	0.0047	0.0046
Skewness	−0.5465	−0.0895	−0.0683	−0.6794	−0.2783	−0.1616	−0.1584
Kurtosis	7.8713	13.781	8.5486	11.126	6.6457	12.192	15.237
JB	[0.0000]	[0.0000]	[0.0000]	[0.0000]	[0.0000]	[0.0000]	[0.0000]
ADF-test	[0.0000]	[0.0000]	[0.0000]	[0.0000]	[0.0000]	[0.0000]	[0.0000]

Note: [] is Maximum Bayes Factor (MBF)

distribution of these markets' return, because the MBF are all zero. Finally, ADF tests indicate that every return series are strongly stationary.

The results of the marginal estimation from GJR-GARCH with skew student-t distribution of error term are presented in Table 2. The result shows that the sum of the estimated parameters $\alpha + \beta$ are close to 1, meaning that the unconditional variance of the error terms is finite where the conditional variance evolves over time and is persistent. In addition, the estimated parameters $\gamma > 0$, meaning negative shocks will have larger effects on volatility than positive shocks or leverage effect. The estimated shape and skewness parameters of the skewed student-t distribution indicate the benefits of employing skewed student-t distribution to approximate the true shape of our returns. We then transform the standardized residuals generated from ARMA(1,1)-GJR-GARCH(1,1) model into the cumulative skewed student-t distribution. These uniform series will be used to model the dependence structure between the marginal distributions of each emerging stock market and advanced stock market.

Table 2. Marginal estimation for GJR-GARCH results.

Index	a_0	a_1	a_2	ω	α	β	γ	Skewness	Shape
China	−0.0005	1.0000	−0.9886	0.00001	0.0454	0.9450	0.0174	0.9525	3.9344
	(0.0003)	(0.0015)	(0.0001)	(0.00001)	(0.0042)	(0.0025)	(0.0103)	(0.0168)	(0.2391)
India	0.0002	−0.1497	0.2077	0.00001	0.0197	0.8987	0.1372	0.9529	6.4631
	(0.0001)	(0.9689)	(0.9602)	(0.00001)	(0.0035)	(0.0194)	(0.0245)	(0.0189)	(0.4279)
Brazil	0.0001	0.6831	−0.7027	0.00001	0.0165	0.9152	0.0884	0.9562	8.2899
	(0.0001)	(0.1685)	(0.1644)	(0.00001)	(0.0106)	(0.0322)	(0.0278)	(0.0174)	(0.5351)
Indonesia	0.0002	−0.7312	0.7586	0.00001	0.0653	0.8623	0.1062	0.9268	4.5023
	(0.0001)	(0.2641)	(0.2519)	(0.00001)	(0.0079)	(0.0153)	(0.0144)	(0.0188)	(0.2518)
Turkey	0.0002	−0.7312	0.7586	0.00001	0.0653	0.8623	0.1062	0.9268	4.5023
	(0.0001)	(0.2641)	(0.2519)	(0.00001)	(0.0079)	(0.0153)	(0.0144)	(0.0188)	(0.2518)
UK	0.00004	0.4217	−0.4542	0.00001	0.0001	0.8942	0.1761	0.8795	8.6310
	(0.00003)	(0.2527)	(0.2485)	(0.00001)	(0.0001)	(0.0280)	(0.0301)	(0.0135)	(0.8467)
USA	0.0001	0.1432	−0.1920	0.00001	0.00002	0.8863	0.1920	0.8967	6.0884
	(0.00004)	(0.4267)	(0.4244)	(0.00001)	(0.01285)	(0.0239)	(0.0436)	(0.0231)	(0.9774)

Note: () denotes standard error

6 Model Comparison and the Estimation Results

The uniform series obtained from the first step of the estimation are then used as input for single-regime and two-regime forms of bivariate Gaussian and Student-t copula discussed in Sect. 2. The lowest value of Bayesian information criterion (BIC) is used as the measure to select the best fit copula model. The estimated BIC values are reported in Table 3. The result shows that ST copula for two regimes copula structure slightly outperforms the one regime copula with respect to 7 out of 10 pairs, suggesting that ST copula is more appropriate choice for the dependence structure between emerging and advanced stock markets. This result confirms us that a one regime copula which is intensively employed in the literature may not be reliable in some financial markets. Interestingly, the MS copula results for two-regime copula of all pairs do not provide a better performance over one regime copula as the higher BIC values are presented. We expect that the filtered probabilities may not well capture the large financial sample data. However, this is just our expectation, we leave this issue for the future study.

Table 3. Model comparison

BIC	Copula		MS-Copula		ST-Copula	
Pair	Gaussian	Student-t	Gaussian	Student-t	Gaussian	Student-t
China - UK	86.836	90.891	127.163	145.378	116.814	**39.569**
China - USA	27.451	**26.902**	69.749	78.244	52.861	26.985
India - UK	542.269	555.344	613.382	638.023	575.151	**498.730**
India - USA	192.765	198.859	232.298	248.407	236.537	**158.152**
Brazil - UK	769.588	816.257	894.698	922.393	783.312	**765.268**
Brazil - USA	**1,324.89**	1,406.24	1,584.74	1,622.97	1,340.36	1,394.55
Indonesia - UK	272.308	279.935	312.435	329.483	288.857	**224.262**
Indonesia - USA	**56.743**	58.256	98.783	113.290	82.388	124.451
Turkey - UK	743.173	779.780	797.024	888.296	741.834	**734.713**
Turkey - USA	319.453	336.324	371.456	411.095	347.123	**281.232**

Note: The bolded value is the minimum of BIC.

In essence, the above empirical comparison analysis presents differing results for the two sets of our sample. Consistent with the existing literature, (see, Pastpipatkul et al. 2016a; Zhu et al. 2016; BenSaïda et al. 2018), the nonlinear copula does not evidently dominate the linear copula model. The evidence of linear dependence in the returns of China-USA, Brazil-USA, and Indonesia-USA market pairs, indicates that the dependence appears unaffected by structural change. Nevertheless, dominance of ST copula for other seven pairs may support the application of the nonlinear dependence structure and smoothly structure changing between the emerging and advanced stock markets.

The parameter estimates and their robust standard errors of student-t copula from one regime or linear student-t copula and two-regime or smooth transition student-t are shown in Table 4 in order to compare their dependence values. The dependence parameters of single regime or linear student-t copula can be classified as low positive value in pairs of China – USA, Indonesia – USA, China – UK, India – USA, Indonesia – UK, and Turkey – USA, and moderate positive value in India – UK, Brazil – UK, Turkey – UK, and Brazil – USA. In the case of smooth transition student-t copula, the estimated $\theta_{S_t=1}$ differs from and higher than $\theta_{S_t=2}$ indicating the difference of dependence between emerging and advanced stock markets in each regime. In addition, threshold parameter (c) and smooth parameter (δ) show the smoothly structural change in the dependence between emerging and advanced stock markets. Most of them have low smooth parameter values (δ) ranging from 2.999 to 4.993 indicating smooth switch threshold copula excepting for higher smooth parameter values of Turkey – USA and India – UK being 19.868 and 18.188 meaning higher speed switch threshold copula.

Table 4. Estimation results

Model	Linear student-t copula	Smooth transition student-t copula			
Pair	θ	$\theta_{S_t=1}$	$\theta_{S_t=2}$	δ	c
China - UK	0.14	0.182	0.011	4.993	0.251
	(0.02)	(0.022)	(0.053)	(0.3523)	(0.083)
China - USA	0.07	0.105	0.009	3.004	0.349
	(0.02)	(0.034)	(0.017)	(1.858)	(0.166)
India - UK	0.36	0.383	0.205	19.868	0.142
	(0.01)	(0.015)	(0.053)	(8.636)	(0.027)
India - USA	0.22	0.274	0.094	3.558	0.317
	(0.01)	(0.027)	(0.065)	(0.024)	(0.042)
Brazil - UK	0.42	0.471	0.336	4.155	0.611
	(0.01)	(0.019)	(0.031)	(0.023)	(0.005)
Brazil - USA	0.54	0.621	0.399	2.999	0.605
	(0.01)	(0.013)	(0.022)	(0.017)	(0.004)
Indonesia - UK	0.26	0.281	0.200	3.054	0.597
	(0.01)	(0.019)	(0.030)	(0.064)	(0.005)
Indonesia - USA	0.11	0.133	0.023	3.889	0.207
	(0.02)	(0.046)	(0.035)	(0.025)	(0.247)
Turkey - UK	0.42	0.469	0.332	4.487	0.582
	(0.01)	(0.015)	(0.024)	(0.097)	(0.003)
Turkey - USA	0.28	0.327	0.055	18.188	0.170
	(0.01)	(0.060)	(0.003)	(3.451)	(0.029)

Note: Degrees of freedom of Student-t copula are not provided. () is standard error

7 Conclusion

This paper aims to study dependence between emerging stock markets and advanced stock markets by proposing a new nonlinear structure copula by combing the copula theory with smooth transition model of Silvennoinen and Teräsvirta (2009) based on Logistic function. Next we compare the performance of nonlinear or two regimes bivariate smooth transition copula with Markov Switching copula and one regime or linear structure copula. The data consists of seven stock price indexes from five emerging stock markets and two advanced stock markets starting from 2004, January to 2018, October. Thus, 10 pairs of linear structure and 20 nonlinear bivariate (MS and ST) copula based ARMA (1, 1)-GJR-GARCH (1, 1) are estimated and compared their performance with BIC. The result shows that ST copula for two regimes copula structure outperforms one regime copula and MS copula. Thus, the nonlinear dependence structure exists. In addition, ST copula is the more appropriate choice for the dependence structure between emerging and advanced stock markets.

Acknowledgement. This research work was partially supported by Chiang Mai University and Faculty of Economics Chiang Mai University.

References

BenSaïda, A., Boubaker, S., Nguyen, D.K.: The shifting dependence dynamics between the G7 stock markets. Quant. Financ. **18**(5), 801–812 (2018)

Cha, B., Oh, S.: The relationship between developed equity markets and the Pacific Basin's emerging equity markets. Int. Rev. Econ. Financ. **9**(4), 299–322 (2000)

Deschamps, P.J.: Comparing smooth transition and Markov switching autoregressive models of US unemployment. J. Appl. Econ. **23**(4), 435–462 (2008)

Glosten, L.R., Jagannathan, R., Runkle, D.E.: On the relation between the expected value and the volatility of the nominal excess return on stocks. J. Financ. **48**(5), 1779–1801 (1993)

Hamilton, J.D.: A new approach to the economic analysis of nonstationary time series and the business cycle. Econometrica **57**, 357–384 (1989)

Joe, H.: Multivariate Models and Dependence Concepts. Chapman & Hall, London (1997)

Joe, H.: Dependence Modeling with Copulas. Chapman & Hall/CRC, London (2014)

Joe, H., Xu, J.J.: The estimation method of inference functions for margins for multivariate models (1996)

Mahfoud, M., Michael, M.: Bivariate archimedean copulas: an application to two stock market indices. BMI Paper (2012)

Manda, K.: Stock market volatility during the 2008 financial crisis. Eur. Financ. Manag. **17**, 789–805 (2010)

Mensah, J.O., Alagidede, P.: How are Africa's emerging stock markets related to advanced markets? Evidence from copulas. Econ. Model. **60**, 1–10 (2017)

Nelsen, R.B.: An Introduction to Copulas. Springer, New York (2007). https://doi.org/10.1007/0-387-28678-0

Pastpipatkul, P., Yamaka, W., Sriboonchitta, S.: Analyzing financial risk and co-movement of gold market, and Indonesian, Philippine, and Thailand stock markets: dynamic copula with Markov-switching. In: Huynh, V.N., Kreinovich, V., Sriboonchitta, S. (eds.) Causal Inference in Econometrics. SCI, vol. 622, pp. 565–586. Springer, Cham (2016a). https://doi.org/10.1007/978-3-319-27284-9_37

Pastpipatkul, P., Maneejuk, P., Sriboonchitt, S.: The best copula modeling of dependence structure among gold, oil prices, and U.S. currency. In: Huynh, V.N., Inuiguchi, M., Le, B., Le, B.N., Denoeux, T. (eds.) Integrated Uncertainty in Knowledge Modelling and Decision Making. LNCS, vol. 9978, pp. 493–507. Springer, Cham (2016b). https://doi.org/10.1007/978-3-319-49046-5_42

Rodriguez, J.C.: Measuring financial contagion: a copula approach. J. Empir. Financ. **14**(3), 401–423 (2007)

Samitas, A., Kenourgios, D., Paltalidis, N.: Financial crises and stock market dependence. In: 14th Annual Conference of the Multinational Finance Society, Thessaloniki, Greece (2007)

Schmidt, T.: Coping with copulas. In: Copulas-From theory to Application in Finance, pp. 3–34 (2007)

Seifoddini, J., Roodposhti, F.R., Kamali, E.: Gold-stock market relationship: emerging markets versus developed markets. Emerg. Mark. J. **7**(1), 17–24 (2017)

Silvennoinen, A., Teräsvirta, T.: Modeling multivariate autoregressive conditional heteroskedasticity with the double smooth transition conditional correlation GARCH model. J. Financ. Econ. **7**(4), 373–411 (2009)

Song, W., Park, S.Y., Ryu, D.: Dynamic conditional relationships between developed and emerging markets. Phys. A Stat. Mech. Appl. **507**(C), 534–543 (2018)

Tripathi, V., Sethi, S.: Integration of indian stock market with major global stock markets. Asian Journal of Business and Accounting **3**(1), 117–134 (2010)

Tripathi, V., Sethi, S.: Inter linkages of Indian stock market with advanced emerging markets. Asia Pac. Finan. Acc. Rev. **1**(1), 34–51 (2012)

Trivedi, P.K., Zimmer, D.M.: Copula modeling: an introduction for practitioners. Found. Trends® Econ. **1**(1), 1–111 (2007)

Wei, S.: China's a-share, b-share, and h-share stock markets and the world financial markets: a cointegration and causality analysis. J. Appl. Bus. Econ. **16**(2), 70–80 (2014)

Wong, W.K., Penm, J., Terrell, R.D., Ching, K.Y.: The relationship between stock markets of major developed countries and Asian emerging markets. J. Appl. Math. Decis. Sci. **8**(4), 201–218 (2004)

Zhu, K., Yamaka, W., Sriboonchitta, S.: Multi-asset portfolio returns: a markov switching copula-based approach. Thai J. Math., 183–200 (2016). Special Issue on Applied Mathematics: Bayesian Econometrics

Mean Absolute Deviation Portfolio Frontiers with Interval-Valued Returns

Songkomkrit Chaiyakan$^{(\boxtimes)}$ (ID) and Phantipa Thipwiwatpotjana (ID)

Department of Mathematics and Computer Science, Faculty of Science,
Chulalongkorn University, Bangkok, Thailand
songkomkrit.c@gmail.com

Abstract. This work discusses the frontiers of mean absolute deviation portfolios arising from the uncertainty of future rates of return. The risk of the overall portfolio is proposed as an objective function to attain a well-diversified portfolio with a predetermined target rate of return. The possible ranges of the target returns are suggested via the strong feasibility of the interval linear system of constraints. No short sales are allowed and a risk-averse investor is assumed to pursue the buy-and-hold strategy. The use of the method is illustrated with the historical returns of S&P 500 stocks, for which the negative correlation condition empirically holds, with a 6-month investment horizon from November 2018 to April 2019. The historical data is collected monthly over the past 4 years from November 2014 to October 2018.

Keywords: Mean absolute deviation portfolio selection model ·
Left hand side uncertainty · Interval linear programming

1 Interval Linear Programming

The problem of portfolio selection with the mean absolute deviation risk can be modeled by linear programming when the expected asset returns are deterministic [1,2]. Practically, the return process is nonstationary. As a result, the expected returns should not be predicted solely by the arithmetic average of historical returns. The distribution of returns over time is also arguable [3]. Due to the disagreement of measurements for these parameters, intervals are employed as representatives. Throughout this work, the theory of interval linear programming is extensively used. This section aims to pave its theoretical frameworks.

1.1 Notations

An interval matrix (or vector) $\boldsymbol{Y}$ is defined by

$$\boldsymbol{Y} = [\underline{Y}, \overline{Y}] = \{Y \mid \underline{Y} \leq Y \leq \overline{Y}\},$$

© Springer Nature Switzerland AG 2019
H. Seki et al. (Eds.): IUKM 2019, LNAI 11471, pp. 222–234, 2019.
https://doi.org/10.1007/978-3-030-14815-7_19

each element of which are compared componentwise. The matrices $\underline{Y}$ and $\overline{Y}$ are its lower and upper bound respectively. Its center and radius matrices are given by

$$Y^c = (\underline{Y} + \overline{Y})/2 \quad \text{and} \quad Y^\Delta = (\overline{Y} - \underline{Y})/2$$

correspondingly. Note that the interval matrix is uncertain but its center and radius are exact.

1.2 Feasibility

The system $Ax = b$ where $A \in \mathbb{R}^{m \times n}$ and $b \in \mathbb{R}^m$ is said to be feasible if it possesses at least one nonnegative solution. When A and b are uncertain, the interval system $\boldsymbol{A}x = \boldsymbol{b}$ is said to be weakly (strongly) feasible if the deterministic system $Ax = b$ is feasible for some (all) $A \in \boldsymbol{A}$ and $b \in \boldsymbol{b}$.

Intuitively, it seems tedious to verify the strong feasibility of the interval system $\boldsymbol{A}x = \boldsymbol{b}$ owing to the uncountability of an interval set. To overcome this difficulty, it may be guessed by considering all end points independently. However, the following theorem addresses this issue with the reduced number of steps by examining the end points, either left ends or right ends for each row, of all intervals embedded in the interval matrix $\boldsymbol{A}$.

Theorem 1 (see [4]). *A system $\boldsymbol{A}x = \boldsymbol{b}$ is strongly feasible if and only if for each $y \in \{\pm 1\}^m$ the system $(A^c - diag(y)A^\Delta)x = b^c + diag(y)b^\Delta$ has a nonnegative solution.*

Sometimes, the strongly feasible solution set may be empty. To further examine the behavior of the system $\boldsymbol{A}x = \boldsymbol{b}$, the weakly feasible solution set may be considered because its element simply satisfies at least one possibility. The following theorem describes its weakly feasible solution set in the exact manner.

Theorem 2 (see [4]). *The weakly feasible solution set to the system $\boldsymbol{A}x = \boldsymbol{b}$ is given by the set $\{x \mid \underline{A}x \leq \overline{b}, \overline{A}x \geq \underline{b}, x \geq 0\}$.*

When the constraint of equality is relaxed to the inequality $\boldsymbol{A}x \leq \boldsymbol{b}$, its weakly feasible solution set is described in the following theorem.

Theorem 3 (see [4]). *The weakly feasible solution set to the system $\boldsymbol{A}x \leq \boldsymbol{b}$ is given by the set $\{x \mid \underline{A}x \leq \overline{b}, x \geq 0\}$.*

1.3 Range of Optimal Values

Consider an interval linear programming problem

$$\begin{aligned} \text{minimize} \quad & c^\mathsf{T}x \\ \text{subject to} \quad & Ax = b, \\ & x \geq 0. \end{aligned} \tag{1}$$

Usually, the primary concern is to compute the set of all weakly optimal solutions to at least one deterministic linear program

$$\text{minimize} \quad c^\mathsf{T} x$$
$$\text{subject to} \quad Ax = b,$$
$$x \geq 0$$

where $A \in \boldsymbol{A}$, $b \in \boldsymbol{b}$ and $c \in \boldsymbol{c}$. Arising from the uncertainties, there is a wide range of these optimal values. Generally, the minimum and maximum of optimal values are not necessarily existent [5]. Instead, its infimum and supremum may be of interest.

Assume the minimum and maximum exist. Both can be obtained by the minimin and maximin problems. Alternatively, the following theorems provide how to calculate the extreme values without the aid of nonlinear programming.

Theorem 4 (see [4,5]). *If the minimum of optimal objective values of the interval linear program* (1) *exists, then it is identical to the optimal value of the following linear program:*

$$\text{minimize} \quad \underline{c}^\mathsf{T} x$$
$$\text{subject to} \quad \underline{A}x \leq \overline{b},$$
$$\overline{A}x \geq \underline{b},$$
$$x \geq 0.$$

Theorem 5 (see [4,5]). *If the maximum of optimal objective values of the interval linear program* (1) *exists, then it is identical to*

$$\max_{z \in \{\pm 1\}^m} f_z$$

where f_z is the optimal value to the following linear program:

$$\text{maximize} \quad (b^c + diag(z)b^\Delta)^\mathsf{T} y$$
$$\text{subject to} \quad (A^c - diag(z)A^\Delta)^\mathsf{T} y \leq \overline{c},$$
$$diag(z)y \geq 0.$$

2   Mean Absolute Deviation (MAD) Portfolio Model

Suppose that, at time T, a risk-averse investor wants to invest proportion w_i of capital in asset i (where $i = 1, 2, \ldots, n$) with the buy-and-hold strategy. During this period, the rate of return of each asset i can be represented by a random variable $R_{i,T+1}$. The overall rate of return of the portfolio becomes

$$R_{T+1} = w_1 R_{1,T+1} + w_2 R_{2,T+1} + \ldots + w_n R_{n,T+1}.$$

When no short sales are allowed, the proportion invested in each asset i cannot be negative: $w_i \geq 0$.

Arising from the return unpredictability, the investor faces risk across the entire portfolio. The risk-averse investor decides to minimize risk with a given target rate of return θ.

2.1   Development of Model with Real-Valued Returns

Konno and Yamazaki [1] introduced the mean absolute deviation (MAD) portfolio model. Using the historical rate of return $\tilde{r}_{i,t}$ for asset i from time 1 to T, the optimal asset allocation w_i^* is achievable by the linear program

$$\text{minimize} \quad \frac{1}{T} \sum_{t=1}^{T} \left| \sum_{i=1}^{n} a_{i,t} w_i \right|$$

$$\text{subject to} \quad \sum_{i=1}^{n} r_i w_i = \theta,$$

$$\sum_{i=1}^{n} w_i = 1,$$

$$w_i \geq 0, \quad i = 1, 2, \ldots, n$$

where $r_i = \mathbb{E}_T[R_{i,T+1}]$, the expected return of asset i at time $T+1$ based on all previous trading information by time T, and $a_{i,t} = \tilde{r}_{i,t} - r_i$ for $i = 1, 2, \ldots, n$ and $t = 1, 2, \ldots, T$. This is equivalent to the linear program

$$\text{minimize} \quad \frac{1}{T} \sum_{t=1}^{T} d_t$$

$$\text{subject to} \quad d_t + \sum_{i=1}^{n} a_{i,t} w_i \geq 0, \quad t = 1, 2, \ldots, T,$$

$$d_t - \sum_{i=1}^{n} a_{i,t} w_i \geq 0, \quad t = 1, 2, \ldots, T,$$

$$\sum_{i=1}^{n} r_i w_i = \theta,$$

$$\sum_{i=1}^{n} w_i = 1,$$

$$w_i \geq 0, \quad i = 1, 2, \ldots, n,$$

$$d_t \geq 0, \quad t = 1, 2, \ldots, T.$$

To reduce the number of constraints, Feinstein and Thapa [2] added the surplus variables $2u_t$ and $2v_t$ to the first two constraints:

$$d_t + \sum_{i=1}^{n} a_{i,t} w_i - 2u_t = 0 \quad \text{and} \quad d_t - \sum_{i=1}^{n} a_{i,t} w_i - 2v_t = 0$$

where $u_t, v_t \geq 0$ for $t = 1, 2, \ldots, T$. Then

$$d_t = u_t + v_t \quad \text{and} \quad \sum_{i=1}^{n} a_{i,t} w_i - u_t + v_t = 0.$$

This leads to the minimization problem

$$\text{minimize} \quad \frac{1}{T} \sum_{t=1}^{T} (u_t + v_t)$$

$$\text{subject to} \quad -u_t + v_t + \sum_{i=1}^{n} a_{i,t} w_i = 0, \quad t = 1, 2, \ldots, T,$$

$$\sum_{i=1}^{n} r_i w_i = \theta, \tag{2}$$

$$\sum_{i=1}^{n} w_i = 1,$$

$$w_i \geq 0, \quad i = 1, 2, \ldots, n,$$

$$u_t, v_t \geq 0, \quad t = 1, 2, \ldots, T.$$

2.2 Model with Interval-Valued Returns

All parameters in the linear program (2) are certain without ambiguity except the conditional expected return $r_i = \mathbb{E}_T[R_{i,T+1}]$ of each asset i also appeared in the term $a_{i,t}$. As discussion earlier in Sect. 1, this parameter should be represented by the interval $r_i = [\underline{r}_i, \overline{r}_i]$ containing the parameter r_i.

To eliminate the dependence of $a_{i,t}$ and r_i in the portfolio model (2), the second constraint is added to the first counterpart. It leads to the interval linear program

$$\text{minimize} \quad \frac{1}{T}\sum_{t=1}^{T}(u_t + v_t)$$

$$\text{subject to} \quad -u_t + v_t + \sum_{i=1}^{n}\tilde{r}_{i,t}w_i = \theta, \quad t = 1, 2, \ldots, T,$$

$$\sum_{i=1}^{n} r_i w_i = \theta, \tag{3}$$

$$\sum_{i=1}^{n} w_i = 1,$$

$$w_i \geq 0, \quad i = 1, 2, \ldots, n,$$

$$u_t, v_t \geq 0, \quad t = 1, 2, \ldots, T.$$

For simplicity, the problem with the objective function $\sum_{t=1}^{T}(u_t + v_t)$ is investigated instead.

The optimization problem (3) in the matrix notation becomes

$$\text{minimize} \quad c^\mathsf{T} x$$

$$\text{subject to} \quad Ax = b, \tag{4}$$

$$x \geq 0$$

where

$$x = \begin{bmatrix} u^\mathsf{T}_{1\times T} & v^\mathsf{T}_{1\times T} & w^\mathsf{T}_{1\times n} \end{bmatrix}^\mathsf{T}, \; c = \begin{bmatrix} 1^\mathsf{T}_{1\times T} & 1^\mathsf{T}_{1\times T} & 0^\mathsf{T}_{1\times n} \end{bmatrix}^\mathsf{T}, \; b = b(\theta) = \begin{bmatrix} 0^\mathsf{T}_{1\times T} & \theta_{1\times 1} & 1_{1\times 1} \end{bmatrix}^\mathsf{T}$$

$$A = A(r) = \begin{bmatrix} -I_{T\times T} & I_{T\times T} & \tilde{R}^\mathsf{T}_{T\times n} \\ 0^\mathsf{T}_{1\times T} & 0^\mathsf{T}_{1\times T} & r^\mathsf{T}_{1\times n} \\ 0^\mathsf{T}_{1\times T} & 0^\mathsf{T}_{1\times T} & 1^\mathsf{T}_{1\times n} \end{bmatrix}, \; \tilde{R} = \begin{bmatrix} \tilde{r}_{1,1} & \cdots & \tilde{r}_{1,t} & \cdots & \tilde{r}_{1,T} \\ \vdots & \ddots & \vdots & \ddots & \vdots \\ \tilde{r}_{i,1} & \cdots & \tilde{r}_{i,t} & \cdots & \tilde{r}_{i,T} \\ \vdots & \ddots & \vdots & \ddots & \vdots \\ \tilde{r}_{n,1} & \cdots & \tilde{r}_{n,t} & \cdots & \tilde{r}_{n,T} \end{bmatrix},$$

$$r = \begin{bmatrix} r_1 & \cdots & r_n \end{bmatrix}^\mathsf{T} = \begin{bmatrix} [\underline{r}_1, \overline{r}_1] & \cdots & [\underline{r}_n, \overline{r}_n] \end{bmatrix}^\mathsf{T} = [\underline{r}, \overline{r}].$$

Note that every matrix belonging to the interval matrix A has a full row rank. In this program, an investor must initialize the target rate of return θ to derive the range of optimal portfolio risks to describe its frontiers for further investigation.

3   Portfolio Frontiers

For single-valued returns, a portfolio frontier [6] is a curve representing the set of all optimal portfolios that offer the lowest risk for a given target rate of return θ. In the case of interval-valued returns, a portfolio frontier is described by the region consisting of all such curves.

3.1   Range of Target Portfolio Returns

Intuitively, a target rate of return θ is valid only when a portfolio under any circumstances can generate the rate of return θ. An investor should determine the possible values of these target values beforehand. In other words, the parameter θ enables the system $\boldsymbol{A}x = b(\theta)$ strongly feasible. The following theorem is employed to establish a criterion for validating a choice of $\theta \in [\underline{\theta}, \overline{\theta}]$.

Theorem 6. *For the interval linear program* (4), *the system of constraints* $\boldsymbol{A}(r)x = b$ *is strongly feasible if and only if each of the systems* $\boldsymbol{A}(\underline{r})x = b$ *and* $\boldsymbol{A}(\overline{r})x = b$ *contains a nonnegative solution.*

Proof. According to Theorem 1, the system $(A^c - z_{T+1}A^\Delta)x = (A^c - \mathrm{diag}(z)A^\Delta)x = b^c + \mathrm{diag}(z)b^\Delta = b$ must have a nonnegative solution for every $z \in \{\pm 1\}^{T+2}$. The term z_{T+1} can take on the values 1 and -1 which correspond to on the left hand side $\boldsymbol{A}(\underline{r})$ and $\boldsymbol{A}(\overline{r})$ respectively. $\qquad\square$

As a result, the parameters $\underline{\theta}$ and $\overline{\theta}$ can be found by minimizing and maximizing the objective function θ respectively subject to each constraint separately specified in Theorem 6. A portfolio has no suitable choice for its target return if $\underline{\theta} > \overline{\theta}$. This problem can occur especially when the collected data on the expected return r is overly inexact, i.e. $\underline{r}_i \ll \overline{r}_i$. Every target value between these two extreme values is attainable in consequence of Theorem 7. The algorithm for computing the range of target portfolio returns is suggested in Algorithm 1.

Theorem 7. *Any portfolio with a target rate of return θ between $\underline{\theta}$ and $\overline{\theta}$ is always feasible.*

Proof. Suppose the two portfolios attain the target rates of return $\underline{\theta}$ and $\overline{\theta}$. By Theorem 6, $\boldsymbol{A}(\underline{r})x^{(1)} = b(\underline{\theta})$ and $\boldsymbol{A}(\underline{r})x^{(2)} = b(\overline{\theta})$ for some nonnegative solutions $x^{(1)}$ and $x^{(2)}$. Since θ is between $\underline{\theta}$ and $\overline{\theta}$, it follows that $\theta = t\underline{\theta} + (1-t)\overline{\theta}$ for some $0 \le t \le 1$. Consider $\boldsymbol{A}(\underline{r})(tx^{(1)} + (1-t)x^{(2)}) = t\boldsymbol{A}(\underline{r})x^{(1)} + (1-t)\boldsymbol{A}(\underline{r})x^{(2)} = tb(\underline{\theta}) + (1-t)b(\overline{\theta}) = b(t\underline{\theta} + (1-t)\overline{\theta}) = b(\theta)$. Hence, the system $\boldsymbol{A}(\underline{r})x = b(\theta)$ has a nonnegative solution, and so does the system $\boldsymbol{A}(\overline{r})x = b(\theta)$ in the similar manner. Theorem 6 implies there always exists a portfolio with the target rate of return θ. $\qquad\square$

Algorithm 1. Determine the range of target portfolio returns

Input: The portfolio selection model (4)
Output: The minimum target $\underline{\theta}$ and the maximum target $\overline{\theta}$
 1: Compute $\underline{\theta}_1 \leftarrow \min\{\theta \mid \boldsymbol{A}(\underline{r})x - b(\theta) = 0, \, x \geq 0, \, \theta \geq 0\}$
 2: Compute $\underline{\theta}_2 \leftarrow \min\{\theta \mid \boldsymbol{A}(\overline{r})x - b(\theta) = 0, \, x \geq 0, \, \theta \geq 0\}$
 3: Set $\underline{\theta} \leftarrow \max\{\underline{\theta}_1, \underline{\theta}_2\}$
 4: Compute $\overline{\theta}_1 \leftarrow \max\{\theta \mid \boldsymbol{A}(\underline{r})x - b(\theta) = 0, \, x \geq 0, \, \theta \geq 0\}$
 5: Compute $\overline{\theta}_2 \leftarrow \max\{\theta \mid \boldsymbol{A}(\overline{r})x - b(\theta) = 0, \, x \geq 0, \, \theta \geq 0\}$
 6: Set $\overline{\theta} \leftarrow \min\{\overline{\theta}_1, \overline{\theta}_2\}$
 7: **if** $\underline{\theta} \leq \overline{\theta}$ **then**
 8:    **return** $\underline{\theta}$ and $\overline{\theta}$
 9: **else**
10:    **return** "No suitable choice for a target portfolio return"
11: **end if**

3.2 Range of Optimal Portfolio Risks

Theorem 8. *Both minimum and maximum of optimal portfolio risks attained by the MAD model* (4) *exist.*

Proof. Obviously, the lowest optimal portfolio risk equals the minimum value of $c^{\mathsf{T}}x$ over the weakly feasible solution set to the constraint $\boldsymbol{A}x = b$ in the MAD model (4). Whenever each asset i has an expected return of $r_i \in [\underline{r}_i, \overline{r}_i]$, the primal problem (4) always possesses an optimal value due to the assumption of strong feasibility imposed on a target portfolio return as aforementioned. The strong duality theorem (see [7] for more details) implies its dual program which is a maximum problem always have an optimal value. With a dual variable y, the maximum value of $b^{\mathsf{T}}y$ over the weakly feasible solution set to the dual constraint becomes the highest optimal portfolio risk. $\qquad\square$

Applying Theorems 4 and 5 to the interval MAD model (4) with $b^{\Delta} = c^{\Delta} = 0$, the lowest risk $\underline{\varrho}$ and the highest risk $\overline{\varrho}$ are obtained by the formula

$$\underline{\varrho} = (1/T) \cdot \min\{c^{\mathsf{T}}x \mid \underline{A}x \leq b, \, \overline{A}x \geq b, \, x \geq 0\}$$
$$\overline{\varrho} = (1/T) \cdot \max\{\varrho_z \mid z \in \{\pm 1\}^{T+2}\}$$

where

$$\varrho_z = \max\{b^{\mathsf{T}}y \mid (A^c - z_{T+1}A^{\Delta})^{\mathsf{T}}y \leq c, \, \mathrm{diag}(z)y \geq 0\}.$$

Yet, z_{T+1} takes on a value of either -1 or 1. The other components of z are unrestricted. The constraint $\mathrm{diag}(z)y \geq 0$ can be reduced to $z_{T+1}y_{T+1} \geq 0$. Hence, the range of optimal portfolio risks can be calculated by Algorithm 2. The symbol e_{T+1} denotes a unit vector whose the $(T+1)^{th}$ component equals 1.

Algorithm 2. Determine the range of optimal portfolio risks

Input: The portfolio selection model (4) with T observations and
   the target portfolio return θ
Output: The lowest risk $\underline{\varrho}$ and the highest risk $\overline{\varrho}$
   1: Compute $\underline{\varrho} \leftarrow (1/T) \cdot \min\{c^\mathsf{T}x \mid A(\underline{r})x \leq b,\ A(\overline{r})x \geq b,\ x \geq 0\}$
   2: Calculate $\overline{\varrho}_1 \leftarrow (1/T) \cdot \max\{b^\mathsf{T}y \mid (A(\underline{r}))^\mathsf{T}y \leq c,\ e_{T+1}^\mathsf{T}y \geq 0\}$
   3: Calculate $\overline{\varrho}_2 \leftarrow (1/T) \cdot \max\{b^\mathsf{T}y \mid (A(\overline{r}))^\mathsf{T}y \leq c,\ -e_{T+1}^\mathsf{T}y \geq 0\}$
   4: Compare $\overline{\varrho} \leftarrow \max\{\overline{\varrho}_1, \overline{\varrho}_2\}$
   5: **return** $\underline{\varrho}$ and $\overline{\varrho}$

---

4   Examples and Numerical Results

Generally, a portfolio can include any types of financial assets. However, only
S&P 500 stocks are considered in this work. All historical data were gathered
in October 2018 from Yahoo! Fiance via the R statistical software. The linear
optimization problems are solved by the dual-simplex method implemented in
MATLAB® software.

4.1   Rates of Return

For an individual stock i listed in the S&P 500 index, the sequence of monthly
log-returns $\eta_{i,t}^k$ (where $1 \leq t \leq 8$ and $1 \leq k \leq 6$) is observed over the past 4 years
from November 2014 to October 2018. Its historical return during the 6-month
period t becomes

$$\widetilde{r}_{i,t} = \exp\left[\sum_{k=1}^{6} \eta_{i,t}^k\right].$$

4.2   Lower Bounds on Expected Returns

This work partly follows the guideline suggested by Kadan and Tang [8]. A
stochastic discount factor (SDF) M_{T+1} discounts the stochastic rate of return
$R_{i,T+1}$ from time T to $T+1$ as if it were equivalent to investing in a risk-free
asset with the deterministic rate of return $R_{f,T}$ fixed at time T. Mathematically,
$\mathbb{E}_T[M_{T+1}R_{i,T+1}] = 1$ (see [9] for more details).

Definition 1 (Negative Correlation Condition). *An asset i satisfies the
negative correlation condition (NCC) if*

$$\mathrm{Cov}_T(M_{T+1}R_{i,T+1}, R_{i,T+1}) \leq 0.$$

Theorem 9. *Let $S_{i,T}$ and $F_{i,T}$ denote the price of asset i at time T and its forward price at time T for delivery at time $T+1$ respectively. An asset i for which the NCC holds possesses its lower bound on expected return via the inequality*

$$\mathbb{E}_T[R_{i,T+1}] \geq R_{f,T} + \frac{2}{S_{i,T}^2} \left[\int_0^{F_{i,T}} \nu_{i,T}^{put}(K)\, dK + \int_{F_{i,T}}^\infty \nu_{i,T}^{call}(K)\, dK \right]$$

across European call and put options on asset i with different strike prices K but the same maturity $T+1$. Both are priced at $\nu_{i,T}^{put}(K)$ and $\nu_{i,T}^{call}(K)$ under the assumption of no arbitrage.

The 6-month U.S. Treasury bill quoted on October 12, 2018 is used as the risk-free rate $R_{f,T}$ which equals 1.0238. Theoretically, the formula of forward price is $F_{i,T} = R_{f,T} S_{f,T}$. The left endpoint rule is employed as a numerical scheme for the integration terms on the right hand side to derive the lower bound on expected return.

4.3   Upper Bounds on Expected Returns

With the buy-and-hold strategy, the investor cannot reinvest the capital gains during the holding period. The returns from reinvestment during the previous 3 periods therefore serves as an appropriate upper bound on expected returns. It can be simply calculated by

$$\bar{r}_i = \max_{6 \leq t \leq 8} \left\{ \exp \left[\sum_{k=1}^6 \max(\eta_{i,t}^k, 0) \right] \right\}.$$

4.4   Frontiers of 9-Asset Portfolio

The sample includes the S&P 500 stocks, for which the negative correlation condition (NCC) empirically holds, across all 9 different sectors as given in Table 1. Their historical returns and estimated bounds on expected returns accrued in the next 6 months from November 2018 to April 2019 are provided in Tables 2 and 3 respectively.

By Algorithm 1, the computed values are $\underline{\theta} = 1.0993 > \overline{\theta} = 1.0961$. This gives no result for a target portfolio return because the obtained bound is overly inexact. Therefore, the new values of the upper bounds are considered depending upon the parameter $\alpha \in [0, +\infty)$ by the equation $\overline{r}^{new}(\alpha) = (1/(\alpha+1)) \cdot \underline{r} + (\alpha/(\alpha+1)) \cdot \overline{r}^{old}$. As the return bounds become more precise with a decreasing value of α, there is more room for the target portfolio return θ as illustrated in Table 4.

The frontier of the portfolio at $\alpha = 6$ is considered. This parameter still provides the realistic bounds as presented in Table 3. By Algorithm 2, its frontiers are shown in Table 5.

Table 1. Examples of S&P 500 stocks across all 9 different sectors.

Ticker	Company	Sector
KMX	CarMax Inc.	Consumer Discretionary
MPC	Marathon Petroleum Corporation	Energy
TROW	T. Rowe Price Group	Financials
DVA	DaVita Inc.	Health Care
ETN	Eaton Corp. Plc.	Industrials
MCHP	Microchip Technology Incorporated	Information Technology
PX	Praxair Inc.	Materials
WY	Weyerhaeuser Company	Real Estate
D	Dominion Energy Inc.	Utilities

Table 2. Historical S&P 500 stock returns over 6-month periods from November 2014 to October 2018.

Ticker	Period 1	Period 2	Period 3	Period 4	Period 5	Period 6	Period 7	Period 8
KMX	1.1337	0.9231	0.8468	1.0434	1.1053	1.0427	0.8627	1.0717
MPC	1.2002	1.2458	1.0589	1.0055	1.1988	1.0756	0.9510	0.8463
TROW	0.9883	1.1984	1.0854	1.0855	0.9237	0.8433	0.8809	0.8375
DVA	1.0448	1.0449	1.0565	1.1957	1.0604	1.0270	0.9515	0.9892
ETN	0.9880	1.0447	1.1079	0.9755	1.0788	1.0545	0.9618	1.0185
MCHP	1.0353	1.2081	1.0370	1.0543	1.0856	0.8946	0.8737	1.0949
PX	0.9725	1.0853	1.0941	0.9123	1.1097	0.9855	0.7747	1.0463
WY	0.9305	0.9308	1.0951	0.9318	1.1316	1.0602	1.0242	0.7618
D	1.2797	1.0994	1.1283	1.0309	1.0914	0.8020	1.0324	1.0986

Table 3. Estimates of expected S&P 500 stock returns accrued in the next 6 months from November 2018 to April 2019 including their adjusted upper bounds with the parameter $\alpha = 6$.

Ticker	Lower bound	Upper bound	Adjusted upper bound (at $\alpha = 6$)
KMX	1.0615	1.1374	1.1266
MPC	1.0446	1.1215	1.1106
TROW	1.0345	1.1036	1.0938
DVA	1.0961	1.0993	1.0989
ETN	1.0372	1.1238	1.1115
MCHP	1.0589	1.1119	1.1044
PX	1.0565	1.1328	1.1220
WY	1.0413	1.1192	1.1081
D	1.0345	1.2001	1.1765

Table 4. Range of target portfolio returns with the parameter $\alpha \in [1, 8]$.

Value of α	Minimum target return	Maximum target return
1	1.0691	1.0961
2	1.0806	1.0961
3	1.0864	1.0961
4	1.0898	1.0961
5	1.0921	1.0961
6	1.0938	1.0961
7	1.0950	1.0961
8	1.0960	1.0961

Table 5. Frontiers of the optimal portfolios

Target return	With interval-valued returns $[\underline{r}, \bar{r}^{new}(6)]$		At returns $\bar{r}^{new}(1.35)$
	Minimum optimal risk	Maximum optimal risk	Optimal risk
1.0938	0.0429	0.1391	0.0429
1.0940	0.0431	0.1351	0.0432
1.0942	0.0432	0.1313	0.0434
1.0944	0.0433	0.1284	0.0437
1.0946	0.0435	0.1256	0.0440
1.0948	0.0436	0.1227	0.0442
1.0950	0.0437	0.1198	0.0445
1.0952	0.0439	0.1169	0.0448
1.0954	0.0440	0.1140	0.0450
1.0956	0.0442	0.1111	0.0453
1.0958	0.0443	0.1082	0.0456
1.0960	0.0444	0.1053	0.0458

5   Discussion and Conclusion

The frontiers of mean absolute deviation portfolios with interval-valued returns are obtainable with the assistance of deterministic linear programming as proposed in Algorithms 1 and 2. A target return considered here is merely limited to a reward at which every possible outcome of portfolio can generate. Therefore, some unexpected results can occur. As illustrated in Table 5, some portfolio frontiers may not exhibit the risk-return tradeoff, thereby being inefficient.

More specifically, when the returns of all 9 stocks are expected to be at the adjusted upper bound $\bar{r}^{new}(6)$ with optimal risks identical to the maximum counterparts (as computed in column 3), the portfolio frontier bahaves erratically because the given range of target returns between 1.0938 and 1.0960 are comparatively low. There exists a higher range of target returns to make the portfolio efficient, but all portfolios under the range of expected returns specified in the given interval $r^{new}(6) = [\underline{r}, \overline{r}^{new}(6)]$ cannot attain these high target rates.

However, when all stock returns are expected to be at the low rates $\overline{r}^{new}(1.35) \in r^{new}(6)$ whose optimal risks (see column 4) are very close to the minimum counterparts (as computed in column 2), the portfolio frontier is efficient because these target returns are sufficiently high to surpass the inefficient frontier.

References

1. Konno, H., Yamazaki, H.: Mean-absolute deviation portfolio optimization model and its applications to Tokyo stock market. Manag. Sci. **37**(5), 519–531 (1991)
2. Feinstein, C.D., Thapa, M.N.: A reformulation of a mean-absolute deviation portfolio optimization model. Manag. Sci. **39**(12), 1552–1553 (1993)
3. Embrechts, P., Frey, R., McNeil, A.: Quantitative Risk Management: Concepts. Techniques and Tools. Princeton University Press, New Jersey (2005)
4. Fiedler, M., Nedoma, J., Ramík, J., Rohn, J., Zimmermann, K.: Linear Optimization Problems with Inexact Data. Springer, New York (2006). https://doi.org/10.1007/0-387-32698-7
5. Hladík, M.: Optimal value range in interval linear programming. Fuzzy Optim. Decis. Mak. **8**(3), 283–294 (2009)
6. Bodie, Z., Kane, A., Marcus, A.J.: Investments. McGraw-Hill Education, New York (2014)
7. Ferris, M.C., Mangasarian, O.L., Wright, S.J.: Linear Programming with Matlab. SIAM, Philadelphia (2007)
8. Kadan, O., Tang, X.: A Bound on Expected Stock Returns. https://doi.org/10.2139/ssrn.3108006. Accessed 22 May 2018
9. Cochrane, J.H.: Asset Pricing. Princeton University Press, New Jersey (2005)

The Impact of Economic Growth, Energy Consumption and Trade Openness on Carbon Emissions: An Empirical Analysis in China

Jianxu Liu[1,3], Zihe Li[2], Changrui Dong[1], and Songsak Sriboonchitta[2,3(✉)]

[1] Faculty of Economics, Shandong University of Finance and Economics,
Jinan, China
[2] Faculty of Economics, Chiang Mai University, Chiang Mai, Thailand
songsakecon@gmail.com
[3] Puey Ungphakon Center of Excellence in Econometrics, Chiang Mai University,
Chiang Mai, Thailand

Abstract. The study aims to test the long-run cointegration relationship and causality among China's carbon emissions, economic growth, energy consumption and trade openness for the period 1971–2013. Autoregressive Distributed Lag (ARDL) model incorporating with structural breaks and Vector Error Correction Model (VECM) Granger causality test are applied in this study. The empirical results reveal that inverted-U shape relationship exists between carbon emissions and economic growth in the long-run, but it doesn't hold in short-run, proving that Environmental Kuznets Curve (EKC) is a long-run phenomenon rather than short-run. Moreover, energy consumption and trade openness are found to have positive impacts on carbon emissions in the long-run and short-run. As for causality test the result showed that bi-directional causal relationship exists between energy consumption and carbon emissions in the long-run. In the short-run, unidirectional causality is found running from trade openness to carbon emissions.

Keywords: Carbon emissions · Environmental Kuznets Curve ·
Trade · Economic growth

1 Introduction

Over the past three decades, the ecological problems caused by global warming have gradually aroused people's vigilance. After 1990s, since Grossman and Krueger [2] proposed the Environmental Kuznets Curve (EKC) inverted U-shape theory, the research on the relationship between environment and economy started to flourish. EKC phenomenon indicates that with the development of economy, the quality of the environment will deteriorating. After reaching a turning point, environmental quality will gradually improving. Nowadays, EKC

© Springer Nature Switzerland AG 2019
H. Seki et al. (Eds.): IUKM 2019, LNAI 11471, pp. 235–244, 2019.
https://doi.org/10.1007/978-3-030-14815-7_20

theory has become an important theoretical basis in guiding researches of environmental protection. A variety of empirical studies is relevant to this theory, such as Lau et al. [5], Ozturk and Acaravci [6] and Shahbaz et al. [8]. The empirical results of these studies have proved the validity of EKC phenomenon in target countries, and have provided a support for environmental protection and energy conservation.

Since China joined the WTO in 2001, foreign trade has become an important pillar of China's economic development, with the deepening of China's trade liberalization. In 2015, China's import and export value of goods reached 3.95 trillion U.S Dollars, trade surplus exceeded 1.72 trillion U.S dollars, ranking the first worldwide (Ministry of Commerce of the PRC, 2015). However, with the expansion of China's foreign trade and economic development, the problem of environmental degradation emerged. Due to technical backwardness, China's exports mainly concentrated in labor-intensive and resource-intensive products. These products consume numerous energy, and bring large amounts of carbon emissions. Around 10.4 billion tons of carbon dioxide was emitted by China in 2015, of which 90% came from fossil fuel consumption (Global Carbon Budget Report, 2016). At present, western countries intend to impose carbon tariffs on some Chinese exports. Once implemented, it will hit China's export trade and delay economic development. Under this grim situation, this paper investigates the existence of Environmental Kuznets curve (EKC) hypothesis in China by using the Chinese sample data from 1971 to 2013. Autoregressive-distributed lag model incorporating with structural breaks and VECM Granger causality test are applied to explore the long-run and short-run dynamic relationship and causal relationship among carbon dioxide emissions, economic growth, energy consumption, and foreign trade of China.

Compared to other EKC studies about China, such as Xu and Lin [10], Jalil and Mahmud [3] and Wang et al. [9], this study considers the structural breakpoints appearing in the series into cointegration, so as to obtain a more reliable results. The contribution of this paper is confirming the existence of EKC. Additionally, energy consumption and trade openness are found to play an important role in effecting carbon emission of China.

2 Data and Model

In order to examine the existence of EKC phenomenon of China, this study applies annual time series data of real GDP per capita, carbon emissions per capita, energy consumption per capita and trade as share of GDP range from 1971 to 2013 (World Development Indicators, 2016) to investigate long run cointegration relationship among these variables. All the variables need to be transformed into nature-log form, so the model can be written as follows:

$$LnC_t = \beta_1 + \beta_2 LnY_t + \beta_3 LnY_t^2 + \beta_4 LnE_t + \beta_5 LnT_t + \varepsilon_t, \qquad (1)$$

where LnC_t, LnE_t, LnY_t, LnY_t^2 and LnT_t denote nature log of carbon emissions per capita, energy consumption per capita, real GDP per capita and its non-linear form, trade openness per capita respectively. According to EKC theory, the sign of β_2 must be positive since economic growth increases carbon emissions, while β_3 is negative, indicating that an advanced economy will reduce carbon emissions. Similarly, β_4 has to be positive to show energy consumption has positive impact on carbon emissions. β_1 and β_5 can be either positive or negative.

3   Methodology

In this study, long run cointegration relationship among China's carbon emissions, energy consumption, economic growth and trade openness is checked by employing Autoregressive distributed lag model (ARDL) proposed by Pesaran et al. [7]. This approach can process the variables which are integrated at $I(0)$, $I(1)$ or mixture of them. The ARDL approach with breakpoints can be estimated using unrestricted error correction (UREC) regressions as follows:

$$LnC_t = \eta_1 + \eta_Y LnY_{t-1} + \eta_{Y2} LnY_{t-1}^2 + \eta_E LnE_{t-1} + \eta_T LnT_{t-1} + \eta_D LnD_{t-1}$$

$$+ \sum_{j=1}^{p} \eta_j \Delta lnC_{t-j} + \sum_{k=0}^{q} \eta_k \Delta lnY_{t-k} + \sum_{l=0}^{r} \eta_l \Delta lnY_{t-l}^2 + \sum_{m=0}^{s} \eta_m \Delta lnE_{t-m} + \sum_{n=0}^{k} \eta_n \Delta lnT_{t-n} + \mu_t, \tag{2}$$

where Δ is difference operator; D is dummy variable for breakpoints; μ_i is random error term.

Once cointegration relationship is confirmed, Vector Error Correction Model (VECM) should be applied to test the causal relationship between the variables. The VECM Granger causality test can be expressed as follows:

$$(1-L)\begin{bmatrix} lnC_t \\ lnY_t \\ lnY_t^2 \\ lnE_t \\ lnT_t \end{bmatrix} = \begin{bmatrix} \tau_1 \\ \tau_2 \\ \tau_3 \\ \tau_4 \\ \tau_5 \end{bmatrix} + \sum_{i=1}^{P}(1-L)\begin{bmatrix} b_{11i} & b_{12i} & b_{13i} & b_{14i} & b_{15i} \\ b_{21i} & b_{22i} & b_{23i} & b_{24i} & b_{25i} \\ b_{31i} & b_{32i} & b_{33i} & b_{34i} & b_{35i} \\ b_{41i} & b_{42i} & b_{43i} & b_{44i} & b_{45i} \\ b_{51i} & b_{52i} & b_{53i} & b_{54i} & b_{55i} \end{bmatrix} \times \begin{bmatrix} lnC_{t-1} \\ lnY_{t-1} \\ lnY_{t-1}^2 \\ lnE_{t-1} \\ lnT_{t-1} \end{bmatrix} + \begin{bmatrix} \delta_1 \\ \delta_2 \\ \delta_3 \\ \delta_4 \\ \delta_5 \end{bmatrix} ECT_{t-1} + \begin{bmatrix} \mu_{1t} \\ \mu_{2t} \\ \mu_{3t} \\ \mu_{4t} \\ \mu_{5t} \end{bmatrix} \tag{3}$$

where $(1-L)$ is difference operator, and ECT_{t-1} is lagged error correction term. Long-run causality among all variables can be tested by t-statistic and significant coefficient of ECT_{t-1} term. F-statistic of Wald-test with all variables involved can measure the short-run causality. Afterwards, CUSUM and CUSUMSQ tests are implied to test the stability of short-run and long-run coefficients.

4   Empirical Results

In this study, Augmented Dickey-Fuller (ADF) and PP unit root tests are implemented to test the stability of series. The results shown in Table 1 reveals that all the series have unit root problem at their levels but stationary at their first differences, which means that all series are integrated at first difference. However, these traditional unit root tests fail to capture the information of unknown time breaks existing in the series, therefore leading to unreliable results. So,

Zivot-Andrews unit root test is elicited to detect the time breaks. The results listed in Table 2 indicate that all series with intercept and trend have structural breakpoints at their levels, but they are stationary at first differences, this evidence implies all the series have the same order, making it possible to employ ARDL model to cointegration.

Table 1. ADF and PP unit root tests results

Variables	ADF unit root test		PP unit root test	
	I(0)	I(1)	I(0)	I(1)
LnC	−2.825	−3.667**	−1.243	−3.784**
LnE	−0.433	−4.385***	−0.08	−4.385***
LnY	0.502	−4.701***	−0.153	−6.251***
LnY	−0.811	−3.932**	0.652	−5.453***
LnT	−1.778	−5.451***	−1.929	−5.093***

Note: *** and ** denote statistical significance at 1% and 5% level, respectively.

Table 2. Zivot-Andrews structural break unit root test results

Variable	Level		1st difference	
	T-statistic	Time break	T-statistic	Time break
LnC	−3.485	1997	−5.237**	2003
LnE	−2.526	2000	−7.014***	1987
LnY	−2.029	1992	−6.223***	1985
LnY	−2.041	2000	−5.214**	1985
LnT	−3.978	2001	−5.778***	2007

Note: *** and ** denote statistical significance at 1% and 5% level, respectively.

Before implementing Autoregressive Distributive Lag modelling (ARDL), an appropriate lag length should be chosen for the model. The study of Lütkepohl [4] found that AIC has superior property for small sample data in confirming lag length compared with other lag length criterion. Our selecting criteria is the minimum AIC values.

It is found that the most suitable lag length is 2. After the lag length is confirmed, we will use F-statistic to examine the cointegration relationship among the carbon emissions, economic growth, energy consumption and trade openness range from 1971 to 2013 in the case of China. The calculated F-statistics are reported in Table 3. It can be seen that once we use carbon dioxide emissions, energy consumption, or GDP (GDP square) as dependent variables, their F-statistic with time breaks included are 3.769, 3.532, and 12.408 (16.787)

respectively, which higher than the upper critical value at 10% and 1% significance level. However, no conclusion can be made when treat trade openness as
dependent variable, because the F-statistic is within the upper and lower bond.
Banerjee et al. [1] mentioned in his study that the error correction term is a
useful way to establish cointegration.

Table 3. The results of ARDL cointegration analysis

Bounds testing to cointegration			Diagnostic tests		
Estimated models	Lag length	F-statistics	R^2	Adj-R^2	D. W test
C = f(E,Y,T)	(2,1,2,0,1,2)	3.769*	0.998	0.998	1.77
2	(1,1,0,0,0,0)	3.532*	0.792	0.749	2.06
3	(2,1,1,1,0,0)	12.408***	0.987	0.983	1.87
4	(2,1,1,1,0,0)	16.787***	0.99	0.986	1.78
5	(2,2,0,2,2,0)	2.811	0.596	0.402	2.14
Critical values for F-statistics (%)			Lower I(0)	Upper I(1)	
1			3.41	4.68	
5			2.62	3.79	
10			2.26	3.35	

Note: ***, ** and * denote statistical significance at 1%, 5% and 10% level,
respectively.

The results of diagnostic tests show that R^2 and adj-R^2 for all linear regression model are quite high, Durbin-Watson (DW) value are close to 2, which
means that all the models are eliminated from the serial correlation problem.
Above all, we can draw a conclusion that long-run relationship exists among the
variables of carbon dioxide emissions, GDP (GDP square), energy consumption
and trade openness.

In order to test whether our long-run relationship is robust or not, we adopt
the Johansen & Juselius cointegration approach, since the lag length of the
variable is already known and also all the variables have been proved to have
same order. So it is feasible to apply Johansen-Juselius approach. From the
results listed in Table 4, we can reject the null hypothesis of r ≤ 1, which means
that at least one group cointegration relationship exists in our variables. This
indicates that long-run relationship among variables in this study is effective and
robust.

After Johansen cointegration approach confirmed the robustness and effectiveness of long-run relationship among the variables, next step is to find out
how much of independent variables impact on carbon dioxide emissions in long-
run and short-run respectively. The long-run estimation results are reported in
Table 5. The results show that all estimated variables in this model are significant and with the expected signs. In the long-run, GDP is a significant and
has a positive sign coefficient. One percent growth in real GDP will increase

Table 4. The result of Johansen-Juselius cointegration

Hypothesis	Trace statistic	Critical value (Prob)	Maximum eigen value	Critical value (Prob)
r = 0	93.261*	69.818 (0.000)	41.151*	33.876 (0.005)
r ≤ 1	52.110*	47.856 (0.018)	28.746*	27.584 (0.035)
r ≤ 2	23.363	29.797 (0.228)	13.695	21.131 (0.390)
r ≤ 3	9.668	15.494 (0.307)	7.663	14.264 (0.414)
r ≤ 4	2.005	3.841 (0.156)	2.005	3.841 (0.156)

Note: * denotes significant at 5% level.

carbon emissions by 0.368%, while the negative statistically significant coefficient of GDP square reveals that after cross the threshold, 1% rise in GDP will decline 0.018% carbon emissions. This evidence support the Environmental Kuznets Curve phenomenon, which describes that in early stages of economic development, the quality of the environment will deteriorate. After reaching a certain threshold, the economic development will improve the quality of the environment.

Table 5. The long-run estimation results

Dependent variable: LNCO2			
Long-run results			
	Coefficient	Std. error	t-Statistic
Constant	−7.573	0.961	−7.881(0.00)***
LNGDP	0.368	0.152	2.418(0.02)**
LNGDPSQ	−0.018	0.011	−2.646(0.01)**
LNENERGY	0.985	0.136	7.225(0.00)***
LNTRADE	0.103	0.038	2.711(0.01)**
BREAK1	−0.194	0.029	−6.573(0.00)***

Note: P-values are in the bracket, *** and ** represent significant at 1% and 5% respectively.

Moreover, it found that energy consumption is the main source of carbon dioxide. Under the condition that other factors remain unchanged, as 1% increase in energy consumption leads to 0.985% rise in carbon dioxide, and the results revealed that trade openness is also an important variable that affects carbon dioxide emissions. The evidence shows that a 0.103% growth in carbon dioxide is associated to 1% increase in trade openness. The results in the short-run are presented in Table 6. The insignificant linear and nonlinear of real GDP per capita imply that Environmental Kuznets Curve phenomenon is untenable. Energy consumption has significant impact on carbon dioxide emissions, 1% increase in energy consumption will lead to 1.105% rise in carbon dioxide. The short-run elasticity of energy consumption is larger than that in long-run, which means that the impact of energy consumption on carbon dioxide in short-run is more obvious. ECM (−1) denotes error correction term. It reflects the adjustment speed from short-run to long-run and it should be negative and statistically significant. The results show that the coefficient of ECM (−1) is −0.694 and statistically significant which suggesting that any deviation in carbon dioxide emissions from short-run to long-run can be corrected by 69.4% each year. The diagnostic tests results imply that our short-run model can disregard the white heteroscedasticity, and autoregressive conditional heteroscedasticity, implying that our model is correctly specified, and has no heteroscedasticity.

Table 6. The short-run estimation results

Dependent variable: D(LNCO2)			
	Coefficient	Std. Error	t-Statistic
D(LNENERGY)	1.015	0.13	7.755***
D(LNGDP)	−0.298	0.319	−0.933
D(LNGDP(-1))	−0.121	0.053	−2.251**
D(LNTRADE)	0.072	0.027	2.646**
D(LNGDPSQ)	0.027	0.027	1.009
D(BREAK1)	−0.079	0.028	−2.846**
D(BREAK1(-1))	0.044	0.029	1.487
ECM(-1)	−0.694	0.159	−4.344***
Diagnostic test	Test-statistic		P-value
χ^2Normal	0.231		0.89
χ^2Serial	0.534		0.592
χ^2ARCH	1.248		0.299
χ^2Remsay	1.723		0.296

Note: *** and ** denotes significant at 10% and 5% respectively.

Finally, CUSUM and CUSUMSQ tests are implemented to test the stability of parameters in error correction model. The plots of CUSUM and CUSUMSQ tests are displayed in Fig. 1. Both plots of CUSUM and CUSUMSQ are within the critical value at 5% significance level; which means that all the parameters in the error correction model are stable.

After the long-run cointegration relationship among variables is determined, then how variables interact with each other should be tested in the following step. In other words, we should test the causal relationship among all variables. With the results of causality among variables, we are able to analyze the causes of environmental degradation and find problems in the industry, and accordingly to develop the appropriate environmental and industrial policies. Based on this, the

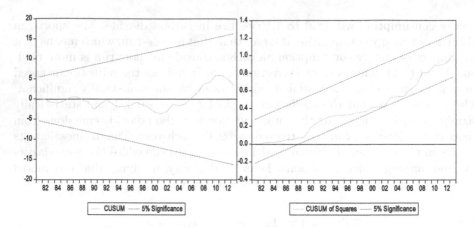

Fig. 1. Plots of CUSUM and CUSUMSQ of recursive residuals.

VECM causality test is applied to find the causal relationship among variables. In this model, long-run causality is expressed by significant t-statistic and negative parameters of error correction term (ECM_{t-1}). The jointly significant LR test on the lagged explanatory variables shows short-run causality.

The long-run and short-run causality results are reported in Table 7. It shows that error correction terms ECM (-1) in VECM model are significant and with negative sign in carbon dioxide equation and energy consumption equation, confirming that long-run causality range from real GDP (real -GDP square), energy consumption, trade openness to carbon dioxide emissions. Similarly, long-run causality can also be found from carbon dioxide, real GDP (GDP square), trade openness to energy consumption.

Table 7. The VECM Granger causality test results

Dependent variable	Short-run $\ln C_t$	$\ln Y_t$	$\ln Y_t^2$	$\ln E_t$	$\ln T_t$	Long-run ECM (-1)
LnC_t	------	2.187 [0.335]	1.875 [0.391]	1.026 [0.598]	7.563*** [0.022]	-0.810*** [-4.719]
LnY_t	0.753 [0.686]	------	2.945 [0.229]	2.541 [0.280]	0.97 [0.615]	-0.417 [-1.117]
LnY_t^2	0.923 [0.630]	3.243 [0.197]	------	2.987 [0.224]	0.907 [0.635]	-0.146 [-1.135]
LnE_t	6.185** [0.045]	1.061 [0.588]	0.952 [0.621]	------	1.977 [0.372]	-0.333** [-2.341]
LnT_t	0.462 [0.793]	4.116 [0.127]	5.592** [0.041]	0.872 [0.646]	------	------

Note: *** and ** represent statistical significant at 1% and 5%.
P-value and T-statistic are shown in the bracket.

According to the results, we can find that unidirectional causality exists in multiple group's variables in the long-run: (1) energy consumption to carbon dioxide emissions. (2) trade openness to carbon dioxide emissions. (3) economic growth to carbon dioxide, this evidence also reinforces the reality of EKC phenomenon exists in China. When energy consumption is used as dependent variable, it can be found that long-run Granger causality runs from (1) economic growth to energy consumption. (2) trade openness to energy consumption (3) carbon dioxide to energy consumption. So here, carbon emissions and energy consumption Granger cause each other. For the short-run, unidirectional Granger causality is found from trade openness to carbon dioxide emissions; carbon dioxide to energy consumption, and economic growth to trade openness.

5 Conclusion

This study attempts to investigate the validity of China's Environmental Kuznets Curve hypothesis by using the annual time series data of carbon emissions, energy consumption, economic growth and trade openness from 1971 to 2013. For this purpose, long-run and short-run cointegration relationships among variables are detected under Autoregressive distributed lag modelling approach. The empirical results suggest that EKC is a long-run phenomenon rather a short-run one in China, and long-run relationship among all variables are robust and effective. Furthermore, the Granger causality test results show that energy consumption and trade openness are two major causes of carbon emissions in China. Based on these results, we conclude that China should make more efforts in energy conservation, clean energy technologies and industry restructuring.

References

1. Banerjee, A., Dolado, J.J., Mestre, R.: Error-correction mechanism tests for cointegration in a single-equation framework. J. Time Ser. Anal. **19**(3), 267–283 (1998)
2. Grossman, G.M., Krueger, A.B.: Environmental Impacts of a North American Free Trade Agreement. National Bureau of Economic Research Working Paper 3914, NBER, Cambridge MA (1991)
3. Jalil, A., Mahmud, S.F.: Environment Kuznets curve for CO2 emissions: a cointegration analysis for China. Energy Policy **37**, 5167–5172 (2009)
4. Lütkepohl, H.: Structural vector autoregressive analysis for cointegrated variables. AStA Adv. Stat. Anal. **90**, 75–88 (2006)
5. Lau, L.S., Chong, C.K., Eng, Y.K.: Investigation of the environmental Kuznets curve for carbon emissions in Malaysia: do foreign direct investment and trade matter? Energy Policy **68**, 490–497 (2014)
6. Ozturk, I., Acaravci, A.: The long-run and causal analysis of energy, growth, openness and financial development on carbon emissions in Turkey. Energy Econ. **36**, 262–267 (2013)
7. Pesaran, M.H., Shin, Y., Smith, R.J.: Bounds testing approaches to the analysis of level relationships. J. Appl. Econ. **16**(3), 289–326 (2001)

8. Shahbaz, M., Mutascu, M., Azim, P.: Environmental Kuznets curve in Romania and the role of energy consumption. Renew. Sustain. Energy Rev. **18**, 165–173 (2013)
9. Wang, S.S., Zhou, D.Q., Zhou, P., Wang, Q.W.: CO2 emissions, energy consumption and economic growth in China: a panel data analysis. Energy Policy **39**, 4870–4875 (2011)
10. Xu, B., Lin, B.: Reducing the CO2 emissions in China's manufacturing industry: evidence from nonparametric additive regression models. Energy **2016**(101), 161–173 (2016)

Machine Learning

NIS-Apriori Algorithm with a Target Descriptor for Handling Rules Supported by Minor Instances

Hiroshi Sakai[1]([⊠]), Kao-Yi Shen[2], and Michinori Nakata[3]

[1] Graduate School of Engineering, Kyushu Institute of Technology,
Tobata, Kitakyushu 804-8550, Japan
sakai@mns.kyutech.ac.jp
[2] Department of Banking and Finance, Chinese Culture University (SCE),
Taipei, Taiwan
kyshen@sce.pccu.edu.tw
[3] Faculty of Management and Information Science, Josai International University,
Gumyo, Togane, Chiba 283-0002, Japan
nakatam@ieee.org

Abstract. For each implication $\tau : Condition\_part \Rightarrow Decision\_part$ defined in table data sets, we see τ is a *rule* if τ satisfies appropriate constraints, i.e., $support(\tau) \geq \alpha$ and $accuracy(\tau) \geq \beta$ for two threshold values α and β $(0 < \alpha, \beta \leq 1)$. If τ is a rule for relatively high α, we say τ is supported by major instances. On the other hand, if τ is a rule for lower α, we say τ is supported by minor instances. This paper focuses on rules supported by minor instances, and clarifies some problems. Then, the NIS-Apriori algorithm, which was proposed for handling rules supported by major instances from tables with information incompleteness, is extended to the NIS-Apriori algorithm with a target descriptor. The effectiveness of the new algorithm is examined by some experiments.

Keywords: Rule generation · Uncertainty · Apriori algorithm · NIS-Apriori algorithm · SQL

1 Introduction

We have been coping with some variations of rule generation related to the Apriori algorithm [1,16], and proposed the NIS-Apriori algorithm for handling tables with definite information (*Deterministic Information Systems: DISs*) and tables with indefinite information (*Non-deterministic Information Systems: NISs*) [9,13]. Furthermore, we recently realized a software tool termed *NIS-Apriori in SQL* [10]. Since SQL has high versatility, the environment yielded by NIS-Apriori based rule generation in SQL will be useful for table data analysis with information incompleteness. The execution logs are uploaded to the web page [11]. In [12–14], we are also considering new topics in conjunction with three-way decisions [17].

© Springer Nature Switzerland AG 2019
H. Seki et al. (Eds.): IUKM 2019, LNAI 11471, pp. 247–259, 2019.
https://doi.org/10.1007/978-3-030-14815-7_21

With such a background, this paper considers two kinds of rules and the problem related to Apriori-based rule generation below:

- An implication τ is a *major rule*, if $support(\tau) \geq \alpha$ and $accuracy(\tau) \geq \beta$ for relatively high α and β.
- An implication τ is a *minor rule*, if $support(\tau) \geq \alpha$ and $accuracy(\tau) \geq \beta$ for lower α and relatively high β.
- Problem: NIS-Apriori-based rule generation will be effective for generating major rules, but it may not be effective for generating minor rules. It is necessary to take measures for minor rule generation.

Major rules reflect the tendency over major instances of the data sets. On the other hand, minor rules reflect the tendency, which strongly holds in minor instances of the data sets. For example, the English alphabet will be the major alphabet in the world, and the Japanese Hiragana and Katakana alphabets are the minor alphabets in the world. However, the major part of publications in Japan consists of the Hiragana, Katakana, and Chinese alphabets, not the English alphabet. Less people in the world understand any Japanese newspaper, but most people in Japan understand it easily. The framework termed *imbalanced data set* [3,4] will be another approach to this issue.

In the application of NIS-Apriori-based rule generation, it will be effective for relatively high α, because the amount of possible implications is usually reduced by using the constraint $support(\tau) \geq \alpha$. However, it is not effective for lower α, because most of implications will satisfy the constraint $support(\tau) \geq \alpha$, and they are still remained as candidates of rules. So, it is very time-consuming for NIS-Apriori-based minor rule generation. For solving this problem, we extend the NIS-Apriori algorithm to that with a target descriptor.

This paper is organized as follows: Sect. 2 surveys the framework of NIS-Apriori-based rule generation, and clarifies the problem of minor rule generation. Section 3 proposes NIS-Apriori-based rule generation with a target descriptor, and Sect. 4 describes the experiments by the implemented system. Section 5 concludes this paper.

2   NIS-Apriori-Based Rule Generation in NISs

This section surveys DIS-Apriori-based rule generation in DISs and NIS-Apriori-based rule generation in NISs, then clarifies the problem related to minor rules.

2.1   DIS-Apriori-Based Rule Generation in DISs

Table 1 is an exemplary DIS ψ_1. We usually predefine a decision attribute *Dec*. In ψ_1, $Dec = price$, and CON is a subset of $\{color, size, weight\}$. In DIS ψ, we term a pair $[A, val_A]$ (an attribute A, an attribute value val_A) a *descriptor*.

Table 1. An exemplary DIS ψ_1 for suitcases. OB (a set of *instances*), AT (a set of *attributes*), VAL_{color} (a set of *attribute values* of color) is $\{red, blue, green\}$, $VAL_{price} = \{high, low\}$.

OB	color	size	weight	price
x_1	red	small	light	low
x_2	red	medium	light	high
x_3	blue	medium	light	high
x_4	red	medium	heavy	low
x_5	red	large	heavy	high
x_6	blue	large	heavy	high

A *rule* is an implication $\tau : \wedge_{A \in CON}[A, val_A] \Rightarrow [Dec, val]$ satisfying below: [1,8,15].

For two threshold values $0 < \alpha, \beta \leq 1.0$,
$support(\tau)(= N(\tau)/N(OB)) \geq \alpha$,
$accuracy(\tau)(= N(\tau)/N(\wedge_{A \in CON}[A, val_A])) \geq \beta$,
Here, $N(*)$ means the amount of instances satisfying the formula $*$,
OB is a set of all instances. We define $support(\tau) = accuracy(\tau) = 0$,
if $N(\wedge_{A \in CON}[A, val_A]) = 0$. (1)

The Apriori algorithm is originally defined for the transaction data sets, and the manipulation of item sets is proposed [1]. However, if we identify each descriptor $[A, val_A]$ with an item, we can similarly apply the Apriori algorithm to rule generation from table data sets. We see the instance x_1 shows an item set and the table ψ_1 is a set of item sets below:

$ItemSet(x_1) = \{[color, red], [size, small], [weight, light], [price, low]\}$,
$Set\_ItemSet(\psi_1) = \{ItemSet(x_1), ItemSet(x_2), \cdots, ItemSet(x_6)\}$.

We term the algorithm handling the above data structure a *DIS-Apriori algorithm* (Algorithm 1). It has the following properties.
(Property 1) The amount of elements in each $ItemSet(x_i)$ is equal to the number of the attributes.
(Property 2) The decision attribute Dec is usually predefined, and the decision part is an element in the set $\{[Dec, val] \mid val$ is a decision attribute value$\}$.
(Property 3) Except (Property 1) and (Property 2), the DIS-Apriori algorithm is almost the same as the Apriori algorithm for the transaction data sets.
We say $\tau' : (\wedge_{A \in CON}[A, val_A]) \wedge [B, val_B]) \Rightarrow [Dec, val]$ is a *redundant* implication for $\tau : \wedge_{A \in CON}[A, val_A] \Rightarrow [Dec, val]$. If we recognize that τ is a rule, we automatically see τ' is also a rule for reducing the amount of rules, namely we handle only a minimal implication as a rule. We have next two additional properties.
(Property 4) If an implication τ' is redundant for τ, $support(\tau') \leq support(\tau)$ always holds.

Algorithm 1. DIS-Apriori algorithm [13]

Require: DIS ψ, the decision attribute Dec, the threshold values α, β.
Ensure: $Rule(\psi)$.
  $Rule(\psi) \leftarrow \{\}$; $i \leftarrow 1$;
  create $SubIMP_i(\subseteq IMP_i)$, where each $\tau_{i,j} \in SubIMP_i$ satisfies $support(\tau_{i,j}) \geq \alpha$;
  while $(|SubIMP_i| \geq 1)$ **do**
    $Rest \leftarrow \{\}$;
    for all $\tau_{i,j} \in SubIMP_i$ **do**
      if $accuracy(\tau_{i,j}) \geq \beta$ **then** add $\tau_{i,j}$ to $Rule(\psi)$;
      else add $\tau_{i,j}$ to $Rest$;
      end if
    end for
    $i \leftarrow i + 1$;
    generate $SubIMP_i(\subseteq IMP_i)$ by using $Rest$, where $\tau_{i,j} \in SubIMP_i$ satisfies
    $support(\tau_{i,j}) \geq \alpha$ and $\tau_{i,j}$ is not redundant for any implication in $Rule(\psi)$;
  end while
  return $Rule(\psi)$

(Property 5) If an implication τ' is redundant for τ, $accuracy(\tau') \leq accuracy(\tau)$ may not hold.

    We also introduce the next IMP_1, IMP_2, $\cdots$, IMP_n.
$IMP_1 = \{\tau : [A, val_A] \Rightarrow [Dec, val]\}$,
    (Any implication with one condition attribute),
$IMP_2 = \{\tau : [A, val_A] \wedge [B, val_B] \Rightarrow [Dec, val]\}$,
    (Any implication with two condition attributes),
$IMP_3 = \{\tau : [A, val_A] \wedge [B, val_B] \wedge [C, val_C] \Rightarrow [Dec, val]\}$,
    (Any implication with three condition attribute).

$$\vdots \qquad \vdots \qquad \vdots \qquad \vdots$$

Here, the subscript i in IMP_i is the amount of descriptors in the condition part, and $|IMP_i|$ is the amount of implications. The DIS-Apriori algorithm makes use of $SubIMP_i(\subseteq IMP_i)$ and Properties 4–5, and it is *sound* and *complete* for the rules defined in the formula (1) [13]. So, any rule defined in the formula (1) can be obtained by Algorithm 1.

2.2 NIS-Apriori-Based Rule Generation in NISs

The table Φ_1 in Table 2 is an exemplary NIS. NISs were proposed by Pawlak [7], Orłowska [6], and Lipski [5] for handling information incompleteness in table data sets. Formerly, information retrieval and question answering were investigated in NISs, and we are recently coping with rule generation from NISs. We replace each non-deterministic information and each ? symbol with a possible value, and we obtain a table with deterministic information. We termed it a *derived DIS* from NIS. Let $DD(\Phi)$ be a set of all derived DISs from Φ. We see (or suppose) an actual DIS ϕ^{actual} exists in $DD(\Phi)$. For Φ_1, $DD(\Phi_1)$ consists of 144 ($=3^2 \times 2^4$) derived DISs. Based on $DD(\Phi)$, we proposed the certain and the possible rules below:

Table 2. An exemplary NIS Φ_1 for suitcases. VAL_{color} (a set of *attribute values* of *color*) is $\{red, blue, green\}$, $VAL_{size} = \{small, medium, large\}$, $VAL_{weight} = \{light, heavy\}$, $VAL_{price} = \{high, low\}$.

OB	color	size	weight	price
x_1	?	small	light	low
x_2	red	?	light	high
x_3	blue	medium	?	high
x_4	red	medium	heavy	low
x_5	$\{red, blue\}$	$\{medium, large\}$	heavy	high
x_6	blue	large	heavy	$\{high, low\}$

Definition 1. [9]

(1) We say τ is a certain rule, if τ satisfies $support(\tau) \geq \alpha$ and $accuracy(\tau) \geq \beta$ in each $\phi \in DD(\Phi)$,

(2) We say τ is a possible rule, if τ satisfies $support(\tau) \geq \alpha$ and $accuracy(\tau) \geq \beta$ in at least one $\phi \in DD(\Phi)$.

Definition 1 seems natural, but we have the computational complexity problem, because the amount of elements in $DD(\Phi)$ increases exponentially. In Φ_1, the amount is 144, and the amount is more than 10^{100} in the Mammographic data set in UCI machine learning repository [2]. For this computational problem, we defined two sets for a descriptor $[A, val]$ below:

$$inf([A, val]) = \{x : instance \mid \text{the value of } x \text{ for } A \text{ is a singleton set } \{val\}\},$$
$$sup([A, val]) = \{x : instance \mid \text{the value of } x \text{ for } A \text{ is a set including } val\},$$
$$inf(\wedge_{A \in CON}[A, val_A]) = \cap_{A \in CON} inf([A, val_A]),$$
$$sup(\wedge_{A \in CON}[A, val_A]) = \cap_{A \in CON} sup([A, val_A]).$$

By using these sets inf and sup, we have solved the computational complexity problem. With respect to an implication τ, the following holds [9].

(Result 1) There is a derived DIS $\psi_{min} \in DD(\Phi)$ satisfying (i) and (ii).

(i) $minsupp(\tau)(= \min_{\psi \in DD(\Phi)}\{support(\tau) \text{ in } \psi\}) = support(\tau)$ in ψ_{min},
(ii) $minacc(\tau)(= \min_{\psi \in DD(\Phi)}\{accuracy(\tau) \text{ in } \psi\} = accuracy(\tau)$ in ψ_{min}.

Thus, τ is a certain rule, if and only if τ is a rule in $\psi_{min} \in DD(\Phi)$, i.e., $minsupp(\tau) \geq \alpha$ and $minacc(\tau) \geq \beta$.

(Result 2) There is a derived DIS $\psi_{max} \in DD(\Phi)$ satisfying (i) and (ii).

(i) $maxsupp(\tau)(= \max_{\psi \in DD(\Phi)}\{support(\tau) \text{ in } \psi\}) = support(\tau)$ in ψ_{max},
(ii) $maxacc(\tau)(= \max_{\psi \in DD(\Phi)}\{accuracy(\tau) \text{ in } \psi\} = accuracy(\tau)$ in ψ_{max}.

Thus, τ is a possible rule, if and only if τ is a rule in $\psi_{max} \in DD(\Phi)$, i.e., $maxsupp(\tau) \geq \alpha$ and $maxacc(\tau) \geq \beta$.

(Result 3) Each formula of four criterion values, $minsupp(\tau)$, $\cdots$, $maxacc(\tau)$, is expressed by using inf and sup sets. This calculation does not depend on the amount of $DD(\Phi)$. (We omit the formulas for them. The details are in [9, 13]). Thus, certain rule generation and possible rule generation does not depend upon the amount of elements in $DD(\Phi)$.

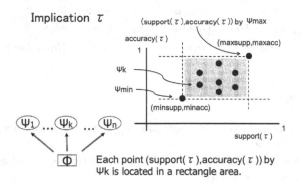

Fig. 1. Each point $(support(\tau), accuracy(\tau))$ by $\psi \in DD(\Phi)$ is located in a rectangle area [13].

Based on the above results, we have a chart in Fig. 1 for each implication τ. We apply the above three results to the DIS-Apriori algorithm in Algorithm 1, and we proposed the NIS-Apriori algorithm. Namely, in certain rule generation $minsupp(\tau)$ and $minacc(\tau)$ are employed instead of $support(\tau)$ and $accuracy(\tau)$ in Algorithm 1. In possible rule generation, $maxsupp(\tau)$ and $maxacc(\tau)$ are employed in Algorithm 1. Therefore, the time complexity of the NIS-Apriori algorithm is more than twice time complexities of the DIS-Apriori algorithm. However, it is possible to calculate criterion four values in polynomial order time, so the NIS-Apriori algorithm does not depend upon the amount of elements in $DD(\Phi)$. The NIS-Apriori algorithm is also sound and complete [13]. Without three results, it will be hard to handle Mammographic data set which has more than 10^{100} derived DISs. Thus, we insist that NIS-Apriori-based rule generation is a significantly new framework supported by possible world semantics.

We recently implemented the NIS-Apriori algorithm in SQL [10], and the execution logs are uploaded to the web page [11]. Figure 2 shows the obtained certain and possible rules ($\alpha = 0.05$ and $\beta = 0.8$) from the Mammographic data set. There are 960 instances and five attributes *assess*, *age*, *shape*, *margin*, *density* (*assess* was added by physicians).

2.3 Problem on Minor Rule Generation

We clarify the problem on minor rule generation by using the Mammographic data set. We employ five α values (0.25, 0.1, 0.05, 0.01, 0.001) and one $\beta = 0.8$. We see the cases I ($\alpha = 0.25$ and $\beta = 0.8$) and II ($\alpha = 0.1$ and $\beta = 0.8$) define major rules, and we see the cases IV ($\alpha = 0.01$ and $\beta = 0.8$) and V ($\alpha = 0.001$ and $\beta = 0.8$) define minor rules.

```
mysql> select * from c1_rule;
+------------+------+----------+------------+---------+--------+
| att1       | val1 | deci     | deci_value | minsupp | minacc |
+------------+------+----------+------------+---------+--------+
  age          80     severity   1            0.056     0.806
  assess       5      severity   1            0.317     0.869
  margin       1      severity   0            0.329     0.859
  end_attrib   NULL   NULL       NULL         NULL      NULL
+------------+------+----------+------------+---------+--------+
4 rows in set (0.00 sec)

mysql> select * from p2_rule;
+------------+------+--------+------+----------+------------+---------+--------+
| att1       | val1 | att2   | val2 | deci     | deci_value | maxsupp | maxacc |
+------------+------+--------+------+----------+------------+---------+--------+
  age          40     assess   4      severity   0            0.094     0.882
  age          50     assess   4      severity   0            0.115     0.821
  age          70     margin   4      severity   1            0.064     0.847
  age          70     shape    4      severity   1            0.108     0.874
  assess       4      margin   3      severity   0            0.061     0.831
  margin       3      shape    4      severity   1            0.069     0.805
  end_attrib   NULL   NULL     NULL   NULL       NULL         NULL      NULL
+------------+------+--------+------+----------+------------+---------+--------+
7 rows in set (0.00 sec)
```

Fig. 2. The obtained certain rules and possible rules ($\alpha = 0.05$ and $\beta = 0.8$) from the Mammographic data sets. Each certain rule satisfies $support \geq 0.05$ and $accuracy \geq 0.8$ in every $\psi \in DD(\Phi_{mammo})$, where the amount of elements ψ is more than 10^{100}.

Table 3 shows the execution time and the amount of rules, where tables $c1rule$, $c2rule$, and $c3rule$ store certain rules obtained from IMP_1, IMP_2, and IMP_3 in Algorithm 1, respectively. Tables $c1rest$, $c2rest$, and $c3rest$ store implications in $Rest$ in Algorithm 1. Tables $p1rule$, $p2rule$, $p3rule$, $p1rest$, $p2rest$, and $p3rest$ store possible rules and implications. In case V, each implication τ satisfies the constraint $support(\tau) \geq 0.001$, so we need to consider any implication. We cannot reduce the amount of implications by using (Property 4) in Sect. 2.1.

3 NIS-Apriori Algorithm with a Target Descriptor

In this section, we extend the NIS-Apriori algorithm to the *NIS-Apriori algorithm with a target descriptor* (tNIS-Apriori). Generally in Apriori-based rule generation, the decision attribute is predefined and any attribute value is considered. In the tNIS-Apriori algorithm, we consider predefined descriptor $[Dec, val]$ in Algorithms 2–3.

Table 3. The comparison of major rule generation and minor rule generation in the Mammographic data set. $|c1rule|$ is the amount of implications in the table $c1rule$. MIN means the minimum amount of implications so as to be one rule, namely $960 \times \alpha$.

Case	α	MIN	$exec\_time$	$\|c1rule\|$ $\|p1rule\|$	$\|c1rest\|$ $\|p1rest\|$	$\|c2rule\|$ $\|p2rule\|$	$\|c2rest\|$ $\|p2rest\|$	$\|c3rule\|$ $\|p3rule\|$	$\|c3rest\|$ $\|p3rest\|$
I	0.25 (*major rule*)	240	19.9 (sec)	2	4	0	2	0	0
				2	4	0	2	0	0
II	0.10 (*major rule*)	96	76.3 (sec)	2	14	4	11	0	1
				5	13	2	11	0	2
III	0.05	48	164.9 (sec)	3	20	6	22	0	8
				9	22	6	29	0	12
IV	0.01 (*minor rule*)	10	540.6 (sec)	6	34	4	75	8	66
				15	34	19	112	7	98
V	0.001 (*minor rule*)	1	889.9 (sec)	7	43	2	192	45	378
				18	36	48	190	44	351

In Algorithm 1, we consider a set $SubIMP_i (\subseteq IMP_i)$, whose element $\tau_{i,j}$ takes $[Dec, \_]$ ($\_$ means any $val \in VAL_{Dec}$) as its decision part. However, Algorithms 2–3, we fix an attribute value $val \in VAL_{Dec}$ and consider a set $SubIMP_{i,[Dec,val]} (\subseteq SubIMP_i \subseteq IMP_i)$, whose element $\tau_{i,j}$ takes $[Dec, val]$ as its decision part. In tNIS-Apriori based rule generation, we have the following advantage and disadvantage.

(Advantage)

The NIS-Apriori algorithm tries to find all rules whose decision attribute is Dec, but the tNIS-Apriori algorithm tries to find all rules whose decision part is $[Dec, val]$. In the Mammographic data set, the decision attribute values are 0 (*benign*) and 1 (*malignant*). So, we apply the tNIS-Apriori algorithm to two decisions $[Dec, 0]$ and $[Dec, 1]$. The execution time by the NIS-Apriori algorithm is usually more time-consuming than that of the tNIS-Apriori algorithm.

(Disadvantage)

In order to have all rules, we need to repeat the execution for each $[Dec, val]$. So, the user's manipulation may be confused, if the amount of decision attribute values is large.

4   Some Experiments

We employed Windows desktop PC (3.60 GHz), and revised the SQL procedures $step1$, $step2$, and $step3$ in the NIS-Apriori algorithm to the SQL procedures $tstep1$, $tstep2$, and $tstep3$ in the tNIS-Apriori algorithm. For example in the Mammographic data set, the next procedure

```
step1('severity',960,0.05,0.8)
```
(Find all implications satisfying $support(\tau) \geq 0.05$, $accuracy(\tau) \geq 0.8$.)

is changed to two procedures below:

```
tstep1('severity','0',960,0.05,0.8), tstep1('severity','1',960,0.05,0.8)
```
(Find all implications with decision value 0) and (Find all implications with decision value 1).

Table 4 shows the comparison of the execution time on the Mammographic data set [2]. Of course, it took less execution time for each of $[severity, 0]$ and $[severity, 1]$ by the tNIS-Apriori algorithm. However, total execution time SUM is worse than that by the NIS-Apriroi algorithm. In this example, the NIS-Apriori algorithm seems better than the tNIS-Apriori algorithm.

Algorithm 2. NIS-Apriori algorithm with a Target Descriptor (Certain rule generation part)

Require: NIS Φ, the descriptor $\underline{[Dec, val]}$, the threshold values α, β.
Ensure: $Certain\_Rule(\Phi)$.
▷ Each changed part from Algorithm 1 is underlined.

$Certain\_Rule(\Phi) \leftarrow \{\}$; $i \leftarrow 1$;
create $SubIMP_{i,[Dec,val]}(\subseteq IMP_i)$ ($\tau_{i,j} \in SubIMP_{i,[Dec,val]}$ and the decision part of $\tau_{i,j}$ is $[Dec, val]$), and $minsupp(\tau_{i,j}) \geq \alpha$ holds;
while $(|SubIMP_{i,[Dec,val]}| \geq 1)$ **do**
    $Rest \leftarrow \{\}$;
    for all $\tau_{i,j} \in SubIMP_{i,[Dec,val]}$ **do**
        if $\underline{minacc(\tau_{i,j})} \geq \beta$ **then** add $\tau_{i,j}$ to $Certain\_Rule(\Phi)$;
        else add $\tau_{i,j}$ to $Rest$;
        end if
    end for
    $i \leftarrow i + 1$;
    generate $SubIMP_{i,[Dec,val]}(\subseteq IMP_i)$ by using $Rest$, where
    $\tau_{i,j} \in SubIMP_{i,[Dec,val]}$ satisfies $minsupp(\tau_{i,j}) \geq \alpha$ and $\tau_{i,j}$ is not redundant
    for any implication in $Certain\_Rule(\Phi)$;
end while
return $Certain\_Rule(\Phi)$

Algorithm 3. NIS-Apriori algorithm with a Target Descriptor (Possible rule generation part)

Require: NIS Φ, the descriptor $\underline{[Dec, val]}$, the threshold values α, β.
Ensure: $Possible\_Rule(\Phi)$.
▷ In possible rule generation, we replace $\underline{minsupp}$ and $\underline{minacc}$ in Algorithm 2 with $\underline{maxsupp}$ and $\underline{maxacc}$, respectively. The other part is the same as Algorithm 2.

Table 4. The execution time (sec) of the tNIS-Apriori algorithm for the Mammographic data set. The column SUM indicates the summation of two cases.

Case	NIS-Apriori (sec)	tNIS-Apriori		
		SUM (sec)	$Dec = 0$ (*benign*) (sec)	$Dec = 1$ (*malignant*) (sec)
I	19.9	22.8	11.1	11.7
II	76.3	77.8	33.5	44.2
III	164.9	185.0	92.3	92.7
IV	540.6	615.9	280.3	335.6
V	889.9	1112.3	495.6	616.7

Table 5. The execution time (sec) of the tNIS-Apriori algorithm for the Congressional Voting data set. The column SUM indicates the summation of two cases.

Case	NIS-Apriori (sec)	tNIS-Apriori		
		SUM (sec)	$Dec = dem(ocrat)$ (sec)	$Dec = rep(ublic)$ (sec)
I	424.1	308.7	27.3	281.4
II	1620.8	1369.5	444.1	925.4
III	3065.5	2616.3	988.6	1627.7
IV	5281.4	5802.9	1806.9	3996.0
V	Ceased	7620.0	1999.5	5620.5

Table 5 shows the comparison of the execution time on the Congressional Voting data set [2]. This data set consists of 435 instances, 17 attributes, each attribute value is either *yes* or *no*. The decision attribute value is either *democrat* or *republic*. Since there are 392 missing values, $DD(\Phi_{congress})$ consists of about 10^{120} ($\risingdotseq 2^{392}$) derived DISs. In case IV, the total execution time SUM is slightly larger, but in case V, we ceased the execution by the NIS-Apriori algorithm, because of its too long execution time. In this example, the tNIS-Apriori algorithm is essential. We have to choose the tNIS-Apriori algorithm for handling the case V.

4.1   Discussion

Of course, the execution time of two algorithms depends upon the details of the algorithms and the characteristics of the data sets. The most time-consuming part of Algorithms 1–3 is 'to generate $SubIMP_i$ by using $Rest$'. We generate $SubIMP_i$ instead of using IMP_i. This strategy is based on Property 4 in Sect. 2.1. For major rule generation, the amount of $|SubIMP_i|$ is generally small.

Table 6. The execution time (sec) of the DIS-Apriori algorithm with a target descriptor for the Car Evaluation data set. The column SUM indicates the summation of four cases.

Case	$Dec = any$	SUM	$Dec = unacc$	$Dec = acc$	$Dec = good$	$Dec = vgood$
Instances	1728	1728	1210	384	69	65
Ratio	100%	100%	70%	22%	4%	4%
I	8.30	13.7	8.09	2.01	1.79	1.81
II	38.26	33.83	15.48	14.87	1.79	1.69
III	255.23	177.37	15.23	158.57	1.72	1.85
IV	3343.52	2014.33	23.71	1513.74	240.09	213.08
V	6004.56	4043.02	25.09	2103.51	1065.04	849.38

Actually, in Mammographic data set, we focus on implications occurring more than 240 times or 96 times. The amount of such implications is small. For minor rule generation, the amount of $|SubIMP_i|$ generally becomes large. In the case V in the Mammographic data set, we need to focus on implications occurring 1 time, namely the *support* constraint is meaningless. We cannot remove any implications satisfying $accuracy(\tau) < \beta$.

In the generation of $SubIMP_i$, we actually pick up all condition descriptors appearing in $Rest$ at first, then we add each of them to implications in $Rest$ and we remove implications which are not in the original table. This manipulation is the most complicated part in the SQL procedure. For example, in certain rule generation (case V) from the Mammographic data set, $|Rest| = 192$ (implications with two condition descriptors) and 20 condition descriptors are picked up. The amount of the candidates of implications is 1068. From these 1068 implications, we generate $SubIMP_{3,[Severity,0]}$ which consists of 987 implications. This seems large amount of implication, however the amount of IMP_3 is huge, because there are about 9600 implications (there are 960 instances and the selection of 3 attributes is $_5C_3 = 5*4*3/3*2*1 = 10$ cases). Even though there are the same implications in 9600 implications, the amount of IMP_3 is much larger than that of $SubIMP_{3,[Severity,0]}$.

The above manipulation seems to be related to the amount of decision attribute values. In the Pittsburgh Bridges data set [2], there are six decision attribute values, and the execution of the NIS-Apriori algorithm was ceased, because of too long execution time. In case V, $|SubIMP_3|$ in certain rule generation exceeds 10000 implications, and the tNIS-Apriori was essential in this case, too.

In DISs, we also executed the DIS-Apriori algorithm with a target descriptor. Table 6 shows the results of rule generation in DIS, the Car Evaluation data set [2]. In Case I, II, and III, the SUM of four execution times is almost the same as the execution time of $Dec = any$. However in Case IV and V, the summation of four execution times is reduced to about 2/3 of $Dec = any$. In the Balance Scale data set and the Phishing data set [2], we similarly had the same results.

5 Concluding Remarks

This paper proposed the tNIS-Apriori algorithm, which is a NIS-Apriori algorithm with a target descriptor. The merits are the following.

(1) For a fixed decision attribute values, tNIS-Apriori algorithm works much better than NIS-Apriori algorithm.
(2) The tNIS-Apriori algorithm is effective for minor rule generation. Actually in Table 5, the NIS-Apriori algorithm cannot generate rules, but tNIS-Apriori algorithm did them.

The NIS-Apriori is suitable for major rule generation, however it is time-consuming for minor rule generation, because the next properties.

(a) If $support(\tau) < \alpha$ holds, we can decide any redundant implication of τ is not a rule (Property 4 in Sect. 2.1).
(b) If $support(\tau) \geq \alpha$ and $accuracy(\tau) < \beta$, this τ is not a rule, but some redundant τ' may satisfy $support(\tau') \geq \alpha$ and $accuracy(\tau') \geq \beta$ (Property 5 in Sect. 2.1).
(c) If we employ the lower threshold value α, most of implications do not satisfy the above (a) and we can not apply the above (a). Furthermore, most of implications satisfy the above (b). Thus, we need to consider large number of redundant implications as candidates of rules.

In order to solve this weak point, we proposed the tNIS-Apriori algorithm. By handling the specified decision descriptor in the tNIS-Apriori algorithm, the candidates of rules are reduced. Thus, we showed the possibility of NIS-Apriori-based minor rule generation. The analysis of the bottlenecks for the execution time, the improvement of the procedures in SQL, and the evaluation with experiments are still in progress now.

References

1. Agrawal, R., Srikant, R.: Fast algorithms for mining association rules in large databases. In: Proceedings of VLDB 1994, pp. 487–499. Morgan Kaufmann (1994)
2. Frank, A., Asuncion, A.: UCI machine learning repository. University of California, School of Information and Computer Science, Irvine, CA (2010). http://mlearn.ics.uci.edu/MLRepository.html
3. Grzymala-Busse, J., Stefanowski, J., Wilk, S.: A comparison of two approaches to data mining from imbalanced data. J. Intell. Manuf. **16**, 565–573 (2005)
4. He, H., Garcia, E.A.: Learning from imbalanced data. IEEE Trans. Knowl. Data Eng. **21**(9), 1263–1284 (2009)
5. Lipski, W.: On databases with incomplete information. J. ACM **28**(1), 41–70 (1981)
6. Orłowska, E., Pawlak, Z.: Representation of nondeterministic information. Theor. Comput. Sci. **29**(1–2), 27–39 (1984)
7. Pawlak, Z.: Systemy Informacyjne: Podstawy Teoretyczne, WNT (1983). (in Polish)
8. Pawlak, Z.: Rough Sets: Theoretical Aspects of Reasoning About Data. Kluwer Academic Publishers, Boston (1991)

9. Sakai, H., Wu, M., Nakata, M.: Apriori-based rule generation in incomplete information databases and non-deterministic information systems. Fundam. Inform. **130**(3), 343–376 (2014)

10. Sakai, H., Liu, C., Zhu, X., Nakata, M.: On NIS-apriori based data mining in SQL. In: Flores, V., et al. (eds.) IJCRS 2016. LNCS, vol. 9920, pp. 514–524. Springer, Cham (2016). https://doi.org/10.1007/978-3-319-47160-0_47

11. Sakai, H.: Execution Logs by RNIA Software Tools (2016). http://www.mns.kyutech.ac.jp/~sakai/RNIA

12. Sakai, H., Nakata, M., Yao, Y.: Pawlak's many valued information system, non-deterministic information system, and a proposal of new topics on information incompleteness toward the actual application. In: Wang, G., Skowron, A., Yao, Y., Ślęzak, D., Polkowski, L. (eds.) Thriving Rough Sets. SCI, vol. 708, pp. 187–204. Springer, Cham (2017)

13. Sakai, H., Nakata, M., Watada, J.: NIS-Apriori-based rule generation with three-way decisions and its application system in SQL. Inf. Sci. (2018). https://doi.org/10.1016/j.ins.2018.09.008

14. Shen, K.Y., Sakai, H., Tzeng, G.H.: Comparing two novel hybrid MRDM approaches to consumer credit scoring under uncertainty and fuzzy judgments. Int. J. Fuzzy Syst. (2018). https://doi.org/10.1007/s40815-018-0525-0

15. Skowron, A., Rauszer, C.: The discernibility matrices and functions in information systems. In: Intelligent Decision Support - Handbook of Advances and Applications of the Rough Set Theory, pp. 331–362. Kluwer Academic Publishers (1992)

16. Sarawagi, S., Thomas, S., Agrawal, R.: Integrating association rule mining with relational database systems: alternatives and implications. Data Min. Knowl. Discov. **4**(2), 89–125 (2000)

17. Yao, Y.Y.: Three-way decisions with probabilistic rough sets. Inf. Sci. **180**, 314–353 (2010)

Utilization of Imprecise Rules
for Privacy Protection

Masahiro Inuiguchi$^{(\boxtimes)}$ and Keisuke Washimi

Graduate School of Engineering Science, Osaka University,
Toyonaka, Osaka 560-8531, Japan
inuiguti@sys.es.osaka-u.ac.jp

Abstract. In this paper, we utilize imprecise rules for privacy protection in the publication of data sets. We assume that data sets show the classification results with more than two classes. First k-anonymous imprecise rules are induced. Using several k-anonymous imprecise rules explaining an object, the object is replaced with several imprecise patterns corresponding to the k-anonymous imprecise rules which are common in at least k objects. In this way, privacy protected data tables are composed. The proposed data table is investigated by numerical experiments from its usefulness in rule induction as well as from its privacy protection ability. The results show that the proposed method will be satisfactorily useful.

Keywords: Rule induction · Imprecise rules · Privacy protection ·
Data anonymization

1 Introduction

In order to protect the privacy in data publication, many privacy preservation techniques have been proposed. The k-anonymity [1,2], ε-differential privacy [3,4], cryptography [5], l-diversity [6], t-closeness [6] and so on have been proposed. Many potential attacks for privacy invasion have been considered and the privacy protection techniques have been investigated. On the other hand, the privacy protection techniques deteriorate the quality of the original data. The balancing the privacy protection and the quality preservation is a difficult issue.

The data privacy has been considered in data mining (see [6]). Rough set approaches [7] provide useful tools for data mining. Nevertheless, the data privacy has not yet considered well in rough set approaches. A few approaches [8–11] using rough sets have been proposed. An anonymization approach utilizing attribute reduction to enhance the usability of anonymous data is proposed in [10]. The concept of k-anonymity is introduced into rule induction in [11] and it is shown that making rules imprecise is useful to obtain k-anonymous rules. Moreover, classification of induced rules is considered to obtain privacy protected data tables [12].

© Springer Nature Switzerland AG 2019
H. Seki et al. (Eds.): IUKM 2019, LNAI 11471, pp. 260–270, 2019.
https://doi.org/10.1007/978-3-030-14815-7_22

In this paper, we investigate a privacy preserving method using imprecise rules [13] assuming a situation that data is published or exchanged among cooperative data analysts. We also assume that there are more than two classes in the data set (decision table). Imprecise rules can conclude the membership only to a union of classes while usual rules (called precise rules) can conclude the membership to a class. We utilize the k-anonymous rule induction method proposed in [11]. For preserving the privacy, we delete inconsistent and individualistic data and replace precise data with imprecise ones. For the replacement, we use k-anonymous imprecise rules supported by the original data. Lumping k objects matching an imprecise rule together as an imprecisely described pattern. Then the proposed anonymized table is composed of imprecisely described patterns. Because we utilize k-anonymous imprecise rules induced from the original data, we may induce similar rules from the anonymized table. By numerical experiments, we examine the proposed approach from viewpoints of the preservation of classification accuracy and the protection ability for decision attribute values when a few condition attribute values are revealed by attackers.

This paper is organized as follows. In next section, rough set approach and imprecise rule induction method are briefly reviewed. In Sect. 3, the proposed data anonymization is described. The numerical experiments and the results are described in Sect. 4. Numerical experiments and their results are shown Some concluding remarks are given in Sect. 5.

2 Rough Set Approach and Imprecise Rule Induction

A decision table is defined by a four-tuple $DT = \langle U, C \cup \{d\}, V, f \rangle$, where U is a finite set of objects, C is a finite set of condition attributes, d is a decision attribute, $V = \bigcup_{a \in C \cup \{d\}} V_a$ with attribute value set V_a of attribute $a \in C \cup \{d\}$ and $f : U \times C \cup \{d\} \to V$ is called an information function which is a total function. By decision attribute value $v_j^d \in V_d$, class $D_j \subseteq U$ is defined by $D_j = \{u \in U \mid f(u, d) = v_j^d\}$, $j = 1, 2, \ldots, p$. Using condition attributes in $A \subseteq C$, we define equivalence classes $[u]_A = \{x \in U \mid f(x, a) = f(u, a), \forall a \in A\}$.

The lower and upper approximations of an object set $X \subseteq U$ under condition attribute set $A \subseteq C$ are defined by

$$A_*(X) = \{x \in U \mid [x]_A \subseteq X\}, \quad A^*(X) = \{x \in U \mid [x]_A \cap X \neq \emptyset\}. \quad (1)$$

Suppose that members of X can be described by condition attributes in A. If $[x]_A \cap X \neq \emptyset$ and $[x]_A \cap (U - X) \neq \emptyset$ hold, the membership of x to X or to $U - X$ is questionable because objects described in the same way are classified into two different classes. Otherwise, the classification is consistent. From these points of view, each element of $A_*(X)$ can be seen as a consistent member of X while each element of $A^*(X)$ can be seen as a possible member of X. The pair $(A_*(X), A^*(X))$ is called the rough set of X under $A \subseteq C$.

In rough set approaches, the attribute reduction, i.e., the minimal attribute set $A \subseteq C$ satisfying $A_*(D_j) = C_*(D_j)$, $j = 1, 2, \ldots, p$, and the minimal length rule induction, i.e., inducing rules about the membership to D_j with minimal conditions which can differ members of $C_*(D_j)$ from non-members, are investigated well. In this paper, we use a rule induction algorithm, MLEM2 [14]. By this algorithm, we obtain minimal set of rules with minimal conditions which can explain all objects in lower approximations of X under a given decision table. By MLEM2, we obtain rules of the form of "if $v_1^L \leq f(u, a_1) \leq v_1^R$, $v_2^L \leq f(u, a_2) \leq v_2^R$, ... and $v_s^L \leq f(u, a_s) \leq v_s^R$ then $u \in X$". MLEM2 generalizes numerical/ordinal condition attribute values to interval values. For each D_j, rules about its membership are induced in MLEM2. Using all rules for all D_j, $j = 1, 2, \ldots, p$, we build a classifier in the same way as that of LERS [14].

In the same way as the induction method for rules about D_i, we can induce rules about a union of D_i's (see [15]). Namely, LEM2-based algorithms can be applied to the induction of rules about a union of D_i's (imprecise rules). We note that imprecise rules can be induced when the number of classes is larger than two, i.e., $p > 2$. Moreover, in the same way, we can build a classifier by induced rules about $Z_s = \bigcup_{j \in \{i_1, i_2, \ldots, i_l\}} D_j$. A rule about a union of l classes in its conclusion is called an l-imprecise rule. The classification of a new object u under rules about unions of D_i's is done by the following procedure:

1. When u matches to at least one of the conditions of the rule, we calculate

$$\hat{S}(D_i) = \sum_{\substack{\text{matching rule } r \\ \text{for } Z \supseteq D_i}} Stren(r) \times Spec(r), \tag{2}$$

where r is called a *matching rule* if the condition part of r is satisfied with u. The strength $Stren(r)$ is the total number of objects in the given dataset correctly classified by rule r. The specificity $Spec(r)$ is the total number of condition attributes in the condition part of rule r. Z is a variable showing a union of classes. For convenience, when there is no matching rules about $Z \supseteq D_i$, we define $\hat{S}(D_i) = 0$. If there exists D_j such that $\hat{S}(D_j) > 0$, the class D_i with the largest $\hat{S}(D_i)$ is selected. If a tie occurs, class D_i with smallest index i is selected from tied classes.

2. When u does not match totally to any rule, for each D_i, we calculate

$$\hat{M}(D_i) = \sum_{\substack{\text{partially matching} \\ \text{rules } r \text{ for } Z \supseteq D_i}} Mat\_f(r) \times Stren(r) \times Spec(r), \tag{3}$$

where r is called a *partially matching rule* if a part of the premise of r is satisfied. The matching factor $Mat\_f(r)$ is the ratio of the number of matched conditions of rule r to the total number of conditions of rule r. Then the class D_i with the largest $\hat{M}(D_i)$ is selected. If a tie occurs, class D_i with smallest index i is selected from tied classes.

Consider a situation that the publication of induced rules are requested. Such a rule publication may be important and necessary for showing the fair and reasonable treatment of objects as well as for knowledge exchange. If the strength $Stren(r)$ is quite small, in other words, r is supported only a few objects, attribute values shown in r identify a few objects. If sensitive data are included in r, this identification invades the privacy. From the viewpoint of data privacy, we cannot publish such rules invading the privacy. On the other hand, hiding such rules may bring an insufficient knowledge exchange and a sense of distrust because the classification of some objects cannot be explained well by the published rules.

The strength $Stren(r)$ of an imprecise rule r is not less than a corresponding precise rule r' because the premise and conclusion of r are weaker than those of r'. By replacing rules with quite small strength scores with imprecise rules, we may protect the data privacy. Moreover, as shown in [15], a set of many imprecise rules can preserve the classification accuracy.

If a rule r satisfies $Stren(r) \geq k$, r is said to be a k-anonymous rule. Inuiguchi et al. [11] proposed an induction method for k-anonymous rules by utilizing imprecise rules. The procedure is given as follows:

(i) Let $\mathcal{R}$ be the set of k-anonymous rules and initialize $\mathcal{R} = \emptyset$. Let $l = 1$.
(ii) Induce a set $\mathcal{S}_1$ of precise rules by MLEM2.
(iii) Select rules $r \in \mathcal{S}_l$ satisfying $Stren(r) \geq k$ and put them in $\mathcal{R}$.
(iv) If $\mathcal{S}_l - \mathcal{R} \neq \emptyset$ and $l < n$, update $l = l + 1$. Otherwise, terminate this procedure.
(v) Define object set B by objects match $r \in \mathcal{S}_l$ such that $Stren(r) < k$.
(vi) Induce a set $\mathcal{S}_l$ of l-imprecise rules for each possible union Z_s of l classes D_i, $i \in \{1, 2, \ldots, p\}$ by MLEM2 inputting $B \cap Z_s$ as a set of objects uncovered by presently induced rules. Return to (iii).

3   The Proposed Data Anonymization

We propose a data anonymization method for decision table based on k-anonymous rules. A decision table is replaced with a table of imprecise patterns each of which is matched by at least k objects in the original decision table, where a pattern stands for a combination of condition and decision attribute values. An imprecise pattern implies a pattern with imprecise attribute values. In a k-anonymous decision table, for each object, there exist at least $(k - 1)$ other objects taking the same pattern. Therefore, the proposed method does not generate a k-anonymous decision table. Because each imprecise pattern in the proposed table has at least k objects in the original decision table (we call this property "k-commonality"), the proposed table preserves the data privacy.

The proposed table is produced by using k-anonymous rules. The k-anonymous rules are induced by the method proposed described in Sect. 2. The proposed k-commonization procedure for a given decision table $DT = \langle U, C \cup \{d\}, V, f \rangle$ under a set $\mathcal{R}$ of k-anonymous rules is as follows:

(i) Let $cT = \langle \langle P, C \cup \{d\}, V, \rho \rangle \rangle$ be a commonized table corresponding to DT, where $\mathcal{P}$ is a set of patterns and $\rho : \mathcal{P} \times C \cup \{d\} \to \bigcup_{a \in C \cup \{d\}} 2^{V_a}$, 2^{V_a} is the power set of V_a, $a \in C \cup \{d\}$. Initialize $\mathcal{P} = \emptyset$ and $\tilde{U} = U$.

(ii) For each object $u \in \tilde{U}$, obtain a minimal set $R(u) \subseteq \mathcal{R}$ of k-anonymous rules such that $Cl(u|R(u)) = Cl(u|\mathcal{R})$, where $Cl(u|R)$ is a set of the estimated values for decision attribute value of u under set of rules R. Execute the following routine:

(ii-a) If $R(u) = \emptyset$, terminate this procedure for $u \in \tilde{U}$.

(ii-b) For each rule $r \in R(u)$, we select k objects matching r and produce an imprecise pattern pt by the procedure (s1)–(s3) described in what follows. Update $\mathcal{P} = \mathcal{P} \cup \{pt\}$.

In (ii), if the class of object u is estimated well by rules in $\mathcal{R}$, $Cl(u|\mathcal{R})$ becomes a singleton. Otherwise, $Cl(u|\mathcal{R})$ becomes a set of multiple decision attribute values or an emptyset. $Cl(u|\mathcal{R}) = \emptyset$ implies that there is no matching rule in $\mathcal{R}$ for object u.

The number of elements of $\mathcal{P}$ can be larger than U. Namely, we obtain a larger table cT than the original decision table DT. The existence of k objects for $r \in R(u)$ is guaranteed by the k-anonymity of r when $R(u) \neq \emptyset$. For the calculation of a minimal set $R(u)$ at step (ii), we apply the following procedure:

(ii-c) Let $M(u) \subseteq \mathcal{R}$ be a set of k-anonymous rules that u matches. Initialize $R(u) = \emptyset$.

(ii-d) If $M(u) = \emptyset$, go to (ii-f).

(ii-e) Select a rule $r \in \mathcal{R}$ randomly. Let $cl(r)$ be a set of decision attribute values in the conclusion of r. Update $R(u) = R(u) \cup \{r\}$ and $M(u) = M(u) - \{r \mid cl(r) \subseteq Cl(u|R(u))\}$. Return to (ii-d).

(ii-f) If $R(u) = \emptyset$, terminate this procedure. Otherwise, for each $r \in R(u)$, we execute the following conditional updating: if $Cl(u|R(u)) = Cl(u|R(u) - \{r\})$, update $R(u) = R(u) - \{r\}$.

(ii-g) Terminate this procedure.

In this paper, the following routine is applied to the selection of k objects matching r at (ii-b) of the proposed procedure:

(s1) Initialize OB by the set of objects in U matching r except u, and $O(r) = \emptyset$.

(s2) Select an object u' from OB which maximizes the number of same condition attribute values as u. If a tie occurs, select the first one among them. Update $O(r) = O(r) \cup \{u'\}$.

(s3) If $|O(r)| < k$, update $Ob = OB - \{u'\}$ and return to (s5), where $|Y|$ shows the cardinality of set Y.

Let $O(r)$ the set of k objects selected by the routine describe above. The composition of pt at (ii-b) of the proposed procedure is explained as follows:

(s4) The decision attribute value set of pt is defined by that in the conclusion of r.

(s5) If a condition attribute $a \in C$ appears in the premise of r, the attribute value set is defined by the value set specified in the premise of r.

Table 1. A simple example

<table>
<tr><td colspan="5">(a) A decision table</td></tr>
<tr><th>obj</th><th>occupation</th><th>age</th><th>sex</th><th>income</th></tr>
<tr><td>u_1</td><td>salesperson</td><td>41</td><td>male</td><td>high</td></tr>
<tr><td>u_2</td><td>salesperson</td><td>34</td><td>female</td><td>high</td></tr>
<tr><td>u_3</td><td>professor</td><td>40</td><td>male</td><td>normal</td></tr>
</table>

(b) 2-common pattern table

pat	occupation	age	sex	income
pt_1	salesperson	[34, 41]	male or female	high

Here 'obj' and 'pat' stand for object and pattern, respectively.

(s6) If a condition attribute $a \in C$ does not appear in the premise of r and it is a numerical/ordinal attribute, the attribute set value is defined by the unique minimal interval covering $\{f(u, a) \mid u \in O(r)\}$.

(s7) If a condition attribute $a \in C$ does not appear in the premise of r and it is a nominal attribute, the attribute set value is defined by the set $\{f(u, a) \mid u \in O(r)\}$.

Example 1. From the simple decision table given in Table 1(a), let us obtain a 2-common pattern table. First we obtain a unique 2-anonymous rule, if occupation is salesperson then income is high. Two objects u_1 and u_2 are selected as objects matching this rule. Then we obtain the 2-common pattern table shown in Table 1(b).

4   Numerical Experiments

4.1   Outline

In order to examine the usefulness of the k-common pattern table, we apply the proposed approach to six data sets obtained from UCI Machine Learning Repository [16]. The six data sets shown in the first column of Table 2 are used for the examination purpose and not necessarily include sensitive data. In Table 2, $|U|$, $|C|$, $|V_d|$ and AT mean the number of objects in the given data table, the number of condition attributes, the number of classes and attribute type, respectively. We execute two numerical experiments. One is to evaluate the privacy preservation of decision attribute value against the revelation of a few condition attribute values. The other is to evaluate the data usability of k-common pattern tables in building classifiers.

In the experiment about the data usability, we apply a 10-fold cross validation method. Namely we partition the data-set into 10 elementary sets and 9 elementary sets are used for training data set and the remaining subset is used for checking data set. Changing the combination of 9 elementary sets, we obtain 10 different evaluations. We calculate the averages and the standard deviations in each evaluation measure. We execute this procedure 10 times with different divisions. In the experiment about the privacy preservation, the same partitions as used in 10-fold cross validation are utilized. All combinations of 10 elementary sets of a partition compose decision tables. In this way, we obtain 10 decision tables for each partition. Because we have 10 partitions, we obtain 100 decision tables to be k-commonalized for each data set.

Table 2. The results of the privacy preservation experiment

Data set	q	DT	$5\text{-}cT$	$10\text{-}cT$	$15\text{-}cT$
car	1	0±0	0±0	0±0	0±0
($|U| = 1,728$,	2	0.1875±0	0.1787±0.0217**	0.1806±0.0195**	0.1844±0.0136*
$|C| = 6$, $|V_d| = 4$,	3	0.1875±0	0.1852±0.0060**	0.1856±0.0055**	0.1866±0.0043*
AT: ordinal)	4	0.4584±0.0007	0.4573±0.0026**	0.4575±0.0025**	0.4579±0.0021*
ecoli	1	0.3716±0.0219	0±0**	0±0**	0±0**
($|U| = 336$,	2	0.9698±0.0043	0.0001±0.0005**	0.0002±0.0012**	0.0005 ± 0.0032**
$|C| = 7$, $|V_d| = 8$,	3	0.9726±0.0040	0.0072±0.0059**	0.0054±0.0059**	0.0041 ± 0.0056**
AT: numerical)	4	0.9726±0.0040	0.0075±0.0053**	0.0052±0.0055**	0.0040 ± 0.0055**
glass	1	0.9234±0.0099	0±0**	0±0**	0±0**
($|U| = 214$,	2	1±0	0.0114±0.0117**	0.0069±0.0104**	0.0099±0.0117**
$|C| = 9$, $|V_d| = 6$,	3	1±0	0.0677±0.0387**	0.0406±0.0275**	0.0522±0.0317**
AT: numerical)	4	1±0	0.2461±0.0897**	0.1586±0.0547**	0.1871±0.0528**
hayes-roth	1	0±0	0±0	0±0	0.3500±0.4530**
($|U| = 159$,	2	0.25±0	0.2675±0.0359**	0.2616±0.0425**	0.1483±0.2008**
$|C| = 4$, $|V_d| = 3$,	3	0.5526±0.0297	0.4550±0.0488**	0.4665±0.0644**	0.0400±0.0627**
AT: numerical)	4	0.8938±0.0095	0.7747±0.0291**	0.6891±0.0329**	0.0143±0.0245**
iris	1	0.4698±0.0313	0.1151±0.1132**	0.1860±0.0955**	0.2043±0.0839**
($|U| = 150$,	2	0.9205±0.0107	0.3188±0.0851**	0.3562±0.0718**	0.3627±0.0595**
$|C| = 4$, $|V_d| = 3$,	3	1±0	0.9053±0.0695**	0.9319±0.0449**	0.9377±0.0365**
AT: numerical)	4	1±0	0.9725±0.0151**	0.9570±0.0178**	0.9537±0.0149**
zoo	1	0±0	0±0	0±0	0±0
($|U| = 101$,	2	0.3333±0	0.3333±0	0.3333±0	0.3333±0
$|C| = 16$, $|V_d| = 7$,	3	0.4120±0.0474	0.4040±0.0397	0.3440±0.1061**	0.2660±0.1060**
AT: nominal)	4	0.7342±0.0534	0.5795±0.0443**	0.5697±0.0353**	0.5723±0.0317**

4.2 Privacy Preservation

We examine the privacy preservation by the proposed k-common pattern table. We assume q condition attribute values of an object in the original table are known by an attacker. Under this assumption, we calculate the probability that the decision attribute value is precisely known from the revealed condition attribute values. To estimate the decision attribute value, we take simply the intersection of sets of decision attribute values of the patterns include the revealed condition attribute values. We apply the same estimation to the original decision table. In both cases, we count combinations of q condition attribute values existing in the original data from which the decision attribute value is uniquely estimated. This number is denoted by $N\text{-}uni$. Then we calculate ratio $N\text{-}uni/N\text{-}all$ for each of decision table and k-common pattern tables, where $N\text{-}all$ is the number of all combinations of q condition attribute values existing in the original data.

The results of the privacy preservation experiment is shown in Table 2. In columns DT (original decision table), $5\text{-}cT$ (5-common pattern table), $10\text{-}cT$ (10-common pattern table) and $15\text{-}cT$ (15-common pattern table) of Table 2, the ratios of the uniquely estimated cases are shown in the style of $av \pm sd$, where av

is the average and sd is the standard deviation. Asterisk $*$ and double asterisk $**$ mean that the ratio of uniquely estimated cases in k-common pattern table is significantly different from that in the original decision table with significance level $\alpha = 0.05$ and $\alpha = 0.01$, respectively.

As shown in Table 2, generally speaking, we found that the privacy is preserved more in the proposed k-common pattern tables. In data sets 'ecoli', 'glass' and 'iris', the advantage of the proposed tables is remarkable. On the other hand, in data sets 'car' and 'zoo', the advantage is not very strong. This fact can be understood the ratios of the uniquely estimated cases are not very big even in the original decision table when q is small. A strange phenomenon is observed in 15-common pattern tables for data set 'hayes-roth'. Namely the ratio decreases as the number q of revealed condition attribute values increases. It is usual that the ratio increases as the number q of revealed condition attribute values increases because the precision usually increases as obtained information increases. This strange phenomenon is caused by the fact only a small number of imprecise rules are induced from the original decision table and thus the patterns in 15-common pattern tables include q revealed condition attribute values decreases as q increases. We note that we count combinations of q condition attribute values whose decision attribute values are uniquely estimated to obtain $N\text{-}uni$ and we did not take care about the correctness of the estimation. Moreover, in the proposed method, we try to keep the classification ability of k-common pattern tables, so that the advantage can be vanishing as q becomes very large.

4.3   Data Usability in Classification

We examine the data usability of k-common pattern tables in classification. To this end, we induce rules from k-common pattern tables in a similar way to MLEM2 and check the classification accuracy of the classifier based on induced rules. Because a k-common pattern table includes imprecise attribute values, some modification is necessary for the lower approximation of a union of classes and rule induction algorithm and the induced rules are imprecise rules. Therefore, the classifier is built by the classification method described in Sect. 2 using the induced imprecise rules.

The lower approximation of a union Z_s of classes in a k-common pattern table $cT = \langle\langle \mathcal{P}, C \cup \{d\}, V, \rho \rangle\rangle$ using condition attribute set $A \subseteq C$ is defined by the following set of patterns:

$$A_*(Z_s) = \{pt \in \mathcal{P} \mid |[pt]|_A \subseteq Z_s\}, \tag{4}$$

where $|[pt]|_A$ and Z_s are defined by

$$|[pt]|_A = \{pt' \in \mathcal{P} \mid \rho(pt', a) \subseteq \rho(pt, a), \; \forall a \in A\}, \quad Z_s = \{pt \in \mathcal{P} \mid \rho(pt, d) \subseteq Z_s\}. \tag{5}$$

To induce imprecise rules from $cT = \langle\langle \mathcal{P}, C \cup \{d\}, V, \rho \rangle\rangle$, we modify MLEM2. For explaining the modification when we induce imprecise rules about Z_s. For a nominal attribute a, we collect all tuples $(a, \bigcup_{pt \in W} \rho(pt, a))$ such that $W \in Z_s$

Table 3. Data usability in classification

data set / item		DT	$5\text{-}cT$	$10\text{-}cT$	$15\text{-}cT$
car	no. rules	$57.22{\pm}1.7411$	$63.56{\pm}3.5590$	$74.66{\pm}3.5643$	$89.76{\pm}4.948$
	accuracy	$0.9867{\pm}0.0097$	$0.8864{\pm}0.031**$	$0.8978{\pm}0.0320**$	$0.8978{\pm}0.0300**$
ecoli	no. rules	$35.89{\pm}2.0293$	$1078.28{\pm}115.2843$	$1155.42{\pm}60.8827$	$1012.59{\pm}52.3200$
	accuracy	$0.7755{\pm}0.0620$	$0.6666{\pm}0.095**$	$0.6590{\pm}0.0810**$	$0.6167{\pm}0.0850**$
glass	no. rules	$25.38{\pm}1.4952$	$111.99{\pm}20.7048$	$209.68{\pm}13.2928$	$187.95{\pm}8.9937$
	accuracy	$0.6844{\pm}0.1005$	$0.5442{\pm}0.1061**$	$0.5164{\pm}0.1206**$	$0.4834{\pm}0.1054**$
hayes-roth	no. rules	$23.17{\pm}1.4075$	$20.62{\pm}2.3400$	$13.95{\pm}1.5256$	$1.96{\pm}1.5028$
	accuracy	$0.8131{\pm}0.0783$	$0.7931{\pm}0.1106$	$0.6375{\pm}0.1463**$	$0.3956{\pm}0.1182**$
iris	no. rules	$7.4{\pm}0.7211$	$6.33{\pm}1.0959$	$5.27{\pm}1.0474$	$5.02{\pm}0.7208$
	accuracy	$0.9286{\pm}0.0550$	$0.9273{\pm}0.0760$	$0.9180{\pm}0.0878$	$0.9213{\pm}0.0920$
zoo	no. rules	$9.67{\pm}0.5487$	$142.13{\pm}14.1263$	$183.81{\pm}16.3736$	$89.76{\pm}10.4393$
	accuracy	$0.9583{\pm}0.0663$	$0.8611{\pm}0.1022**$	$0.7940{\pm}0.1249**$	$0.6891{\pm}0.1242**$

as elementary condition candidates. For a numerical or ordinal attribute a, we calculate all midpoints v^L (resp. v^R) of two consecutive values among all lower (resp. upper) bounds of $\rho(pt, a)$, $a \in C$ and collect all tuples $(a, [v^L, +\infty))$ (resp. $(a, (-\infty, v^R])$) as elementary condition candidates. For a tuple $t = (a, Q)$ ($Q \subseteq V_a$) and a set of tuples, we use $||t|| = \{pt \in \mathcal{P} \mid \rho(pt, a) \subseteq Q\}$ and $||[T]|| = \bigcap_{t \in T} ||t||$ instead of $[t]$ and $[T]$, respectively. By those modifications, a minimal set of minimal length of imprecise rules explaining all patterns in $C_*(Z_s)$ is obtained by MLEM2. Moreover, a classifier can be build by using the classification method under imprecise rules described in Sect. 2.

The classification accuracy of the classifier using imprecise rules induced from k-common pattern table by MLEM is calculated by 10 times run of 10 cross validation method for $k = 5, 10, 15$. For comparison, the classification accuracy of the classifier using precise rules induced from the original decision table by MLEM is also calculated. The results are shown in Table 3. In Table 3, "no. rules" and "accuracy" stand for the number of induced rules and the classification accuracy, respectively. Those values are shown in the style of $av \pm sd$, where av is the average and sd is the standard deviation. Asterisk $*$ and double asterisk $**$ mean that the ratio of uniquely estimated cases in k-common pattern table is significantly different from that in the original decision table with significance level $\alpha = 0.05$ and $\alpha = 0.01$, respectively.

As shown in Table 3, the classification accuracy of the classifier using rules induced from the k-common pattern table is evidently smaller than that from the original decision table. This is because there is no guarantee the original rules used to build a k-common pattern table are induced from the k-common pattern table. Moreover, for objects which cannot be covered by any k-anonymous rules, their decision attribute values will not be estimated correctly. Thus, the deterioration of classification accuracy is not always avoidable. Nevertheless, the deterioration of classification accuracy is less than 12% in four data sets, 15% in five data sets and 21% in all data sets when $k = 5$. As we utilize imprecise rules, the number of rules are remarkably increased in three data sets. In data set 'hayes-roth', because only a few 15-anonymous rules exist, the classification accuracy of 15-cT is very small.

5 Conclusion

In this paper, we proposed a data anonymization method based on k-anonymous imprecise rules. In this method, the data matches a k-anonymous rule in the original table is replaced with imprecise patterns. Then a table of imprecise patterns is obtained. We examined the usefulness of the proposed method by numerical experiments. The results show that the proposed method will be satisfactorily useful. In our future study, we will modify the selection of k objects taking care that the resulting attribute value sets do not match rules about different unions of classes. Moreover, we will investigate a data anonymization method using rules induced from decision tables with two classes.

Acknowledgment. This work was partially supported by JSPS KAKENHI Grant Number 18H01658.

References

1. Samarati, P.: Protecting respondents' identities in microdata release. IEEE Trans. Knowl. Data Eng. **13**(6), 1010–1027 (2001)
2. Sweeney, L.: K-Anonymity: a model for protecting privacy. Int. J. Uncertain. Fuzziness Knowl. Based Syst. **10**(5), 557–570 (2002)
3. Dwork, C.: Differential privacy: a survey of results. In: Agrawal, M., Du, D., Duan, Z., Li, A. (eds.) TAMC 2008. LNCS, vol. 4978, pp. 1–19. Springer, Heidelberg (2008). https://doi.org/10.1007/978-3-540-79228-4_1
4. Dwork, C., Roth, A.: The algorithmic foundations of differential privacy. Found. Trends Theor. Comput. Sci. **9**(3–4), 211–407 (2014)
5. Yakoubov, S., Gadepally, V., Schear, N., Shen, E., Yerukhimovich, A.: A survey of cryptographic approaches to securing big-data analytics in the cloud. In: Proceedings of 2014 IEEE High Performance Extreme Computing Conference, pp. 1–6. IEEE Xplore (2014)
6. Fung, B.C.N., Wang, K., Chen, R., Yu, P.S.: Privacy-preserving data publishing: a survey of recent developments. ACM Comput. Surv. **42**(4), Article 14 (2010)
7. Pawlak, Z.: Rough sets. Int. J. Comput. Inf. Sci. **11**(5), 341–356 (1982)
8. Zhou, Z., Huang, L., Yun, Y.: Privacy preserving attribute reduction based on rough set. In: Proceedings of 2nd International Workshop on Knowledge Discovery and Data Mining, WKKD 2009, pp. 202–206. AAAI, Portland, USA (2009)
9. Rokach, L., Schclar, A.: k-Anonymized reducts. In: Proceedings of 2010 IEEE International Conference on Granular Computing, pp. 392–395. IEEE Xplore (2010)
10. Ye, M., Wu, X., Hu, X., Hu, D.: Anonymizing classification data using rough set theory. Knowl. Based Syst. **43**, 82–94 (2013)
11. Inuiguchi, M., Hamakawa, T., Ubukata, S.: Imprecise rules for data privacy. In: Ciucci, D., Wang, G., Mitra, S., Wu, W.-Z. (eds.) RSKT 2015. LNCS, vol. 9436, pp. 129–139. Springer, Cham (2015). https://doi.org/10.1007/978-3-319-25754-9_12

12. Ohki, M., Inuiguchi, M.: A k-anonymous rule clustering approach for data publishing. J. Adv. Comput. Intell. Intell. Inform. **21**(6), 980–988 (2017)
13. Inuiguchi, M.: Rough set analysis of imprecise classes. In: Wang, G., Skowron, A., Yao, Y., Ślęzak, D., Polkowski, L. (eds.) Thriving Rough Sets: Studies in Computational Intelligence, vol. 708, pp. 157–185. Springer, Cham (2017). https://doi.org/10.1007/978-3-319-54966-8_8
14. Grzymala-Busse, J.W.: MLEM2 - discretization during rule induction. In: Klopotek, M.A., Wierzchon, S.T., Trojanowski, K. (eds.) Intelligent Information Processing and Web Mining, vol. 22. Springer, Heidelberg (2003). https://doi.org/10.1007/978-3-540-36562-4_53
15. Hamakawa, T, Inuiguchi, M.: On the utility of imprecise rules induced by MLEM2 in classification. In: Proceedings of 2014 IEEE International Conference on Granular Computing, pp. 76–81. IEEE Xplore (2014)
16. UCI Machine Learning Repository. http://archive.ics.uci.edu/ml/

Approximate Multiobjective Multiclass Support Vector Machine Restricting Classifier Candidates Based on k-Means Clustering

Keiji Tatsumi$^{(\boxtimes)}$ ⓘ, Takahumi Sugimoto, and Yoshifumi Kusunoki ⓘ

Graduate School of Engineering, Osaka University,
2-1 Yamada-Oka, Suita, Osaka, Japan
{tatsumi,kusunoki}@eei.eng.osaka-u.ac.jp

Abstract. In this paper, we propose a reduction method for the multiobjective multiclass support vector machine (MMSVM), one of alltogether method of the SVM. The method can maintain the discrimination ability, and reduce the computational complexity of the original MMSVM. First, we derive an approximate convex multiobjective optimization problem for the MMSVM by linearizing some constraints, and we secondly restrict the normal vectors of classifier candidates by using centroids obtained from the k-means clustering for each class dataset. The derived problem can be solved by the reference point method based on the centers of gravity of class datasets, in which the geometric margins between all pairs are exactly maximized. Some numerical experiments for benchmark problems show that the proposed method can reduce the computational complexity without decreasing its generalization ability widely.

Keywords: Multiclass classification · Support vector machine ·
Multiobjective optimization · Reference point method ·
k-means clustering

1 Introduction

The binary support vector machine (SVM) [12] is one of popular machine learning methods, which finds a classifier with a high classification ability by maximizing the geometric margin between data and a separating hyperplane. In addition, various extended methods of the binary SVM have been investigated for multiclass classification. In this paper, we focus on all-together (AT) method among the extended ones, especially, the multiobjective multiclass SVM (MMSVM) [7]. It was reported that comparing with the simplest AT method maximizing functional margins, the MMSVM can obtain a classifier with a higher classification rate by maximizing exactly the geometric margin between each class pair. However, it requires a larger amount of computational resources than the simplest AT and other methods [7–9].

© Springer Nature Switzerland AG 2019
H. Seki et al. (Eds.): IUKM 2019, LNAI 11471, pp. 271–283, 2019.
https://doi.org/10.1007/978-3-030-14815-7_23

Therefore, in this paper, we propose a method of reducing the computational complexity without decreasing its generalization ability widely. First, we derive an approximate convex multi-objective optimization problem (MOP) by linearizing some constraints of the original non-convex MOP which is used to find a classifier in MMSVM. Secondly, we restrict the normal vectors of classifier candidates to a space spanned by centroids obtained from the preliminary k-means clustering. Thirdly, we solve the derived MOP by applying the reference point method. Through numerical experiments for benchmark problems, we evaluate performance of classifiers obtained by the proposed method and its computational complexity.

2   Multiclass Classification

The multiclass classification means discriminating data into more than two classes. We assume that a dataset (x^i, y_i), $i = 1, \ldots, l$ are generated by the same distribution $P(x, y)$, where $x^i \in \mathcal{R}^n$ denotes an n-dimensional input, and $y_i \in M := \{1, \ldots, m\}$ denotes a label which the corresponding x^i should be classified into. The aim is finding a classifier $f(x)$ which satisfies $y_i = f(x^i)$, $i = 1, \ldots, l$ and which can correctly classify a new unknown input x from the same distribution. In this paper, we assume that there exists an appropriate feature space F and a corresponding function $\phi : \mathcal{R}^n \to F$. Thus, we mainly discuss a linear classification on F which uses the kernel method.

In the representative SVMs for multiclass classification such as one-against-all (OAA) [2] and all-together (AT) methods [11,12], the following discriminant function is often used:

$$f(x) = \underset{p \in M}{\operatorname{argmax}} \ w^{p^\top} \phi(x) + b^p$$

where w^p, b^p, $p \in M$ denote a weight vector and a bias value, respectively. Thus, the aim is finding appropriate (w^p, b^p), $p \in M$.

2.1   SVM Maximizing Functional Margins

As the simplest AT method, the SVM maximizing the sum of functional margins was proposed in [11,12], which can be straightforwardly derived from the binary SVM.

$$\begin{aligned}
\text{(AT)} \quad &\min \sum_{p \in M} \sum_{q \in M} ||w^p - w^q||^2 \\
&\text{s.t. } (w^p - w^q)^\top \phi(x^i) + (b^p - b^q) \geq 1, \ i \in I_p, q > p, \ p, q \in M, \\
&\quad (w^q - w^p)^\top \phi(x^i) + (b^q - b^p) \geq 1, \ i \in I_q, q > p, \ p, q \in M,
\end{aligned}$$

where $I_p := \{i \in \{1, \ldots, l\} | \ y_i = p\}, p \in M$. Note that maximizing the functional margin in binary SVM can guarantee exact maximization of the distance between data and a separating hyperplane, called *geometric margin*, which can contribute its high generalization ability. On the other hand, in the problem (AT) for the multiclass classification, maximizing functional margins $1/\|w^p - w^q\|$ does not necessarily guarantees the maximization of the geometric margins. Namely, the functional margin for a class pair pq does not necessarily represent the distance between the corresponding separating hyperplane:

$$(w^p - w^q)^\top \phi(x) + (b^p - b^q) = 0$$

and the closest data in classes $\{p, q\}$, as pointed out in [7], which is represented by

$$d_{pq}(w, b) = \min_{i \in I_p \cup I_q} \frac{|(w^p - w^q)^\top \phi(x)^i + (b^p - b^q)|}{\|w^p - w^q\|}, \quad q > p, \; p, q \in M.$$

Thus, it might be difficult to expect the generalization ability similar to the binary SVM. The method of maximizing exactly geometric margins was already proposed in [7]. We introduce it in the next section.

2.2 SVM Maximizing Geometric Margins

In order to maximize exactly the geometric margins, an AT method called MMSVM was already proposed, which was formulated as the following multiobjective optimization problem (MOP) [7]:

$$(\mathrm{M}) \quad \max_{w, b, \sigma} \theta_{12}(w, \sigma), \ldots, \theta_{m-1, m}(w, \sigma),$$
$$\text{s.t. } (w^p - w^q)^\top \phi(x^i) + (b^p - b^q) \geq \sigma_{pq}, \; i \in I_p, q > p, \; p, q \in M,$$
$$(w^q - w^p)^\top \phi(x^i) + (b^q - b^p) \geq \sigma_{pq}, \; i \in I_q, q > p, \; p, q \in M,$$
$$\sigma_{pq} \geq 1, \; q > p, \; p, q \in M,$$

where we define $\theta_{pq}(w, \sigma) = \sigma_{pq}/\|w^p - w^q\|$, $q > p$, $p, q \in M$. Note that (M) has more than two objective functions, and the number of them is that of all combinations of class pairs. In [7], it was shown that at any Pareto optimal solution (w^*, b^*, σ^*) of (M), each of objective functional values $\theta_{pq}(w^*, \sigma^*)$ is equal to the geometric margin $d_{pq}(w^*, b^*)$ of the corresponding class pair [7].

Since in general the optimal solutions of the MOP are often given as a set called *Pareto optimal* solutions, and, in addition, (M) is not convex. The problem is more difficult to solve than the single-objective optimization problem (SOP). However, a method of finding a Pareto optimal solution by solving a convex SOP was introduced, and the kernel method can be easily applied to (M).

Now, let's consider the kernel method for (M). The weight vector w^p of the separating hyperplane is represented as a weighted sum of $\phi(x^i)$ by introducing new decision variables $\alpha_i^p \in R$, $i = 1, \ldots, l$, $p \in M$:

$$w^p = \sum_{i=1}^{l} \alpha_i^p \phi(x^i), \quad p \in M. \tag{1}$$

Then, by defining $K := (\phi(x^1), \ldots, \phi(x^l))^\top (\phi(x^1), \ldots, \phi(x^l))$, $\alpha^p := (\alpha_1^p, \ldots, \alpha_l^p)^\top$, $p \in M$, $\bar{\theta}_{pq}(\alpha, \sigma) := \sigma_{pq}/\sqrt{(\alpha^p - \alpha^q)^\top K (\alpha^p - \alpha^q)}$, $q > p$, $p, q \in M$, $\kappa(x^i) := (k(x^1, x^i), \ldots, k(x^l, x^i))^\top$, $i = 1, \ldots, l$, (M) can be rewritten as

(M2) $\displaystyle\max_{\alpha, b, \sigma} \bar{\theta}_{12}(\alpha, \sigma), \ldots, \bar{\theta}_{(m-1)m}(\alpha, \sigma)$

   s.t. $(\alpha^p - \alpha^q)^\top \kappa(x^i) + (b^p - b^q) \geq \sigma_{pq}$, $i \in I_p$, $q > p$, $p, q \in M$,

   $(\alpha^q - \alpha^p)^\top \kappa(x^i) + (b^q - b^p) \geq \sigma_{pq}$, $i \in I_q$, $q > p$, $p, q \in M$,

   $\sigma_{pq} \geq 1$, $q > p$, $p, q \in M$,

In addition, the discriminant function can be represented by $f(x) = \displaystyle\operatorname*{argmax}_{p \in M} \{\alpha^{p*\top} \kappa(x) + b^{p*}\}$, where $(\alpha^*, b^*, \sigma^*)$ is the Pareto optimal solution of (M2).

As a method of solving (M2), the ε-constraint method is used, which is one of popular scalarization methods for the MOP [4]. In the method, the SOP is derived instead of (M2), in which one of objective functions of (M2) is used as the objective function of the new SOP, and other objective functions are changed into its constraints by an appropriate constant vector ε. In addition, the following transformation was used to solve the SOP [7,8].

Now, we focus on all positive eigenvalues values $\lambda_1, \ldots, \lambda_\tau$ of K, and the corresponding eigenvectors $t^1, \ldots, t^\tau$, where $\tau > 0$ denotes the number of positive eigenvalues. Then, we have that

$$K = [t^1, \ldots, t^\tau] \operatorname{diag}\{\lambda_1, \ldots, \lambda_\tau\}[t^1, \ldots, t^\tau]^\top =: T\Lambda T^\top \tag{2}$$

Then, new decision variables z^p are defined as $z^p := \Lambda^{\frac{1}{2}} T^\top \alpha^p$, $p \in M$, and the following convex SOP is obtained:

(εM2) $\displaystyle\max_{z, b, \sigma} \frac{c_{rs}}{\|z^r - z^s\|}$

   s.t. $\dfrac{\sigma_{pq}}{\|z^p - z^q\|} \geq \varepsilon_{pq}$, $q > p$, $(p, q) \neq (r, s)$, $p, q \in M$,

   $(z^p - z^q)^\top \Lambda^{\frac{1}{2}} \bar{t}^i + (b^p - b^q) \geq \sigma_{pq}$, $i \in I_p$, $q > p$, $p, q \in M$,

   $(z^q - z^p)^\top \Lambda^{\frac{1}{2}} \bar{t}^i + (b^q - b^p) \geq \sigma_{pq}$, $i \in I_q$, $q > p$, $p, q \in M$,

   $\sigma_{pq} \geq 1$, $q > p$, $(p, q) \neq (r, s)$, $p, q \in M$,

   $\sigma_{rs} = c_{rs}$,

where $\bar{t}^{i\top}$ denotes the i-th row vector of T, the constant ε_{pq}, $p, q \in M$ is appropriately selected for the feasibility of (εM2), and class pair rs is appropriately selected. Note that the constraint $\sigma_{rs} = c_{rs}$ with a sufficiently large constant c_{rs} is added so that (εM2) is convex, and a large c_{rs} guarantees that the optimal solution of (εM2) is Pareto optimal [7]. Moreover, (εM2) is a second-order cone programming problem (SOCP), which is a convex problem having the second-order and linear constraints, and which can be effectively solved by using some primal-dual interior method [1]. In addition, numerical experiments showed that the geometric margins of separating hyperplanes constructed by the optimal solution of (εM2) are larger than those obtained by the functional margin method

(AT), and that classifiers of (M2) obtained by (εM2) have better classification ability than (AT) [7–9].

Next, let us evaluate the computational resources required to solve (εM2) and (AT). Since in (εM2), a constant vector ε is determined by the optimal solution of (AT), (εM2) requires solving (AT). In addition, CPU time of solving an SOCP (εM2) is considerably larger than that of (AT) because of many decision variables. Moreover, if the element number l of all datasets is large, the diagonalization of $l \times l$ matrix K also requires a large amount of computational resources. Therefore, in [5], an approximation method for (M) was proposed, in which a single-objective SOCP is derived by introducing the sum of objective functions of (M) and linearizing the right-hand side of the first and second constraints of (M). The SOCP is easily solved due to its convexity, and an feasible solution of (M) can be easily obtained from the optimal solution of the SOCP. The numerical experiments showed that the generalization ability of classifiers obtained by approximation method is better than that of (AT).

In this paper, we derive an approximate MOP by using the same approximation technique, and, furthermore, we restrict the normal vectors of classifier candidates to a space spanned by centroids obtained from a preliminary clustering in order to reduce its computational complexity.

3   Approximate MMSVM

In this section, we introduce the following problem by defining $\delta_{pq} := \sigma_{pq}^2$ and putting a constant upper limit $\rho \geq 1$ on δ_{pq}, $q > p$, $p, q \in M$.

$$
\begin{aligned}
&\min_{w,b,\delta} \ \eta_{12}(w, \delta), \ldots, \eta_{(m-1)m}(w, \delta)\\
\text{(S1)} \quad &\text{s.t. } (w^p - w^q)^\top \phi(x^i) + (b^p - b^q) \geq \sqrt{\delta_{pq}}, \ i \in I_p, \ q > p, \ p, q \in M,\\
&\qquad (w^q - w^p)^\top \phi(x^i) + (b^q - b^p) \geq \sqrt{\delta_{pq}}, \ i \in I_q, \ q > p, \ p, q \in M,\\
&\qquad 1 \leq \delta_{pq} \leq \rho, \ q > p, \ p, q \in M.
\end{aligned}
$$

Here, η_{pq} is defined by $\eta_{pq}(w, \delta) = \|w^p - w^q\|^2 / 2\delta_{pq}$, $q > p$, $p, q \in M$. If ρ is sufficiently large, (S1) can be considered to be equivalent to (M). Then, in order to approximate (S1), we replace the right-hand sides of the first and second constraint inequalities with $(\delta_{pq} + \sqrt{\rho})/(1 + \sqrt{\rho})$ by using a constant ρ in the same way to [5]. Then, we obtain

$$
\begin{aligned}
&\min_{w,b,\delta} \ \eta_{12}(w, \delta), \ldots, \eta_{(m-1)m}(w, \delta)\\
\text{(S2)} \quad &\text{s.t. } (w^p - w^q)^\top \phi(x^i) + (b^p - b^q) \geq \frac{\delta_{pq} + \sqrt{\rho}}{1 + \sqrt{\rho}}, \ i \in I_p, \ q > p, \ p, q \in M,\\
&\qquad (w^q - w^p)^\top \phi(x^i) + (b^q - b^p) \geq \frac{\delta_{pq} + \sqrt{\rho}}{1 + \sqrt{\rho}}, \ i \in I_q, \ q > p, \ p, q \in M,\\
&\qquad 1 \leq \delta_{pq} \leq \rho, \ q > p, \ p, q \in M.
\end{aligned}
$$

Figure 1 shows the relation of $\sqrt{\delta}$ and $(\delta + \sqrt{\rho})/(1 + \sqrt{\rho})$, which shows that $\sqrt{\delta} \geq \frac{\delta + \sqrt{\rho}}{1 + \sqrt{\rho}}$ for any $\delta \in [1, \rho]$. By making use of the property, for any Pareto

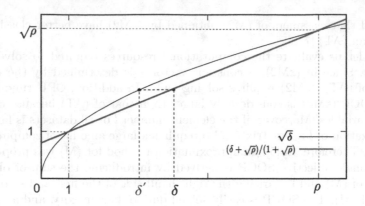

Fig. 1. Approximate affine function

optimal solution (w, b, δ) of (S2), we can obtain a feasible solution (w, b, δ') of (S1) as follows:

$$\delta' = \left(\left(\frac{\delta_{12} + \sqrt{\rho}}{1 + \sqrt{\rho}} \right)^2, \dots, \left(\frac{\delta_{(m-1)m} + \sqrt{\rho}}{1 + \sqrt{\rho}} \right)^2 \right)^\top.$$

Since (S2) can be regarded as convex, it is easier to solve the problem than (M2). In addition, we can show the following properties between solutions of (S1) and (S2): The relation of objective function values at (w, b, δ) and (w, b, δ') is given by

$$\frac{\eta_{pq}(w, b, \delta')}{\eta_{pq}(w, b, \delta)} = \frac{\delta_{pq}}{\delta'_{pq}} \leq \frac{(1 + \sqrt{\rho})^2}{4\sqrt{\rho}}, \quad q > p, \ p, q \in M.$$

Thus, for the Pareto solution or the feasible solution $(\bar{w}, \bar{b}, \bar{\delta})$ of (S1) which dominate (w, b, δ') such that $\eta_{pq}(\bar{w}, \bar{b}, \bar{\delta}) \leq \eta_{pq}(w, b, \delta'), q > p, \ p, q \in M$, we have that

$$\eta_{pq}(w, b, \delta') \leq \frac{(1 + \sqrt{\rho})^2}{4\sqrt{\rho}} \eta_{pq}(\bar{w}, \bar{b}, \bar{\delta}), \quad q > p, \ p, q \in M.$$

The approximate method for MMSVM proposed in this section is called AMMSVM.

4   AMMSVM Based on K-Means Clustering

Next, we introduce a dimension reduction which restricts the weights of (1) of separating hyperplanes to AMMSVM. This method is based on the assumption that the representation of appropriate weights of the separating hyperplanes does not need all datasets such as (1), and thus, the weights can be represented by the weighted sum of a smaller number of data. Namely, instead of using all $\phi(x^i), i = 1, \ldots, l$, the method selects representative points in F for weights w^p of each $p \in M$. Since by using this restriction, the feasible region of the proposed problem is smaller than the original (M2), solving time and used memories can be expected to be widely reduced. At the same time, the proposed method keeps all the constraints of (M2) which guarantee that all training data is correctly classified. Thus, the method is quite different from a reduction method of deleting training data by some preliminary technique [3]. The proposed method uses the k-means clustering [6] with an appropriate number of clusters to each class data $\phi(x^i), i \in I_p$, and centroids of obtained clusters are used as representative points for the class.

4.1   k-Means Clustering

In the proposed method, the k-means clustering is applied to each dataset $\phi(x^i), i \in I_p$ for class p in order to obtain clusters $\{\phi(x^l)\}_{l \in I_p^k}, k = 1, \ldots, c_p$, such that $I_p = \cup_{k=1}^{c_p} I_p^k$, individually, which means minimizing the following function:

$$E^p = \sum_{k=1}^{c_p} \sum_{i \in I_p^k} \|\phi(x^i) - \psi^{p,k}\|^2,$$

where c_p denotes the number of clusters which is appropriately selected for class p. In the numerical experiments at Sect. 5, we set $c_p = \lfloor r|I_p| \rfloor$, and r is a small constant. Centroids of each cluster k are given by $\psi^{p,k} = \sum_{i \in I_p^k} \phi(x^i)/|I_p^k|$. Here, note that the kernel method can be easily applied to the clustering method.

It is well-known that the k-means does not necessarily find the global minimum of E^p. Thus, we executed 20 times k-means clustering and select the centroids of the clustering in which the least E^p was obtained in the numerical experiments.

4.2   Dimension Reduction Based on k-Means Clustering

The centroids obtained by the k-means clustering for the dataset in class p are represented by $\psi^{p,k}, k = 1, \ldots, c_p$. By introducing new decision variables $\beta_{q,k}^p$, $k = 1, \ldots, n_p, p, q \in M$, the weight w_p of AMMSVM for class $p \in M$ are given by

$$w^p = \sum_{q \in M} \sum_{h=1}^{c_p} \beta_{q,h}^p \psi^{q,h} = \Psi \beta^p \tag{3}$$

where matrix Ψ and a decision vector β^p is defined as

$$\Psi := (\psi^{1,1}, \psi^{1,2}, \ldots, \psi^{1,c_1}, \psi^{2,1}, \ldots, \psi^{m,c_m}) \in \mathcal{R}^{c_{all} \times c_{all}},$$
$$\beta^p := (\beta_{1,1}^p, \beta_{1,2}^p, \ldots, \beta_{1,c_1}^p, \beta_{2,1}^p, \ldots, \beta_{m,c_m}^p)^\top \in \mathcal{R}^{c_{all}},$$

and c_{all} is defined as $\sum_{p \in M} c_p$. Then, the discriminant function is represented by

$$f(x) = \underset{p}{\mathrm{argmax}} \left\{ (\Psi \beta^p)^\top \phi(x) + b^p \right\},$$

and the decision variables (β^p, b^p), $q > p$, $p, q \in M$ are determined by solving the following MOP:

(KMS)

$$\min_{\beta, b, \delta} \frac{1}{2} \frac{\|\Psi \beta^1 - \Psi \beta^2\|^2}{\delta_{12}}, \dots, \frac{1}{2} \frac{\|\Psi \beta^{m-1} - \Psi \beta^m\|^2}{\delta_{(m-1)m}}$$

$$\text{s.t.} \quad (\Psi \beta^p - \Psi \beta^q)^\top \phi(x^i) + (b^p - b^q) \geq \frac{\delta_{pq} + \sqrt{\rho}}{1 + \sqrt{\rho}}, \quad i \in I_p, \ q > p, \ p, q \in M,$$

$$(\Psi \beta^q - \Psi \beta^p)^\top \phi(x^i) + (b^q - b^p) \geq \frac{\delta_{pq} + \sqrt{\rho}}{1 + \sqrt{\rho}}, \quad i \in I_q, \ q > p, \ p, q \in M,$$

$$1 \leq \delta_{pq} \leq \rho, \quad q > p, \ p, q \in M.$$

The geometric margins between all class pairs are maximized by solving (KMS) under the restriction (3). Moreover, since a centroid of each cluster is represented by the weighted sum of $\phi(x^i)$, the kernel method can be applied to (KMS).

Now, similarly to (2), we have that $\Psi^\top \Psi = \hat{T} \hat{\Lambda} \hat{T}^\top$, where $\hat{\Lambda} \in \mathcal{R}^{\tau_c \times \tau_c}$ is a diagonal matrix whose diagonal components are all positive eigenvalues of $\Psi^\top \Psi$, $T \in \mathcal{R}^{c_{all} \times \tau_c}$ consists of the corresponding eigenvectors, and τ_c denotes the number of the positive eigenvalues. Then, by introducing decision variables:

$$z^p = \hat{\Lambda}^{\frac{1}{2}} \hat{T}^\top \beta^p, \tag{4}$$

and defining as $\bar{k}^p(x) := \left(\sum_{j \in I_p^1} k(x^j, x)/|I_p^1|, \ \dots, \ \sum_{j \in I_p^{c_p}} k(x^j, x)/|I_p^{c_p}| \right)^\top$ and $\bar{\kappa}(x) := (\bar{k}^1(x), \dots, \bar{k}^m(x))^\top$, we can transform (KMS) to the following MOP:

(KMS2)

$$\min_{z, b, \delta} \frac{1}{2} \frac{\|z^1 - z^2\|^2}{\delta_{12}}, \dots, \frac{1}{2} \frac{\|z^{m-1} - z^m\|^2}{\delta_{(m-1)m}}$$

$$\text{s.t.} \quad (z^p - z^q)^\top \hat{\Lambda}^{-\frac{1}{2}} \hat{T}^\top \bar{\kappa}(x^i) + (b^p - b^q) \geq \frac{\delta_{pq} + \sqrt{\rho}}{1 + \sqrt{\rho}}, \quad i \in I_p, \ q > p, \ p, q \in M,$$

$$(z^q - z^p)^\top \hat{\Lambda}^{-\frac{1}{2}} \hat{T}^\top \bar{\kappa}(x^i) + (b^q - b^p) \geq \frac{\delta_{pq} + \sqrt{\rho}}{1 + \sqrt{\rho}}, \quad i \in I_q, \ q > p, \ p, q \in M,$$

$$1 \leq \delta_{pq} \leq \rho, \quad q > p, \ p, q \in M.$$

We can easily show that any Pareto optimal solution of (KMS) is obtained by solving (KMS2) as follows:

Theorem 1. *For a Pareto optimal solution (z^*, b^*, δ^*) of (KMS2), a solution (β^*, b^*, δ^*) of which β^* is defined as $\beta^{p*} = \hat{T}\hat{\Lambda}^{-\frac{1}{2}} z^{p*}$, $p \in M$ is Pareto optimal for (KMS). Conversely, for a Pareto optimal solution (β^*, b^*, δ^*) of (KMS), (z^*, b^*, δ^*) in which z^* is defined by (4) is Pareto optimal for (KMS2).*

4.3   Solving Based on Reference Point Method

In this subsection, we apply the reference point method to solve (KMS2), which finds a Pareto optimal solution by minimizing the distance between a given reference point and Pareto optimal solutions in the objective space. The following SOP can be derived:

(KMS3)

$$\min_{z,b,r,\delta} \quad \max_{p,q \in M} \{\omega_{pq}(r_{pq} - r^*_{pq})\} + \mu \sum_{p,q \in M} \omega_{pq} r_{pq}$$

$$\text{s.t.} \quad 2 r_{pq} \delta_{pq} \geq \|z^p - z^q\|^2, \quad r_{pq} \geq 0, \ q > p, \ p, q \in M,$$

$$(z^p - z^q)^\top \hat{\Lambda}^{-\frac{1}{2}} \hat{T}^\top \bar{\kappa}(x^i) + (b^p - b^q) \geq \frac{\delta_{pq} + \sqrt{\rho}}{1 + \sqrt{\rho}}, \quad i \in I_p, \ q > p, \ p, q \in M,$$

$$(z^q - z^p)^\top \hat{\Lambda}^{-\frac{1}{2}} \hat{T}^\top \bar{\kappa}(x^i) + (b^q - b^p) \geq \frac{\delta_{pq} + \sqrt{\rho}}{1 + \sqrt{\rho}}, \quad i \in I_q, \ q > p, \ p, q \in M,$$

$$1 \leq \delta_{pq} \leq \rho, \quad q > p, \ p, q \in M.$$

Here, r^*_{pq} is a given reference point which is used as the criterion of minimizing. In (KMS3), the distance between the reference point and Pareto optimal solutions is measured by the augmented Tchebyshev function, in which ω is a weight vector which determines a balance between objective functions, and μ shows the rate of the second term to the first one. The weight vector ω and reference point r^* are selected as the following three kinds of pairs:

$$\begin{array}{lll} \text{(R0)} & \omega_{pq} = 1, & r^*_{pq} = 0, \\ \text{(R1)} & \omega_{pq} = \|g^p - g^q\|^2, & r^*_{pq} = 0, \\ \text{(R2)} & \omega_{pq} = \|g^p - g^q\|^2, & r^*_{pq} = 1/\|g^p - g^q\|^2, \end{array}$$

where g^p denotes the center of gravity of the data set of class p, namely, $g^p = (1/|I_p|) \sum_{i \in I_p} \phi(x^i)$. The selection is based on the idea that the appropriate balance of margins is roughly estimated at the balance of the distances between the centers of gravity. Then, the discriminant function is represented by

$$f(x) = \underset{p}{\operatorname{argmax}} \left\{ z^{*p\top} \hat{\Lambda}^{-\frac{1}{2}} \hat{T}^\top \left(k(x^1, x^i), \dots, k(x^l, x^i) \right)^\top + b^{*p} \right\}. \quad (5)$$

where (z^*, b^*, δ^*) denotes the optimal solution of (KMS3). Although (KMS3) is not smooth, an equivalent smooth SOCP can be easily derived, which means that a Pareto optimal solution of (KMS3) can be easily obtained by solving the SOCP. Moreover, in the numerical experiments, we solve the dual problem of the smooth SOCP problem because its computational time can be expected to be less than that of the original problem.

4.4 Comparison of Computational Complexities

Now, let us compare computational complexities of (εM2) and (KMS3). The constant vector ε used in (εM2), are determined by solutions of (AT), which is a large-scale quadratic optimization problem using all class datasets at once. On the other hand, the calculation of centroids for (KMS3) requires a considerably small amount of computational resources even if it is executed 20 times because the k-means clustering is individually applied to each class dataset.

Next, let us compare the diagonalization of kernel matrices used in (εM2) and the dual problem of (KMS3), and the numbers of the decision variables and constraints of them. The size of matrices diagonalized for (εM2) and (KMS3) are $l \times l$ and $c_{all} \times c_{all}$, respectively. The numbers of decision variables of (εM2) and the dual problem of (KMS3) are $m(m+2\tau+1)/2-1$ and $m(m+\tau_c)$, respectively, and the sizes of constraints of them are $(m-1)(l+m/2)-1$ and $(m-1)(l+3m/2)$), respectively. Since in general, we have that $m \le \tau_c \le c_{all} \ll \tau \le l$, more reduction can be expected if c_{all} is small.

5 Numerical Experiments

We applied the existing methods, AT and MMSVM and the variations of the proposed methods to seven benchmark problems [10], and compared the mean correct classification rate and mean CPU time by using the 10-fold cross-validation, in which hyperparameters were appropriately selected. To solve optimization problems, we used software package MOSEK. As variations of the proposed methods, we used the AMMSVM which does not use the dimension reduction based on the k-means clustering and which is solved by the reference point method, and KMSs in which r was varied in $\{0.05, 0.1, 0.15, 0.2\}$, which are represented by AMM, KMS5, KMS10, KMS15 and KMS20, respectively, and three kinds of reference point and weights, R0, R1 and R2, were used for AMMSVM and each KMS. We used the RBF kernel, namely, $k(x, y) = \exp\left(-\gamma\|x - y\|^2\right)$.

The results are shown in Tables 1 and 2. In Table 1, the numbers in parentheses denote the best hyperparameters of each method: (γ) in AT and MMSVM, and (γ, ρ, μ) in AMMSVM and KMSs. The italic and bold number denote the first and second best classification rate for each problem.

Table 1 shows that MMSVM obtained a high classification rate for many problems, while the classification rates of AMMSVM and KMSs with a large r are equal or slightly smaller than those of MMSVM. The classification rates of KMSs mostly increase as r increases. In particular, although the rates are considerably low if r is small for Ecoli, the highest rate was obtained by the KMS with $r = 0.15$. In addition, AMMSVM and KMSs achieved a higher rate than MMSVM

Table 1. Mean correct classification rate of four methods for benchmark problems

	Wine	Balance	DNA	Car	Dermatology	Zoo	Ecoli
AT	**98.89** (1)	90.72 (5)	96.00 (0.05)	99.31 (5)	97.26 (0.5)	**97.09** (0.5)	80.95 (50)
MMSVM	**98.89** (1)	**97.92** (1)	**95.95** (0.05)	99.48 (5)	98.09 (5.00e−04)	**97.09** (1.00e−04)	80.96 (50)
AMM R0	**98.89** (1,10,10)	100.00 (0.05,10,1e−5)	95.80 (0.05,1,10)	**99.42** (5,10,1e−5)	97.81 (1e−4,100,1e−5)	**97.09** (0.5,1,1e−5)	81.26 (50,1,1e−5)
R1	**98.89** (5,1,1e−5)	100.00 (0.05,100,1e−5)	95.80 (0.05,1,100)	99.36 (1,100,1e−5)	97.81 (1e−4,10,100)	**97.09** (0.5,1,1e−5)	81.26 (50,100,1e−5)
R2	**98.89** (5,1,1e−5)	100.00 (0.05,100,1e−5)	95.80 (0.05,1,100)	99.36 (1,100,1e−5)	97.81 (1e−4,10,100)	**97.09** (0.5,1,1e−5)	81.26 (50,100,1e−5)
KMS5 R0	98.33 (1,100,1e−5)	100.00 (1e−4,1e+3,1e−5)	92.90 (0.05,10,1e−5)	98.15 (0.5,10,100)	98.08 (0.01,100,1e−5)	98.09 (0.05,100,1e−5)	43.8 (500,1,1e−5)
R1	98.33 (10,10,100)	100.00 (1e−4,10,1e−5)	93.40 (0.05,1e+6,1e−5)	98.32 (1,1,100)	**98.36** (0.01,1e+3,10)	98.09 (1e−4,100,10)	43.80 (500,1,1e−5)
R2	**98.89** (10,10,1)	100.00 (1e−4,10,1e−5)	92.95 (0.05,1,10)	98.21 (1,1,10)	**98.36** (0.005,100,0.1)	98.09 (1e−4,100,10)	43.80 (500,1,1e−5)
KMS10 R0	98.33 (5,10,1e−5)	100.00 (0.05,10,1e−5)	94.35 (0.05,1,10)	99.31 (1,1,1e−5)	98.09 sparabreak (0.005,10,10)	98.09 (0.05,100,1e−5)	80.66 (1,1e+6,1e−4)
R1	98.33 (10,1,1e−5)	100.00 (0.05,10,100)	94.30 (0.05,10,1e−5)	99.02 (1,10,1)	98.09 (0.1,1e+5,1e−5)	98.09 (1e−3,1e+6,1e−5)	80.66 (0.1,1e+6,1e−5)
R2	**98.89** (10,1,1)	100.00 (0.05,100,1e−5)	93.75 (0.05,1,1)	99.19 (1,10,100)	**98.36** (0.1,1e+4,1e−5)	98.09 (1e−4,1e+5,1e−4)	78.58 (5,1e+5,1e−4)
KMS15 R0	98.33 (0.5,1e+6,1)	100.00 (0.05,10,1e−5)	94.70 (0.05,1,10)	99.31 (1,100,1)	97.81 (1e−4,100,10)	98.09 (0.1,100,1e−5)	82.75 (10,100,1e−5)
R1	99.44 (5,1,1e−5)	100.00 (0.05,100,1e−5)	94.55 (0.05,1,100)	99.25 (1,100,1e−5)	97.82 (0.1,100,10)	98.09 (0.05,100,1e−5)	81.60 (5,1e+3,1e−5)
R2	98.33 (5,1,1e−5)	100.00 (0.05,100,1e−5)	94.85 (0.05,1,1)	99.25 (1,100,1e−5)	98.09 (1,100,1e−5)	98.09 (0.1,100,1e−5)	79.50 (1,100,1e−5)
KMS20 R0	98.33 (1,1,10)	100.00 (0.05,10,1e−5)	95.00 (0.05,10,1)	99.25 (1,100,100)	97.82 (0.1,100,1e−5)	98.09 (0.1,100,1e−5)	**81.86** (5,1e+4,1e−5)
R1	98.33 (1,10,10)	100.00 (0.05,100,1e−5)	94.75 (0.05,10,1e−5)	99.31 (1,10,10)	98.63 (0.1,10,100)	98.09 (0.1,100,1e−5)	81.56 (10,10,1e−5)
R2	**98.89** (1,10,0.1)	100.00 (0.05,100,1e−5)	94.75 (0.05,1e+5,1e−5)	99.25 (1,10,1)	97.82 (0.1,10,1)	98.09 (0.1,100,1e−5)	80.42 (5,100,1e−5)

for some problems. The superiority of AMMSVM and KMSs is considered to be caused by the diversity of obtained solutions: The reference point method used in AMMSVM and KMSs finds a solution under various kinds of balances of objective functions, while in MMSVM, a single margin is maximized by the ε-constraint method using the optimal solution of (AT). From Table 2, we can see that AMMSVM reduced CPU time than AT or MMSVM without a dimension reduction, and KMSs did more greatly for large-scale problems even if $r = 0.15$ or 0.20. Comparing performance of KMSs using three kinds of reference point and weights, namely, KMSs with R0, R1 and R2, each method obtained a high rate for different problems, though no significant difference was observed.

Table 2. Mean CPU time (sec) of four methods for benchmark problems

	Wine	Balance	DNA	Car	Dermatology	Zoo	Ecoli
AT	0.313	4.567	104.677	258.805	12.566	1.103	21.233
MMSVM	0.802	15.727	854.189	1019.577	5.414	3.206	34.363
AMM R0	0.253	1.484	103.038	184.486	1.934	0.248	11.956
R1	0.441	1.478	103.688	162.613	1.930	0.295	11.625
R2	0.247	1.491	79.488	120.569	1.552	0.317	11.261
KMS5 R0	0.116	0.559	6.392	4.009	0.292	0.356	0.748
R1	0.169	0.614	6.144	3.942	0.645	0.370	0.409
R2	0.105	0.581	4.145	6.344	0.645	0.141	0.458
KMS10 R0	0.291	0.858	8.363	8.644	0.542	0.445	1.055
R1	0.294	0.878	12.692	8.431	0.950	0.123	0.830
R1	0.109	0.853	12.656	12.136	0.759	0.156	0.513
KMS15 R0	0.375	1.361	19.758	25.919	0.550	0.408	0.961
R1	0.373	1.242	15.536	26.998	1.280	0.130	0.923
R2	0.131	1.308	21.891	28.375	1.300	0.494	1.044
KMS20 R0	0.423	1.361	38.100	46.297	0.917	0.481	1.331
R1	0.455	1.828	38.447	48.706	1.538	0.195	1.314
R2	0.120	3.120	37.603	50.661	1.591	0.503	1.889

6   Conclusion

In this paper, we have proposed an approximate method of MMSVM which approximates its non-convex MOP by linearizing the constraints, and a reduction method which restricts the normal vectors of separating hyperplanes by using centroids from the k-means clustering. Through numerical experiments, we have observed that the proposed methods are effective to reduce the computational resources without decreasing the classification ability widely.

References

1. Alizadeh, F., Goldfarb, D.: Second-order cone programming. Math. Program. Ser. B **95**, 3–51 (2003)
2. Bottou, L., et al.: Comparison of classifier methods: a case study in handwriting digit recognition. In: Proceedings of the 12th IAPR International Conference on Pattern Recognition, pp. 77–87 (1994)
3. Boyang, L., Qiangwei, W., Jinglu, H.: A fast SVM training method for very large datasets. In: Proceedings of the 2009 International Joint Conference on Neural Networks, pp. 14–19 (2009)
4. Ehrgott, M.: Multicriteria Optimization. Springer, Heidelberg (2005). https://doi. org/10.1007/3-540-27659-9
5. Kusunoki, Y., Tatsumi, K.: A multi-class support vector machine based on geometric margin maximization. In: Huynh, V.-N., Inuiguchi, M., Tran, D.H., Denoeux, T. (eds.) IUKM 2018. LNCS (LNAI), vol. 10758, pp. 101–113. Springer, Cham (2018). https://doi.org/10.1007/978-3-319-75429-1_9
6. MacQueen, J.B.: Some methods for classification and analysis of multivariate observations. In: Fifth Berkeley Symposium on Mathematics, Statistics and Probability, pp. 281–297 (1967)
7. Tatsumi, K., Tanino, T., Hayashida, K.: Multiobjective multiclass support vector machines maximizing geometric margins. Pac. J. Optim. **6**(1), 115–140 (2010)
8. Tatsumi, K., Kawachi, R., Tanino, T.: Nonlinear extension of multiobjective multiclass support vector machine. In: Proceedings of the IEEE SMC, pp. 1338–1343 (2010)
9. Tatsumi, K., Tanino, T.: Support vector machines maximizing geometric margins for multi-class classification. Off. J. Span. Soc. Stat. Oper. Res. **22**(3), 815–840 (2014)
10. Lichman, M.: UCI machine learning repository (2013). http://archive.ics.uci.edu/ ml
11. Weston, J., Watkins, C.: Multi-class Support Vector Machines. In: Verleysen, M. (ed.) ESANN99, Belgium, Brussels (1999)
12. Vapnik, V.N.: Statistical Learning Theory. Wiley, NewYork (1998)

Generative Cooperative Net for Image Generation and Data Augmentation

Qiangeng Xu[1,2], Zengchang Qin[1,3(✉)], and Tao Wan[4(✉)]

[1] Intelligent Computing and Machine Learning Lab, School of ASEE,
Beihang University, Beijing, China
zcqin@buaa.edu.cn
[2] Department of Computer Science, University of Southern California,
Los Angeles, USA
[3] Keep Labs, Keep Inc., Beijing, China
[4] School of Biological Science and Medical Engineering, Beihang University,
Beijing, China
taowan@buaa.edu.cn

Abstract. How to build a good model for image generation given an abstract concept is one of fundamental problems in computer vision. In this paper, we explore a generative model for the task of generating fictitious images with desired features. We propose the Generative Cooperative Net (GCN) for image generation. The idea is similar to generative adversarial networks (GANs) except that the generators and discriminators are trained to work accordingly. We conducted experiments on hand-written digit generation and facial expression generation. In experimental studies, we found that GCN's two cooperative counterparts (the generator and the classifier) can work together nicely and achieve promising results. Such generative model can be used as a data-augmentation tool. Our experiment of applying this method to an emotion classification task shows that the synthesis images can help to improve classification accuracy. It is easy to set up and generate a very large synthesized dataset.

Keywords: Generative Cooperative Net (GCN) · GAN · Data augmentation

1 Introduction

Generative model for image generation has been an active research area and has been applied to various computer vision applications such as super-resolution [10], image painting [15], manifold learning [20] and semantic segmentation [14]. A wide variety of deep learning approaches employ generative parametric models while some models use the encoder-decoder structure that can reconstruct the image from latent representation [4]. Different from those previous works, in this paper we focus on learning the high-level concepts and generating fictious images with desired features (e.g. shown in Fig. 1). We developed a new

© Springer Nature Switzerland AG 2019
H. Seki et al. (Eds.): IUKM 2019, LNAI 11471, pp. 284–294, 2019.
https://doi.org/10.1007/978-3-030-14815-7_24

generative network model called *Generative Cooperative Net* (GCN). It contains two parts: a generator based on features and a multi-task classifier to distinguish these features. These two parts are trained accordingly and getting better together. A well-trained GCN could provide a highly efficient alternative for data-augmentation because it can generate a large amount of diverse samples with desired labels. In this paper, our major contributions are as follows: (1) We proposed a new generative model, Generative Cooperative Net (GCN) to learn the high-level concepts of images' latent description and generate high-quality fictitious images with desired features. (2) We propose a data augmentation technique of using GCN in the facial emotion classification task.

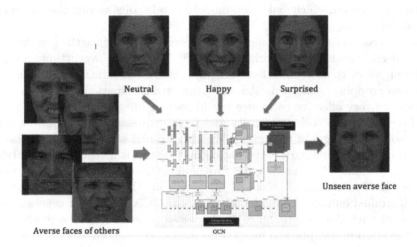

Fig. 1. An example of learning high-level concepts such as facial expression using GCN. The model takes faces of (a) Neutrality, (b) Happiness (c) Surprise of a person as well as facial expressions (including aversion) from other people. We can generate an aversion face of this person by learning the abstract concept of aversion.

2   Related Work

Although the generative model for image generation has been well-studied by using methods such as "Deep Boltzmann Machine" [16], the quality and capability of the generative model was boosted especially after researchers started to "reverse" the top-down deep learning network with up-scale structures such as "de-convolutional layer" introduced by Zeiler *et al.* [19]. One type of the generative networks (e.g. the model introduced in [2]) are parametric and applied supervised learning to project the high level description to the corresponding images. Although such structure can build perfect mapping for the label-image pairs, those networks always suffer from "memorizing" the training set and have limited ability on knowledge transfer.

Another family of deep generative networks are Generative Adversarial Networks (GANs) introduced by Goodfellow *et al.* [3]. Such framework provides an attractive alternative to supervised learning methods and enable the networks to learn the data distribution. Certain models can generate realistic images from the random sample of a Gaussian noise, meanwhile providing limited controllability over the desired features. To extend the controllability, Mirza *et al.* [13] introduced the "Conditional GANs" which can learn a conditional distribution of the input label and the output, that would guide the model to generate desired features. The network still need a large dataset to learn the manifold of the concept. Our research has adopted the supervised learning methods to learn the concept directly. Since we do not need to learn the distribution of a visual concept compared to the models using GANs, the dataset we need is much smaller. On the other hand, the classifier in our framework could facilitate the knowledge transformation, which makes our model a good choice among the supervised generative models.

Image classification has become a mature research area since the fast development of deep convolution networks [1,5,17,18]. To prevent overfitting and data starvation, all of those studies have adopted data augmentation methods such as random cropping, scale variation and Affine transformation. Although those methods are very effective to reduce the impact of images' details, recognition of high-level features such as identity and facial expression still require a sufficient amount of images of the target. We also studied some face augmentation methods including landmark perturbation and synthesis methods on hairstyles, glasses, poses and illuminations [12]. Although those methods could provide new combination of images, most of them rely on the specific 3D knowledge and well-studied human-face model. Instead, our GCN model can generate realistic images with desired attributes to do data-augmentation by a much easier implementation and applied to many other tasks.

3   Method and Model Architecture

Our goal is to generate fictitious images with desired features. To better achieve this goal, we designed the model with two components, a generator and a classifier. The generator takes a set of high-level feature vectors as inputs, and output the corresponding images.

$$D = \{(f_1^1, f_1^2, \ldots, f_1^M), (f_2^1, f_2^2, \ldots, f_2^M), \ldots, (f_N^1, f_N^2, \ldots, f_N^M)\}$$

where M is the amount of features each image has and N denotes the amount of images we have in the training set. In terms of the facial expression generation task, the features we used in training are: people's identity (one-hot encoding vector in 70 dimensions), the category of his or her expression and the transformation (rotation angle) being applied. Two fully connected layers were used for each feature to extent up to 256 dimensions, then concatenated those layers of each individual feature and applied another two other fc-layers on them. We then reshape the vector of the fifth fc-layer to a 8×8 matrix with 256 channels. From here, we applied fractional-stride convolution to replace up-pooling strategy used in most generative models. Each deconvolutional layer's stride has been set to 2. According to the output size relationship:

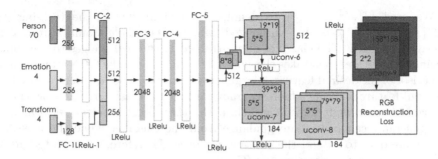

Fig. 2. The architecture of the generator in GCN

$$size_{out} = (size_{in} - 1) \times stride + kernel_{size} - 2 \times pad$$

We modify each layers to meet the target dimension of the final output layers (in case of face expression generation which is 158×158, we have 4 such layers). The details of the generator's structure is shown in Fig. 2. We also tested different types of activation layers and found out the LeakyReLU with negative slop of 0.1 could outperform RelU, PRelU and other commonly-used activation layers.

The classifier's structure is based on the AlexNet introduced by Alex et al. [1]. We could in fact use more complicated deep network such as VGG [17] or even the 152 layer ResNet introduced by He et al. [5]. However, after considering the memory usage, training speed and most importantly, the gradient degradation problem in deep learning (17 layers of the generator plus layers of the classifier), we decided to stick with AlexNet. We made some modification to the AlexNet to adapt to our task, two fc-layers and a softmax layer were assigned for each feature separately, which makes it a multi-task classifier. After exploring the activation layers, we found out LeakyReLu layers with negative slop of 0.2 would outperform the original ReLu layers. Besides, we also reduced the stride of first convolutional layer since our image resolution is smaller than the ImageNet. The structure of our classifier is shown in Fig. 3.

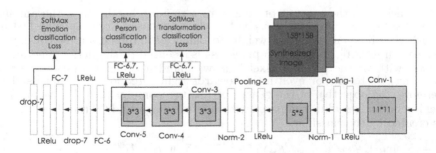

Fig. 3. Structure of the classifier: a multi-task AlexNet.

We used $\{I_r^1, I_r^2, \ldots, I_r^N\}$ to denote the real images in the training set and $\{I_s^1, I_s^2, \ldots, I_s^N\}$ as the synthesized images. The whole training process is described in Fig. 4. The classifier is pre-trained in the real images to recognize specific features and then takes the synthesis images while the GCN training as a supervisor of the desired feature. The generator tries to generate images based on the input feature and gets a high accuracy through the classifier. These two parts are trained accordingly which can help each other to get a better performance. The generator can have more guidance from the classifier while the classifier gains more data to extend the categorical space. Therefore, the architecture is called Generative Cooperative Net (GCN).

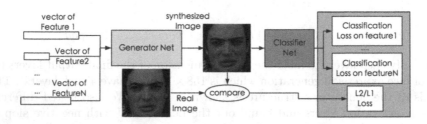

Fig. 4. The loss inside the green box will be back-propagated to the classifier and the generator accordingly. (Color figure online)

3.1   Objective Function

Since each image has multiple features, we need to test on each of them and adjust the weights to get an all-rounded result (e.g. a smiling face of person No. 34 should look like person No. 34 and can be recognized as smiling). Instead of using multiple classifier, we only used one multi-labeled classification network, which produced cross entropy loss on each feature. The objective of the classifier is:

$$\min_W -\frac{1}{N}\left[\sum_{i=1}^{N}\sum_{f=1}^{M}\sum_{j=1}^{K_f} Weight_f\left\{y_f^{(i)}=j\right\}\log\frac{e^{\theta_{f_j}^T x_f^{(i)}}}{\sum_{l=1}^{K_f}e^{\theta_{f_l}^T x_f^{(i)}}}\right] \tag{1}$$

In which N is the amount of images in a running batch, M is the amount of features for one sample and K_f is the class dimension for a specific feature f. Each classification loss has a weight to balance the dimension. According to the research in [2,6], the pixel to pixel loss is essential to the detail generation. Besides, without the pixel to pixel loss, we found it difficult for generator to create sensible results or even move towards right direction in initial batches. Therefore, the objective of the generator has two components: (1) the pixel to pixel Euclidean loss while compared with the real image, and (2) the classification loss created by the classifier.

$$\min_W \frac{1}{N}\{\sum_{i=1}^{N}||Pixel_r^i - Pixel_s(f_{i_1},\ldots,f_{i_M})^i||^2 - \left[\sum_{i=1}^{N}\sum_{f=1}^{M}\sum_{j=1}^{K_f} Weight_f\left\{y_f^{(i)}=j\right\}\log\frac{e^{\theta_{f_j}^T x_f^{(i)}}}{\sum_{l=1}^{K_f}e^{\theta_{f_l}^T x_f^{(i)}}}\right]\}$$
$$\tag{2}$$

3.2 Image Augmentation

We conducted a data-augmentation experiment by using a trained GCN. Since we observed that the GCN is also capable to synthesize two people's face, we generated all combination of two people's face under every emotion. We then used them to form a much larger dataset and performed emotion recognition training on it. In the end, we compared the accuracy achieved on the synthesized dataset with the accuracy trained on the original dataset. Augmented data can help the model to boost performance by injecting more data to balance class distribution, that is very common in many real-world problems.

4   Experimental Studies

4.1   Dataset

We here conducted two experiments to explore the task of generating fictitious images with desired features. In the first experiment, we used the "Karolinska Directed Emotional Faces (KDEF)" [11], which is a data set of 4900 pictures of human facial expressions of emotion. The dataset contains 70 individuals, each displaying 7 different emotional expressions and each expression being photographed (twice) from 5 different angles. To simplify the generative task, we only selected the front views and picked 4 emotions: neutrality, aversion, happiness and surprise. Moreover, We used OpenCV's default frontal face detection cascade script to further crop the image and re-sized them to the resolution of 158×158. We also rotated each image by 0, 90, 180 and 270° and conducted mirror operation to each degree, which result in 7 different transformations from every original image.

As for the second generation experiment, we used the MNIST dataset [9] to train and generate handwriting digits. We select 100 different examples for each digit, and coloured each grey levels image to red, green and blue versions. Same as the first experiment, we also augmented the dataset by rotating each image by 0, 90, 180 and 270°.

4.2   Model Training

Both of our generation experiments are implemented in Caffe [7]. We used Adam [8] with base learning rate of 0.0002. For the face expression generation, we found the momentum β_1 as 0.9, β_2 as 0.99 and $\epsilon = 10^{-8}$ would make the training most stable. Meanwhile, the digit generation task seems to favor momentum β_2 at 0.995. Both tasks have batch size of 64 and the learning rates are divided by 2 after every 1000 batches. We trained both tasks up to 40000 batches and found out the loss have become very stable. Since the handwriting digit is only 28×28, we reduced the deconvolutional layers in generator and the convolutional layers in classifier to fit the dimension.

To test the capability of GCN on generating fictitious images, we deliberately selected 10 people that would either miss several transformations or miss

1 facial expression. Then, we tested the trained generator by inputting the fictitious expression for those people and evaluated the generated image. As for the MNIST dataset, we only picked out several transformations or colors from a handwriting digit to form our test set, since there is no other high-level features like emotion for this task. We found setting the loss of emotion and identity to be 10 and euclidean loss to be 1 could produce the best result because of the better balance achieved between perception and details. We also explored other weights combination such as 100 for the classification and 1 for the Euclidean loss, which made both the generator and classifier hard to converge.

To explored GCN's capability of data-augmentation, we trained a GCN by selecting 65 people with one session for all 5 emotions. Then we deployed the well-trained GCN and input 0.5 on each person in every combination of two people. We then, got $65 \times 65 \times 5$ synthesized images and filtered out images with low quality. We used a single-task emotion classifier to train on the synthesized and the original dataset. The classifier has the same structure of the GCN's classifier. We used Adam with base learning rate of 0.0002, momentum β_1 as 0.9 and β_2 as 0.999. Both datasets have been trained for 40000 batches with batch size of 64 and the weight decay of 0.0005. The test set included the faces of these people in another session and the faces of other people under these 5 emotions.

4.3   Result Analysis

As for the face generation task, if a target feature combination is existed in the training set, the reconstruction result looks very promising. The synthesized images are almost identical to the real images (Fig. 5(a)) and would only miss out some minor details. For the handwriting figure generation task, the reconstructions are fairly good as well, except for some noisy pixels. (Fig. 5(b)).

Low Level Feature Generation. The low level features we want to generate include the rotational transformation for KDEF and both rotational and coloring transformation on MNIST. For facial expression generation, we took the 90° rotation of a person's averse face out of the training set. We then, fast-forward with the input vector of this person's identity, the averse facial expression and the 90° rotation. Even the network has never seen the image, our results show that the concept of rotation could be learned through training (Fig. 6(a)). In the digit generation experiment, we first only picked out the red color of a figure "2" and tried to generate the red "2" after training. We also took out all three colors of a digit "7" to test whether GCN can generate the image while 2 features are both missing. As we can see in Fig. 6(b), the results on the "missing color" is better than result on the "missing transformation". It is possible that mathematically, the rotation transformation relationship has more complexity than the color concept, thus more difficult for GCN to learn the relations.

High Level Feature Generation. The reason we chose an expression generation task for our research is because human facial expression of emotion could

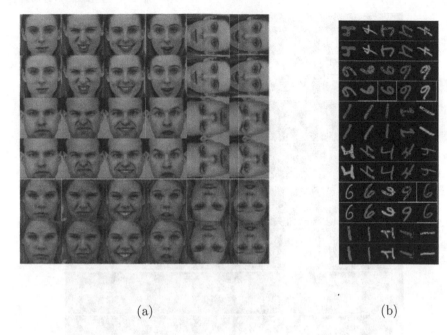

(a)   (b)

Fig. 5. Existing objects reconstruction: (a) The 1st, 3rd and 5th rows are the synthesized faces and the 2nd, 4th and 6th are the real faces in the training set. (b) The odd rows are the synthesized digits and the even rows are the corresponding real digits

provide two "easy to recognize" high-level features, the emotion and the face identity. Here we select 2 people to test each emotion. We deliberately took out the original images and transformations of this facial expression of the 2 people and tested the network's ability to generate this fictitious expression for them.

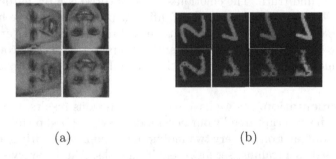

(a)   (b)

Fig. 6. Low level features generation: (a) Images on the first row are ground truth images with 90° and 180° rotation, which have been taken out of the training set. Images on the second row are the synthesized images. (b) The first column shows the generation result of a 90° rotated red digit, when the training set only miss the red color. The other three columns show the generation of a 180° rotated "7" when all colors has been taken out for this rotation. (Color figure online)

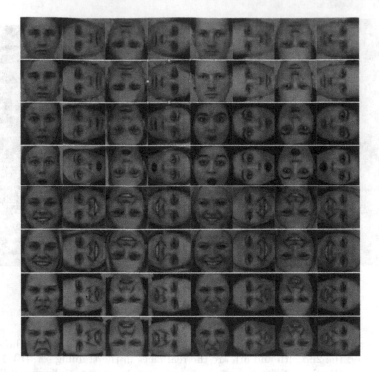

Fig. 7. High level features generation: the 1st, 3rd, 5th and 7th rows are the synthesized images and the 2nd, 4th, 6th and 8th rows are their corresponding real images. The first two rows show the neutral faces, while the second two rows show the surprised faces, the third two rows are the happy faces and the last two rows are the averse faces.

Inspecting the result shown in Fig. 7, we can see the emotions such as happiness and aversion could be nicely learnt and the synthesized images are very close to the ground truth. The emotion such as surprise is a bit harder to train, since in KDEF, the facial expressions of different people on surprise have larger variance (the degree of the mouth expansion and raising eye brows are very different among individuals). More importantly, the person's identity could be well preserved which significantly increased the credibility of the synthesized image.

Image Augmentation. As we have seen from previous results, the "fictitious emotion" can be well captured by our new model. "synthesized people" we generated using a combination of every two individual could have a fairly good quality, and can be used as a training set for recognition tasks. Almost every synthesized image can preserve the emotion concept of its parent images and look like both people (Fig. 8 shows an example of averse face generation). By combining people with each others at the ratio of 0.5 to 0.5, we can finally obtain 21125 faces from the original 325 faces. After training with the same hyper-parameter settings, the emotion classification accuracy can boost from 92% on the original dataset, to 94% on the synthesized dataset.

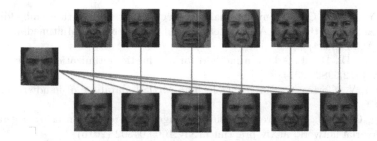

Fig. 8. An illustration of generating 6 synthesized faces by combining one averse face of one person with averse faces of 6 different people. The synthesized faces can also be used in other recognition tasks such as emotion recognition.

5   Conclusions and Future Work

In this paper, we proposed the generative cooperative Net (GCN) model which can generate fictitious images with desired high-level visual concepts. Unlike the GANs, the GCN model does not have adversarial modules: the generator and classifier work cooperatively to minimize the objective function. Besides, the GCN model can used for data augmentation. Since the synthesized images are not simply the linear combination of the original images, it could provide an unique transformation on the target concept and preserve the concepts we hope to keep. In our case study of using the KDEF data, it transforms the person identity but keep the emotion. The GCN model looks like an alternative structure to the GANs, but it could be actually incorporated in the GANs framework and performed solely on its generator's training phase. Our future work would focus on this integration and investigate a more difficult task such as using small dataset to learn and enable emotion transformation for any new faces. We will also test more datasets to investigate how to fuse high-level concepts in other problems but not only image generation.

References

1. Alex, K., Sutskever, I., Geoffrey, H.E.: ImageNet classification with deep convolutional neural networks. In: Proceedings of the NIPS, pp. 1097–1105 (2012)
2. Dosovitskiy, A., Springenberg, J.T., Brox, T.: Learning to generate chairs with convolutional neural networks. In: Proceedings of the CVPR, pp. 1538–1546 (2015)
3. Goodfellow, I., et al.: Generative adversarial nets. In: Proceedings of the NIPS, pp. 2672–2680 (2014)
4. Goodfellow, I.J.: NIPS 2016 tutorial: generative adversarial networks. arXiv preprint arXiv:1701.00160 (2016)
5. He, K., Zhang, X., Ren, S., Sun, J.: Deep residual learning for image recognition. In: The IEEE Conference on Computer Vision and Pattern Recognition (CVPR), June 2016
6. Isola, P., Zhu, J.Y., Zhou, T., Efros, A.A.: Image-to-image translation with conditional adversarial networks. ArXiv e-prints, November 2016

7. Jia, Y., et al.: Caffe: convolutional architecture for fast feature embedding. In: Proceedings of the 22nd ACM International Conference on Multimedia, pp. 675–678. ACM (2014)

8. Kingma, D., Ba, J.: Adam: a method for stochastic optimization. arXiv preprint arXiv:1412.6980 (2014)

9. LeCun, Y., Cortes, C., Burges, C.J.: The MNIST database of handwritten digits (1998)

10. Ledig, C., et al.: Photo-realistic single image super-resolution using a generative adversarial network. arXiv preprint arXiv:1609.04802 (2016)

11. Lundqvist, D., Flykt, A., Öhman, A.: The Karolinska Directed Emotional Faces (KDEF). CD ROM from Department of Clinical Neuroscience, Psychology section, Karolinska Institutet, pp. 91–630 (1998)

12. Lv, J.J., Shao, X.H., Huang, J.S., Zhou, X.D., Zhou, X.: Data augmentation for face recognition. Neurocomputing **230**, 184–196 (2017)

13. Mirza, M., Osindero, S.: Conditional generative adversarial nets. arXiv preprint arXiv:1411.1784 (2014)

14. Noh, H., Hong, S., Han, B.: Learning deconvolution network for semantic segmentation. In: The IEEE International Conference on Computer Vision (ICCV), December 2015

15. Pathak, D., Krähenbühl, P., Donahue, J., Darrell, T., Efros, A.: Context encoders: feature learning by inpainting (2016)

16. Salakhutdinov, R., Hinton, G.: Deep Boltzmann machines. In: Artificial Intelligence and Statistics, pp. 448–455 (2009)

17. Simonyan, K., Zisserman, A.: Very deep convolutional networks for large-scale image recognition. CoRR abs/1409.1556 (2014)

18. Szegedy, C., et al.: Going deeper with convolutions. In: The IEEE Conference on Computer Vision and Pattern Recognition (CVPR), June 2015

19. Zeiler, M.D., Krishnan, D., Taylor, G.W., Fergus, R.: Deconvolutional networks. In: Proceedings of the CVPR, pp. 2528–2535 (2010)

20. Zhu, J.-Y., Krähenbühl, P., Shechtman, E., Efros, A.A.: Generative visual manipulation on the natural image manifold. In: Leibe, B., Matas, J., Sebe, N., Welling, M. (eds.) ECCV 2016. LNCS, vol. 9909, pp. 597–613. Springer, Cham (2016). https://doi.org/10.1007/978-3-319-46454-1_36

Optimal Classifier Parameter Status Selection Based on Bayes Boundary-ness for Multi-ProtoType and Multi-Layer Perceptron Classifiers

Yuya Tomotoshi[1], David Ha[1], Emilie Delattre[2], Hideyuki Watanabe[3], Xugang Lu[4], Shigeru Katagiri[1(✉)], and Miho Ohsaki[1]

[1] Graduate School of Science and Engineering, Doshisha University, Kyoto, Japan
`skatagir@mail.doshisha.ac.jp`
[2] University of Mons, Mons, Belgium
[3] Advanced Telecommunications Research Institute International, Kyoto, Japan
[4] National Institute of Information and Communications Technology, Kyoto, Japan

Abstract. Recently, we proposed a new method to select an optimal classifier parameter status (value) using our new criterion that is referred to as *uncertainty measure* and directly evaluates the *Bayes boundary-ness* of estimated boundaries. The utility of the method was shown in a task of selecting an optimal Gaussian kernel width, which closely approximates the linear Bayes boundary in the feature space produced by Support Vector Machine classifier. In this paper, we apply the method to two types of classifiers whose class boundaries are basically nonlinear: Multi-ProtoType (MPT) classifier and Multi-Layer Perceptron (MLP) classifier. From experiments using a synthetic dataset and four real-life datasets, we show that our method can provide an optimal (in size and value) classifier parameter status, which basically corresponds to the nonlinear Bayes boundary in given feature spaces, for MPT and MLP classifiers.

Keywords: Bayes boundary · Bayes error · Classifier status selection

1 Introduction

The ultimate goal of the statistical approach to pattern classifier development is to find an optimal classifier parameter status (value) that leads to the *Bayes boundary*, which corresponds to the Bayes error, or the minimum classification error probability [1]. Essentially, the Bayes boundary is defined over an infinite number of pattern samples, each represented in some feature space; however, its

This work was supported in part by JSPS KAKENHI No. 18H03266 and MEXT's Program for Strategic Research Foundation at Private Universities (2014–2018), called "Driver-in-the-Loop".

© Springer Nature Switzerland AG 2019
H. Seki et al. (Eds.): IUKM 2019, LNAI 11471, pp. 295–307, 2019.
https://doi.org/10.1007/978-3-030-14815-7_25

estimation is rather difficult in real-life datasets where it is generally unknown and only a finite number of samples are available.

To resolve the above difficulty, such data-driven methods as Hold-Out (HO), Cross Validation (CV), and REGularization (REG) have been widely applied to classifier training, and also various statistical criteria/frameworks like AIC and SRM have been investigated [2]. However, methods like a simple combination of HO and REG are not necessarily reliable in the Bayes boundary estimation because of its dependency on sample-splitting for training, validation, and testing [3]. The CV method has been one of the standards for achieving the goal boundary, but it often requires long training time and is less applicable to large-scale datasets. Moreover, statistics-based criteria/frameworks cannot be directly applied because of their strict theoretical assumptions.

As an alternative to the CV method, we proposed a new method to select an optimal classifier parameter status that corresponds to the Bayes boundary from the boundaries estimated by a one-shot training of classifier parameters, which is done over the entire set of given pattern samples [4,5]. The method aims to directly evaluate the estimated boundaries, based on *Bayes boundary-ness*, not via the conventional course of estimating either sample distribution functions or classification error probability. Here, the Bayes boundary-ness is the specific nature of Bayes boundary that the on-boundary samples of boundary-forming classes have equal posterior probabilities, or in other words, are *uncertain*; it is represented by *uncertainty measure*, which indicates a high value for the Bayes boundary and a low value otherwise. In theory, the method can be applied to any reasonable shape of Bayes boundary and any type of classifier. Its utility was shown in a task of selecting an optimal Gaussian kernel width of Support Vector Machine (SVM) classifier, which drew a linear boundary in a kernel-mapped high-dimensional feature space [5]. Moreover, we showed, based on theoretical analyses, that the uncertainty measure could be statistically unbiased and of small variance, if it was computed using only samples sufficiently close to the Bayes boundary [6]. However, the general applicability of the method has not yet been fully investigated.

Taking into account the above situation, we evaluate in the paper our Bayes-boundary-ness-based method for finding an optimal classifier parameter status in nonlinear Bayes boundary cases where Multi-ProtoType (MPT) classifier and Multi-Layer Perceptron (MLP) classifier are adopted. We first introduce a classifier development task (Sect. 2), and we summarize the Bayes-boundary-ness-based method (Sect. 3). Next, we report experiment evaluation results for the MPT and MLP classifiers over one synthetic dataset and four real-life datasets (Sect. 4). In the evaluation, we adopt the classification error probability estimated by the CV method as a reference and evaluate our method in comparison to that reference. Finally, we conclude the paper (Sect. 5).

2   Classifier Development

Given a pattern sample $x \in \mathcal{X}$, where $\mathcal{X}$ is a D-dimensional feature-represented sample space (feature space), we consider a task of classifying x into one of J classes $(C_1, \ldots, C_J)$, based on the following classification decision rule:

$$C(x) = C_k \text{ iff } k = \arg\max_j g_j(x; \Lambda), \tag{1}$$

where $C(\ )$ is a classification operator, Λ is a set of classifier parameters to be optimized, $g_j(x; \Lambda)$ is a discriminant function for C_j, which represents the degree of confidence to which the classifier assigns x to class C_j. A higher value for $g_j(x; \Lambda)$ represents higher confidence.

In the statistical approach to the above classification, an ultimate goal of classifier development is to find the status Λ^* that achieves the optimal Bayes boundary (B^*) that corresponds to the minimum classification error probability condition, or in other words, Bayes error. As discussed in Sect. 1, various methods have been investigated to reach this goal (e.g. [1,2]). Among them, the most reliable, high-road-like methods are based on such resampling as the CV method and the bootstrap method. For example, in the CV method, a given dataset is first split into M data subsets (folds), where M is an integer; then, classifier parameters are trained on $(M - 1)$ folds and tested on the remaining testing fold; the training and testing are repeated M times while switching the testing fold (as a result, the $(M - 1)$ training folds). Here, the training is conducted with various methods based on the maximum likelihood distribution function estimation and the minimization of such losses as the classification error count loss and the cross-entropy loss. The CV method basically provides an accurate estimate of the Bayes error; however, its run is rather time-consuming and its application to large-scale datasets is less practical. Moreover, the CV method does not directly provide the optimal parameter set that corresponds to the Bayes boundary because its estimate of the Bayes error is an average of error rates, each individually estimated by one of M trained classifiers. To obtain a single set (status) of optimal classifier parameters, we therefore need some further procedure. A motivation for developing the Bayes-boundary-ness-based method was to avoid these disadvantages of CV, i.e. the time-consuming repetition of training/testing and the need for a further procedure.

At a given sample location x, one dominant discriminant function score $g_j(\cdot)$ enables the classifier to assign x to class C_j. When the two highest scores are equal, the classifier cannot decide between the two corresponding classes, and this is when x lies on the estimated boundary between the two classes. The concept of boundary directly extends to three or more classes; however, equality between three or more scores at a sample location is less likely. Therefore, the rest of this paper assumes boundaries between two classes. We denote $B_{iy}(\Lambda)$ as the estimated boundary that separates C_y and C_i. $B_{iy}(\Lambda)$ and $B_{yi}(\Lambda)$ are interchangeable since they represent the same boundary.

3   Optimal Classifier Parameter Status Selection Based on Bayes Boundary-ness

3.1   Procedure Outline

In our proposed method, we evaluate the Bayes boundary-ness using the uncertainty measure that should indicate its highest value on the Bayes boundary and lower values otherwise.

Given a set of trained classifier parameter statuses $\{\Lambda\}$, we compute the value of the uncertainty measure denoted as $U(\Lambda)$ for each status and select the status Λ^* that most closely approximates the Bayes boundary B^*. Based on the definition of $U(\Lambda)$, we can expect that $U(\Lambda)$ closely represents the Bayes boundary-ness, or a certain similarity between an estimated boundary $B(\Lambda)$ and the Bayes boundary B^*. The higher the value of $U(\Lambda)$, the closer it is to B^*.

The Bayes boundary B^* fits *on-boundary* samples in the feature space whose posterior probabilities are equal. In this case, its corresponding uncertainty measure $U(\Lambda^*)$ should obviously be computed using those on-boundary samples. Similarly, $U(\Lambda)$ should fundamentally be computed using on-boundary samples on its corresponding boundary $B(\Lambda)$. However, in practical situations where only a finite number of samples are available, it is not necessarily possible to keep on-boundary samples for any estimated boundary. Therefore, the practical computation of $U(\Lambda)$ requires some approximation of the on-boundary samples.

Based on the above requirement, our status selection method consists of the following two steps: Step 1 *near-boundary* sample selection and Step 2, the approximated computation of uncertainty measure using selected near-boundary samples. We will describe them in detail in the following subsections.

3.2   Step 1: Near-Boundary Sample Selection

Overview. We assume that a training sample set $\mathcal{T}$, which consists of a finite number of samples, is available. To obtain near-boundary samples, of which set is denoted by $\mathcal{N}_B(\Lambda)$, we next perform the following three procedures: (1) dividing $\mathcal{T}$ into subsets $\{\mathcal{T}_{sub}\}$, each containing the training samples that form a class boundary between the two classes whose discriminant functions are higher than others; (2) generating virtual samples, called *anchors*, each lying on $B(\Lambda)$; (3) generating near-boundary sample set $\mathcal{N}_B(\Lambda)$ by selecting the nearest neighbor of each anchor $(\boldsymbol{x}_a)$ from $\mathcal{T}_{sub}$, denoted by $\mathrm{NN}(\boldsymbol{x}_a, \mathcal{T}_{sub})$, as a near-boundary sample.

All three of these procedures are simply implemented by repeating double loops, one for the sample pair selection and the other for the anchor generation and near-boundary sample selection (Algorithm 1).

Training Sample Splitting. As mentioned in Sect. 2, we assume that a class boundary is formed by the two highest-value discriminant functions, or their corresponding class parameters, even in multi-class classification. Based on this

Algorithm 1. Selection of near-boundary sample set $\mathcal{N}_B(\Lambda)$

Input: Classifier parameters Λ that were trained on $\mathcal{T}$
Output: $\mathcal{N}_B(\Lambda)$

1   **if** $\forall \boldsymbol{x}, \boldsymbol{x} \in \hat{C}_0$ *or* $\forall \boldsymbol{x}, \boldsymbol{x} \in \hat{C}_1$ **then** $\mathcal{N}_B(\Lambda) \leftarrow \emptyset$;
2   **else**
3   **while** *stop criterion not met* **do**
4   Randomly pick $\boldsymbol{x}, \boldsymbol{x}' \in \mathcal{T} : C(\boldsymbol{x}) \neq C(\boldsymbol{x}')$
5   Find $\alpha_0 \in [0, 1] : g_0(\alpha_0 \boldsymbol{x} + (1 - \alpha_0)\boldsymbol{x}'; \Lambda) = g_1(\alpha_0 \boldsymbol{x} + (1 - \alpha_0)\boldsymbol{x}'; \Lambda)$
6   Find $t_a = \mathrm{NN}(\alpha_0 \boldsymbol{x} + (1 - \alpha_0)\boldsymbol{x}', \mathcal{T}_{sub})$
7   If $t_a \notin \mathcal{N}_B(\Lambda)$ then add to $\mathcal{N}_B(\Lambda)$
8   **end**
9   **end**
10 **return** $\mathcal{N}_B(\Lambda)$

---

assumption, we first divide training samples $\mathcal{T}$ into subsets $\{\mathcal{T}_{sub}\}$, each only containing the two-class samples that form a class boundary (Fig. 1). Moreover, for efficient computation in the successive procedures, we represent the sample subsets in a matrix form; if training sample $\boldsymbol{x}$ is assigned to C_i and C_j, whose discriminant functions $g_i(\boldsymbol{x}; \Lambda)$ and $g_j(\boldsymbol{x}; \Lambda)$ indicate the first and second highest values, respectively, it is registered in element A_{ij} of matrix A.

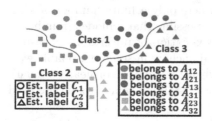

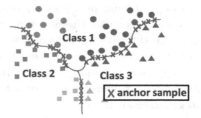

Fig. 1. Graphic illustration of training sample subgrouping into subsets, each only containing two-class samples related to a class boundary.

Fig. 2. Graphic illustration of anchor generation.

Anchor Generation. Naturally, a segment connecting two different-class samples crosses a class boundary at least once. Therefore, we can basically generate an anchor, which is located on a boundary, on the connecting segment. Based on this understanding, we next try to generate anchors on $B_{ij}(\Lambda)$ using samples in A_{ij} and A_{ji}. Picking one sample from A_{ij} and another from A_{ji}, we make a segment by connecting the two; then, we search the anchor that should lead to the discriminant function equality between C_i and C_j. If the size of A_{ji} is equal to 0, we produce anchors by using elements in $A_{ij} \times A_{ki}(k \neq j)$. Our actual search is done by finding α ($\in [0, 1]$) that satisfies $g_i(\alpha \boldsymbol{x} + (1 - \alpha)\boldsymbol{x}'; \Lambda) = g_j(\alpha \boldsymbol{x} + (1 - \alpha)\boldsymbol{x}'; \Lambda)$, where $\{\boldsymbol{x}, \boldsymbol{x}'\}$ is two picked samples (Fig. 2).

Near-Boundary Sample Selection. For each anchor, we select its nearest neighbor sample from $\mathcal{T}_{sub}$ as a near-boundary sample (Fig. 3). That is, we prepare a near-boundary sample as an approximation of on-boundary sample, which should be ideally used for computing uncertainty measure $U(\Lambda)$.

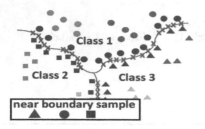

Fig. 3. Graphic illustration of near-boundary sample selection via anchors.

Fig. 4. Graphical illustration of kNN-based entropy computation for every partition along an estimated boundary.

Because the uncertainty measure should be computed in theory using an infinite number of samples, we might need as many anchors (or their corresponding near-boundary samples) as possible. However, the number of possible anchors has its upper bound in a real-life situation where only a finite number of samples are available; also, the situation would be possible that a modest number of anchors are sufficient to obtain a good approximation of uncertainty measure. Therefore, we control the number of generated anchors in some reasonable way. One simple way is to stop the generation when the number of near-boundary samples does not increase or a preset number of training sample pairs are used up to generate anchors.

3.3 Step 2: Uncertainty Measure Computation

Overview. The uncertainty measure would be to quantitatively represent how close to the Bayes boundary an estimated class boundary is. However, our goal is to find the Bayes boundary or its close approximation among estimated boundaries, or in other words, to evaluate if an estimated boundary is close to the Bayes boundary or not; therefore, it will be sufficient that the uncertainty measure indicates its maximum value around the Bayes boundary and indicates lower values otherwise.

Among many possibilities for implementing such uncertainty measure, we first adopt the following Shannon entropy $H_{ij}(\boldsymbol{x})$ to compute $U(\Lambda)$:

$$H_{ij}(\boldsymbol{x}) = -(P(C_{(1)}|\boldsymbol{x})\log P(C_{(1)}|\boldsymbol{x}) + P(C_{(2)}|\boldsymbol{x})\log P(C_{(2)}|\boldsymbol{x})), \quad (2)$$

where $C_{(1)}$ and $C_{(2)}$ correspond to the classes with the two highest class posterior probabilities, and index i and j indicate the classes with the two highest discriminant function scores. Importantly, in multi-class classification tasks, the classes with the two highest class posterior probabilities can be different from the classes with the two highest discriminant function scores. To deal with this situation, we introduce the following normalization rule:

1. If $\{(1), (2)\} \neq \{i, j\}$, then set $H_{ij}(\boldsymbol{x})$ to 0.
2. Else normalize $P(C_{(1)}|\boldsymbol{x})$ and $P(C_{(2)}|\boldsymbol{x})$ so that their sum amounts to 1. Note that $P(C_{(1)}|\boldsymbol{x}) + P(C_{(2)}|\boldsymbol{x}) = 1$ does not necessarily hold.

Next, we define a boundary uncertainty score for sub-boundary $B_{ij}(\Lambda)$, which is a part of $B(\Lambda)$, as follows:

$$U_{ij}(\Lambda) = \frac{1}{N_{ij}^B} \sum_{\boldsymbol{x} \in N_{ij}^B(\Lambda)} H_{ij}(\boldsymbol{x}), \tag{3}$$

where N_{ij}^B is the number of samples in $N_{ij}^B(\Lambda)$, which is a subset of $\mathcal{N}_B(\Lambda)$ that contains only the near-boundary samples of C_i and C_j.

Finally, a multi-class uncertainty measure score $U(\Lambda)$ is simply the mean over several boundary uncertainty scores $U_{ij}(\Lambda)$:

$$U(\Lambda) = \frac{1}{N^B} \sum_{i,j \in J} N_{ij}^B U_{ij}(\Lambda), \tag{4}$$

where N^B is the number of near-boundary samples in $\mathcal{N}_B(\Lambda)$.

k-Nearest-Neighbor-Based Entropy Estimation. To compute entropy $H(\boldsymbol{x})$, we adopt the posterior probability estimation using the k Nearest Neighbor (kNN) method, focusing on its advantage, i.e. a general and model-free nature.

For given sample $\boldsymbol{x}$, we select its k nearest neighbors from $\mathcal{N}_B(\Lambda)$ and use it to estimate class posterior probabilities. For example, if k_i neighbors belong to C_i ($\sum_{i=1}^{J} k_i = k$), we estimate posterior probabilities as $\hat{P}(C_i|\boldsymbol{x}) = k_i/k$. One reason we use here only the near-boundary samples as the neighbors of the given sample is that we aim to estimate the entropy (or posterior probability) along $B(\Lambda)$. The validity of this approach becomes clearer when $B(\Lambda)$ is the Bayes boundary B^*: the posterior probabilities of all of the on-boundary samples along B^* share the common nature that their values are 0.5. We elaborated the importance of using only the near-boundary samples in the posterior probability estimation in the symptotic convergence analysis [6].

When applying the kNN method, the determination of k itself is an issue to investigate. However, because we do not need accurate posterior probability estimates in our procedure but only the estimates that are sufficient for distinguishing the Bayes boundary from others, some simple determination would be acceptable. Moreover, the density of training samples generally depends on locations in the feature space. Therefore, k should be determined based on a local sample density; we can use a large value for k to increase the estimation reliability when the density is high, and we should use a small value to avoid the degradation in estimation.

To implement the above consideration in our proposed method, we introduce the following adaptive clustering procedure that recursively splits $\mathcal{N}_B(\Lambda)$ into two small partitions until the number of the partition members gets between N_m and N_M ($N_m < N_M$), where N_m and N_M are preset thresholds for avoiding unreliable estimations due to the "few samples" problem and the "distant-from-boundary samples" problem, respectively (Fig. 4). The adopted dichotomy clustering is affected by initialization. To solve this instability, we ran the clustering R times while changing the initial settings, and averaged the entropy score over the R entropy estimates.

4   Experimental Evaluation

4.1   Classifiers

In this paper, we adopted two types of classifiers, i.e. MPT classifier and MLP classifier, which are different from an SVM-based classifier that was already used for the evaluation of our Bayes-boundary-ness-based method [5].

The parameter statuses to be trained (optimized) of the MPT classifier are the prototype locations in the feature space, or in other words, the values of prototype (vector) elements, and the number of prototypes per class. We trained the prototype locations using k-Means Clustering (kMC) and jointly selected the optimal per-class number and locations (values) of the prototypes by applying our proposed method.

The parameter statuses to be trained of the MLP classifier are basically the network weights, the number of layers, and the number of nodes for each layer. For simplicity, we set the number of layers as four (one input layer, two hidden layers, and one output layer) and assumed the common number of nodes to the two hidden layers. We then set the activation function to sigmoid, trained the network weights using the standard Limited-memory BFGS (L-BFGS) loss minimization with the cross-entropy loss, and selected with one sweep the optimal number of hidden-layer nodes and the optimal values of the trained network weights using our proposed method. The implementation details of the classifiers are available online.[1]

4.2   Datasets

We conducted evaluations using a 5-class synthetic dataset (referred to as "5-class GMM"), which we made by producing samples by a single two-dimensional Gaussian function for each class, and four real-life datasets from the UCI Machine Learning Repository,[2] "Abalone," "Letter Recognition," "MNIST (test)," and "Landsat Satellite." In particular, the Abalone dataset used was

---

[1] https://scikit-learn.org.
[2] https://archive.ics.uci.edu/ml/index.php.

a custom version that grouped the original class categories into "young", "medium", and "old"; the MNIST (test) dataset was the "test" subset of the original MNIST set. We summarize the characteristics of these datasets in the following:

- 5-class GMM: $N = 1000, D = 2, J = 5$ (Synthetic data)
- Abalone: $N = 4,177, D = 7, J = 3$ (Custom version)
- Letter Recognition: $N = 20,000, D = 16, J = 26$
- MNIST (test): $N = 10,000, D = 784, J = 10$
- Landsat Satellite: $N = 6,435, D = 36, J = 7$

Here, N, D, and J denote the number of samples available, the dimension of samples (vectors), and the number of classes, respectively.

The Bayes boundary for real-life data is usually unknown. However, to evaluate our method, we need it as a reference, or standard for comparison. We estimated the Bayes boundary for each dataset via a Bayes error estimation based on the Cross Validation (CV) data-handling scheme; taking into account the number of samples, we adopted 40-fold CV for the larger datasets of MNIST and Letter Recognition; 80-fold CV for the Abalone and Landsat Satellite datasets. The reference Bayes boundary then corresponds to the minimum of the CV-based classification error probability. In contrast, we estimated the reference Bayes boundary for a synthetic dataset by generating a large number of samples (50,000) based on an adopted distribution function. The importance here is that our method selects the Bayes boundary from many boundaries, each being trained using the entire set of given samples in one shot, even for real-life dataset.

4.3 Hyperparameters

Because our method does not aim at an accurate estimate of the Bayes error, we can expect that model status selection results are not sensitive to the hyperparameter values in the method. The values of our hyperparameters like R and N_M should basically be small based on the theoretical and computational requirements of our method; therefore, in the experiments of this section, we simply set the hyperparameter values as follows: $R = 10$, $N_m = 4$, $N_M = 8$, and 20 for the maximum number of dichotomies in the anchor generation.

4.4 Results on Synthetic Datasets

We first evaluate the quality of Step 1, by applying the MPT classifier to the 5-class GMM data (Fig. 5). In Fig. 5, the right graph plots scattering samples, the left shows class boundaries drawn by the MPT classifier with one prototype per class, and the center shows class boundaries drawn by the MPT classifier with 50 prototypes per class. Moreover, the left and center graphs show selected near-boundary samples (purple dots). From the graphs, we find that the method successfully selected the near-boundary samples and also that the

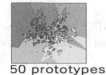

1 prototype   50 prototypes   GMM 5-class

Fig. 5. Evaluation of Step 1 using MPT classifier with 1 prototype (per class) or 50 prototypes over 5-class GMM dataset.

right-size classifier (one prototype per class) accurately draws the Bayes boundary and the overlarge-size classifier (50 prototypes per class) seriously overfits training samples.

Next, we show the classifier status selection results gained with our proposed method in the form of comparison between the uncertainty measure scores and the classification error probability estimates (Figs. 6 and 7); here, we obtained the classification error probability estimates using the 50,000 testing samples, which are independent from the 1,000 training samples. In Fig. 6, we draw the uncertainty measure score curve (blue) and the estimated classification error probability curve (red) for the MPT classifier; the MLP classifier in Fig. 7. In these figures, the left vertical axis represents the classification error probability estimates (LT_e), and the right vertical axis represents the uncertainty measure score ($-U$); for clarification, we use the sign-reversed uncertainty measure score $-U$. Moreover, the horizontal axis in Fig. 6 represents the number of prototypes per class; the number of hidden-layer nodes in Fig. 7.

In these figures, the minimum (valley point) of the red curve should ideally fit that of the blue curve: the minimum of the red curve corresponds to the Bayes error and the minima of the two curves predicate the Bayes boundary. The figures clearly show such ideal matching; from Fig. 6, we can select the optimal set of prototypes trained over the training samples using the optimal number of prototypes, and the optimal set of network weights trained using the optimal number of hidden nodes from Fig. 7.

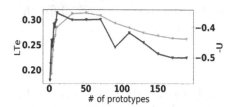

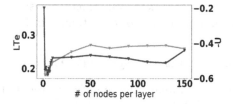

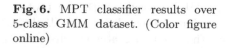

Fig. 6. MPT classifier results over 5-class GMM dataset. (Color figure online)

Fig. 7. MLP classifier results over 5-class GMM dataset. (Color figure online)

4.5   Results on Real-Life Datasets

For the real-life datasets, we estimated the classification error probabilities for both of the training samples and the testing samples using the CV scheme; the minimum of the classification error probabilities for testing samples corresponds to the Bayes error. We compare our uncertainty measure curve (blue) and the classification error probability estimate (over testing samples) curve (green), with an additional comparison in the error probability estimate curves between the training sample (orange) and the testing samples (Figs. 8, 9, 10, 11, 12, 13, 14 and 15).

In the top graphs of each figure, we can clearly see the over-learning phenomenon that appears in the monotonically decreasing trend of the orange curve. From the green curves, we find that their minima don't necessarily appear in the middle region of the horizontal axis. A possible reason for this phenomenon is that the increase in the number of prototypes or in the number of hidden-layer nodes does not always lead to an increase in the complexity of the trained class boundaries. This phenomenon in the curve shape is different from that obtained using SVM classifiers [5]. However, regardless of the green-curve shape, the minimum of the green curve predicates the Bayes boundary in these figures.

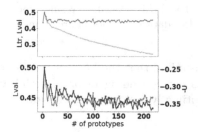

Fig. 8. MPT classifier results for Abalone dataset. (Color figure online)

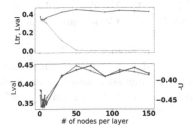

Fig. 9. MLP classifier results for Abalone dataset. (Color figure online)

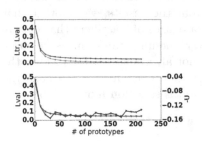

Fig. 10. MPT classifier results for Letter Recognition dataset. (Color figure online)

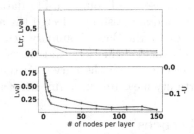

Fig. 11. MLP classifier results for Letter Recognition dataset. (Color figure online)

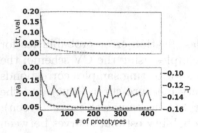
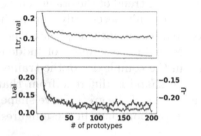

Fig. 12. MPT classifier results for MNIST dataset. (Color figure online)

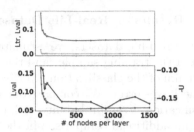
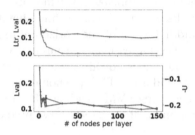

Fig. 13. MLP classifier results for MNIST dataset. (Color figure online)

Fig. 14. MPT classifier results for Landsat dataset. (Color figure online)

Fig. 15. MLP classifier results for Landsat dataset. (Color figure online)

From the bottom graphs in these figures, we also find that the blue curve of our uncertainty measure scores closely follows its corresponding error probability curve (green) and its minimum provides an optimally trained classifier status.

5   Conclusion

We reported experimental results of evaluating our Bayes-boundary-ness-based classifier status selection method with MPT and MLP classifiers over one synthesis dataset and four real-life datasets. From the results, we showed that our method successfully selected the appropriate sizes of classifiers that closely approximate the Bayes boundary-ness. However, the uncertainty measure curve contains small oscillation, and therefore it is not always suited for picking the best selection from the curve. Further improvements are needed to make the uncertainty measure curve smoother and the status selection more reliable.

References

1. Bishop, C.: Pattern recognition and machine learning (2006)
2. Duda, R.O., Hart, P.E.: Pattern classification and scene analysis (1973)
3. Shiraishi, H., Watanabe, H., Katagiri, S., Xugang, L., Hori, C., Ohsaki, M.: Relation between data grouping and robustness to unseen data in large geometric margin minimum classification error training. IEICE Technical report, PRMU2014-101, pp. 177–182, January 2015

4. Ha, D., Maes, J., Tomotoshi, Y., Watanabe, H., Katagiri, S., Ohsaki, M.: A classification-uncertainty-based criterion for classification boundary. IEICE Technical report, vol. 117, no. 442, PRMU2017-166, pp. 121–126, February 2018
5. Ha, D., et al.: Optimal classifier model status selection using Bayes boundary uncertainty. In: IEEE Workshop on Machine Learning for Signal Processing (2018)
6. Ha, D., Watanabe, H., Tomotoshi, Y., Delattre, E., Katagiri, S.: Optimality analysis of boundary-uncertainty-based classifier selection method. In: ACM International Conference on Signal Processing and Machine Learning (2018)

Averaged Logits: An Weakly-Supervised Approach to Use Ratings to Train Sentence-Level Sentiment Classifiers

Xiaoyu Tang, Wei Ou[(✉)], and Van-Nam Huynh

Japan Advanced Institute of Science and Technology,
Asahidai 1-1, Nomi, Ishikawa, Japan
ouwei@jaist.ac.jp

Abstract. As an important aspect of review mining, sentence-level sentiment classification has received much attention from both academia and industry. Many recently developed methods, especially the ones based on deep learning models, have centred around the task. In a majority of the existing methods, training sentence-level sentiment classifiers require sentence-level sentiment labels, that are usually expensive to obtain. In this research, we propose a novel approach, named 'Averaged logits', that uses the prevalently available ratings, instead of sentence-level sentiment labels to train the classifiers. In the approach, the rating of a review is assumed to be the 'average' of the sentiments of the individual sentences. We experiment with this idea under the framework of the recurrent neural network model. The results show that, the performance of the proposed approach is close to that of the traditional SVM and Naive Bayes classifiers trained by labelled sentences when their training sizes are approximately equal, and close to that of the neural network based classifiers trained by labelled sentences when the proposed approach uses approximately 5 times more training samples.

Keywords: Sentence-level sentiment analysis · Review data ·
Neural networks

1 Introduction

Product reviews posted by previous shoppers have been serving as an important source of information that helps on-line consumers or vendors to make their decisions. However, reading reviews to understand why the authors like or dislike a target product or service is usually very time-demanding, even frustrating when the volume of the reviews is large. In recent years, developing methods that can automatically extract customer opinions to ease the review-understanding process, has been a very active topic in both academia and industry.

An important aspect of the opinion mining process is sentence-level sentiment classification. Currently, existing methods for the classification task are usually based on supervised machine learning models [3,5,11,12,16,18,21], that require a

© Springer Nature Switzerland AG 2019
H. Seki et al. (Eds.): IUKM 2019, LNAI 11471, pp. 308–319, 2019.
https://doi.org/10.1007/978-3-030-14815-7_26

large number of labelled sentences to train the classifiers. In practice, the labelled data is not easy to obtain as the labelling process takes intense human labor and even linguistic expertise. This is especially true when the review sentences are full of informal and ambiguous language. The difficulties of the sentence labelling process can be reflected by the fact that only few datasets containing sentence-level sentiment labels are publicly available on the internet.

One possible way to reduce or avoid the labelling efforts is to take use of the prevalently available ratings, that can be regarded as a natural form of document level sentiment labels, to train the classifiers. Ratings provide a corse hint of the sentiment orientations of sentences in each review. For instances, a 5-star on Amazon indicates the sentences are prone to be positive whereas 1-star the sentences negative. Therefore, they can serve as a form of weak supervision signal for the classification task.

However, there exists a problem in training sentence-level sentiment classifiers with ratings: the sentiments of some individual sentences may be very different from the sentiment of the containing review. In other words, the true sentiment orientations of individual sentences can be easily misrepresented by the document-level ratings, that would inevitably result in harmful impacts to the classifiers. Because of the difficulty, the ratings have been ignored in a majority of the existing methods and only exploited in few rare works [22,27,31,33]. Even in those rare works, the ratings are not used alone as the supervision signal, instead, they either serve as the complementary signal to sentence-level labels, or work with opinion lexicons and linguistic rules, that also take human labor and linguistic knowledge to build.

This paper makes use of the ratings in the sentiment classification task. However, we aim to use ratings as the only supervision signal and avoid the laborious efforts for sentence-level labelling. We propose a novel approach, called 'Averaged logits' based on Recurrent Neural Network (RNN), that takes individual sentences as the inputs, and uses the averages of the resulting logit vectors of sentences from the same reviews to compute the sentiment distributions of the containing reviews. The cross-entropy between the resulting sentiment distributions and the ratings are used as the cost function. In the experiment, we trained the proposed approach on the public available MRS (Movie Review Scale) dataset [20] containing around 4000 reviews, and test on 2 review sentence datasets: SST-1 (Sentiment Tree Bank) and SST-2 [23]. We compare the performance of the proposed approach against a number of existing models using sentence-level sentiment labels as the supervision signal. The results show that, the performance of the proposed approach is close to that of the traditional SVM and Naive Bayes classifiers trained by labelled sentences when their training sizes are approximately equal, and close to that of the neural network based classifiers trained by labelled sentences when the proposed approach uses approximately 5 times more training samples.

2 Related Work

This section first presents a brief introduction to existing methods that use labelled sentences to train sentence-level sentiment classifiers, then present a few rare methods that exploit document-level sentiment labels to train sentence-level sentiment classifiers.

2.1 Sentence-Level Sentiment Classification Based on Labelled Sentences

A critical phase in generic text classification problem is extracting effective input features to represent the raw text. This is especially true for sentence-level sentiment classification, as sentences are usually very short and noisy. Based on how the features are extracted, the existing methods are divided into two categories: (1) lexicon-based methods whose feature extraction process involves the use of external opinion lexicons; (2) machine-learning based methods that exploit the statistical patterns or contextual structures of sentences to define the features.

Lexicon Based Methods. Lexicon based models require lexicons consisting of opinion words or phrases. Those methods assume that opinions words and phrases are the dominating indicator for sentiment classification, therefore, the input features of those methods are usually derived based on the presence or absence of the opinion words in a text fragment. A big array of models, ranging from rule based algorithms [4,15,26], to unsupervised [13,14,32] and supervised machine learning models [2,8,19,24,30], have been proposed to build sentiment classifiers with the help of sentiment lexicons.

A problem with the lexicon based methods is that the opinion lexicons are manually built, therefore, the scope and size of a lexicon is usually very limited. There exists a high possibility that the words of an opinionated sentence has no overlap with the opinion lexicon. Researchers have proposed to automatically expand the lexicons by iteratively searching WordNet or other dictionaries for synonyms and antonyms of the opinion words as the addition to the lexicons [1,7,9,28]. The shortcoming of this approach is that the added words from the external dictionaries can be unrelated to the training corpus, therefore, cannot reflect the domain knowledge and corpus-specific statistics. To address the problem, Liu et al. [4,6] proposed to search the training corpus, instead of external dictionaries, for words that are semantically or syntactically related to the opinion words. However, it is usually not easy to determine the sentiment orientation of the newly added words by the semantic or syntactical connection, therefore, external linguistic knowledge and intense labor are still needed to help make the decision.

Machine Learning Based Methods. Compared with lexicon based methods, machine learning based methods do not involve opinion lexicons in the feature extraction process. Generic text features, such part-of-speech tags, tf-idf weights, term frequencies, etc., are very popular in those methods [3,5,11,12,16,18,21]. Inevitably, those features are much more noisy than the features generated from opinion lexicons. To compensate for the downside, those methods usually use complex model structures or large datasets. This is especially true in recent years as many newly developed methods are built upon deep learning models and big data. Socher et al. [23] proposed the recursive neural network, in which the authors first use the tree structure grammars to parse an input sentence into sub-phrases, then combine the embedding vectors of words in each sub-phrase through a set of pre-defined combination functions as the compositional representation for the sub-phrase. The sub-phrase representations are passed to a DNN model to predict the sentiment of each phrase. By fitting the DNN, not only the sentiment classifier, but also the phrase representations, can be learned. Tai et al. [25] proposed the Tree-LSTM, in which the authors apply the LSTM model on a sentence, and combine the hidden and memory cell states of words in each sub-phrase as the representation of the sub-phrase. Kim [10] stacks the embedding representations of words in each sentence and applies the convolutional neural network on the stacked feature matrix to train a sentiment classifier. Wang et al. [29] combine the expressiveness of LSTM and CNN by proposing a hybrid model, that first applies 1-dimensional convolutional filters on the embedding feature matrixes of the input sentences, then passes the resulting feature maps to RNN units to get deep sentence representations for sentiment classification.

2.2 Exploiting Document-Level Sentiment Labels for Sentence-Level Sentiment Classification

Besides the methods mentioned in the Introduction section, there also exist the following methods that exploit document-level sentiment labels to train sentence-level sentiment classifiers. In those models, document-level sentiments are usually used as a form of weak supervision, that has to work with other knowledge, such as opinion lexicons, human linguistic expertise, sentence-level sentiment labels, etc., to train sentence-level classifiers. Qu et al. [22] proposed a multi-expert model that makes use of local syntactic patterns of sentences, and opinion lexicons to build a set of rule-based base predictors, and predict the sentiment labels based on the votes of the base predictors. Yang et al. [33] proposed a CRF model that uses sentence-level sentiment labels as the main supervision signal, and use overall ratings as a form of posterior regularisation to keep sentence-level sentiments and document-level sentiments consistent. Täckström et al. [27] proposed a Hidden CRF model, that treats the sentence-level sentiments of a review as latent variables, and the overall ratings are observable variables conditioning on the latent variables. Opinion lexicons are used in the method to define the feature functions for the CRF model. Wu et al. [31] proposed the SSWS model that uses two levels of features: document-level sentiments and word-level sentiments, along with a predefined set of linguistic rules, to train a linear sentence-level sentiment classifier.

As previously mentioned, those methods also require intense human labor and linguistic expertise to build the opinion lexicons and linguistic rules. We differ our research from those existing work by that we focus on generalised, end-to-end approach to train the sentence-level sentiment classifiers, to avoid the laborious efforts to build the lexicons or the external linguistic expertise.

3   Averaged Logit

This section introduces our approach that uses recurrent neural network as the learning model, and the review ratings as the ground truth labels for sentence-level sentiment classification. In this section, a brief description of the generic application of RNN in sentence-level sentiment classification is presented first, followed by the introduction of the proposed approach.

3.1   RNN for Sentence-Level Sentiment Classification

The many-to-one structured RNN is usually used for sentiment classification. In this structure, each word position of a sentence is treated as a time step, and the embedding or one-hot vector of the word at each position is fed into the RNN unit of the corresponding time step. The activation of the last RNN unit in the sequence is further fed to a few fully connected MLP layers, and the logit vector of the last fully connected layer is used to compute the probability distribution of the sentiment classes. A pictorial description of the structure is shown in Fig. 1. A mathematical description of the process is shown as follows.

Assuming there are N words in a sentence, the representation of each word n is denoted as v_n, the mapping function of the RNN sequence as f_r, then the output of the RNN sequence a_r can be expressed as:

$$a_r = f_r(v_1, v_2, ...v_N) \tag{1}$$

Let f_c be the mapping function of the fully connected layers, that maps a_r to the logit vector of the last fully connected layer:

$$z = f_c(a_r) \tag{2}$$

The logit vector z is used to compute the probability of the ith sentiment class s_i by the softmax function:

$$p(s_i) = \frac{\exp(z_i)}{\sum_j (\exp(z_j))} \tag{3}$$

The cross entropy between $p(s)$ and the one-hot vector of the ground truth sentiment label is usually used as the loss. The back-propagation algorithm can be used to derive the parameters of the RNN units and the fully connected layers.

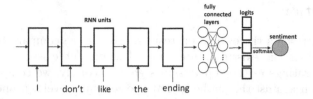

Fig. 1. The application of RNN in sentiment classification

3.2   RNN Models Based on Averaged Logits

The proposed model is built upon the basic RNN models, however, it makes a difference by modelling the compositional effects of the sentiments of individual sentences belonging to the same reviews. Assuming there are M sentences in a review, each of these sentences is fed into the RNN network. Let $z^{(m)}$ be the resulting logit of each sentence m. In the proposed approach, the logit vectors of the M sentences are combined into a synthesised logit vector $\bar{z}$ by their mean:

$$\bar{z} = \frac{1}{M} \sum_m z^{(m)} \tag{4}$$

The synthesised logit vector $\bar{z}$ is used to compute the probability of each possible rating r_i

$$p(\hat{r}_i) = \frac{\exp(\bar{z}_i)}{\sum_j (\exp(\bar{z}_j))} \tag{5}$$

The loss of the proposed network is the cross-entropy between $p(\hat{r})$ and the one-hot vector of the ground truth rating. A pictorial illustration of the proposed model is shown in Fig. 2.

In the test phase, given a unseen sentence, the resulting logit vector of the sentence is used to determine the sentiment label of the sentence.

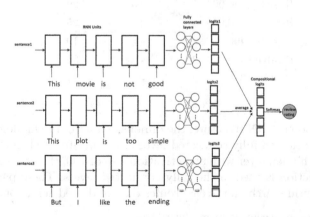

Fig. 2. A pictorial illustration of the sentiment composition approach

4   Evaluation

This section presents the evaluation results. Firstly, we compare the proposed approach against the existing sentence-level sentiment classification methods that also use ratings as the supervision signal. Secondly, we compare the proposed approach against the methods that use sentence-level sentiment labels as the supervision signal.

In the experiment, the word-to-vector (W2V) model [17] is used to represent the text. Instead of training dataset specific W2V models, the publicly available and general-purpose W2V model[1] pre-trained on the Google News corpus is used on all tasks. Two types of RNN: GRU and LSTM are chosen as the building blocks of the proposed approach. The Tensorflow package[2] is used to implement the proposed approach.

4.1   Comparison Against Models Using Ratings as Supervision Signal

Wu et al. [31] used a dataset of around 650,000 reviews covering 3 domains, book, dvd, and electronics, to train sentence-level sentiment classifiers, and used a small review sentence set manually annotated with binary sentiment labels to evaluate the performance of the classifiers. The same training and test sets are also used in this paper and their statistics are shown in Table 1.

Table 1. Statistics of the 3-domain dataset. R-ele: reviews for electronics, R-dvd: reviews for dvds, R-bks: reviews for books. RS: the review sentences for the three domains. $|V|$: the size of the data. Num(C): number of classes, Avglen(R): average number of sentences of the reviews, Avglen(S): average number of words of the review sentences

	$|V|$	Num(C)	Avglen(R)	Avglen(S)
R-ele	23,000	5	8	22
R-dvd	120,000	5	10	29
R-bks	500,000	5	11	25
RS-ele	400	2	-	19
RS-dvd	420	2	-	31
RS-bks	350	2	-	21

In the proposed RNN structure, the dimensionality of the hidden state vector is set to 200. 2 hidden fully connected layers are concatenated to the RNN unit sequence, one hidden layer with an activation size of 100, the other 30. The Relu activation function is used in the fully connected layers. These parameters are decided via a grid search on the electronics data and used across on all domains.

---

[1] https://code.google.com/archive/p/word2vec/.
[2] https://www.tensorflow.org/.

Table 2. Performance comparison among models supervised by ratings

	Book	DVD	Electronics
HCRF	0.7021	0.7566	0.7615
ME	0.7207	0.7846	0.7865
SSWS	0.7482	0.8082	0.8017
LSTM (Averaged logits)	0.719	0.777	0.764
GRU(Averaged logits)	0.730	0.786	0.772

Since the ground truth ratings of the reviews in the training set are on a 5-star scale, the predicted sentence-level sentiment labels are also integers on that scale. For binary sentiment classification, the predicted scores are converted into 'positive' if they are over 3 stars, and 'negative' otherwise. The Macro-average F1 score is used to measure the classification performance. The proposed model is compared with the existing models introduced in Sect. 2 that also use ratings to train sentence-level classifiers: the multi-expert model (ME) proposed by [22], the hidden conditional random field (HCRF) model proposed by [27], and the SSWS model proposed by [31]. The overall evaluation results are shown in Table 2. As the results indicate, the performance of the proposed model is close to that of HCRF and ME, slightly weaker than SSWS. As introduced in Sect. 2, all the 3 existing models utilise sentiment lexicons, linguistic expertise, and manually crafted decision rules, whereas the proposed model does not require those external knowledge.

Table 3. Statistics of the movie review and sentence datasets. Num(C): number of classes, $|V|$: data size, Avglen(R): average number of sentences of the reviews, Avglen(S): average number of words of the sentences.

	$|V|$	Num(C)	Avglen(R)	Avglen(S)
SMR	5000	5[a]	11	28
CSMR	55,000	5	8	21
SST-1	2210	5	-	18
SST-2	1821	2	-	19

[a]In the SMR dataset, the original rating of each review takes a decimal value in the range of $[0, 1]$. We converted those ratings into integers falling in the range of $[0, 5)$ by the following scheme in this experiment: 0 if rating ≤ 0.2; 1 if $0.2 <$ rating ≤ 0.4; 2 if $0.4 <$ rating ≤ 0.6; 3 if $0.6 <$ rating ≤ 0.8; 4 if rating > 0.8

4.2 Comparison Against Models Supervised by Sentence-Level Sentiments

The majority of the papers introducing sentence-level sentiment classification models have used the following datasets for evaluation:

- SST1: Stanford Sentiment Treebank, it is an extension of MR but provides following finer-scaled labels: very negative, negative, neutral, positive, very positive [23].
- SST2: Same as SST1 but each sentence is annotated with the binary positive/negative sentiment label.

Those datasets contain only stand-alone review sentences, therefore, cannot be used to train the proposed classifier. Instead, the scale movie review dataset (SMR) [20] containing 5000 movie reviews is used as the training set. We train GRU and LSTM classifiers by the proposed approach on SMR once, and evaluate them on the test subsets of SST-1 and SST-2.

Furthermore, since the proposed approach does not require manual efforts for data annotation, it is easy to obtain a very large training set by crawling reviews from the web. In the experiment we crawled around 50,000 reviews[3] from Rotten Tomato, and combined them with the SMR set to form a new dataset, called 'CSMR', as another training set. Statistics of SMR, SST-1, SST-2, and CSMR are shown in Table 3.

Table 4. Performance comparison between the proposed approach and existing models trained with sentence-level sentiment labels

Method type	Model	SST-1	SST-2
Recursive [23]	Basic	0.432	0.824
	Matrix-vector	0.444	0.829
CNN [10]	Static	0.455	0.868
	Non-static	0.480	0.872
RNN [25]	LSTM	0.464	0.849
	GRU	0.447	0.837
Traditional [23]	SVM	0.407	0.794
	Naive Bayes	0.410	0.818
Averaged logit trained on SMR	GRU	0.412	0.809
	LSTM	0.401	0.802
Averaged logit trained on CSMR	GRU	0.477	0.867
	LSTM	0.463	0.861

The ground-truth sentiment labels of the reviews in the SMR set are on a 5-point scale, therefore, the predicted labels are also scores on the 5-point scale.

---

[3] All the reviews were written by 'audience' with ratings on a 5-point scale.

We convert them into the binary format for the classification task on STS-2 with the same scheme mentioned in the previous subsection. The proposed approach is compared against the existing models trained on the training partitions of SST-1 and SST-2 under the supervision of sentence-level sentiments. The performance comparison is shown in Table 4.

The results indicate that, the performance of the proposed approach trained on SMR is close to that of the traditional SVM and Naive Bayes classifiers trained on SST-1 and SST-2, and weaker than that of the neural network models trained on SST-1 and SST-2 by margins of 2%–7%. Though the margins are not trivial, they come at the cost of expensive manual labor for the sentence labelling process.

When the proposed approach is trained on the CSMR dataset whose size is around 5 times larger than SST-1 and SST-2, the test accuracy of the proposed approach on SST-1 reaches 0.477, 0.867 on SST-2, that can rival the performance of the best of the neural network models trained on the two sentence datasets.

5   Conclusion

In review mining, a majority of existing methods for sentence-level sentiment classification are based on supervised machine learning models, that require a large number of labelled sentences to train the classifiers. In practice, the labelled sentence datasets are usually very expensive to obtain. In this paper, we propose a novel approach, called 'Averaged logits', that uses ratings, instead of sentence-level sentiment labels, to train the sentence-level sentiment classifiers. Evaluation results show that, the performance of the proposed approach is close to that of the traditional SVM and Naive Bayes classifiers trained by labelled sentences when their training sizes are approximately equal, and close to that of the neural network based classifiers trained by labelled sentences when the proposed approach uses approximately 5 times more training samples. In the future, we plan to crawl a larger movie review dataset to train the classifiers to have better understanding of the true potential of the proposed approach.

References

1. Blair-Goldensohn, S., Hannan, K., McDonald, R., Neylon, T., Reis, G.A., Reynar, J.: Building a sentiment summarizer for local service reviews. In: WWW Workshop on NLP in the Information Explosion Era, vol. 14, pp. 339–348 (2008)
2. Choi, Y., Cardie, C.: Learning with compositional semantics as structural inference for subsentential sentiment analysis. In: Proceedings of the 2008 Conference on Empirical Methods in Natural Language Processing, Honolulu, Hawaii, pp. 793–801. Association for Computational Linguistics, October 2008. http://www.aclweb.org/anthology/D08-1083
3. Cui, H.: Comparative experiments on sentiment classification for online product reviews, p. 6

4. Ding, X., Liu, B., Yu, P.S.: A holistic lexicon-based approach to opinion mining. In: Proceedings of the 2008 International Conference on Web Search and Data Mining, WSDM 2008, pp. 231–240. ACM, New York (2008). http://doi.acm.org/10.1145/1341531.1341561

5. Gamon, M.: Sentiment classification on customer feedback data: noisy data, large feature vectors, and the role of linguistic analysis, p. 7

6. Ganapathibhotla, M., Liu, B.: Mining opinions in comparative sentences. In: Proceedings of the 22nd International Conference on Computational Linguistics-Volume 1, pp. 241–248. Association for Computational Linguistics (2008)

7. Hu, M., Liu, B.: Mining and summarizing customer reviews. In: Proceedings of the Tenth ACM SIGKDD International Conference on Knowledge Discovery and Data Mining, pp. 168–177. ACM (2004)

8. Kennedy, A., Inkpen, D.: Sentiment classification of movie reviews using contextual valence shifters. Comput. Intell. **22**(2), 110–125 (2006)

9. Kim, S.M., Hovy, E.: Determining the sentiment of opinions. In: Proceedings of the 20th International Conference on Computational Linguistics, p. 1367. Association for Computational Linguistics (2004)

10. Kim, Y.: Convolutional neural networks for sentence classification. arXiv:1408.5882 [cs], August 2014

11. Kouloumpis, E., Wilson, T., Moore, J.D.: Twitter sentiment analysis: the good the bad and the omg! In: ICWSM, vol. 11(538–541), p. 164 (2011)

12. Lafferty, J., McCallum, A., Pereira, F.C.: Conditional random fields: probabilistic models for segmenting and labeling sequence data (2001)

13. Li, F., Huang, M., Zhu, X.: Sentiment analysis with global topics and local dependency. In: AAAI, vol. 10, pp. 1371–1376 (2010)

14. Lin, C., He, Y.: Joint sentiment/topic model for sentiment analysis, p. 10

15. Lu, C.Y., Lin, S.H., Liu, J.C., Cruz-Lara, S., Hong, J.S.: Automatic event-level textual emotion sensing using mutual action histogram between entities. Expert Syst. Appl. **37**(2), 1643–1653 (2010). http://www.sciencedirect.com/science/article/pii/S0957417409006046

16. Martineau, J., Finin, T.: Delta TFIDF: an improved feature space for sentiment analysis. In: ICWSM, vol. 9, p. 106 (2009)

17. Mikolov, T., Chen, K., Corrado, G., Dean, J.: Efficient estimation of word representations in vector space. arXiv preprint arXiv:1301.3781 (2013)

18. Mullen, T., Collier, N.: Sentiment analysis using support vector machines with diverse information sources. In: Lin, D., Wu, D. (eds.) Proceedings of EMNLP 2004, Barcelona, Spain, pp. 412–418. Association for Computational Linguistics, July 2004

19. Ng, V., Dasgupta, S., Arifin, S.M.N.: Examining the role of linguistic knowledge sources in the automatic identification and classification of reviews. In: Proceedings of the COLING/ACL 2006 Main Conference Poster Sessions, Sydney, Australia, pp. 611–618. Association for Computational Linguistics, July 2006. http://www.aclweb.org/anthology/P/P06/P06-2079

20. Pang, B., Lee, L.: Seeing stars: exploiting class relationships for sentiment categorization with respect to rating scales. In: Proceedings of the 43rd Annual Meeting on Association for Computational Linguistics, pp. 115–124. Association for Computational Linguistics (2005)

21. Pang, B., Lee, L., Vaithyanathan, S.: Thumbs up? Sentiment classification using machine learning techniques. arXiv:cs/0205070, May 2002

22. Qu, L., Gemulla, R., Weikum, G.: A weakly supervised model for sentence-level semantic orientation analysis with multiple experts. In: Proceedings of the 2012 Joint Conference on Empirical Methods in Natural Language Processing and Computational Natural Language Learning, Jeju Island, Korea, pp. 149–159. Association for Computational Linguistics, July 2012. http://www.aclweb.org/anthology/D12-1014

23. Socher, R., et al.: Recursive deep models for semantic compositionality over a sentiment treebank. In: Proceedings of the 2013 Conference on Empirical Methods in Natural Language Processing, Seattle, Washington, USA, pp. 1631–1642. Association for Computational Linguistics, October 2013. http://www.aclweb.org/anthology/D13-1170

24. Taboada, M., Brooke, J., Tofiloski, M., Voll, K., Stede, M.: Lexicon-based methods for sentiment analysis. Comput. Linguist. **37**(2), 267–307 (2011)

25. Tai, K.S., Socher, R., Manning, C.D.: Improved semantic representations from tree-structured long short-term memory networks. arXiv:1503.00075 [cs], February 2015

26. Turney, P.D.: Thumbs up or thumbs down? Semantic orientation applied to unsupervised classification of reviews. arXiv:cs/0212032, December 2002

27. Täckström, O., McDonald, R.: Discovering fine-grained sentiment with latent variable structured prediction models. In: Clough, P., et al. (eds.) ECIR 2011. LNCS, vol. 6611, pp. 368–374. Springer, Heidelberg (2011). https://doi.org/10.1007/978-3-642-20161-5_37

28. Valitutti, A., Strapparava, C., Stock, O.: Developing affective lexical resources. PsychNology J. **2**(1), 61–83 (2004)

29. Wang, X., Jiang, W., Luo, Z.: Combination of convolutional and recurrent neural network for sentiment analysis of short texts. In: Proceedings of COLING 2016, the 26th International Conference on Computational Linguistics: Technical Papers, Osaka, Japan, pp. 2428–2437. The COLING 2016 Organizing Committee, December 2016. http://aclweb.org/anthology/C16-1229

30. Wilson, T., Wiebe, J., Hoffmann, P.: Recognizing contextual polarity: an exploration of features for phrase-level sentiment analysis. Comput. Linguist. **35**(3), 399–433 (2009). http://www.mitpressjournals.org/doi/10.1162/coli.08-012-R1-06-90

31. Wu, F., Zhang, J., Yuan, Z., Wu, S., Huang, Y., Yan, J.: Sentence-level sentiment classification with weak supervision, pp. 973–976. ACM Press (2017). http://dl.acm.org/citation.cfm?doid=3077136.3080693

32. Xianghua, F., Guo, L., Yanyan, G., Zhiqiang, W.: Multi-aspect sentiment analysis for Chinese online social reviews based on topic modeling and HowNet lexicon. Knowl.-Based Syst. **37**, 186–195 (2013)

33. Yang, B., Cardie, C.: Context-aware learning for sentence-level sentiment analysis with posterior regularization. In: Proceedings of the 52nd Annual Meeting of the Association for Computational Linguistics (Volume 1: Long Papers), Baltimore, Maryland, pp. 325–335. Association for Computational Linguistics, June 2014. http://www.aclweb.org/anthology/P14-1031

Convolutional Neural Networks with Interpretable Kernels

Vojtech Molek$^{(\boxtimes)}$ ⓘ and Irina Perfilieva$^{(\boxtimes)}$ ⓘ

Institute for Research and Applications of Fuzzy Modeling,
Centre of Excellence IT4Innovations, University of Ostrava,
30. dubna 22, Ostrava, Czech Republic
{vojtech.molek,irina.perfilieva}@osu.cz

Abstract. We are focused on the theoretical background of convolutional neural networks. In particular, we examine the problem whether semantic meaning can be assigned to convolutional kernels in the first layers and how this fact can simplify the learning procedure. In this respect, we prove the suitability and efficiency of the F-transform kernels. We describe various experiments that support our claim.

Keywords: F-transform · Convolutional kernels ·
Convolutional neural networks · Interpretability

1 Motivation and Formulation of the Problem

The field of Deep learning (DL) has recently gained a lot of attention from both public and scientific communities. This fact can be attributed to the recent significant progress in the field [1,9]. The progress was partially enabled by the technological advent of graphics processor units (GPUs) and increasing amount of labeled data, crucial for a *supervised* learning.

In the past, neural networks did not contain many layers and neurons due to the technological limitations and were referred to as a *shallow*. The technological progress allowed to create, train and use networks with many layers and neurons [10], so-called deep networks, thus the term *deep learning*.

We will be focusing on the theoretical analysis of deep-learning methods and their realization, ConvNets in particular. We consider their applications to classification and image recognition. We remind [1] that deep-learning methods are *representation-learning* methods with multiple levels of representation. On each next level, a new representation is created as a result of a non-linear transformation of data on the preceding level (starting from the raw input). The representation is based on a set of features that allows making an approximation reconstruction.

Learning of features is the most significant aspect of a ConvNet. What kind of features are extracted is determined by convolutional kernels (filters) in each

Supported by University of Ostrava.

H. Seki et al. (Eds.): IUKM 2019, LNAI 11471, pp. 320–332, 2019.
https://doi.org/10.1007/978-3-030-14815-7_27

convolutional layer. As a result of a training procedure, the kernels are highly dependent on the training dataset. In this sense, a ConvNet learns those features and filters that are *most relevant (discriminative) for the given task*. In general, relevant features describe an object or even allow the "inverse" process - approximate reconstruction of the original object from its features. From the point of view of the further successful recognition, this requires two things: (a) robust reconstruction of an image from the extracted features; (b) completeness of the training dataset with respect to certain (called rigid) transformations. In this contribution, our main concern is the first requirement, while the second one is an aside result of the proposed technique.

We propose to "help" a ConvNet and start its training from already preselected kernels in its first (convolutional) layer. Our motivation is based on the above-given preamble, which can be summarized as follows: *any convolution-based transformation which has the inverse form can be selected as a pool of kernels that are used in convolution layers of a ConvNet*. The reasonable convolution-based transformation that we are pushing forward by this contribution is based on a fuzzy partition and is known as the *higher degree F-transform* [5,6]. Moreover, due to the universality of the F-transform and its ability to perform reconstruction with arbitrary precision (for all functions from the corresponding Hilbert space), it is independent of any training dataset. In its convolutional form, the higher degree F-transform extracts the features that characterize the following universal properties of any functional object. They are: average values of a function range, its first and higher derivatives within a certain area. If a functional object is an image, then these universal properties/features are: average values of brightness, edginess, convexity and so on. The degree of averageness depends on the size of the corresponding area (kernel). In this respect, we say that the F-transform provides with the interpretable convolutional kernels.

Last, but not least, the F-transform significantly reduces noise and by this, it returns a smooth/robust reconstruction on its inverse step [4]. Below, we provide general information regarding the F-transform.

The F-transform has two modes: *forward (direct)* transforms original data into the so-called *component* space, and *inverse* transforms backward. The inverse transform approximates the original function with arbitrary precision. Furthermore, besides the conventional integral form, the direct F-transform can be represented with the help of convolution, which makes it suitable for kernels in a ConvNet.

To strengthen our motivation and confirm our consideration, we show the theoretical details of the F-transform (Sect. 2) and the results of the two experiments with various ConvNets. In the first experiment (see Sect. 3.1), we take the LeNet-5 and replace the kernels in its first layer by the second degree F-transform kernels. The obtained *FT-based LeNet-5* (we will refer to it as to FTNet) is applied to the database MNIST in such a way that the weights in the first layer (F-transform kernels) are not updated during learning. The results are compared with those with full learning and random selection of weights. The second experiment (see Sect. 3.2) analyses several ConvNets trained on the database ImageNet with the purpose to find analogies between their convolution kernels in the first layer and the F-transform kernels.

The structure of the paper is as follows: in Sect. 2, we give details of the higher degree F-transform [6]. In Sect. 3, we explain the details of our experiments.

2   The F-transform of a Higher Degree (F^m-transform)

In this section, we recall the main facts (see [6] for more details) about the higher degree F-transform and specifically F^2-transform - the technique, which will be used in the proposed below ConvNet with the FT kernels (FTNet).

2.1   Fuzzy partition

The F-transform is the result of a convolution of an object function (image, signal, etc.) and a generating function of what is regarded as a *fuzzy partition* of a universe.

Definition 1. *Let $n > 2$, $a = x_0 = x_1 < \ldots < x_n = x_{n+1} = b$ be fixed nodes within $[a, b] \subseteq \mathbb{R}$. Fuzzy sets $A_1, \ldots, A_n : [a, b] \to [0, 1]$, identified with their membership functions defined on $[a, b]$, establish a fuzzy partition of $[a, b]$, if they fulfill the following conditions for $k = 1, \ldots, n$:*

1. *$A_k(x_k) = 1$;*
2. *$A_k(x) = 0$ if $x \in [a, b] \setminus (x_{k-1}, x_{k+1})$;*
3. *$A_k(x)$ is continuous on $[x_{k-1}, x_{k+1}]$;*
4. *$A_k(x)$ for $k = 2, \ldots, n$ strictly increases on $[x_{k-1}, x_k]$ and for $k = 1, \ldots, n-1$ strictly decreases on $[x_k, x_{k+1}]$;*
5. *for all $x \in [a, b]$ holds the Ruspini condition*

$$\sum_{k=1}^{n} A_k(x) = 1. \tag{1}$$

The elements of fuzzy partition $\{A_1, \ldots, A_n\}$ are called *basic functions*.

In particular, an h-uniform fuzzy partition of $[a, b]$ can be obtained using the so called *generating function*

$$A : [-1, 1] \to [0, 1], \tag{2}$$

which is defined as an even, continuous and positive function everywhere on $[-1, 1]$ except for on boundaries, where it vanishes. Basic functions $A_2, \ldots, A_{n-1}$ of an h-uniform fuzzy partition are rescaled and shifted copies of A in the sense that for all $k = 2, \ldots, n - 1$;

$$A_k(x) = \begin{cases} A(\frac{x - x_k}{h}), & x \in [x_k - h, x_k + h], \\ 0, & \text{otherwise.} \end{cases}$$

Below, we will be working with one particular case of an h-uniform fuzzy partition that is generated by the triangular shaped function A^{tr} and its h-rescaled version A_h^{tr}, where

$$A^{tr}(x) = 1 - |x|,\ x \in [-1, 1],\ \text{and}\ A_h^{tr}(x) = 1 - \frac{|x|}{h},\ x \in [-h, h].$$

A fuzzy partition generated by the triangular shaped function A^{tr} will be referred to as *triangular shaped*.

2.2   Space $L_2(A_k)$

Let us fix $[a, b]$ and its h-uniform fuzzy partition $A_1, \dots, A_n$, where $n \geq 2$ and $h = \frac{b-a}{n-1}$[1]. Let k be a fixed integer from $\{1, \dots, n\}$, and let $L_2(A_k)$ be a set of square-integrable functions $f : [x_{k-1}, x_{k+1}] \rightarrow \mathbb{R}$. Denote $L_2(A_1, \dots, A_n)$ a set of functions $f : [a, b] \rightarrow \mathbb{R}$ such that for all $k = 1, \dots, n$, $f|_{[x_{k-1}, x_{k+1}]} \in L_2(A_k)$. In $L_2(A_k)$, we define an *inner product* of f and g

$$\langle f, g \rangle_k = \int_{x_{k-1}}^{x_{k+1}} f(x)g(x)d\mu_k = \frac{1}{s_k} \int_{x_{k-1}}^{x_{k+1}} f(x)g(x)A_k(x)dx,$$

where

$$s_k = \int_{x_{k-1}}^{x_{k+1}} A_k(x)dx.$$

The space $(L_2(A_k), \langle f, g \rangle_k))$ is a *Hilbert space*. We apply the Gram-Schmidt process to the linearly independent system of polynomials $\{1, x, x^2, \dots x^m\}$ restricted to the interval $[x_{k-1}, x_{k+1}]$ and convert it to an orthogonal system in $L_2(A_k)$. The resulting orthogonal polynomials are denoted by $P_k^0, P_k^1, P_k^2, \dots, P_k^m$.

Example 1. Below, we write the first three orthogonal polynomials P^0, P^1, P^2 in $L_2(A)$, where A is the generating function of a uniform fuzzy partition, and $\langle \cdot, \cdot \rangle_0$ is the inner product:

$$P^0(x) = 1,$$
$$P^1(x) = x,$$
$$P^2(x) = x^2 - I_2,\ \text{where}\ I_2 = h^2 \int_{-1}^{1} x^2 A(x)dx,$$

If generating function A^{tr} is triangular shaped and h-rescaled, then the polynomial P^2 can be simplified to the form

$$P^2(x) = x^2 - \frac{h^2}{6}. \tag{3}$$

We denote $L_2^m(A_k)$ a linear subspace of $L_2(A_k)$ with the basis $P_k^0, P_k^1, P_k^2 \dots, P_k^m$.

---

[1] The text of this and the following subsection is a free version of a certain part of [6] where the theory of a higher degree F-transform was introduced.

2.3 F^m-transform

In this section, we define the F^m-transform, $m \geq 0$, of a function f with polynomial components of degree m. Let us fix $[a, b]$ and its fuzzy partition $A_1, \ldots, A_n$, $n \geq 2$.

Definition 2 ([6]). *Let $f : [a, b] \rightarrow \mathbb{R}$ be a function from $L_2(A_1, \ldots, A_n)$, and let $m \geq 0$ be a fixed integer. Let F_k^m be the k-th orthogonal projection of $f|_{[x_{k-1}, x_{k+1}]}$ on $L_2^m(A_k)$, $k = 1, \ldots, n$. We say that the n-tuple $(F_1^m, \ldots, F_n^m)$ is an F^m-transform of f with respect to $A_1, \ldots, A_n$, or formally,*

$$F^m[f] = (F_1^m, \ldots, F_n^m).$$

F_k^m is called the kth F^m-transform component of f.

Explicitly, each kth component is represented by the mth degree polynomial

$$F_k^m = c_{k,0}P_k^0 + c_{k,1}P_k^1 + \cdots + c_{k,m}P_k^m, \tag{4}$$

where

$$c_{k,i} = \frac{\langle f, P_k^i \rangle_k}{\langle P_k^i, P_k^i \rangle_k} = \frac{\int_a^b f(x)P_k^i(x)A_k(x)dx}{\int_a^b P_k^i(x)P_k^i(x)A_k(x)dx}, \quad i = 0, \ldots, m.$$

Definition 3. *Let $F^m[f] = (F_1^m, \ldots, F_n^m)$ be the direct F^m-transform of f with respect to $A_1, \ldots, A_n$. Then the function*

$$\hat{f}_n^m(x) = \sum_{k=1}^n F_k^m A_k(x), \quad x \in [a, b], \tag{5}$$

is called the inverse F^m-transform of f.

The following theorem proved in [6] estimates the quality of approximation by the inverse F^m-transform in a normed space L_1.

Theorem 1. *Let $A_1, \ldots, A_n$ be an h-uniform fuzzy partition of $[a, b]$. Moreover, let functions f and A_k, $k = 1, \ldots, n$ be four times continuously differentiable on $[a, b]$, and let $\hat{f}_n^m$ be the inverse F^m-transform of f, where $m \geq 1$. Then*

$$\|f(x) - \hat{f}_n^m(x)\|_{L_1} \leq O(h^2),$$

where L_1 is the Lebesgue space on $[a + h, b - h]$.

2.4 F^2-transform in the Convolutional Form

Let us fix $[a, b]$ and its h-uniform fuzzy partition $A_1, \ldots, A_n$, $n \geq 2$, generated from $A : [-1, 1] \rightarrow [0, 1]$ and its h-rescaled version A_h, so that $A_k(x) = A(\frac{x-x_k}{h}) = A_h(x - x_k)$, $x \in [x_k - h, x_k + h]$, and $x_k = a + kh$. The F^2-transform of a function f from $L_2(A_1, \ldots, A_n)$ has the following representation

$$F^2[f] = (c_{1,0}P_1^0 + c_{1,1}P_1^1 + c_{1,2}P_1^2, \ldots, c_{n,0}P_n^0 + c_{n,1}P_n^1 + c_{n,2}P_n^2), \tag{6}$$

where for all $k = 1, \ldots, n$,

$$P_k^0(x) = 1, \ P_k^1(x) = x - x_k, \ P_k^2(x) = (x - x_k)^2 - I_2, \tag{7}$$

where $I_2 = h^2 \int_{-1}^{1} x^2 A(x) dx$, and coefficients are as follows:

$$c_{k,0} = \frac{\int_{-\infty}^{\infty} f(x) A_h(x - x_k) dx}{\int_{-\infty}^{\infty} A_h(x - x_k) dx}, \tag{8}$$

$$c_{k,1} = \frac{\int_{-\infty}^{\infty} f(x)(x - x_k) A_h(x - x_k) dx}{\int_{-\infty}^{\infty} (x - x_k)^2 A_h(x - x_k) dx}, \tag{9}$$

$$c_{k,2} = \frac{\int_{-\infty}^{\infty} f(x)((x - x_k)^2 - I_2) A_h(x - x_k) dx}{\int_{-\infty}^{\infty} ((x - x_k)^2 - I_2)^2 A_h(x - x_k) dx}. \tag{10}$$

In [6,7], it has been proved that

$$c_{k,0} \approx f(x_k), \ c_{k,1} \approx f'(x_k), \ c_{k,2} \approx f''(x_k), \tag{11}$$

where $\approx$ is meant up to $O(h^2)$.

Without going into technical details, we rewrite (8)–(10) into the following discrete representations

$$c_{k,0} = \sum_{j=1}^{l} f(j) g_0(ks - j), \ c_{k,1} = \sum_{j=1}^{l} f(j) g_1(ks - j), \ c_{k,2} = \sum_{j=1}^{l} f(j) g_2(ks - j), \tag{12}$$

where $k = 1, \ldots, n$, $n = \lfloor \frac{l}{s} \rfloor$, s is the so called "stride" and g_0, g_1, g_2 are normalized functions that correspond to generating functions A_h, $(x A_h)$ and $((x^2 - I_2) A_h)$. It is easy to see that if $s = 1$, then coefficients $c_{k,0}$, $c_{k,1}$, $c_{k,2}$ are results of the corresponding discrete convolutions $f \star g_0$, $f \star g_1$, $f \star g_2$. Thus, we can rewrite the representation of F^2 in (6) in the following vector form:

$$F^2[f] = ((f \star_s g_0)^T \mathbf{P}^0 + (f \star_s g_1)^T \mathbf{P}^1 + (f \star_s g_2)^T \mathbf{P}^2), \tag{13}$$

where $\mathbf{P}^0$, $\mathbf{P}^1$, $\mathbf{P}^2$ are vectors of polynomials with components given in (7), and $\star_s$ means that the convolution is performed with the stride s, $s \geq 1$.

3   FT-Based LeNet-5: Architecture and Efficiency

The ConvNet with the F-transform kernels was proposed in [3] where it was named as FTNet. In this section, we remind the essential details and results of [3] proposal. Moreover, we present the results of new experiments.

As a baseline network, we use a restricted version of the LeNet-5 with the architecture shown in Table 1.

In [3], we modified the LeNet-5 by replacing convolution-type units in the first and third convolution layers C_1 and C_3 by the similar units which realize

Table 1. The baseline network architecture.

Parameters	Conv C_1	Max pool S_2	Conv C_3	Max pool S_4	F-Con FC_5	Output
Input shape	$28 \times 28 \times 1$	$28 \times 28 \times 32$	$14 \times 14 \times 32$	$14 \times 14 \times 64$	3136	500
Output shape	$28 \times 28 \times 32$	$14 \times 14 \times 32$	$14 \times 14 \times 64$	$7 \times 7 \times 64$	500	10
Window size	5×5	2×2	$5 \times 5 \times 32$	2×2	-	-
Stride	1×1	2×2	1×1	2×2	-	-
Kernels	32	-	64	-	-	-
Activation	ReLU	-	ReLU	-	ReLU	Softmax

Table 2. FT-Net architecture.

Hyper-parameter	C_1	S_2	C_3	S_4	FC_5	FC_6
Kernel size	5×5	-	5×5	-	-	-
# feature maps	8	-	64	-	-	-
Stride	1×1	2×2	1×1	2×2	-	-
Pooling size	-	2×2	-	2×2	-	
# FC units	-	-	-	-	500	10

convolution with the F-transform kernels according to (12) adapted to functions of two variables. Together with negative versions of kernels, we have eight feature maps in the layer C_1. Each of C_1 feature maps is processed with all eight kernels again in C_3 - altogether, we have sixty-four feature maps in C_3.

The details of the FTNet architecture are given below in Table 2. All experiments were realized on MNIST [2] dataset available at Yan LeCun website[2]. MNIST dataset consists of 60 000 28×28 pixel 8 bit grey-scale training images and 10 000 testing 8 bit grey-scale images with same resolution of the hand written digits.

3.1 Efficiency of FTNet

For this contribution, we made several experiments in order to obtain a comparative analysis of the FTNet and LeNet-5 efficiencies (in terms of training time and accuracy of recognition). In the 1st experiment, we excluded the first convolutional layer of the FTNet from parameter update procedure to preserve its weights. This allows observing the network response to features extracted with the help of the F-transform kernels. Our conjecture was as follows: *Features extracted by convolution operation with the F-transform kernels are sufficient to capture essential features of the dataset MNIST.*

In comparison with the baseline network, the FTNet has four times fewer kernels: 32 versus 8. All 8 kernels are normalized so that their respective sum equals to either 0 (kernels with negative numbers) or 1 (kernels with positive

---

[2] http://yann.lecun.com/exdb/mnist.

numbers) to be consistent with other layers initialization (normal distribution $\mathcal{N}(0, 1)$). Table 3 shows the result of training the network for one epoch. The accuracy subscript indicates the median[3] of accuracy at i^{th} iteration. The training is realized with softmax output layer, cross entropy loss function, using gradient descent with momentum as optimization method[4] and weights decay.

Table 3. Training time and accuracy of the two different first layer initializations.

1st conv layer	Training time [sec]	Accuracy$_0$ [%]	Accuracy$_{500}$ [%]	Accuracy$_{10^4}$ [%]
Random init	53.81	11.55	97.20	99.08
F-transf. init	37.27	14.49	96.56	98.94

The results in Table 3 confirms the conjecture and shows that the F-transform kernels are sufficiently descriptive and have a relatively low number of learning parameters. Using the F-transform kernels in ConvNet architecture supports the F-transform theory as a basis of neural networks and allows users to leverage this knowledge in order to raise the level of interpretability.

In the 2nd experiment, we examined the impact of the F-transform kernels on network behavior. With this purpose, we performed the hyperparameters optimization via grid search.

The grid search is an optimization method that systematically searches an optimal combination through a selected subspace of the hyperparameters space. In this experiment, we optimized the following parameters for first two convolutional and pooling layers: convolutional kernel size $\mathcal{D}$, presence and type of the subsampling $\mathcal{S}$, layer weights trainability $\mathcal{T}$, and a form of the layer weights initialization $\mathcal{I}$. The observed hyperparameters values are shown in Table 4.

Table 4. The observed hyperparameters values: $\mathcal{D}$, $\mathcal{T}$ and $\mathcal{I}$ are associated with first and second conv layer; $\mathcal{S}$ is associated with both first and second convolutional and pooling layer.

$\mathcal{D}$	$\mathcal{S}$	$\mathcal{T}$	$\mathcal{I}$
3×3	None	Trainable	Static kernels
5×5	2×2 pooling	Non-trainable	Kernels from $\mathcal{N}(0, 1)$
-	2×2 stride	-	-

The value of the hyperparameter $\mathcal{T}$ controls whether the first convolutional layer and/or second convolutional layer are excluded from the weights updating

---

[3] We performed several independent training runs.
[4] Hyperparameters values were left at their respective default values - as they are set in the Caffe framework example.

procedure; $\mathcal{I}$ controls whether the first convolutional layer and/or the second convolutional layer are initialized with random kernels sampled from $\mathcal{N}(0,1)$ or static (conventional or F-transform) kernels.

- **Conventional kernels** - Gaussian kernel, its negative, $Sobel_x$, $Sobel_y$, Laplace and its negative.
- **F-transform kernels** - F^0-transform and its negative, F_x^1-transform, F_y^1-transform, F^2-transform and its negatives.

Note: the second convolutional layer uses the same kernels as in the first layer. By this, it performs two successive convolutions with all possible kernel combinations. In this experiment, the baseline network was initialized with only 6 kernels in order to analyze the network behavior with respect to either fixed kernels or randomly initialized. We evaluated the network behavior in terms of the loss function values. The grid search over 576 possible combinations leads to the following three groups of the network configurations depending on the loss function values:

- The lowest value ≈ 1.5 of the loss function was achieved on the subset of hyperparameters where subsampling (in either pooling or stride form) was used in both first and second convolutional layers irrespective of the selection of kernels. We will further refer to this configuration as to *Configuration 1*.
- The second lowest value ≈ 6 of the loss function was achieved on the subset of hyperparameters where subsampling was used in only one (first or second) convolutional layer irrespective of the selection of kernels.
- The highest value ≈ 25 of the loss function was achieved on the subset of hyperparameters where subsampling was not used at all.

These conclusions implicitly confirm that for the dataset MNIST, the F-transform kernels are suitable in the same extent as the randomly learned ones. To make this observation better supported, we compared the (static and updated by learning) F-transform kernels in the first layer of the two FTNets with the *Configuration 1*. The result is demonstrated in Fig. 1 together with the visualized difference between the corresponding kernels. The following comments can be made: the FTNet updated by learning F-transform kernels did not modify the F^2-transform kernels at all, whereas the F^0-transform and F^1-transform kernels were slightly rotated without any changes of their shapes. This explicitly confirms the suitability of the F-transform for feature extraction. Only small corrections are generally needed to fully capture the nature of the dataset. This is attributed to those numerals in MNIST that were painted with non-horizontal/vertical lines. Therefore, we can use the F-transform kernels for first layer initialization of ConvNets in order to decrease the training time.

Another important conclusion (regarding this experiment with MNIST) is that the most influential operation is subsampling. It decreases the size of data being processed by a network and therefore, decreases the number of free parameters in a network. So the obvious rationale is that the decreasing number of parameters increases the speed of convergence to the feasible solution.

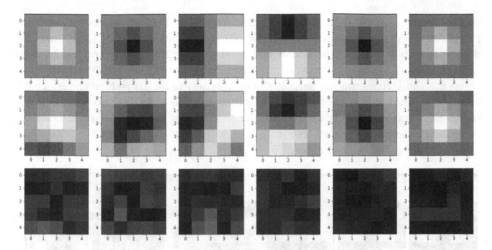

Fig. 1. Top row: visualization of the F-transform kernels. Middle row: F-transform kernels after 100 epoch learning of the FTNet. Bottom row: difference between the top and middle rows.

In our 3^d experiment we were focused on the invariance with respect to different kind of transformations. In particular, we were interested in the invariance with respect to rotations where we examined whether the features extracted with static (not updated by learning) kernels remain unchanged. For this purpose, we created new manually painted inputs (numerals from 0 to 9 are shown in Fig. 2) in the spirit of MNIST and rotated them by 5° up to 355°.

The results of this experiment showed that the network with the *Configuration 1* demonstrates satisfactory invariance with respect to rotations within the interval $[-30°, +30°]$.

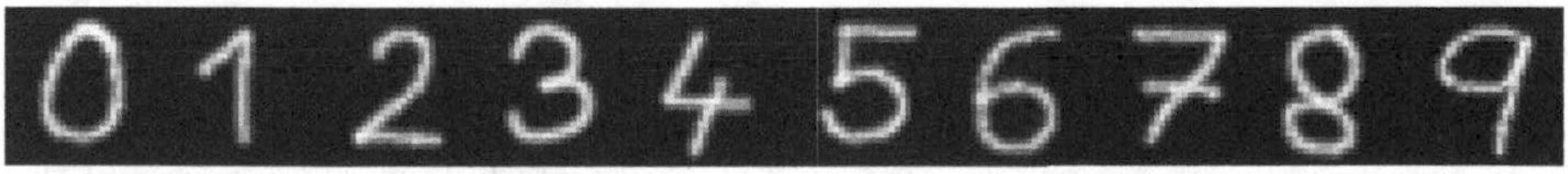

Fig. 2. Manually painted numerals used in rotation invariance test.

3.2   Semantic Meaning of Principal Kernels in Convolutional Layers

In this subsection, we tackle the problem of interpretability from the opposite angle. Instead of initializing a network with predefined kernels, we examine the first convolutional layers and the (already trained) kernels corresponding to them taken from several networks. The purpose is to find general semantic meanings of kernels and through them compare with the F-transform kernels. We selected 6 networks: VGG16, VGG19, InceptionV3, MobileNet, ResNet, and AlexNet, as the representative examples of ConvNets. All of the networks were trained on ImageNet [8], using the same training dataset, consisting of ≈1.2M RGB images[5] with various resolution (usually downsampled to 256×256). We extracted kernels from the first convolutional layer of all considered networks and analyzed whether there are similarities among kernels across the networks. To reduce the space of kernels, we first apply the hierarchical clustering on every network kernel set separately, and then, look for the similarities among clusters. The *medoids* of the found clusters are shown in Fig. 3.

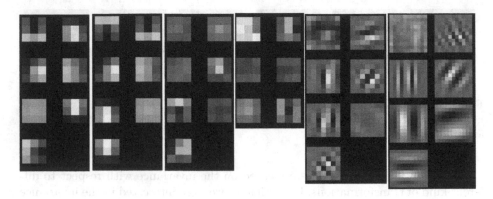

Fig. 3. From left to right: the medoids of the clusterized kernels from the first convolutional layer of VGG19, VGG16, InceptionV3, MobileNet, ResNet and AlexNet.

We observed that the extracted clusters contain similar elements (kernels) across the different networks that share one of the following characteristics:

– **Gaussian-like**;
– **Edge detection (with various angle specifications)**;
– **Texture detection**;
– **Color blobs**.

If we compare the semantic meaning of the extracted clusters (in terms of the above-given characteristics) with that of the F-transform kernels in the FTNet, then we see the coincidence in the first two items from the above-given list. To be more precise, the F^0-transform kernels are the Gaussian-like, and the F^1-transform kernels are (horizontal or vertical) edge detectors. This again supports our conclusion regarding the suitability of the F-transform kernels in the first layers of ConvNets.

The more general conclusion relates to the disclosed semantic meanings of convolutional kernels. This knowledge is helpful for optimal and non-exhaustive design of ConvNets.

---

[5] ImageNet dataset differs, depending on the year of ILSVRC competition.

4   Conclusion and the Future Work

In the proposed contribution, we have been focused on representation learning and the way to improve it. The network gradually extracts features from raw input to later approximate reconstruction. We have shown similar properties can be found in fuzzy logic technique - F-transform. We have introduced ConvNet architecture with F-transform kernels - FTNet.

FTNet was compared with the baseline, LeNet-5 - like, network with randomly initialized weights. FTNet was learning faster due to the decreased number of convolutional kernels. The kernels and their impact was tested with grid search through hyperparameters. The most influential hyperparameter with respect to loss function value was subsampling (pooling or stride). Both FTNet and baseline network have performed comparably on rotated data. By letting FTNet to training F-transform kernels, we obtained F-transform kernels specialized for feature extraction from MNIST. The specialized kernels had the same shape as original F-transform kernels and contained only small modification such as small rotations. These results confirm the suitability of FTNet for MNIST and usage of F-transform kernels for general feature extraction.

To further evaluate what type of kernels are being learned we have extracted kernels from 6 popular pre-trained networks. The kernel clusters were divided into 4 groups matching classic image processing techniques (gradient detection, texture detection, color detection, and blurring). F-transform kernels match with these groups and so it is adequate to use F-transform kernels for initialization and pre-training.

There are several open questions for future research. How to properly expand F-transform kernels for more complex datasets? What kernels are redundant and how to properly reduce the number of kernels by other means than F-transform kernels substitution? How influential are kernel rotations to network accuracy and how many of rotated kernels should be included?

Acknowledgment. The work was supported from ERDF/ESF "Centre for the development of Artificial Intelligence Methods for the Automotive Industry of the region" (No. CZ.02.1.01/0.0/0.0/17_049/0008414).

References

1. LeCun, Y., Bengio, Y., Hinton, G.: Deep learning. Nature **521**(7553), 436–444 (2015)
2. LeCun, Y., Bottou, L., Bengio, Y., Haffner, P.: Gradient-based learning applied to document recognition. Proc. IEEE **86**(11), 2278–2324 (1998)
3. Molek, V., Perfilieva, I.: Convolutional neural networks with the F-transform kernels. In: Rojas, I., Joya, G., Catala, A. (eds.) IWANN 2017. LNCS, vol. 10305, pp. 396–407. Springer, Cham (2017). https://doi.org/10.1007/978-3-319-59153-7_35
4. Porfilieva, I., Holčapek, M., Kreinovich, V.: A new reconstruction from the F-transform components. Fuzzy Sets Syst. **288**, 3–25 (2016)
5. Perfilieva, I.: Fuzzy transforms: theory and applications. Fuzzy Sets Syst. **157**(8), 993–1023 (2006)

6. Perfilieva, I., Danková, M., Bede, B.: Towards F-transform of a higher degree. In: IFSA/EUSFLAT Conference, pp. 585–588 (2009)
7. Perfilieva, I., Kreinovich, V.: Fuzzy transforms of higher order approximate derivatives: a theorem. Fuzzy Sets Syst. **180**(1), 55–68 (2011)
8. Russakovsky, O., et al.: ImageNet large scale visual recognition challenge. Int. J. Comput. Vision **115**(3), 211–252 (2015)
9. Schmidhuber, J.: Deep learning in neural networks: an overview. Neural Netw. **61**, 85–117 (2015)
10. Szegedy, C., Ioffe, S., Vanhoucke, V.: Inception-V4, inception-ResNet and the impact of residual connections on learning. CoRR abs/1602.07261 (2016). http://arxiv.org/abs/1602.07261

Machine Learning Applications

Ties Between Mined Structural Patterns in Program and Their Identifier Names

Yoshiki Mashima[1(✉)], Sachio Hirokawa[2], and Kazuhiro Takeuchi[3]

[1] Graduate School of Engineering, Osaka Electro-Communication University,
Osaka, Japan
mi17a004@oecu.jp
[2] Research Institute for Information Technology, Kyushu University,
Fukuoka, Japan
hirokawa@cc.kyushu-u.ac.jp
[3] Faculty of Information and Communication Engineering,
Osaka Electro-Communication University, Osaka, Japan
takeuchi@osakac.ac.jp

Abstract. Identifier names in readable and maintainable source codes are always descriptive. These names are given based on the implicit knowledge of experienced programmers. In this paper, we propose a structural pattern mining method based on support vector machines (SVM) for source codes. We extract 1,000 method names in object-oriented source codes collected from online software repositories and create 1,000 datasets labeled by positive and negative class. The structural features used for the input feature vectors to the SVM learning are designed for representing partial characteristics in the abstract syntax tree (AST) parsed from a source code. Applying this method, we made an F1 score list of the 1,000 method names, which shows the degree of patterning of each name, by using our structural features. From the list, we confirmed structural patterns were strongly associated with specific method names. A qualitative evaluation of method names was also conducted by mapping the structural feature vector of each program example to the two-dimensional plane in the same way as a previous major study. From the evaluation, we confirmed that the contrasting structure among the programs corresponds to the names given to programs. Furthermore, we show examples of visualization of structural patterns using structural features extracted by feature selection.

Keywords: Software engineering · Implicit knowledge extraction ·
Pattern mining · Support vector machines

1 Introduction

Source code is normally associated with natural language (NL) in various aspects of software development, such as specification documents, comments, and reviews. In particular, programmers need to name identifiers in source codes,

© Springer Nature Switzerland AG 2019
H. Seki et al. (Eds.): IUKM 2019, LNAI 11471, pp. 335–346, 2019.
https://doi.org/10.1007/978-3-030-14815-7_28

such as variables, parameters, functions, and classes. Giving descriptive names to identifiers is important for making codes readable and maintainable. This naming is generally based on the experience and common sense of programmers.

In recent years, it has become possible to obtain an enormous number of source codes from online repositories such as GitHub. From such available open source repositories, research to relate natural language to source codes has been promoted. The associated linkage between NL and source codes contributes to the advancement of software development tools. For example, Gvero et al. proposed a program development support tool "AnyCode", which accepts free-form sentences mixed with English and Java language as inputs [7]. AnyCode, based on statistical usage patterns such as the Java syntax and type, generates executable statements written in the Java language. On the other hand, a technique to summarize source codes in natural language has also been attempted [8]. Such studies are helpful for understanding and searching for source codes, which are important tasks not only for software engineering but also for the applications of natural language processing.

In this paper, we propose a method for finding the source code structure strongly associated with method names. Programs written in the object-oriented language are generally developed in class units consisting of multiple data structures and methods. Because a highly structured source code is created by combining many classes and methods, the implicit knowledge in writing programs must be hidden behind the naming of the methods. We use method names in the set of source codes to create and label training data for machine learning problems. The training data is a set of examples that consist of programs named with a specific method name (positive examples) and those with other methods names (negative examples) and maintains a balance between the numbers of examples of two classes. We train a binary classification model with support vector machines (SVM) using the data [6]. We also apply a feature selection method to the SVM classification problem and extract structural features from programs with specific method names. Those structural features are designed to represent the features of the subtree of the abstract syntax tree (AST). We regard this task as a kind of graph mining method to find structural patterns from source codes.

2   Background

To create an application such as AnyCode, it is necessary to combine multiple fundamental modules such as source code search and suggestion. The widespread use of a large source code repository (e.g. GitHub) contributes to improve these fundamental modules. For example, for the code suggestion application, Raychev et al. calculate and apply a stochastic language model from the source code repository [10]. Nguyen et al. improve the predictability of the token-based n-gram language model by introducing semantic annotations in source codes [9].

Studies on applications that associate source code with NL such as source code search and code suggestion have also been undertaken. Iyer et al. propose a data-driven approach to generate NL summaries of source codes [8]. In addition, Allamanis et al. study to capture the problem of predicting method names from a given partial source code as an extreme summary of the source code [2,4]. These studies respond to the fact that the neural network method in natural language processing generation began to show remarkable results.

We focused on the naming of identifier names strongly related to program structures and functions. We constructed a synonym dictionary for method naming by applying singular value decomposition (SVD) to make groups of names from source code repositories [12]. Based on this dictionary, we also express a whole program by a labeled graph and predict functions of the partial code named by a specific method name from the graph structure [13].

In this paper, we apply the method proposed by Adachi et al. to program pattern mining of the source codes by associating the structural patterns of the source code to method names [1,11]. The approach proposed in this paper is close to the approach of Alon et al. [5]. Specifically, Alon et al. use the set of paths on AST (Fig. 1) parsed from the target source code as representing the source code features [5]. They claim that the path-based representation of AST is common to many programming languages. As an evaluation of the representation, they experimented with the prediction task of variable names and method names in programs in four languages such as JavaScript, Java, Python, C# in both CRF based learning, and Word2vec based learning. The proposed method in this paper also uses a set of paths on AST as a feature representation, which can be input to the SVM in a way different from that of Alon et al.

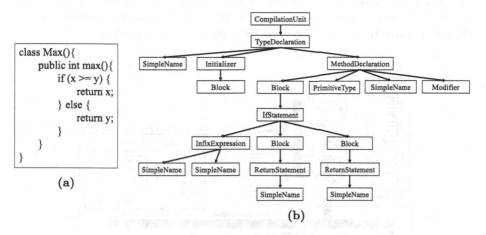

Fig. 1. An example of an abstract syntax tree corresponding to a source code; (a) a program and (b) an abstract syntax tree.

Allamanis et al. proposed to represent both syntactic and semantic structures of the source code using graphs [3]. They used graph-based deep learning methods to reason over program structures and experimented predicting the name of a variable, given its use as an application of their model. Their program representation showed the advantages of modeling known structure based on comparison to methods that use less structure. The difference between their work and our proposal in this paper is that our method deals with structural feature extraction from AST as a binary classification problem using SVM.

3   Algorithm

3.1   Main Components of Our Proposal

Our proposal consists of the following three viewpoints (Fig. 2). In Steps (1)–(3), using the features extracted from an AST, a source code is converted to a vector representation. In Steps (4)–(7), the feature set is selected, and the SVM is trained and evaluated. In Steps (8)–(9), a feature set with the highest binary classification result is acquired and visualized in a graph pattern.

Feature Representation. We propose a vector representation by extracting paths composed of nonterminal symbols from the AST.

Feature Selection. Extracting useful knowledge and patterns from huge data such as documents is a challenging task. Adachi et al. proposed a feature selection method that utilizes SVM for extracting useful knowledge from documents [1,11]. They build an SVM that performs binary classification with a binary feature as input. The feature selection is based on the output values of the SVM, which they refer to as SVM scores.

Visualization. With the feature selection by the SVM, a feature set (path set) is extracted for characterizing a program pattern. To assess a selected feature set, we visualize the selected optimal path set as a graph pattern.

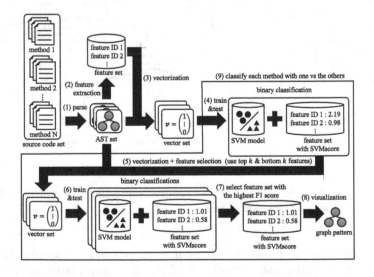

Fig. 2. Our proposal graph mining method overview

3.2   Feature Representation

In this paper, the paths that appear in an AST are treated as the features of the tree. This section explains our idea that the representation of source code is figured as a set of paths in its AST. First, we define the AST used in our proposal. Next, we delineate a set of paths in its AST (AST-path), which is followed by a vector representation by using the AST-path.

Definition 1 (Abstract syntax tree). *An AST for a source code c is a tuple $\langle N, V \rangle$, where N is a set of nonterminal nodes and V is a set of edges composed of nonterminal nodes.*

Definition 2 (AST-path). *An AST-path of length k is a tuple $\langle S, s \rangle$, where S is a sequence of the form: $n_1 n_2 ... n_{k-1}$, where for $i \in [1...k]$: $n_i \in N$ are nonterminals, and s is a SVM score.*

Definition 3 (vector representation). *A vector representation $v = \langle x_1, x_2, ..., x_n \rangle$ is an n-dimensional vector, where n is a number of AST-paths and x_n is a boolean value.*

Based on these definitions, Algorithm 1 shows how vectors are generated for a source code. Given feature set F, source code src, and path length k as the input, a vector V (path-vector) is the output. The AST is traversed with depth-first search and all nodes are visited (line 5). When the length of the *Stack* exceeds the path length k, the nodes are acquired as the feature from the top of the *Stack* (line 7). We check whether the feature exists in the dictionary and add its *id* to vector V if there is a feature (lines 9, 12).

3.3   Feature Selection

It is natural to select features to extract important them. In this section, we describe the feature selection method. We apply the SVM score proposed by Adachi et al. and evaluate the features. They calculated the SVM score by using the SVM perf implementing a linear kernel and we define the SVM score as follows:

Definition 4 (SVM score). *The SVM score is the output value of the trained SVM model after inputting one of the features used them at training, e.g., when one of AST-paths is $f_1 = \langle 1, 0, 0, ..., 0 \rangle$, we calculate the SVM score s as the follows: $s = w f_1^T + b$.*

Algorithm 2 shows our proposed feature selection method using the SVM score. We sort features in the descending order based on the SVM score (line 2). According to the absolute value, a number of k positive and negative features are acquired, respectively (lines 4, 5).

Algorithm 1. Source code to vector representation

Input: feature set F, source code src, and path length k.
Output: path-vector $\mathcal{V}$.
1: $\mathcal{V} \leftarrow \emptyset$
2: Stack $\leftarrow \emptyset$
3: $AST \leftarrow$ parse(src)
4: Traverse AST in pre-order (depth-first order)
5: **for** all nodes N traversed **do**
6:      Stack $\leftarrow$ add(N)
7:      **if** length(Stack) $\geq k$ **then**
8:          $f \leftarrow$ getPathDescendingOrder(Stack, k)
9:          **if** isExist(f, F) **then**
10:             $id \leftarrow F[f]$
11:         **end if**
12:         $\mathcal{V} \leftarrow \mathcal{V} \cup id$
13:     **end if**
14:     **if** isBacktrack(N) **then**
15:         Stack $\leftarrow$ remove(N)
16:     **end if**
17: **end for**

Algorithm 2. Feature Selection with SVM

Input: feature set with SVMscore F, and a number of select feature k.
Output: selected feature set SF.
1: $SF \leftarrow \emptyset$
2: $F \leftarrow$ sortSVMscore(F)
3: **for** all features i in k **do**
4:      $SF \leftarrow SF \cup$ getTopFeature(i) // positive feature
5:      $SF \leftarrow SF \cup$ getBottomFeature(i) // negative feature
6: **end for**

3.4 Visualization

In the general graph mining method, the frequency information of the appearing graph structures is used. We select features that contribute to binary classification using the SVM and construct a graph pattern from the acquired feature set. Because features are composed of path fragments extracted from the ASTs, it is assumed that an optimal feature set constitutes a part of the characteristic graph pattern. We define our proposed graph pattern composed of the feature set as follows:

Definition 5 (graph pattern). *A graph pattern is a tuple $\langle N, V, \phi \rangle$ where N is a set of nonterminal nodes, V is a set of edges composed of nonterminal nodes, and $\phi : V \rightarrow X$ is a function that maps an edge to an associated value.*

Algorithm 3 shows how to construct and to visualize the graph pattern. This algorithm receives as the input that the feature set with the most highest SVM score F as the input, and outputs a graph pattern. The threshold is 70% of the maximum SVM score, and we decided it based on the graph size visually judge-able (lines 4, 5). The collected features are divided for each edge, and the sum of the SVM scores is assigned to each edge (line 8). Thus, the graph pattern is constructed with the important feature set contributing to the binary classification.

Algorithm 3. Visualize Feature Set (Graph Pattern)

Input: feature set with the most highest SVMscore F.
Output: graph pattern G.
1: $G \leftarrow \emptyset$
2: $h \leftarrow \text{getHighScore}(F)$
3: **for** all features f in F **do**
4:    **if** $\text{getScore}(f) \geq 0.7 * h$ **then**
5:       $V \leftarrow \text{getEdges}(f)$
6:       **for** all edge v in V **do**
7:          **if** $\text{isExist}(v, G)$ **then**
8:             $G \leftarrow G \cup \text{sumScore}(v, G[v])$
9:          **else**
10:             $G \leftarrow G \cup v$
11:          **end if**
12:       **end for**
13:    **end if**
14: **end for**

---

4 Experiments

4.1 Experimental Setup

The research questions we assumed are the following:

RQ 1. Can we identify some specific method names from the others by their own structural pattern?
RQ 2. Can the proposed method extract a graph pattern characterizing a method name?

Approaching these research questions, we define the similarity between methods based on the selected feature set.

Definition 6 (graph similarity). *The similarity of two graph patterns g_1 and g_2 is defined as the follows:*

$$similarity(g_1, g_2) = \sum_{i \in V_{g1} \cap V_{g2}} \frac{1}{2} \left(\phi_{g1}(i) + \phi_{g2}(i) \right) \tag{1}$$

The data corpus used in these experiments is the source code set described in the Java language extracted by Allamanis et al. from GitHub. We collected 1,000 methods with a high frequency of occurrence and 1,000 samples for each method from the corpus (total one million samples). In the binary classification method, 1,000 samples included in the target method are taken as positive examples and five samples, from each method except for the target method, are taken as negative examples (4995 samples in total). In addition, the experiments were designed that all methods were included in the training data and verification data, and we performed 5-fold cross-validation. The evaluation method is a general F1 score used for binary classification.

4.2   Experimental Results

Figure 3 shows an overview of the binary classification results by using optimal features and all features. The number of optimal features is 1,000. In many cases, the identification result is a high score, and it is confirmed that the optimal feature is better than all features. Table 1 is an example of the binary classification result. It shows the top 40 and the bottom 40 method names. Almost all the top 40 method names are more than two words, whereas almost all the bottom 40 method names are single words. Therefore, it is suggested that the detailed method names may include their specific pattern in their program structure. These results provide evidence for an affirmative answer to **RQ1**. For some method names, we can identify a specific program structure from its corresponding name.

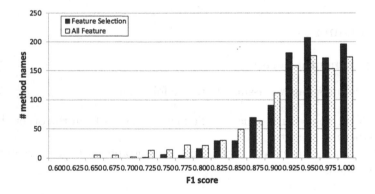

Fig. 3. A histogram of F1 scores obtained from each method by using optimal features and all features. The horizontal axis is the division of the F1 score and the vertical axis is the number of method names corresponding to the division. The range of the division is 0.025. A significant number of binary classification results are the high score. Further, in the case of the optimal features, the method names identified are higher.

We conducted a qualitative evaluation of method names by mapping the structural feature vector of each program example to the two-dimensional plane

Table 1. A binary classification list of F1 scores using optimal features sorted in the descending order: (a) top 40 names and (b) bottom 40 names. The larger the number of words contained in the method name, the higher the binary classification result.

<table>
<tr><td colspan="3" align="center">(a)</td><td colspan="3" align="center">(b)</td></tr>
<tr><th>Rank</th><th>Method name</th><th>F1</th><th>Rank</th><th>Method name</th><th>F1</th></tr>
<tr><td>1</td><td>eStaticClass</td><td>1.0000</td><td>961</td><td>complete</td><td>0.8120</td></tr>
<tr><td>2</td><td>tryConsume</td><td>1.0000</td><td>962</td><td>apply</td><td>0.8091</td></tr>
<tr><td>3</td><td>eIsSet</td><td>1.0000</td><td>963</td><td>find</td><td>0.8075</td></tr>
<tr><td>4</td><td>eInverseRemove</td><td>1.0000</td><td>964</td><td>bind</td><td>0.8074</td></tr>
<tr><td>5</td><td>createFollowerAfterReturn</td><td>1.0000</td><td>965</td><td>perform</td><td>0.8070</td></tr>
<tr><td>6</td><td>writeQName</td><td>1.0000</td><td>966</td><td>activate</td><td>0.8066</td></tr>
<tr><td>7</td><td>writeQNameAttribute</td><td>1.0000</td><td>967</td><td>compile</td><td>0.8053</td></tr>
<tr><td>8</td><td>writeQNames</td><td>1.0000</td><td>968</td><td>register</td><td>0.8024</td></tr>
<tr><td>9</td><td>getPullParser</td><td>1.0000</td><td>969</td><td>select</td><td>0.8013</td></tr>
<tr><td>10</td><td>isReaderMTOMAware</td><td>1.0000</td><td>970</td><td>handle</td><td>0.8011</td></tr>
<tr><td>11</td><td>buildPartial</td><td>1.0000</td><td>971</td><td>init</td><td>0.8007</td></tr>
<tr><td>12</td><td>removeSemanticListeners</td><td>1.0000</td><td>972</td><td>evaluate</td><td>0.8006</td></tr>
<tr><td>13</td><td>isApplicableAndPattern</td><td>1.0000</td><td>973</td><td>getData</td><td>0.8005</td></tr>
<tr><td>14</td><td>addSemanticListeners</td><td>1.0000</td><td>974</td><td>getValue</td><td>0.7987</td></tr>
<tr><td>15</td><td>getEditTextValidator</td><td>1.0000</td><td>975</td><td>connect</td><td>0.7982</td></tr>
<tr><td>16</td><td>performDirectEditRequest</td><td>1.0000</td><td>976</td><td>add</td><td>0.7980</td></tr>
<tr><td>17</td><td>getChildBySemanticHint</td><td>1.0000</td><td>977</td><td>filter</td><td>0.7959</td></tr>
<tr><td>18</td><td>eSet</td><td>0.9995</td><td>978</td><td>generate</td><td>0.7921</td></tr>
<tr><td>19</td><td>eUnset</td><td>0.9995</td><td>979</td><td>render</td><td>0.7914</td></tr>
<tr><td>20</td><td>_Fields</td><td>0.9995</td><td>980</td><td>login</td><td>0.7891</td></tr>
<tr><td>21</td><td>findByThriftId</td><td>0.9995</td><td>981</td><td>refresh</td><td>0.7887</td></tr>
<tr><td>22</td><td>findByThriftIdOrThrow</td><td>0.9995</td><td>982</td><td>validate</td><td>0.7873</td></tr>
<tr><td>23</td><td>fieldForId</td><td>0.9995</td><td>983</td><td>initialize</td><td>0.7867</td></tr>
<tr><td>24</td><td>dynamicQueryCount</td><td>0.9995</td><td>984</td><td>search</td><td>0.7854</td></tr>
<tr><td>25</td><td>createFollower</td><td>0.9995</td><td>985</td><td>delete</td><td>0.7849</td></tr>
<tr><td>26</td><td>parseDelimitedFrom</td><td>0.9995</td><td>986</td><td>save</td><td>0.7844</td></tr>
<tr><td>27</td><td>getSerializedSize</td><td>0.9995</td><td>987</td><td>start</td><td>0.7792</td></tr>
<tr><td>28</td><td>buildParsed</td><td>0.9995</td><td>988</td><td>get</td><td>0.7767</td></tr>
<tr><td>29</td><td>getThriftFieldId</td><td>0.9990</td><td>989</td><td>convert</td><td>0.7766</td></tr>
<tr><td>30</td><td>setBeanIdentifier</td><td>0.9990</td><td>990</td><td>send</td><td>0.7743</td></tr>
<tr><td>31</td><td>mergeFrom</td><td>0.9990</td><td>991</td><td>open</td><td>0.7705</td></tr>
<tr><td>32</td><td>onDialogEvent</td><td>0.9990</td><td>992</td><td>check</td><td>0.7654</td></tr>
<tr><td>33</td><td>handleNotificationEvent</td><td>0.9990</td><td>993</td><td>process</td><td>0.7533</td></tr>
<tr><td>34</td><td>getEditText</td><td>0.9990</td><td>994</td><td>execute</td><td>0.7481</td></tr>
<tr><td>35</td><td>dynamicQuery</td><td>0.9990</td><td>995</td><td>update</td><td>0.7388</td></tr>
<tr><td>36</td><td>getAccessibleEditPart</td><td>0.9990</td><td>996</td><td>build</td><td>0.7367</td></tr>
<tr><td>37</td><td>eGet</td><td>0.9985</td><td>997</td><td>prepare</td><td>0.7321</td></tr>
<tr><td>38</td><td>createRouteBuilder</td><td>0.9985</td><td>998</td><td>load</td><td>0.7311</td></tr>
<tr><td>39</td><td>createDefaultEditPolicies</td><td>0.9985</td><td>999</td><td>create</td><td>0.7302</td></tr>
<tr><td>40</td><td>fetchByPrimaryKey</td><td>0.9985</td><td>1000</td><td>resolve</td><td>0.7198</td></tr>
</table>

in the same way as Allamanis et al. did [2]. Figure 4 shows a classification map between getter methods (e.g., getId) and setter methods (e.g., setString). Regardless of the projection algorithm being linear or nonlinear, it is confirmed that it can separate the getter methods from the setter methods. This result emphasizes the usefulness of our proposal as a vector representation of AST and feature selection for the representation. Allamanis et al. employed terminal symbols in AST as the feature to predict method names. In contrast, we did not use terminal symbols in AST; instead, we used paths that consisted only of nonterminal symbols in it as features. In Fig. 4, we obtained the same results as Allamanis did.

Figure 5 shows an example of a graph pattern in the getter method (getId) and the setter method (setString). The getId method has an IfStatement and a pair of a ReturnStatement and a NumberLiteral. This is generally the structure required to obtain the necessary numeric id. Furthermore, a setString method has a pair of a SingleVariableDeclaration and a SimpleType, and a pair of a ReturnStatement and BooleanLiteral. This is necessary to confirm if the typed values could be stored. These graph patterns are the most important patterns according to the defined Algorithm 3. It is suggested that these patterns represent each method name well.

Table 2 shows an example of similar method names for the getter method (getId) and the setter method (setString). The getId method result is the method representing the meaning "get value", such as getVersion, getLocation, and getDescription. The setString method result is the method representing the meaning "store value", such as updateStatus, store, and receive. These results provide some evidence for an affirmative answer to **RQ2**.

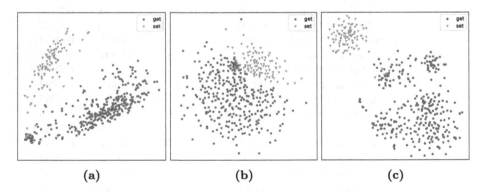

(a)	(b)	(c)

Fig. 4. Plotting getter and setter methods on a 2D linear space using (a) principal component analysis (PCA), (b) multidimensional Scaling (MDS), and a 2D non-linear projection using (c) t-SNE.

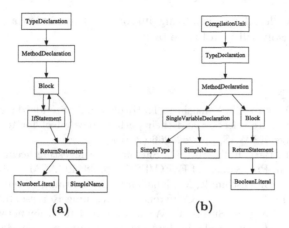

Fig. 5. Visualised graph patterns of (a) getId method and (b) setString method. These graph patterns consist of structures that can be inferred from method names.

Table 2. List of similar method names inferred for (a) the getId method and (b) the setString method. The similarity measure was calculated based on our proposed structural feature.

<table>
<tr><td colspan="2" align="center">(a)</td><td colspan="2" align="center">(b)</td></tr>
<tr><td>rank</td><td>similar method name</td><td>rank</td><td>similar method name</td></tr>
<tr><td>1</td><td>getVersion</td><td>1</td><td>complete</td></tr>
<tr><td>2</td><td>getLocation</td><td>2</td><td>updateStatus</td></tr>
<tr><td>3</td><td>getDescription</td><td>3</td><td>store</td></tr>
<tr><td>4</td><td>getStatus</td><td>4</td><td>dispatch</td></tr>
<tr><td>5</td><td>getKey</td><td>5</td><td>move</td></tr>
<tr><td>6</td><td>getConfig</td><td>6</td><td>receive</td></tr>
<tr><td>7</td><td>getContext</td><td>7</td><td>begin</td></tr>
<tr><td>...</td><td>...</td><td>...</td><td>...</td></tr>
<tr><td>997</td><td>refreshUnderline</td><td>997</td><td>performDirectEditRequest</td></tr>
<tr><td>998</td><td>performDirectEditRequest</td><td>998</td><td>getEditTextValidator</td></tr>
<tr><td>999</td><td>writeQNames</td><td>999</td><td>refreshUnderline</td></tr>
</table>

5   Conclusion

In this paper, we examined ties between mined structural patterns in source codes and their identifier names. The main difference between the recent major works such as Alon et al. and Allamanis et al. and our proposal is that our method constructs the learning data for the binary classification for a specific method name. Using our method, one can apply an elaborated feature selection technique to the classification modeling based on the supervised data to know the significance of the structural features. One of the advantages of our method, compared to aforementioned works, is that it is possible to assess the structural features in more detail by visualizing important features as a graph. Further detailed comparison will be a future task.

References

1. Adachi, Y., Onimura, N., Yamashita, T., Hirokawa, S.: Standard measure and SVM measure for feature selection and their performance effect for text classification. In: Proceedings of iiWAS, pp. 262–266 (2016)
2. Allamanis, M., Barr, E.T., Bird, C., Sutton, C.: Suggesting accurate method and class names. In: Proceedings of ESEC/FSE, pp. 38–49. ACM (2015)
3. Allamanis, M., Brockschmidt, M., Khademi, M.: Learning to represent programs with graphs. In: International Conference on Learning Representations (2018)
4. Allamanis, M., Peng, H., Sutton, C.: A convolutional attention network for extreme summarization of source code. In: International Conference on Machine Learning, pp. 2091–2100 (2016)
5. Alon, U., Zilberstein, M., Levy, O., Yahav, E.: A general path-based representation for predicting program properties. In: Proceedings of the 39th ACM SIGPLAN Conference on PLDI, pp. 404–419. ACM (2018)
6. Cortes, C., Vapnik, V.: Support vector networks. Mach. Learn. **20**(3), 273–297 (1995)
7. Gvero, T., Kuncak, V.: Synthesizing java expressions from free-form queries. In: Proceedings OOPSLA, pp. 416–432 (2015)
8. Iyer, S., Konstas, I., Cheung, A., Zettlemoyer, L.: Summarizing source code using a neural attention model. In: Proceedings AMACL, pp. 2073–2083. ACL (2016)
9. Nguyen, T.T., Nguyen, A.T., Nguyen, H.A., Nguyen, T.N.: A statistical semantic language model for source code. In: Proceedings ESEC/FSE, pp. 532–542. ACM (2013)
10. Raychev, V., Vechev, M., Yahav, E.: Code completion with statistical language models. In: Proceedings of the 35th ACM SIGPLAN Conference on PLDI, pp. 419–428. ACM (2014)
11. Sakai, T., Hirokawa, S.: Feature words that classify problem sentence in scientific article. In: Proceedings iiWAS, pp. 360–367 (2012)
12. Yamashita, H., Takeuchi, K., Hashimoto, K.: Word usage in programming codes for software repository mining. In: Proceedings ACIS, pp. 351–357 (2014)
13. Yamashita, H., Takeuchi, K., Hashimoto, K.: Resolving functional ambiguities in labeled graph representation of programs: an application of dictionary construction based on software repository mining. In: Proceedings KICSS, pp. 536–545 (2015)

Big Data and Machine Learning for Economic Cycle Prediction: Application of Thailand's Economy

Chukiat Chaiboonsri and Satawat Wannapan[✉]

Puey Ungphakorn Centre of Excellence in Econometrics, Faculty of Economics,
Chiang Mai University, Chiang Mai, Thailand
chukiat1973@gmail.com, lionz1988@gmail.com

Abstract. Since traditional econometrics cannot guarantee that the parametric estimation based on some of time-series variables provides the best solution for economic predictions. Interestingly, combining with mathematics, statistics, and computer science, the big data analysis and machine learning algorithms are becoming more and more computationally highlighted. In this paper, 29 yearly collective factors, which are qualitative information, quantitative trends, and social movement activities, are employed to process in three machine learning algorithms such as k-Nearest Neighbors (kNN), Tree models and random forests (RF), and Support vector machines (SVM). Technically, collective variables using in this paper were observed from the source agents who successfully accumulated data details from trends of the world for easily accessing, for instance, Google Trends or World Bank Database. With advanced artificial calculations, the empirical result is very precise to real situations. The predicting result also clearly shows Thailand economy would be very active (peak) in the upcoming quarters. Consequently, this advanced artificial learning successfully done in this paper would be the new approach to helpfully provide policy recommendations to authorities, especially central banks.

Keywords: Macroeconomics · Machine learning · Big data ·
Econometric forecast

1 Introduction

It is inevitable to state that the world is closed to be completely united by an enormous amount of data. Every day we talk, criticize, analyze, and even mobilize. These activities create a huge category of information, which is interestingly implied to humans' evolutional thinking. However, the problem is information, frequently referred as "big data". It is extremely vast and difficult to observe and predict simultaneously. Additionally, the study on big data calculations essentially needs more than one science and one innovation. Consequently, it seems to be very rare that the big data analysis is considered into the area of social science, especially in fiscal and monetary economics.

Since more than one hundred years economists have been relied on only the traditional estimating assumption, which is to fix non-considerable variables to be

© Springer Nature Switzerland AG 2019
H. Seki et al. (Eds.): IUKM 2019, LNAI 11471, pp. 347–359, 2019.
https://doi.org/10.1007/978-3-030-14815-7_29

constant. Graphically, this assumption is just one of sciences used for data analyses, which is displayed in Fig. 1. It is not enough and makes a cause for many researchers have to put more assumptions to guarantee that their estimating models can computationally fit for data. Furthermore, this unmitigated mistake is being still employed to many econometric predictions and used to provide policy recommendations to authorities, especially central banks. Accordingly, to escape from it, Machine Learning (ML) is invented to efficiently combine with big data analyses, and this would be the solution for clarifying and forecasting the economy's trends, hidden signs, and crises in the modern era of academic researches.

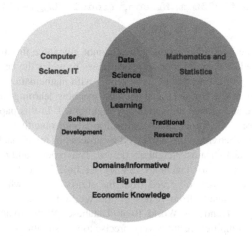

Fig. 1. The conceptual framework of research

Considering into central banks, it is obvious their outcomes are also microeconomic decisions and interactions. These responsibilities came with the collection and access to a wealth of new data sources, which moves central banks into the realm of big data (Chakraborty and Joseph 2017). Historically, applying data science investigations and machine learning coined in 1959 by Samuel (1959) is rare in econometrics for economic researches. For central banks, however, these data computing analyses are becoming more highlighted. For example, the paper of Bholat (2015) was proposed to situate topics within the context of strategic plans and initiatives and summarized the article linking central banks' emerging interest in Big Data approaches. The letter of Bank of Italy proposed by Signorini (2018) was mentioned the actual and potential value of big data is more and more crucial for economic researches. Moreover, the academic issue stated by Daniel Hinge (2017) was expressed that machine learning may not yet be at the stage where central bankers are being replaced with robots, but the field is recently bringing powerful tools to bear on big economic questions. Accordingly, there is no reason why big data and machine learning cannot be the possibly suitable solution for monetary-econometric researches in Thailand.

2   The Objective and Scope of Research

The fundamental aim of this paper is to computationally predict Thailand's economic structural trends by applying big data and machine learning. Interestingly, mixed observations such as qualitative survey details and time-trend data series during 2004 to 2017 are being employed to do an econometric estimation by artificial intelligent approaches. All of variables are described and presented in Table 1. Technically, collective variables using in this paper were observed from the reliable source agents who successfully accumulated information from trends of the world for convenient accessing, for instance, Google Trends, World Bank Database, or YouTube Searching Engine.

Table 1. The details of collective information used to data science analyses reference from the Thailand's Key Macroeconomic (Identified by BOT) and Big data from Google Trends database

Variable Detail	Symbol	Time-series range	Source
GDP (growth rate)	RBC_GDP	2004–2017	The World Bank Database
Population (growth rate)	POP	2004–2017	The World Bank Database
Unemployment (% change)	UN_EM	2004–2017	The World Bank Database
Industrial value added (% per GDP)	IND	2004–2017	The World Bank Database
Agricultural value added (% per GDP)	AGR	2004–2017	The World Bank Database
Gross domestic investment (% per GDP)	GDI	2004–2017	The World Bank Database
Service imports (% per GDP)	SER_IM	2004–2017	The World Bank Database
Service exports (% per GDP)	SER_EX	2004–2017	The World Bank Database
Military expenditures (% per GDP)	MIL	2004–2017	The World Bank Database
Public debt (% per GDP)	PU_DEP	2004–2017	The World Bank Database
Consumer price index (% per annual)	CPI	2004–2017	The World Bank Database
Foreign direct investment (US$)	FDI	2004–2017	The World Bank Database
Available lands (km^2)	LAND	2004–2017	The World Bank Database
Preserved forests (km^2)	FOR	2004–2017	The World Bank Database

(continued)

Table 1. (*continued*)

Variable Detail	Symbol	Time-series range	Source
International reserves	INT	2004–2017	The World Bank Database
Fixed interest rate	FIX_I	2004–2017	The World Bank Database
Exchange rate (Baht per US$)	EX	2004–2017	The World Bank Database
Official development assets (US$)	ODA	2004–2017	The World Bank Database
Values of stock market trading	MKT	2004–2017	The World Bank Database
Overall Thailand economic situations	Big_data 1	2004–2017	Google shopping database
Investment situations	Big_data 2	2004–2017	Google shopping database
Stock market situations	Big_data 3	2004–2017	Google shopping database
Thailand business movements	Big_data 4	2004–2017	Google shopping database
Employments	Big_data 5	2004–2018	Google shopping database
Thailand international trades and investments	Big_data 6	2004–2019	News search
Thailand agricultural situations	Big_data 7	2004–2020	New search
Thailand industrial situations	Big_data 8	2004–2017	Image search
Thailand banking situations	Big_data 9	2004–2018	Google shopping database
Thailand political atmospheres	Big_data 10	2004–2019	YouTube search
Thailand social atmospheres	Big_data 11	2004–2020	YouTube search

3   Theory and Methodology

3.1   The Fundamental Concept of Machine Learning in Data Science

With complex components to perform the machine learning computation, there are essential approaches described as follows: firstly, decision tree learning is used to make a decision tree as a predictive model, which maps observations about an item in order to conclude about the item's target value. Second, an artificial neural network (ANN) learning algorithm is employed to model complex relationships between inputs and outputs or to capture the statistical structure in an unknown joint probability distribution between collective variables. Third, inductive logic programming (ILP) is an approach used to rule learning using logic programming as a uniform representation for input samples, background knowledge, and hypotheses. Forth, support vector machines (SVMs) are an array of related supervised learning methods used for classification and regression. Fifth, clustering analysis is a method of unsupervised learning, and a common technique for statistical data analysis. Sixth, Bayesian networks are a belief network that presents a set of random variables and their conditional independencies via a directed acyclic graph (DAG). This modern statistics can efficiently performs algorithm exist for inference and learning. Seventh, representation learning (RL) is employed to disentangle the underlying factors of variation that explain the observed data (Bengio 2009). Eighth, a genetic algorithm (GA) is a search heuristic that mimics the process of natural selection, which is used to improve the performance of genetic and evolutionary algorithms (Zhang et al. 2011). Ninth, rule-based machine learning is a contrast model to other machine learners that commonly verifies a singular model for predicting (Bassel et al. 2011). Rule-based machine learning approaches include learning classifier systems, association rule learning, and artificial immune systems. Tenth, feature selection approach is employed to select an optimal subset of relevant features for use in model construction.

3.2 Algorithms for Machine Learning Analyses

(a) k-Nearest Neighbors (kNN)
k-Nearest Neighbors (kNN) is a non-parametric method for both classification and regression problems (Chakraborty and Joseph 2017). Data is performed as its k-nearest to others in the feature space. For solving the problem, data observations are assigned to the majority class of its nearest observations, for example, the mean value of its nearest neighbors in the regression model. The performance of k-NN can be analyzed by focusing on the error rate of miss-classified examples for classification or squared errors for regression problems, respectively. The output value for a single sample x_i is achieved as following the steps, which are neighbor selection and value assignment. The former is the calculation of the distance x_i to all other points in the feature space. The stage determines its k closet neighbors $\{x_j\}_i^k$. Euclidean distance is commonly used as a distance measure. The latter is conducted to assign the output y_i the class membership (i.e. $y_i \in \{C_1, C_2, \ldots, C_c\}$ $\forall i = 1, 2, \ldots, n$, where c is a number of class layers) by the majority vote of its k nearest neighbors. The k-NN regression, y_i, is delegated to be the average value of its single nearest neighbor, $y_i = 1/k$ $\sum_{j=1; x_i \in \{x_j\}_i}^{k} x_j$. One can be also considered as a distance-weighted average.

(b) Support vector machines (SVM)
Support vector machines (SVM) are a powerful technique for both classification and regression type problems (Chakraborty and Joseph 2017). Generally, two-class classification problems are modeled by logistic regressions (Logit model), the position to hyperplane in the feature space is projected to a (0,1) interval, which can be defined as probabilities of class membership. Computationally, An SVM will try to find a decision boundary to separate these two classes by the maximal margin, represented by the gray area (as shown in Fig. 2). The separating boundary (black line), which maximizes the

margin, is, however, not straightforward in the general case. Thus, the concept behind SVMs is now two-fold. First, to solve a presentation of the feature spaces in which the data is linearly separable and, second, to identify the points in the input space which define, or support, the maximal margin, the support vectors. As seen in Fig. 2, the separating boundary, decision rule and error function for two-class classification can be expressed as

$$
\begin{array}{ll}
\left|x_{SV}^T \cdot \beta\right| \equiv 1 & \text{support vectors} \\
h(x_i, \beta) = sign\left(x_i^T.\beta\right) & \text{hypothesis} \\
ERR(X, Y, \beta) = -\frac{1}{2m}\sum_{i=1}^{m}\left(|y_i - h(x_i, \beta)|\right) & \text{error function.}
\end{array}
$$

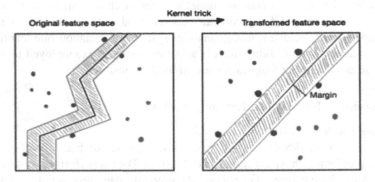

Fig. 2. Schematic presentation of a two-class (green and red dots) support-vector classifier in a two-dimensional feature space. (Color figure online)

The coefficients β define the hyperplane satisfying the decision boundary equation. The support vectors lying on the boundary of the gray area and the target values, y_i, take the values $y_i \in -1, 1$. The coefficients β can be provided by solving the (dual) optimization problem:

$$
L(\alpha) = \sum_{i=1}^{m} \alpha_i - \frac{1}{2}\alpha^T.H.\alpha,
$$

$$
\text{with } \beta = \sum_{i=1}^{m} \alpha_i y_i x_i, \; H_{ij} \equiv y_i y_j x_i^T .x_j, \tag{1}
$$

under the constraints $\alpha.Y = 0$ and $\alpha_i \geq 0$. Considering into the Lagrangian $L(\alpha)$, it focuses though there is a large number of free parameters α_i, namely one for each observation x_i. The input data X enters the Lagrangian only via an inner product, resulting in a scalar. This allows to define transformations $T(.)$ and inner products via a so-called *kernel*,

$$K(\hat{x}, x') = \langle T(x), T(x') \rangle. \tag{2}$$

Commonly, the *radial basis function* (Gaussian kernel) relied on the polynomial expansion of the exponential function is employed to be the choice for a kernel function,

$$K(\hat{x}, x') = \exp\left[-\gamma \|x - x'\|^2\right]. \tag{3}$$

(c) Tree models and random forests

Tree models are one of popular non-parametric machine learning techniques for regression and classification problems. Conceptually, the training dataset based on the input features until assignment criterion with respect to the target variable is consecutively divided into a data basket (leaf). The purpose is to minimize the entropy $H(Y|X)$ (objective function) within areas of the baskets conditioned on the features X. At the beginning, the full set X of m observations is identified for conditioning on the features x, which leads to the highest information gain (I), and this can be expressed as follows:

$$I(Y|x) = H(Y|X) - \sum_{v \in \{x\}} \frac{|X_v|}{|m|} H(Y\}X_v) \qquad \text{information gain (classification)}$$

$$H(Y|X) = -\sum_{c=1}^{C} p(Y = c|X) \log(p(Y = c|X)) \quad \text{entropy (classification)}$$

$$H(Y|X) = \frac{1}{m} \sum_{j=1}^{k} \sum_{i=1}^{m_j} \left(y_i - \mu_j|_{x_i \in X_j}\right)^2 \qquad \text{MSE (regression)},$$

where $p(Y = c|X)$ is the connected frequency of class c observations in X. $|X_v|$ is the set of observations which take on each value. In a regression setting, the entropy can be replaced by the mean squared error (MSE) and splits are performed along the dimensions which most reduce the error (Galton 1907). A schematic representation of a tree model with two features, x_1 and x_2, is displayed in Fig. 3.

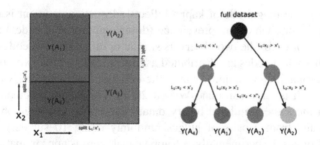

Fig. 3. Schematic representation of a tree model. Left: the target data space is systematically segregated by the tree model based on the input features. Right: a tree representation of the final model. The tree is grown from the top to the bottom. (Modified from Chakraborty and Joseph 2017)

3.3 Model Validation

Similar to econometric models, it is difficult to give the best result from the beginning when doing the computation of machine learning systems. Adjustments and robustness checking are essential in most cases. In this paper, Cohen's kappa coefficient is employed to validate the best prediction from machine learning algorithms. Cohen's kappa coefficient (κ) is a statistic tool measuring inter-rater agreement for qualitative (categorical) items. It is mostly implied to be a more robust measurement than simple percent agreement calculation. However, the complexity of Cohen's kappa interpretation is the obstruction to use this coefficient. Some researchers have suggested that it is conceptually simpler to evaluate disagreement between items (Pontius and Millones 2011). The calculation of Cohen's kappa measurements is to verify the agreement between two raters who each classify N items into C mutually exclusive categories. The coefficient (κ) is defined as:

$$\kappa \equiv \frac{p_0 - p_e}{1 - p_e} = 1 - \frac{1 - p_0}{1 - p_e}, \tag{4}$$

where p_0 is the relative observed agreement among raters, which identifies to accuracy, and p_e is the hypothetical probability of chance agreement. The observed data is used to calculate the probabilities of each randomly observable seeing in each category. If (κ) equals one, this implies that the raters are in complete agreement. On the other hand, if (κ) equals zero then here is no agreement among the raters other than what would be expected by chance. Cohen's kappa coefficient is possible to be negative. This implies that there is no effective agreement between the two raters or the agreement is worse than random. For categories k, a number of items N and n_{k_i} is the number of time raters i predicted category k,

$$p_e = \frac{1}{N^2} \sum_k n_{k1} n_{k2}. \tag{5}$$

Interestingly, the magnitude of kappa reflects adequate agreement is depended on two important factors, which are prevalence (doing equiprobable codes) and bias (the marginal probabilities for the two observers similar or different). For codes, Kappas are obviously higher when codes are distributed asymmetrically by the two observers. This is contrast to probability variations, which the effect of bias is greater when Kappa is small than when it is large (Sim and Wright 2005). The Kappa calculation shown in Fig. 4 is from the same simulated binary data. Each point on the graph is computationally generated from a pairs of judges randomly rating 10 subjects for having a diagnosis of X or not. Experimentally, a Kappa equals zero is approximately equivalent to an accuracy = 0.5.

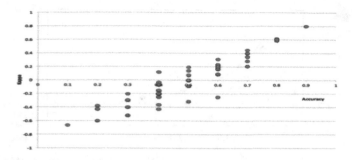

Fig. 4. Kappa accuracy and coefficient presentation

4 Empirical Results

4.1 Descriptive Information

First of all, taking consideration into Thailand economic trends, yearly collective GDP during 2004 to 2017 is displayed in Fig. 5. The classification of the trend obviously shows that Thailand economy is dramatically fluctuated. This fluctuation causes the prediction to become more difficult, doing traditionally econometric tools alone cannot precisely provide the best model. In particularly, general means are not reliable anymore. Consequently, the optimal algorithm calculation called Newton Method is employed to extend the ability of data explanations.

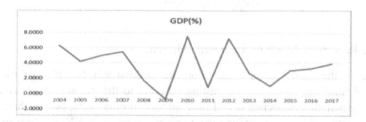

Fig. 5. The presentation of economic trends in Thailand during 2004 to 2017

To mathematically solve the issue, the basic idea of Newton's method is typically used as a basis on linearization. Graphically, the convergence processing of Newton's method is approximated by its tangent line, which is already gone closer to the root at x* (as shown in Fig. 6). Empirically, the result represents that the real middle value of data observations is quite different from the mean (Seen details in Table 2).

From the result of the optimal value, the economic trend can be clarified as a period of real business cycles. The peak period is defined at the level, which is higher than the optimal value (3.55%). Expansion is in the interval between 2.1% and 3.5%. Recession belongs to the interval between 0% and 2%, and fall is defined as the level below 0%. The details are demonstrated in Table 3.

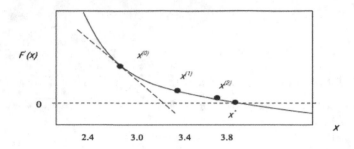

Fig. 6. Presentation the convergence processing in Newton-optimal approach (Modified from Satawat et al. 2018)

Table 2. The comparison between the optimal value and normal mean of yearly Thailand GDP

Variables (Growth rate)	General mean	Optimal value (Newton's method)
GDP	3.67%	3.55%

Table 3. The demonstration of real business cycles in Thailand economic trends

Growth rate	Description
>3.55%	Peak
2.1% to 3.5%	Expansion
0% to 2%	Recession
<0%	Fall

4.2   The Results of Machine Learning Algorithm

This stage is the model comparison. For the comparison of different models, an approach of training, validation and testing suitable to the given dataset are adopted. Three different such approaches would raise an alert if either one, two or three or more outliers are detected within the four model features such as k-NN, random forest, and SVM, respectively. The results are shown in Table 4. In this computation, the random forest (rf) is the best model and contains the highest parametric values when selecting by both the values of accuracy and Kappa coefficient, which are 0.7 and 0.4167, respectively. Moreover, this comparison result is graphically displayed in Fig. 7.

Obviously, as the random forest model is chosen to be the algorithm predictor for Thailand macroeconomic variables, the forecasting result strongly confirms that Big Data using in this computational paper can provide a very sensibility in a similar way to the real situation, more precisely than traditional econometric estimations. The details are represented in Table 5. Empirically, the predicting result clearly shows Thailand economy would be very active (peak) in the upcoming quarters. In other words, the rate of economic expansion would be 4.5%, higher than the optimal value (3.55%) during the third quarter to fourth quarter.

Table 4. Comparative performance statistics of various machine learning models for predicting Thailand macroeconomic variables

Method	Cross-validation	The final values	Accuracy	Kappa's coefficient
Random forest (rf)	Cohen's kappa	k = 9	0.7	0.4167
k-NN (kn)	Cohen's kappa	c = 0.25 sigma = 1.286084e-18	0.6	0
SVM (svm)	Cohen's kappa	mtry = 15	0.5417	0

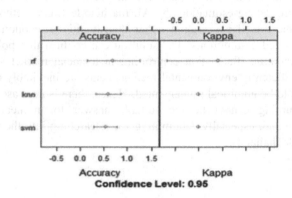

Fig. 7. The model selection in machine learning systems by Cohen's kappa

Table 5. The overall prediction from three machine learning algorithms

Years	RBC_GDP (Current values)	RBC_GDP (Description)	RF*** (Prediction)	KNN (Prediction)	SVM (Prediction)
2018(Q1)	(4.8%)	Peak	*Peak*	*Peak*	*Peak*
2018(Q2)	(4.2%–4.7%)	Peak	*Peak*	*Peak*	*Peak*
2018(Q1–Q4)f	(4.5%)	Peak	*Peak*	*Peak*	*Peak*

5 Conclusion

The huge challenge to apply the advanced computational method, as called "machine learning algorithms" for predicting the big data in macroeconomic variables is successfully done in this paper. Since it is totally different from traditionally parametric estimations and is more powerful. The machine learning systems can capture an enormous amount of informative details in databases, including qualitative data, quantitative factors, and even time-series trends. In this paper, 29 time-series variables regarding to Thailand economic structures during 2004 to 2017 were included to predict the sensible upcoming trend.

With the minimization of modeling assumptions, machine learning systems can efficiently calculate both stationary and non-stationary data. Methodologically, the data was processed into two sections, algorithm calculations and model selections.

Empirically, the random forest algorithm is the best model selected by Cohen's kappa coefficient. Consequently, the predicting result clearly shows Thailand economy would be very active (peak) in the upcoming quarters (2018q3 to 2018q4).

To interpret this computational result and apply to policy recommendations, it is obvious that the big data using in this paper dose not includes only economic factors, but it contains social variables such as political activities, environmental issues, and even IT social behaviors. With machine learning algorithms, the result which is concluded that Thailand economy would be the sunshine time is extremely reliable much more than traditional econometric models for time-series forecasting. Since the complexity in multi-processing estimations by AI, machine learning methods can explain the outliners in the mixed observation rather than traditional econometric methods, which inevitably need assumptions. The solution can confirm that political stability, social network updating, fluctuation controlling in financial market systems, online news, and even efficiently environmental managements are inevitably connected, and these points have to be empirical implemented. Hence, there is no reason why big data and machine learning cannot be the suitable answer for monetary-econometric researches in Thailand, especially data mining for researching into the responsibilities of BOT (Bank of Thailand).

References

Bassel, G.W., Glaab, E., Marquez, J., Holdsworth, M.J., Bacardit, J.: Functional network construction in Arabidopsis using rule-based machine learning on large-scale data sets. Plant Cell 23(9), 3101–3116 (2011)

Bholat, D.: Big data and central banks. Q. Bull. Q1, pp. 86–93 (2015). https://www.researchgate.net/publication/276101527_Big_Data_and_central_banks

Bishop, C.M.: Pattern Recognition and Machine Learning. Springer, New York (2006). ISBN 0-387-31073-8

Chakraborty, C., Joseph, A.: Machine learning at central banks. Staff Working Paper No. 647. Bank of England (2017)

Galton, F.: Vox populi. Nature 75, 450–451 (1907)

Hinge, D.: Big Data in Central Banks. Published by Infopro Digital Services Ltd, Central Banking Publications, London (2017)

Landis, J.R., Koch, G.G.: The measurement of observer agreement for categorical data. Biometrics 33(1), 159–174 (1977). https://doi.org/10.2307/2529310

Pontius, R., Millones, M.: Death to Kappa: birth of quantity disagreement and allocation disagreement for accuracy assessment. Int. J. Remote Sens. 32, 4407–4429 (2011)

Samuel, A.L.: Some studies in machine learning using the game of checkers. In: Levy, D.N.L. (ed.) Computer Games, pp. 335–365. Springer, New York (1988). https://doi.org/10.1007/978-1-4613-8716-9_14

Sim, J., Wright, C.C.: The Kappa statistic in reliability studies: use, interpretation, and sample size requirements. Phys. Ther. 85, 257–268 (2005)

Signorini, L.F.: Harnessing big data & machine learning technologies for central banks. The Printing and Publishing Division, Bank of Italy, Rome (2018)

Wannapan, S., Chaiboonsri, C., Sriboonchitta, S.: Macro-econometric forecasting for during periods of economic cycle using bayesian extreme value optimization algorithm. In: Kreinovich, V., Sriboonchitta, S., Chakpitak, N. (eds.) TES 2018. SCI, vol. 753, pp. 706–723. Springer, Cham (2018). https://doi.org/10.1007/978-3-319-70942-0_51

Zhang, J., et al.: Evolutionary computation meets machine learning: a survey. IEEE **6**(4), 68–75 (2011)

A Comparative Study on SOM-Based Visualization of Potential Technical Solutions Using Fuzzy Bag-of-Words and Co-occurrence Probability of Technical Words

Yasushi Nishida(✉) and Katsuhiro Honda

Graduate School of Engineering, Osaka Prefecture University,
Sakai, Osaka 599-8531, Japan
nishida@iao.osakafu-u.ac.jp, honda@cs.osakafu-u.ac.jp

Abstract. Self-Organizing Maps (SOM) is a powerful tool in visualizing mutual connection among various objects. In a previous work, SOM-based visualization was applied for revealing potential technical solutions varied in Japanese patent documents, in which meaningful pairs of technical words are implied in SOMs. Before application, text documents were quantified into numerical vectors considering co-occurrence frequency among technical words in sentences, and then, SOMs were constructed summarizing word features of co-occurrence probability vectors or correlation coefficient vectors. Recently, a fuzzy bag-of-words model was proposed for handling sparse characteristics of word feature values and shown to be useful in document classification. In this paper, a comparative study on utilizing fuzzy bag-of-words in conjunction with previous feature values is performed with the goal of revealing potential technical solutions varied in patent documents.

Keywords: Patent documents · Technical solution ·
Fuzzy bag-of-words

1 Introduction

Self-Organizing Maps (SOM) [1] is a powerful tool in visualizing mutual connection among various objects and has been utilized in many fields. In a previous work [2], SOM-based visualization was utilized in patent document analysis with the goal of supporting inspiration of potential solving means by extracting representative words characterizing past technical solutions in Japanese patent document data. Meaningful pairs of technical words were expected to be implied in SOMs for revealing potential technical solutions varied in the sentences of *means for solving the problem* section in each publication of patent applications.

Before typical document analysis applications, text documents are often quantified into numerical vectors based on the Bag-of-Words (BoW) concept

© Springer Nature Switzerland AG 2019
H. Seki et al. (Eds.): IUKM 2019, LNAI 11471, pp. 360–369, 2019.
https://doi.org/10.1007/978-3-030-14815-7_30

considering frequencies of basis terms (keywords) in each document [3]. Term-frequency-inverse document frequency (TF-IDF) [4] can be also considered for weighting the frequency degrees so as to degrade the responsibility of common terms. In [2], a BoW model was generated considering co-occurrence frequency among technical words in sentences, and then, SOMs were constructed summarizing word features of co-occurrence probability vectors or correlation coefficient vectors. The experimental results successfully demonstrated that co-occurrence probability vectors were utilized for revealing connections, which were directly utilized in existing patents, while correlation coefficient vectors were expected to be useful for revealing potential connections, which can be utilized in future developments.

Recently, a Fuzzy Bag-of-Words (FBoW) [5] model was proposed for handling sparse characteristics of word feature values. In general document analysis tasks, it is often the case that a document only contains a very small portion of all basis terms and we may fail to reveal intrinsic connection among basis terms from sparse BoW vectors because the conventional BoW counts the frequencies of terms with the hard mapping function. FBoW is a fuzzy extension of the conventional BoW based on a fuzzy mapping function such that it introduces vagueness in the matching between words and the basis terms. Fuzzy mapping enables a word semantically similar to a basis term to be activated in the BoW model and was shown to be useful in document classification.

From the comparative viewpoint, co-occurrence-based word features of Ref. [2] are expected to have similar characteristics to FBoW representation but they are constructed with different processes. In this paper, a comparative study on utilizing FBoW in conjunction with previous feature values is performed with the goal of revealing potential technical solutions varied in patent documents. The remaining parts of this paper are organized as follows: Sect. 2 briefly review the process of SOM-based visualization of potential technical solutions in patent documents and the FBoW model. The results of a comparative experiment are presented in Sect. 3 and a summary conclusion is given in Sect. 4.

2   Brief Review on SOM-Based Visualization of Potential Technical Solutions and Fuzzy Bag-of-Words Model

2.1   Preprocessing

In [2], 6213 publications of patent applications in Japan were extracted as patent document data to be analyzed. These publications of patent applications were stored by a patent search tool on May 9, 2017 based on "AND" search under the following two conditions:

Condition 1: Num. of *information provision* is more than 1.
Condition 2: Application date is January 1, 2005 or later.

Information provision is a systematic process for providing evidence information. A third party can provide the examiner of the patent office such information that a patent may have no novelty or no inventive step of the patented invention. In many cases, *information provision* is awakened from competitors, who want to prevent the right of the corresponding invention. In this sense, the patent having *information provision* is expected to be "an invention considered to be a threat for competitors", and the above condition was adopted for extracting many useful inventions in various technical fields.

With the goal of finding potential technical solution means, the sentences of *means for solving the problem* section in each publication of patent applications were quantified based on the BoW concept as follows: The sentences of patent documents were divided into word units by performing morphological analysis utilizing MeCab software, where the MeCab version of mecab-ipadic-NEologd [6] was used. Here, each sentence was regarded as a datum, which is characterized by a BoW vector with the frequencies of occurrences of nouns, and then, we had total 55358 sentences.

In order to eliminate meaningless terms, words having the occurrence frequency of 30% or more are regarded as general words and are filtered out. Words having the frequency of 0.3% or less are also filtered out because such rare words may contain some typos or be coined words. Additionally, in order to reduce the dimensionality, 100 words having the highest TF-IDF scores were selected as representative words.

Finally, the numerical data matrix to be analyzed is composed of 55358 × 100 elements with 55358 sentences and 100 terms.

2.2   SOM-Based Visualization of Connection Among Technical Terms

In order to intuitively represent the mutual connection among technical terms, SOMs are constructed considering the features of 100 representative terms, where the characteristic features among technical words are summarized into term × term feature matrices. In [2], two different types of feature summarization were adopted.

Co-occurrence Probability Vectors. We can expect that meaningful pairs of elements (words) such as an important constituent and its behavior often co-occur in the *means for solving the problem* in the same patent documents. Therefore, word co-occurrence information is available for revealing mutual connection among technical words in various technical solution means varied in conventional patent documents.

In order to focus on co-occurrence frequency among technical words in sentences, word feature vectors in the original BoW matrix are processed into co-occurrence probability vector among technical words. That is, a data matrix to be analyzed is constructed by calculating the vectors for all representative words, where each row corresponds to a co-occurrence probability vector containing co-occurrence degrees with the other representative words. Technical words having similar co-occurrence characteristics are expected to have similar vectors.

Each element of a co-occurrence probability vector is calculated as follows: Let the number of sentences be L, the number of co-occurrences of word w_i and word w_j in sentence l be $N_{ij}(l)$, and the total appearances of word w_i in all sentences be M_i. Then, the co-occurrence probability P_{ij} of words w_i and w_j is given as:

$$P_{ij} = \frac{\sum_{l=1}^{L} N_{ij}(l)}{\sum_{l=1}^{L} M_i(l)} \tag{1}$$

Co-occurrence probability degrees P_{ij} are merged into a matrix P_{word} considering all n representative words, where row elements correspond to co-occurrence probability vectors $\boldsymbol{P_{w1}}, ..., \boldsymbol{P_{wn}}$ of the representative words as:

$$P_{word} = \begin{pmatrix} P_{11} & P_{12} & \cdots & P_{1n} \\ P_{21} & P_{22} & \cdots & P_{2n} \\ \vdots & & \ddots & \vdots \\ P_{n1} & P_{n2} & \cdots & P_{nn} \end{pmatrix} = \begin{pmatrix} \boldsymbol{P_{w1}} \\ \boldsymbol{P_{w2}} \\ \vdots \\ \boldsymbol{P_{wn}} \end{pmatrix} \tag{2}$$

Correlation Coefficient Vectors. The above feature vectors are further processed for revealing intrinsic connection among technical words, which is not directly found in certain patent documents but can be varied in related documents. While each element of word co-occurrence probability vectors represent the frequency of the combinatorial use of representative words in conventional patents, it is expected that the correlation coefficient of co-occurrence probability vectors can be utilized for finding the potential combination of technical word pairs. Therefore, a high correlation coefficient of two co-occurrence probability vectors may imply a potential pair of technical words because they may be related to common technical issues even when the co-occurrence probability of the two words is not so high.

Correlation coefficient C_{ij} of two co-occurrence probability vectors $\boldsymbol{P_{wi}}$ and $\boldsymbol{P_{wj}}$ of representative words w_i and w_j can be expressed as follows:

$$C_{ij} = \frac{\sum_{x=1}^{n}(P_{ix} - \overline{P_{wi}})(P_{jx} - \overline{P_{wj}})}{\sqrt{\sum_{x=1}^{n}(P_{ix} - \overline{P_{wi}})^2}\sqrt{\sum_{x=1}^{n}(P_{jx} - \overline{P_{wj}})^2}} \tag{3}$$

where $\overline{P_{wi}}$ is the arithmetic mean of co-occurrence probabilities of word w_i.

The correlation coefficient for each word is summarized into a symmetric matrix C_{word}, where row elements of correlation coefficient matrix C_{word} correspond to correlation coefficient vectors $\boldsymbol{C_{w1}}, ..., \boldsymbol{C_{wn}}$ of the representative words.

$$C_{word} = \begin{pmatrix} C_{11} & C_{12} & \cdots & C_{1n} \\ C_{21} & C_{22} & \cdots & C_{2n} \\ \vdots & & \ddots & \vdots \\ C_{n1} & C_{n2} & \cdots & C_{nn} \end{pmatrix} = \begin{pmatrix} C_{w1} \\ C_{w2} \\ \vdots \\ C_{wn} \end{pmatrix} \tag{4}$$

2.3　Fuzzy Bag-of-Words

In order to tackle the sparse feature of the conventional BoW model, Fuzzy BoW was proposed considering the fuzzy mapping function [5]. As is implied in Sect. 2.1, the conventional BoW model derives the frequency of each representative term by counting the number of exact word matching. Assume that $A_{t_i}(w)$ is a mapping function and the term frequency of a basis term t_i is calculated by summing up $A_{t_i}(w)$ for all words w in a sentence. Exact word matching is equivalent to employing the following membership function:

$$A_{t_i}(w) = \begin{cases} 1, \text{ if } w \text{ is } t_i \\ 0, \text{ otherwise} \end{cases} \tag{5}$$

On the other hand, the FBoW model adopts semantic matching or fuzzy mapping to project the words occurred in documents to the basis terms. To implement semantic matching, a fuzzy membership function is given as follows:

$$A_{t_i}(w) = \begin{cases} \cos(W[t_i], W[w]), \text{ if } \cos(W[t_i], W[w]) > 0 \\ 0, \hspace{3.5cm} \text{otherwise} \end{cases} \tag{6}$$

where $W[w]$ denote word embeddings for word w, and are utilized in estimating mutual semantic similarities among words. In [5], word2vec [7] was adopted in word embeddings, and the fuzzy membership degree that measures the similarity between attribute (words in documents) and set (basis terms in BoW space) was approximated by their cosine similarity score.

Then, the numerical vector representation z of a document under fuzzy BoW model is given by

$$z = xH, \tag{7}$$

where x is a vector composed of the number of occurrence of words and H is composed of fuzzy memberships of Eq. (6) as $H = \{A_{t_i}(w_j)\}$. Supported by the fuzzy representation, elements of the original (sparse) BoW matrix are extended to the similarity-weighted sum of occurrence frequencies of their related words, and the FBoW representation is expected to have richer information than the original (sparse) BoW representation.

In the next section, the utility of the FBoW model is compared in SOM-based visualization of potential technical solutions by adopting the FBoW vectors instead of the conventional BoW vectors in Sect. 2.1

3   Comparative Experiments

Comparative experiments were performed with the goal of intuitively extracting element pairs for acceleration of innovation through visualization of the relationships among technical elements of the solution means.

Before analysis, 6213 publications of patent applications in Japan were decomposed into 59048 sentences, and 941 words (nouns) were extracted through morphological analysis by MeCab [6]. In the followings, 100 representative words, which most frequently appeared, were used for analysis. Then, the BoW representation was constructed as a 59048×100 matrix, whose 96.9% elements are zero.

In the comparative approach of [2], the original (sparse) BoW representation was used as co-occurrence information, which also has 96.9% zero elements. Then, the co-occurrence probability vectors or the correlation coefficient vectors to be inputted into SOM were constructed.

On the other hand, in the FBoW-based approach, the original (sparse) BoW representation was extended into a FBoW representation considering the 100-dimensional word embeddings $W[w]$ for 100 representative words given by word2vec, which was performed with window size 5 in the Continuous Bag-of-Words (CBOW) model. Supported by such fuzzy imputation, the FBoW matrix had 8.3% zero elements only. Then, the co-occurrence probability vectors or the correlation coefficient vectors to be inputted into SOM were constructed. Here, be noted that the word2vec representation was used only for estimating mutual semantic similarities among words, i.e., Eq. (6), in FBoW matrix construction and was not utilized in the following SOM implementation.

SOM was implemented in Python Programming Language with the TensorFlow [8] framework, where the parameters were set as follows: SOM map size is 40×40, the learning rate coefficient α is 0.4, the neighborhood radius is 20, the learning frequency is 500 times, and the number of learning is 500. The initial values of SOM is given with random values drawn from uniform distributions, and the similarity degrees are calculated with Euclidean distances.

3.1   Conventional SOM Visualization

First, the conventional model of [2] was used for constructing SOM visualization with the original (sparse) BoW representation. The following sections introduce SOMs having most popular characteristics in 10 trials with different initializations.

The SOMs constructed with co-occurrence probability vectors and correlation coefficient vectors are shown in Figs. 1 and 2, respectively [2]. Some intuitively typical connections are emphasized with enclosures for illustrative purposes.

As emphasized in Fig. 1, the co-occurrence probability-based approach is useful for revealing the direct connections utilized in existing patents such as "飲料, 殺菌, 濃度 (beverage, sterilization, density)" and "温度, 時間 (temperature, time)".

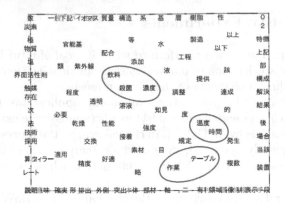

Fig. 1. SOM of co-occurrence probability vectors [2]

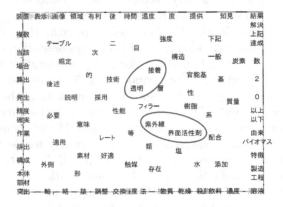

Fig. 2. SOM of correlation coefficient vectors [2]

On the other hand, the correlation coefficient-based approach is useful for revealing some intrinsic connections, which were concealed in Fig. 1. For example, "紫外線, 界面活性剤 (ultraviolet, surfactant)" seem not to have a direct connection but we could find their intrinsic connections with paint technology in some patent documents although the word was filtered out in preprocessing. Possible connection among "接着, 透明 (glueing, transparency)" is also implied.

Beside such useful features of revealing potential connections, the conventional model may fail to find other promising connections because the original BoW representation is often so sparse that some intrinsic connections can be concealed by many zero elements. Then, in the followings, the effects of utilizing other approaches such as FBoW representation are compared.

3.2 SOM of word2vec Representation

In FBoW construction, word2vec representation was utilized only in word similarity calculation, while word2vec representation itself is also a candidate of rich
representation of word features. Then, before applying FBoW representation to
SOM visualization of co-occurrence tendencies, for a comparison, SOM of the
word2vec representation is investigated as shown in Fig. 3. The word2vec representation is expected to be useful for revealing the general characteristics of
word relations in the patent documents, and then, we can find plausible connection such as "上記, 下記 (as stated above, as follows)" and "材料, 素材 (material,
ingredient)", which could not be found in Figs. 1 and 2.

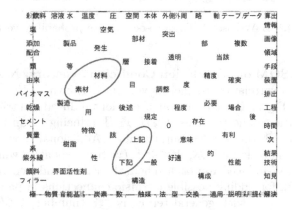

Fig. 3. word2vec-induced SOM

However, with the goal of finding the connection in a certain solution mean,
the word2vec-induced SOM may present only trivial relations and we should
further investigate considering the intrinsic co-occurrence tendencies. Therefore,
in the following, FBoW-induced SOM visualization is applied.

3.3 FBoW-Based SOM Visualization

In this subsection, the FBoW-based model was used for constructing SOM visualization.

FBoW-Based SOM of Co-occurrence Probability Vectors. The FBoW-
induced SOM with co-occurrence probability vectors is shown in Fig. 4. As
emphasized in enclosures, such typical connections as "水, セメント (water,
cement)" are newly implied while the previous connections of Fig. 1 were concealed. It may be because the FBoW-induced SOM located many general terms
such as "説明 (description)", "適用 (apply)" and "採用 (adopt)" in the center
area, which appear in almost all patent documents. That is, the fuzzy imputation model of FBoW can contribute to revealing some intrinsic but important
connections while many general terms are also overemphasized.

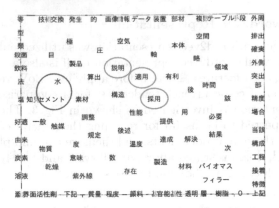

Fig. 4. FBoW-based SOM of co-occurrence probability vectors

FBoW-Based SOM of Correlation Coefficient Vectors. Next, the FBoW-based SOM with correlation coefficient vectors is shown in Fig. 5. In the similar manner to Fig. 2, the correlation coefficient-induced contributes to finding such intrinsic connections as "接着, 透明 (glueing, transparency)" and "紫外線, 界面活性剤 (ultraviolet, surfactant)". Additionally, we can also find such a typical connection as "水, セメント (water, cement)", which was newly found in the FBoW-induced approach. Then, the FBoW-induced approach seems to be more useful for finding possible intrinsic connections in paint technology.

However, in Fig. 5, such general terms as "数 (number)", "次 (next)" and "後述 (see below)" were also located in the center area. In order to improve the interpretability of the center area, which is supposed to present most typical features, other word2vec representations can be adopted by rejecting the influences of such non-technical words, in future works.

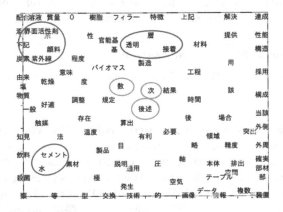

Fig. 5. FBoW-based SOM of correlation coefficient vectors

4   Conclusion

This paper presented a comparative study on SOM-based visualization of technical words in patent documents for revealing potential solving means, which can be used in potential innovation. Besides the conventional approach of utilizing BoW representations, FBoW representation was newly adopted in constructing the word vectors to be used as SOM inputs.

Supported by fuzzy imputation of sparse BoW representation, the FBoW-induced SOM contributed to finding intrinsic potential connections of technical terms. However, many general terms were tend to locate in the center area and might conceal other potential connections.

A possible future work is to improve the interpretability of the center area of SOM by modifying word2vec or FBoW representations so that the influences of non-technical words are rejected. In [5], the FBoW model was combined with clustering for reducing the dimensionality of FBoW representations, where similar words are merged into clusters. Such clustering-based summarization may be a promising way of improving the FBoW-induced SOM visualization. Combination with the TF-IDF concept may also be a potential approach. Another direction of future study can be noise term rejection in SOM implementation.

Acknowledgment. This work was achieved through the use of large-scale computer systems at the Cybermedia Center, Osaka University.

References

1. Kohonen, T.: Self-Organizing Maps, 3rd edn. Springer, Heidelberg (2000). https://doi.org/10.1007/978-3-642-56927-2
2. Nishida, Y., Honda, K.: Visualization of potential technical solutions by self-organizing maps and co-cluster extraction. In: Joint 10th International Conference on Soft Computing and Intelligent Systems and 19th International Symposium on Advanced Intelligent Systems, pp. 820–825 (2018)
3. Lan, M., Tan, C.L., Su, J., Lu, Y.: Supervised and traditional term weighting methods for automatic text categorization. IEEE Trans. Pattern Anal. Mach. Intell. **31**(4), 721–735 (2009)
4. Salton, G., Buckley, C.: Term-weighting approaches in automatic text retrieval. Inf. Process. Manage. **24**(5), 513–523 (1988)
5. Zhao, L., Mao, K.: Fuzzy bag-of-words model for document representation. IEEE Trans. Fuzzy Syst. **26**(2), 794–804 (2018)
6. Sato, T.: Neologism dictionary based on the language resources on the web for MeCab (2015). https://github.com/neologd/mecab-ipadic-neologd
7. Mikolov, T., Chen, K., Corrado, G., Dean, J.: Efficient estimation of word representations in vector space. In: International Conference Learning Representations (2013). https://arxiv.org/pdf/1301.3781.pdf
8. Abadi M., et al.. TensorFlow. large-scale machine learning on heterogeneous systems. https://www.tensorflow.org/

Fuzzy Co-clustering for Categorization of Subjects in Questionnaire Considering Responsibility of Each Question

Katsuhiro Honda$^{(\boxtimes)}$, Ruixin Yang, Seiki Ubukata, and Akira Notsu

Graduate School of Engineering, Osaka Prefecture University,
Sakai, Osaka 599-8531, Japan
{honda,subukata,notsu}@cs.osakafu-u.ac.jp

Abstract. Categorization of subjects is a basic approach for summarizing the results of various questionnaires. Co-clustering is a technique for simultaneous co-clustering of mutually familiar objects and items such that each co-cluster is formed by the subject group with their typical questions. This research considers such a situation that a questionnaire is designed for finding some pre-defined categories, which are characterized by some typical questions, but some questions may not be necessarily fit to the target categories. Then, fuzzy co-clustering is performed in conjunction with evaluation of the responsibility of each question for the target categorization. The proposed fuzzy memberships are constructed under a hierarchical structure of category characterization and question evaluation. The characteristic feature of the proposed method is demonstrated through numerical experiments with an artificial data set.

Keywords: Fuzzy clustering · Co-clustering · Questionnaire analysis

1 Introduction

In general survey researches, categorization of subjects is a basic approach for summarizing the results of various questionnaires. With the goal of revealing the characteristics of subjects, a questionnaire with many questions can be performed such that each subject is designed for selecting *yes* or *no* with some typical features of the pre-defined categories. Then, the questionnaire result can be summarized into a rectangular relational matrix of *subjects* (object) × *questions* (items).

Co-clustering is a technique for simultaneous clustering of mutually familiar objects and items such that each co-cluster is formed by the subject group with their typical questions. Fuzzy co-clustering has been utilized in many application fields such as document-keyword analysis [1], customer-product analysis [2] and gene expression analysis [3], where the goal is to estimate two different types of cluster memberships of objects and items such that pairs of objects and items have large membership values in same clusters when their cooccurrence degrees

© Springer Nature Switzerland AG 2019
H. Seki et al. (Eds.): IUKM 2019, LNAI 11471, pp. 370–379, 2019.
https://doi.org/10.1007/978-3-030-14815-7_31

are high. Fuzzy clustering for categorical multivariate data (FCCM) [4] and its variants [1,5] adopted the aggregation degree of co-clusters as the objective function to be maximized with fuzzification penalty terms and optimized two types of memberships based on the fuzzy c-means (FCM) [6]-like iterative algorithm.

Although the conventional fuzzy co-clustering handled all items (questions) with equal responsibilities, some questionnaires can be performed for categorizing subjects into some pre-defined categories, whose characteristics were already known. This research considers such a situation that a questionnaire is designed for finding some pre-defined categories, which are characterized by some typical questions. If all questions were well constructed such that each subject exactly belongs to a certain pre-defined category and each question is exactly related only to a designed category, we can find the subject category by simply counting the questions, which the subject matched. However, in general surveys, some questions may not be necessarily fit to the target categories and some subjects may belong to the boundaries of categories.

Then, this paper proposes a novel fuzzy co-clustering model, which can estimate not only the familiar pairs of *subjects* (object) × *questions* (items) but also the responsibility of each question for the target categorization. The proposed fuzzy memberships are constructed under a hierarchical structure of category characterization and question evaluation, where two different types of item memberships are defined such as category typicality and intra-category responsibility while object membership is remained like the conventional fuzzy co-clustering. The characteristic feature of the proposed method is demonstrated through numerical experiments with an artificial data set.

The remaining parts of this paper are organized as follows: Sect. 2 briefly review the conventional fuzzy co-clustering and Sect. 3 proposed a novel fuzzy co-clustering model. The experimental results are presented in Sect. 4 and a summary conclusion is given in Sect. 5.

2   FCM-Type Fuzzy Co-clustering

Assume that we have $n \times m$ cooccurrence information $R = \{r_{ij}\}$ among n objects and m items, where r_{ij} ($r_{ij} > 0$) represents the connectivity degree among object i and item j. For example, in document-keyword analysis, r_{ij} can be the frequency of keyword j in document i. The goal of fuzzy co-clustering is to find the co-clusters of familiar pairs of objects and items by simultaneously estimating fuzzy memberships of objects u_{ci} and items w_{cj} such that mutually familiar objects and items tend to have large memberships in the same cluster considering the aggregation degree of each co-cluster.

The sum of aggregation degrees to be maximized is defined as [4]:

$$L = \sum_{c=1}^{C} \sum_{i=1}^{n} \sum_{j=1}^{m} u_{ci} w_{cj} r_{ij}, \tag{1}$$

where C is the number of clusters. Here, object memberships u_{ci} have a similar role to those of FCM under the same condition, such that $\sum_{c=1}^{C} u_{ci} = 1$ implies u_{ci} to be identified with the probability of belongingness of object i to cluster C.

If item memberships w_{cj} also obey a similar condition of $\sum_{c=1}^{C} w_{cj} = 1$, the aggregation criterion has a trivial maximum of $u_{ci} = w_{cj} = 1, ^\forall i, j$ in a particular cluster c. Then, in order to avoid the trivial solution, w_{cj} are forced to be exclusive in each cluster, such that $\sum_{j=1}^{m} w_{cj} = 1$, and so, w_{cj} represent the relative typicalities of items in each cluster.

In [4], the linear aggregation criterion of Eq. (1) is non-linearized with respect to u_{ci} and w_{cj} by entropy-based penalties [7] for fuzzification of two-types of memberships as:

$$L_{fccm} = \sum_{c=1}^{C} \sum_{i=1}^{n} \sum_{j=1}^{m} u_{ci} w_{cj} r_{ij}$$

$$-\lambda_u \sum_{c=1}^{C} \sum_{i=1}^{n} u_{ci} \log u_{ci} - \lambda_w \sum_{c=1}^{C} \sum_{j=1}^{m} w_{cj} \log w_{cj}, \qquad (2)$$

where λ_u and λ_w are the fuzzification weights for object and item memberships, respectively. The larger the weight $\lambda.$ is, the fuzzier the partition is.

As an alternative approach, Kummamuru et al. [1] extended FCCM by introducing the quadric term-based fuzzification mechanism [8] instead of the entropy-based fuzzification, so that it can handle larger data sets. Another fuzzy co-clustering approach was induced by the statistical co-clustering of multinomial mixture models (MMMs), whose pseudo-log-likelihood function has a similar form to the FCCM criterion. Since the soft partition of MMMs is also achieved by an entropy-based penalty, Honda et al. [5] proposed an MMMs-induced fuzzy co-clustering with an adjustable fuzziness penalty.

3 Fuzzy Co-clustering for Categorization of Subjects in Questionnaire

3.1 Proposed Objective Function

If the goal of a questionnaire survey is to categorize subjects into some pre-defined categories, questions should be designed for connecting each subject to typical features of the categories such that a certain subject group is expected to select *yes* for the questions related to the typical features. In such cases, co-clustering may be performed considering connections among subjects and categories rather than among subjects and each question. However, if a questionnaire is not designed with enough qualities, some questions may not have clear connection with the target category and should be eliminated from analysis. Therefore, the responsibility of each question should also be evaluated.

In this section, a novel fuzzy co-clustering model is proposed considering a hierarchical structure of category characterization and question evaluation. Assume that we can expect that each subject is drawn from one of p pre-defined categories and category k ($k = 1, \ldots, p$) is characterized by m_k typical features. For example, subjects can be categorized into p-types of personalities, where a type of personality (type k) is characterized with m_k behavioral pattern. For category type k, the answers of n subjects are summarized into $n \times m_k$ cooccurrence information $R_k = \{r_{ij}^k\}$ among n objects and m_k items, where r_{ij}^k ($r_{ij}^k > 0$) represents the typicality degree of object i with item j in type k. Then, they are merged into p matrices $R_1, \ldots, R_p$ for all p-types.

Here, two different types of item memberships are defined such as category typicality and intra-category responsibility. Each cluster can be composed of subjects from some categories, where their features are characterized by intra-category questions.

z_{ck} is the membership degree of category k in cluster c and is utilized for finding co-clusters of familiar pairs of subjects and categories. Additionally, w_{cj}^k is the membership degree of question j of category k in cluster c and is utilized for ignoring meaningless questions in category k. Then, the proposed objective function is defined as follows:

$$L = \sum_{c=1}^{C} \sum_{i=1}^{n} \sum_{k=1}^{p} \sum_{j=1}^{m_k} u_{ci} z_{ck} w_{cj}^k r_{ij}^k$$

$$-\lambda_u \sum_{c=1}^{C} \sum_{i=1}^{n} u_{ci} \log u_{ci} - \lambda_w \sum_{c=1}^{C} \sum_{k=1}^{p} \sum_{j=1}^{m_k} w_{cj}^k \log w_{cj}^k$$

$$-\lambda_z \sum_{c=1}^{C} \sum_{k=1}^{p} z_{ck} \log z_{ck}, \tag{3}$$

where λ_z is the fuzzification penalty for category memberships z_{ck}.

Following the FCCM concept, each membership is calculated under the following constraints:

$$\sum_{c=1}^{C} u_{ci} = 1, \ \forall i, \quad \sum_{k=1}^{p} \sum_{j=1}^{m_k} w_{cj}^k = 1, \ \forall c, \quad \sum_{k=1}^{p} z_{ck} = 1, \ \forall c. \tag{4}$$

We can expect that subject i has large membership u_{ci} in cluster c when he/she belongs to type k with large z_{ck}, whose characteristic is featured by question j with large w_{cj}^k. In this way, subject categorization is performed by considering the responsibility of each question in a questionnaire survey.

3.2 Updating Rules

Considering the optimal condition for each memberships, the updating rules for three-types of fuzzy memberships are given as follows:

$$u_{ci} = \frac{\exp\left(\frac{1}{\lambda_u} \sum_{k=1}^{p} \sum_{j=1}^{m_k} z_{ck} w_{cj}^k r_{ij}^k\right)}{\sum_{\ell=1}^{C} \exp\left(\frac{1}{\lambda_u} \sum_{k=1}^{p} \sum_{j=1}^{m_k} z_{\ell k} w_{\ell j}^k r_{ij}^k\right)}, \tag{5}$$

$$w_{cj}^k = \frac{\exp\left(\dfrac{1}{\lambda_w}\sum_{i=1}^{n} u_{ci}z_{ck}r_{ij}^k\right)}{\sum_{\ell=1}^{p}\sum_{s=1}^{m_k}\exp\left(\dfrac{1}{\lambda_w}\sum_{i=1}^{n} u_{ci}z_{c\ell}r_{is}^\ell\right)},\tag{6}$$

and

$$z_{ck} = \frac{\exp\left(\dfrac{1}{\lambda_z}\sum_{i=1}^{n}\sum_{j=1}^{m_k} u_{ci}w_{cj}^k r_{ij}^k\right)}{\sum_{\ell=1}^{p}\exp\left(\dfrac{1}{\lambda_z}\sum_{i=1}^{n}\sum_{j=1}^{m_\ell} u_{ci}w_{cj}^\ell r_{ij}^\ell\right)}.\tag{7}$$

The FCM-like algorithm iteratively updates the tree-types of fuzzy memberships until convergence.

4   Experimental Results

4.1   Experimental Data Set

In this study, a comparative experiment was performed with a toy questionnaire data set, which is composed of answers for 12 questions given by 15 subjects and is designed such that the larger the value of the answer is, the greater the degree of fitness to the question becomes. Figure 1 shows the experimental data set to be analyzed, where the goal is to partition the subjects into three types, each

		Question (Yes : 3、 No : 1、 Neither : 2)											
		type 1				type 2				type 3			
		1-1	1-2	1-3	1-4	2-1	2-2	2-3	2-4	3-1	3-2	3-3	3-4
o b j e c t	1	3	2	3	1	2	1	1	1	1	1	2	2
	2	2	3	3	3	1	1	1	2	1	2	1	3
	3	3	3	2	2	1	1	2	1	2	1	1	1
	4	3	2	3	1	2	1	1	3	1	1	2	3
	5	2	3	3	3	1	2	1	2	1	2	1	2
	6	1	1	2	2	3	2	3	1	2	1	1	1
	7	1	2	1	3	2	3	3	3	1	1	1	2
	8	2	1	1	1	3	3	2	2	1	1	2	1
	9	1	1	2	3	3	2	3	1	2	1	1	3
	10	1	2	1	2	2	3	3	3	1	2	1	2
	11	2	1	1	1	1	1	2	2	3	2	3	1
	12	1	1	1	2	1	2	1	3	2	3	3	3
	13	1	1	2	1	2	1	1	1	3	3	2	2
	14	2	1	1	3	1	1	2	3	3	2	3	1
	15	1	2	1	2	1	2	1	2	2	3	3	3

Fig. 1. Toy questionnaire data

of which is supposed to be characterized by four typical questions. For example, objects 1 to 5 are supposed to be in type 1, where they are generally typical in questions 1 to 4. In the same manner, objects 6 to 10 and 11 to 15 are related to type 2 and type 3, respectively.

Besides object categorization, this experiment has another purpose of evaluating the typicality of each question for the type under the co-clustering concept. In Fig. 1, the fourth question is a meaningless item, which is irrelevant to type classification, and so, it should be ignored in object partitioning. Therefore, this experiment aims not only to partition objects into the types but also to evaluate the significance of each question considering the typicality in each type under the goal of designing an optimal questionnaire with rejection of meaningless questions.

4.2   Conventional FCCM Partition

First, for a comparison, the data set of Fig. 1 was processed by the conventional FCCM algorithm, where all questions are equally handled without category information and the goal is to extract co-clusters of familiar object and item pairs by estimating object memberships u_{ci} and item memberships w_{cj}^k. This situation is equivalent to the proposed method, where all type memberships z_{ck} are fixed as $z_{ck} = 1$ without considering type classification.

Table 1 shows the derived fuzzy memberships u_{ci} and w_{cj}^k, where fuzzification penalties were set as $\lambda_u = 0.6$ and $\lambda_w = 0.95$, respectively. Object memberships u_{ci} indicate that 15 objects were fairly partitioned into the designed 3 types. On the other hand, item memberships w_{cj} imply that meaningless questions (Questions 4, 8, and 12) worked like noise such that they were slightly connected to types, which were not designed for their target types. In other words, when questionnaires are not properly designed and contain meaningless questions, which do not fit type features, the conventional method may not work well for category characterization.

Table 1. Fuzzy memberships u_{ci} and w_{cj}^k in FCCM ($\lambda_u = 0.6$, $\lambda_w = 0.95$)

(a) Object membership u_{ci}

type k		k=1					k=2					k=3			
object i	1	2	3	4	5	6	7	8	9	10	11	12	13	14	15
cluster c 1	0.794	0.868	0.611	0.778	0.839	0.141	0.007	0.113	0.162	0.074	0.064	0.061	0.275	0.058	0.085
2	0.064	0.061	0.275	0.059	0.085	0.794	0.868	0.611	0.778	0.839	0.141	0.07	0.113	0.162	0.074
3	0.141	0.07	0.113	0.162	0.074	0.064	0.061	0.275	0.059	0.085	0.794	0.868	0.611	0.778	0.839

(b) Item membership w_{cj}

type k		k=1				k=2				k=3		
item j	1	2	3	4	5	6	7	8	9	10	11	12
cluster c 1	0.127	0.146	0.581	0.031	0.003	0.001	0.001	0.011	0.001	0.003	0.002	0.088
2	0.001	0.003	0.002	0.088	0.127	0.146	0.581	0.031	0.003	0.001	0.001	0.011
3	0.003	0.001	0.001	0.011	0.001	0.003	0.002	0.088	0.127	0.146	0.581	0.031

4.3 Partitions by Proposed Algorithm

Next, the proposed method was implemented for object-item partitioning in conjunction with evaluation of responsibility of each question.

Plausible Result. Table 2 shows the derived fuzzy memberships, which seems to be performed with plausible parameter setting.

Table 2. Fuzzy membership u_{ci}, w_{cj}^k and z_{ck} in the proposed method ($\lambda_u = 0.6$, $\lambda_w = 0.95$ $\lambda_z = 0.7$)

(a) Object membership u_{ci}

type k	k=1					k=2					k=3				
object i	1	2	3	4	5	6	7	8	9	10	11	12	13	14	15
cluster c 1	**0.813**	**0.895**	**0.683**	**0.803**	**0.885**	0.131	0.006	0.096	0.136	0.057	0.055	0.043	0.219	0.060	0.057
2	0.055	0.043	0.219	0.060	0.057	**0.813**	**0.895**	**0.683**	**0.803**	**0.885**	0.131	0.006	0.096	0.136	0.057
3	0.131	0.006	0.096	0.136	0.057	0.055	0.043	0.219	0.060	0.057	**0.813**	**0.895**	**0.683**	**0.803**	**0.885**

(b) Item membership w_{cj}^k

type k	k=1				k=2				k=3			
item j	1	2	3	4	5	6	7	8	9	10	11	12
cluster c 1	**0.151**	**0.173**	**0.644**	*0.029*	0.000	0.000	0.000	0.000	0.000	0.000	0.000	0.000
2	0.000	0.000	0.000	0.000	**0.151**	**0.173**	**0.644**	*0.029*	0.000	0.000	0.000	0.000
3	0.000	0.000	0.000	0.000	0.000	0.000	0.000	0.000	**0.151**	**0.173**	**0.644**	*0.029*

(c) Type membership z_{ck}

type k	k=1	k=2	k=3
cluster c 1	**1.000**	0.000	0.000
2	0.000	**1.000**	0.000
3	0.000	0.000	**1.000**

From the object partition viewpoint, we can see that 15 objects were fairly partitioned into the designed three clusters such that all objects have their maximum u_{ci} in plausible clusters, and type memberships z_{ck} imply that each cluster is properly related only to the designed type. On the other hand, from the item partition viewpoint, item membership w_{cj}^k shows that the fourth question of each type (Questions 4, 8 and 12) was rejected with a smaller membership than the well-designed ones. Additionally, item memberships of other types became much smaller than those of Table 1, i.e., item memberships were estimated considering not only the typicality of each question itself but also the plausibility of the corresponding target types. In the target type, the responsibility of the required question was emphasized while that of noise questions was suppressed. So, the typicality of questions were fairly evaluated under the hierarchical structure of category characteristics.

This result indicates the following two aspects: (i) We have surely three designed types of subjects under the co-cluster concept. (ii) The questionnaire used in this experiment is not necessarily designed well and some questions are not fit to the types. In this way, the proposed method is useful not only for confirming the type existence but also for evaluating the quality of the questionnaire.

Table 3. Fuzzy membership u_{ci}, w_{cj}^k and z_{ck} in the proposed method ($\lambda_u = 0.5$, $\lambda_w = 0.5$ $\lambda_z = 0.7$)

(a) Object membership u_{ci}

type k	$k=1$					$k=2$					$k=3$				
object i	1	2	3	4	5	6	7	8	9	10	11	12	13	14	15
cluster c 1	**0.861**	**0.853**	**0.827**	0.401	**0.852**	0.084	0.020	0.022	0.084	0.020	0.114	0.017	0.430	0.019	0.137
2	0.021	0.129	0.075	0.543	0.129	0.017	**0.828**	0.146	0.016	**0.828**	**0.752**	**0.963**	0.087	**0.957**	**0.756**
3	0.117	0.017	0.097	0.054	0.018	**0.898**	0.150	**0.831**	**0.898**	0.150	0.132	0.018	**0.481**	0.022	0.105

(b) Item membership w_{cj}^k

type k	$k=1$				$k=2$				$k=3$			
item j	1	2	3	4	5	6	7	8	9	10	11	12
cluster c 1	0.056	**0.146**	**0.795**	0.004	0.000	0.000	0.000	0.000	0.000	0.000	0.000	0.000
2	0.000	0.000	0.000	0.000	0.000	0.000	0.000	**0.999**	0.000	0.000	0.000	0.000
3	0.000	0.000	0.000	0.000	**0.851**	0.015	**0.132**	0.000	0.000	0.000	0.000	0.000

(c) Type membership z_{ck}

type k	$k=1$	$k=2$	$k=3$
1	**1.000**	0.000	0.000
cluster c 2	0.000	**1.000**	0.000
3	0.000	**1.000**	0.000

Parameter Settings. In order to study the influences of fuzzification penalties, the proposed method was also implemented with other parameter settings.

Table 3 shows the result with more crisp setting than the previous one, where two fuzzification penalties (λ_u and λ_w) had smaller values than Table 2. Because of crisp partition-like features, each cluster was characterized by only few items in Table 2-(b), and cluster 2 was connected with a noise question (Question 8). As a result, the object partition (Table 2-(a)) was severely violated from the designed types and the type responsibility (Table 2-(c)) failed to catch type 3. Therefore, fuzzy partition seems to contribute to fair evaluation of the questionnaire design.

Next, Table 4 shows the result, where only the fuzzification penalty for the newly added membership (λ_z) was changed from Table 2. Even though λ_z became more than double of Table 2, type memberships z_{ck} were almost equivalent to the previous ones as shown in Table 4-(c). However, from the subtables Table 4-(a) and (b), u_{ci} and w_{cj}^k had slightly fuzzier values than Table 2, i.e., λ_z can also contribute to fuzzification of object and item memberships.

Table 4. Fuzzy membership u_{ci}, w_{cj}^k and z_{ck} in the proposed method ($\lambda_u = 0.6$, $\lambda_w = 0.95$ $\lambda_z = 1.5$)

(a) Object membership u_{ci}

type k	$k=1$					$k=2$					$k=3$				
object i	1	2	3	4	5	6	7	8	9	10	11	12	13	14	15
cluster c 1	**0.812**	**0.895**	**0.683**	**0.802**	**0.885**	0.131	0.060	0.096	0.136	0.057	0.055	0.043	0.219	0.060	0.057
2	0.055	0.043	0.219	0.060	0.057	**0.812**	**0.895**	**0.683**	**0.802**	**0.885**	0.131	0.060	0.096	0.136	0.057
3	0.131	0.060	0.096	0.136	0.057	0.055	0.043	0.219	0.060	0.057	**0.812**	**0.895**	**0.683**	**0.802**	**0.885**

(b) Item membership w_{cj}^k

type k	$k=1$				$k=2$				$k=3$			
item j	1	2	3	4	5	6	7	8	9	10	11	12
cluster c 1	**0.151**	**0.173**	**0.644**	*0.030*	0.000	0.000	0.000	0.000	0.000	0.000	0.000	0.000
2	0.000	0.000	0.000	0.000	**0.151**	**0.173**	**0.644**	*0.030*	0.000	0.000	0.000	0.000
3	0.000	0.000	0.000	0.000	0.000	0.000	0.000	0.000	**0.151**	**0.173**	**0.644**	*0.030*

(c) Type membership z_{ck}

type k	$k=1$	$k=2$	$k=3$
cluster c 1	**1.000**	0.000	0.000
2	0.000	**1.000**	0.000
3	0.000	0.000	**1.000**

By the way, when λ_z becomes slightly larger than $\lambda_z = 1.5$, the cluster structure of Table 4 was drastically broken such that one cluster was equally shared by all cluster, i.e., very fuzzy partition with $z_{ck} = 1/C$. Type memberships z_{ck} seem to work well with almost crisp selection of type.

Then, it is quite important to choose the optimal penalty setting and a potential future work is to develop the method of selecting a better penalty parameters.

5   Conclusion

In this paper, a novel co-clustering-based method was proposed for simultaneously performing extraction of object-item co-clusters and evaluation of the typicality of each question in questionnaire data analysis for subject categorization. The conventional co-clustering model of FCCM was extended to the task by introducing type-related item memberships and type memberships.

In an experiment with an artificial data set, objective classification clustering was carried out in consideration of rejection of meaningless questions in questionnaire data analysis for type classification. Compared with the conventional FCCM method, the proposed method achieved better clustering, where it is possible to realize which questions are exactly more important in each type.

A future work is to develop the mechanism for properly selecting parameter sets because the experimental results implied proposed method works well only with proper parameter settings. As another future work, we also have a plan of applying the proposed method to a real world questionnaire data set related to such fields as life and educational psychology, and develop methods for setting parameters according to data.

Acknowledgment. This work was supported in part by JSPS KAKENHI Grant Number JP18K11474.

References

1. Kummamuru, K., Dhawale, A., Krishnapuram, R.: Fuzzy co-clustering of documents and keywords. In: 2003 IEEE International Conference on Fuzzy Systems, vol. 2, pp. 772–777 (2003)
2. Honda, K., Notsu, A., Ichihashi, H.: Collaborative filtering by sequential user-item co-cluster extraction from rectangular relational data. Int. J. Knowl. Eng. Soft Data Paradigms **2**(4), 312–327 (2010)
3. Liu, Y., Wu, S., Liu, Z., Chao, H.: A fuzzy co-clustering algorithm for biomedical data. PLoS ONE **12**(4), e0176536 (2017)
4. Oh, C.-H., Honda, K., Ichihashi, H.: Fuzzy clustering for categorical multivariate data. In: Joint 9th IFSA World Congress and 20th NAFIPS International Conference, pp. 2154–2159 (2001)
5. Honda, K., Oshio, S., Notsu, A.: Fuzzy co-clustering induced by multinomial mixture models. J. Adv. Comput. Intell. Intell. Inform. **19**(6), 717–726 (2015)
6. Bezdek, J.C.: Pattern Recognition with Fuzzy Objective Function Algorithms. Plenum Press, New York (1981)
7. Miyamoto, S., Ichihashi, H., Honda, K.: Algorithms for Fuzzy Clustering. Springer, Heidelberg (2008). https://doi.org/10.1007/978-3-540-78737-2
8. Miyamoto, S., Umayahara, K.: Fuzzy clustering by quadratic regularization. In: 1998 IEEE International Conference Fuzzy Systems and IEEE World Congress Computational Intelligence, vol. 2, pp. 1394–1399 (1998)

Extracting Access Patterns with Hierarchical Latent Tree Analysis: An Empirical Study on an Undergraduate Programming Course

Leonard K. M. Poon[✉][iD]

The Education University of Hong Kong, Hong Kong SAR, China
kmpoon@eduhk.hk

Abstract. Students commonly access learning resources with online systems nowadays. The access logs of those learning resources potentially provide rich information for course instructors to understand students. However, when there are many learning resources, the insights on students could be obscured by the large amount of unprocessed log data in its raw form. In this paper, we propose to extract the access patterns using a recently proposed method for topic modeling. The method, named hierarchical latent tree analysis, can capture co-occurrences of access to learning resources and group related learning resources together. We further show a way to consider the access periods during a course from the log data to reflect different behaviors of the students. We empirically test the proposed method using the log data captured in an undergraduate programming course with 63 students. We evaluate the performance of the proposed method based on the task for predicting students' scores. The regression model built on the features extracted by the proposed method shows significantly better performance over three baseline models. We further demonstrate meaningful insights on the student behaviors can be obtained from the extracted access patterns.

Keywords: Access pattern extraction ·
Hierarchical latent tree analysis · Log analysis ·
Probabilistic graphical models · Educational data mining

1 Introduction

Students commonly access learning resources with some online systems nowadays. The learning resources may include lecture notes, exercises, and quizzes used in a course. The access logs of those learning resources potentially provide rich information for course instructors to understand students.

To have more detailed information on the specific resources students have used, the access may be recorded on a finer scale. For example, student access

© Springer Nature Switzerland AG 2019
H. Seki et al. (Eds.): IUKM 2019, LNAI 11471, pp. 380–392, 2019.
https://doi.org/10.1007/978-3-030-14815-7_32

may be recorded for individual pages of lecture notes or individual questions of quizzes. By considering such a fine scale, a course may contain over a hundred of learning resources. With many items, the insights on students could be obscured by the large amount of unprocessed log data in its raw form.

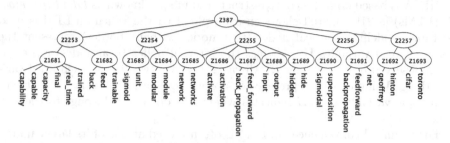

Fig. 1. A subtree of a latent tree model obtained by HLTA on a text data set built from NIPS abstracts. HLTA was originally proposed for topic modeling.

In this paper, we propose to extract access patterns using a recently proposed method named hierarchical latent tree analysis (HLTA). The method was originally proposed for topic modeling, but we show how we can use it for log analysis in this paper. HLTA can capture co-occurrences of access to learning resources and group related learning resources together. A straightforward application of HLTA for log analysis may consider whether a student has accessed a particular learning resource regardless of the time of access. However, whether an access occurs before a lesson, during a lesson, or during the examination period may indicate different learning attitudes of students. Hence, we further propose a way to incorporate the information of access periods to better reflect the differences.

To evaluate the effectiveness of our proposed method, we conducted an empirical study based on the access log data collected from an undergraduate programming course with 63 students. We perform regression analysis using the extracted access patterns as explanatory variables to predict the students' examination scores. We also discuss the access patterns discovered by HTLA.

The remaining of the paper is organized as follows. Section 2 introduces HLTA. Section 3 explains how access patterns can be extracted using HLTA and how access periods can also be considered in the analysis. Section 4 describes how to use the access patterns for regression. Section 5 evaluates the proposed method using the log data collected in an undergraduate programming course. Section 6 discusses related work and Sect. 7 concludes this paper.

2   Hierarchical Latent Tree Analysis

Hierarchical latent tree analysis (HLTA) is a recently proposed method for topic modeling [6]. The task of topic modeling can be described as follows. Consider a collection $\mathcal{D} = \{d_1, \ldots, d_N\}$ of N documents. Suppose P words are used in the vocabulary $\mathcal{V} = \{w_1, \ldots, w_P\}$. Each document d can be represented by the bag-of-words model using a vector $d = (c_1, \ldots, c_P)$, where c_i denotes the count of word w_i occurring in document d. Topic modeling aims to detect a number K of topics $z_1, \ldots, z_K$ among documents $\mathcal{D}$, where K can be given or learned from data. A topic is often characterized by the representative words.

HLTA is based on a class of tree-structured models known as *latent tree models* (LTMs) [8,21]. Figure 1 shows an example of LTMs. When an LTM is used for topic modeling, the leaf nodes (text nodes without border) represent the observed word variables $\boldsymbol{W}$, whereas the internal nodes (oval nodes) represent the unobserved topic variables $\boldsymbol{Z}$. All variables have two states. Each word variable $W_i \in \boldsymbol{W}$ indicates the presence or absence of word $w_i \in \mathcal{V}$ in a document. Each topic variable $Z_i \in \boldsymbol{Z}$ indicates whether a document belongs to the i-th topic.

For technical convenience, an LTM is often rooted at one of its latent nodes. It can then be regarded as a Bayesian network [10] with edges directed away from the root. The numerical information of the model includes a marginal distribution for the root and a conditional distribution for each edge. For example, edge $Z1685 \rightarrow \texttt{network}$ in Fig. 1 is associated with probability $P(\texttt{network}|Z1685)$. The conditional distribution associated with each edge characterizes the probabilistic dependence between the two nodes that the edge connects. The product of all those distributions defines a joint distribution over all the latent variables $\boldsymbol{Z}$ and observed variables $\boldsymbol{W}$. Denote the parent of a variable X as $pa(X)$ and let $pa(X)$ be an empty set when X is the root. Then the LTM defines a joint distribution over all observed and latent variables as follows:

$$P(\boldsymbol{W}, \boldsymbol{Z}) = \prod_{X \in \boldsymbol{W} \cup \boldsymbol{Z}} P(X|pa(X)). \tag{1}$$

Given a document d, the values of the binary word variables $\boldsymbol{W}$ are observed. We also use $d = (w_1, \ldots, w_P)$ to denote those observed values. Whether a document d belongs to a topic $Z \in \boldsymbol{Z}$ can be determined by the probability $P(Z|d)$. The LTM allows each document to belong to multiple topics.

For topic modeling, an LTM has to be learned from the document data $\mathcal{D}$. The structural learning method builds LTMs level by level. Thus, it also known as hierarchical latent tree analysis. Intuitively, the co-occurrence of words is captured by the latent variables at the lowest level, whose co-occurring patterns are captured by the latent variables at higher levels. Consider the model in Fig. 1 as an example. The terms "back-propagation" and "feed-forward" tend to co-occur in NIPS abstracts. Their co-occurrence is captured by the latent variable $Z1687$. Similarly, variable $Z1688$ captures the co-occurrence of "input" and "output". Since those four terms also tend to co-occur, their co-occurring patterns are in turn captured by variable $Z2255$ at one higher level.

To extract the topic hierarchy from an LTM, the tree structure in the model is followed and each internal node is used to represent a topic. A topic is characterized by the words most relevant to it. Specifically, the mutual information (MI) [9] is computed between a topic variable and each of its descendent word variables. At most seven descendent words with the highest MI are then picked to characterize the topic. Conceptually, the topics at higher levels are more general and those at lower levels are more specific in the topic hierarchy.

3   Extracting Access Patterns

In this paper, we propose to use HLTA to extract patterns of user access from log data. We assume the user identity, file requested, and time of access are recorded in the server log. A sample of the log is shown in Fig. 2. We describe two ways to extract access patterns in this section.

```
202.45.xx.xxx - user1 [09/Oct/2017:17:05:00 +0000] "GET /courses/IIT1082/lab/operators.html
     HTTP/1.1" 200 7180 "https://ltm.eduhk.hk/courses/IIT1082/lab/index.html" "Mozilla/5.0
     (Windows NT 6.1; Trident/7.0; rv:11.0) like Gecko"
220.241.xx.xxx - user2 [10/Oct/2017:10:24:03 +0000] "GET /courses/IIT1082/lab/index.html
     HTTP/1.1" 200 28493 "https://moodle.eduhk.hk/course/view.php?id=29321" Mozilla/5.0
     (Windows NT 6.1) AppleWebKit/537.36 (KHTML, like Gecko) Chrome/61.0.3163.100 Safari/537.36"
```

Fig. 2. A sample log file on a Web server. This study uses the user identity, page requested, and time of access to extract access patterns of the students.

3.1   Basic Version

In the basic version, HLTA is used to capture which items are accessed by the same users. Each item is associated with a binary variable, indicating whether the item has been accessed by a user. Suppose there are P items in the system and N users have used the system. A user can then be represented by a P-dimensional vector d, where each element can be either 0 or 1. The vector indicates whether the user has accessed each of the P items. A data set $\mathcal{D} = \{d_1, \ldots, d_N\}$ is then be built from the log to represent the access by the N users. It is used as input to HLTA to capture the co-occurring access to items by the same users.

HLTA produces an LTM. An example of the models resulting from HLTA for log analysis is shown in Fig. 3. In the model, the leaf nodes represent the binary observed variables indicating whether an item has been accessed by a user. The internal nodes represent the latent variables symbolizing the access patterns discovered. The latent variables have two states. They indicate whether a user has exhibited a certain access pattern.

The access patterns can be comprehended more easily by looking at the pattern tree as shown in Fig. 4. The pattern tree is extracted from the model shown in Fig. 3. Each line represents an access pattern. It shows which items are usually accessed by the same users. The items are ordered by their distinctiveness on each line. Only the most distinguishing items are shown. The number at the beginning of each line denotes the fraction of users that has exhibited the access pattern. Although HLTA supports multiple levels of patterns, only a single level has been discovered in this example.

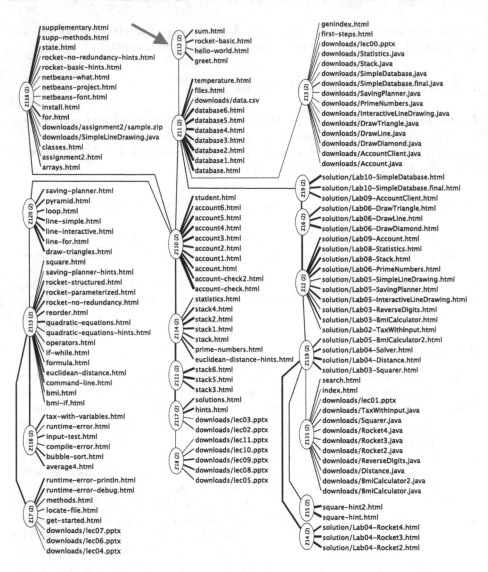

Fig. 3. The latent tree model obtained from the access log of students in an undergraduate programming course using HLTA. The leaf nodes (text nodes without border) represent the user access to individual learning resources. The internal nodes (oval nodes) represent the access patterns discovered.

We use the access pattern shown at the line pointed by an arrow in Fig. 4 for illustration. The access pattern shows that the items `greet.html`, `rocket-basic.html`, `sum.html`, and `hello-world.html` are typically accessed by the same group of users. The number at the beginning of the line denotes that 84.1% of users have shown this access pattern. That part of the pattern tree

```
1
 └0.788 quadratic-equations.html rocket-no-redundancy.html rocket-parameterized.html reorder.html rocket-structured.html comman
 ├0.643 line-for.html line-interactive.html draw-triangles.html line-simple.html saving-planner.html loop.html pyramid.html
 ├0.462 account-check.html account4.html account3.html account5.html account-check2.html account1.html account2.html
 ├0.376 database2.html database1.html database4.html database6.html database5.html database.html database3.html
 ├0.092 downloads/SimpleDatabase.final.java downloads/DrawDiamond.java downloads/SimpleDatabase.java downloads/DrawLine.j
6
 └0.841 greet.html rocket-basic.html sum.html hello-world.html  ◄──────────
 ├0.440 solution/Lab10-SimpleDatabase.html solution/Lab09-AccountClient.html solution/Lab10-SimpleDatabase.final.html
 ├0.470 solution/Lab06-DrawDiamond.html solution/Lab06-DrawTriangle.html solution/Lab06-DrawLine.html
 ├0.418 solution/Lab08-Statistics.html solution/Lab06-PrimeNumbers.html solution/Lab08-Stack.html solution/Lab05-InteractiveLine
 ├0.326 solution/Lab04-Distance.html solution/Lab03-Squarer.html solution/Lab04-Solver.html solution/Lab05-BmiCalculator2.html
11
 └0.420 square-hint2.html square-hint.html
 ├0.322 solution/Lab04-Rocket3.html solution/Lab04-Rocket2.html solution/Lab04-Rocket4.html
 ├0.059 downloads/ReverseDigits.java downloads/Rocket4.java downloads/Distance.java downloads/BmiCalculator.java downloads/f
 ├0.487 for.html supp-methods.html supplementary.html state.html rocket-basic-hints.html classes.html arrays.html
 ├0.516 stack2.html stack1.html stack4.html stack.html prime-numbers.html euclidean-distance-hints.html statistics.html
16
 └0.443 stack5.html stack6.html stack3.html
 ├0.364 hints.html solutions.html downloads/lec02.pptx downloads/lec03.pptx
 ├0.896 downloads/lec09.pptx downloads/lec08.pptx downloads/lec05.pptx downloads/lec10.pptx downloads/lec11.pptx
 ├0.816 input-test.html tax-with-variables.html bubble-sort.html average4.html compile-error.html runtime-error.html
 └0.787 runtime-error-debug.html runtime-error-println.html methods.html locate-file.html downloads/lec04.pptx downloads/lec06.ppt
```

Fig. 4. Access pattern tree extracted from the model shown in Fig. 3. Each line represents an access pattern. It shows which items (files) are usually accessed by the same users (students). The floating numbers at the beginning of lines indicate the fraction of users that have exhibited the corresponding access patterns.

is extracted from the latent variable $Z112$ in Fig. 3. The fraction is obtained by computing the marginal probability $P(Z112)$ from the joint distribution defined by the model (Eq. 1).

3.2 Extended Version with Access Periods

In the above basic version, the time information in the log data is ignored. In the extended version, we divide the time of access into a certain number of periods. Then we use a binary variable for each pair (w, p) of item w and period p to indicate whether a user has accessed the item w during the period p. Each item can be associated with a different number of periods depending on its nature. Denote the number of periods associated with item w by q_w. Each user is then represented by a Q-dimensional vector, where $Q = \sum_{i=1}^{P} q_{w_i}$ and P is the number of items. For example, if each item is associated with three periods, then $3P$ binary variables will be used as the observed variables in HLTA.

4   Regression with Access Patterns

The access patterns of a user can be used as extracted features for other tasks. Specifically, suppose there are R_l latent variables that represent the access patterns at level l in a pattern tree. We can choose a level l of patterns depending on the level of granularity of the access patterns needed. We can then use the probabilities of the latent variables $Z_1, \ldots, Z_{R_l}$ at level l as explanatory variables in regression. Let S be the response variable. The regression model can be written as:

$$s_i = \beta_0 + \beta_1 P(Z_1 = 1|d_i) + \cdots + \beta_{R_l} P(Z_{R_l} = 1|d_i) + \epsilon_i, \quad i = 1, \ldots, N,$$

where β_j is a regression coefficient or the intercept term, ϵ_i is the error term, and N is the number of users. The posterior probability $P(Z_j = 1|d_i)$ for the pattern j and user i can be computed by clique tree propagation [10] based on the observed access values d_i.

5   Empirical Evaluation

We conducted an empirical study using the access log of the students in an undergraduate programming course. The course included two class sessions with a total of 63 students. It consisted of 13 teaching weeks with a 3-hour lesson each week. We considered the student access from one week before the semester to the end of the semester in the study.

During each lesson, around one hour was reserved for students to do programming exercises. The exercises were provided on webpages. The website also included lecture notes, supplementary examples, and solutions. Altogether there are 146 online files. Student access to those files was logged on the server. We assume that each file represented a unit of learning resources of interest.

5.1   Student Access Patterns

Basic Access Data. We first consider the access patterns extracted using the basic version of the proposed method. The model obtained from HLTA is shown in Fig. 3 and the pattern tree extracted is shown in Fig. 4.

The model obtained contains 20 latent variables, meaning that 20 access patterns have been discovered. Most of them are connected to variables corresponding to the related online files. For example, variable $Z13$ is connected to 12 variables related to the access to Java files. This access pattern indicates that students who have accessed some of those Java files usually have accessed the other Java files. The pattern is shown on the fifth line in the pattern tree. We see that only 9% of students have shown this pattern. The low access rate can be explained by the fact that the same Java programs could also be found on the HTML pages. If we look at the patterns on lines 7 to 10 in the pattern tree, we see that around 33% to 47% of students have accessed the solutions online.

The pattern tree also shows other interesting patterns. For example, it shows co-occurring access to the pages `greet.html`, `rocket-basic.html`, `sum.html`, and `hello-world.html`. Those pages correspond to four separate programming exercises in the first lesson. This explains why those pages are usually accessed together with high access rate (84%). Line 3 (`account-check.html ...`) in the pattern tree shows that the different steps of a longer programming exercise were usually accessed by the same students. Line 4 (`database2.html ...`) shows a similar pattern on the steps of another longer exercise. Since those exercises were used near the end of the semester, fewer students had accessed them.

Data with Access Periods. While the above pattern tree captures interesting access patterns, it ignores the time information in the access log. For a course, whether a student accesses a learning resource before a lesson or during the examination period may indicate signally different attitudes of the students. Therefore, we now analyze the access log using the extended version of the proposed method to consider the periods of access by the students as well.

In particular, some resources are associated with a particular lesson. We consider six periods for each of those resources. We consider the period from one hour before the lesson to four hours after the lesson as *during* the lesson. The time before the lesson period is considered as *before* the lesson. The time after the lesson period but before the examination period is considered as *after* the lesson. The *examination* period is defined as those 168 hours (seven days) before the starting time of the examination. We also consider the 168 hours before the submission deadline of the course project as the *project* period. The other access time is considered as belonging to an *unspecific* period. For resources not associated with a particular lesson, we consider only the examination period, the project period, and the unspecific period for each of them. After taking account of the access periods, the input data to HLTA contained 813 attributes. HLTA took a few seconds to run on an Intel Xeon E5-2650v3 CPU using 4 CPU cores.

```
1
├0.353 rocket-structured.html-exam euclidean-distance.html-exam stack1.html-exam quadratic-equations.html-exam command-lir
│  ├0.375 stack1.html-exam line-for.html-exam stack.html-exam statistics.html-exam loop.html-exam stack2.html-exam arrays.htn
│  ├0.324 rocket-structured.html-exam euclidean-distance.html-exam quadratic-equations.html-exam command-line.html-exam rc
│  ├0.389 pyramid.html-exam prime-numbers.html-exam draw-triangles.html-exam
│  ├0.357 bmi-if.html-exam if-while.html-exam saving-planner.html-exam
6
├0.262 reorder.html-exam methods.html-exam bmi.html-exam formula.html-exam operators.html-exam square.html-exam locat
│  ├0.255 stack4.html-exam stack3.html-exam stack5.html-exam stack6.html-exam
│  ├0.174 tax-with-variables.html-exam input-test.html-exam average4.html-exam bubble-sort.html-exam first-steps.html-exam ru
├0.315 account6.html-exam account-check.html-exam account-check2.html-exam account5.html-exam account1.html-exam acco
│  ├0.333 solution/Lab06-PrimeNumbers.html-exam solution/Lab05-SavingPlanner.html-exam solution/Lab10-SimpleDatabase.html-e
11
├0.288 line-for.html-after draw-triangles.html-after stack2.html-after stack3.html-after loop.html-after stack5.html-after stack1.html
├0.428 statistics.html-during arrays.html-during draw-triangles.html-during line-for.html-during stack1.html-during stack.html-durin
├0.500 rocket-no-redundancy.html-during quadratic-equations.html-during rocket-parameterized.html-during rocket-structured.htn
├0.045 downloads/TaxWithInput.java-after solution/Lab09-AccountClient.html-during square.html-project locate-file.html-project dr
├0.474 bmi.html-during square.html-during compile-error.html-during first-steps.html-during operators.html-during locate-file.html-
16
├0.517 tax-with-variables.html-before runtime-error.html-before compile-error.html-before average4.html-before input-test.html-bet
├0.562 square.html-after reorder.html-after locate-file.html-after bmi.html-after operators.html-after formula.html-after
├0.222 solution/Lab04-Distance.html-after solution/Lab05-BmiCalculator2.html-after solution/Lab04-Solver.html-after solution/Labt
├0.169 student.html-before files.html-before database5.html-before database4.html-before account4.html-before account1.html-br
├0.045 rocket-basic-hints.html-exam database6.html-exam square-hint.html-exam database1.html-exam database.html-exam star
```

Fig. 5. The pattern tree obtained by HLTA after considering also the access periods. The access periods are denoted by *before* a lesson, *during* a lesson, *after* a lesson, *exam*ination period, *project* period, and *unspecific* period. The last part of the name of each access variable indicates the access period.

The pattern tree obtained is shown in Fig. 5. The tree has two levels of patterns. The top level (level 2) contains 13 patterns. The bottom level (level 1) contains 86 patterns. Some of the level-1 patterns are shown on lines 2 to 8. The level-2 patterns can be regarded as more general than the level-1 ones.

The patterns on lines 1, 9, 10, and 20 show patterns of access during the examination period. They correspond to different resources and have different access rate. Line 12 (`statistics.html-during ...`) shows a pattern of access during the lesson. Lines 13 and 15 also show similar patterns. On the other hand, lines 16 (`tax-with-variables.html-before ...`) and 19 (`student.html-before ...`) show that two patterns of access to the resources before the lesson.

The above results show that the information provided by the access patterns have been enriched by considering the access periods. The patterns show not only which resources have been accessed by the same students, but also when the access has occurred.

5.2   Prediction of Student Scores

To further evaluate the usefulness of the extracted access patterns, we use them in a regression model to predict the examination scores. The basic access patterns and the patterns with access periods are used in two separate regression models. For the basic version, all the 20 access patterns are used as explanatory variables in the model. For the version with the access periods, only the 13 level-2 patterns are used. The regression analysis is conducted using the R language [20]. We also consider regression models obtained with the stepwise model selection by AIC (using the routine `stepAIC`).

We consider three baseline models for comparison. The first one is the regression model with only the intercept term. It essentially uses the mean score to predict the students' scores. The second one uses the scores of other assessment items in the course as explanatory variables, including weekly reflection writing, two programming assignments, and a course project. The third baseline model is obtained by performing `stepAIC` on the second model.

The prediction performance of the regression models is measured by the root mean square error (RMSE). The evaluation is conducted with 5-fold cross validation. The mean and the standard deviation of the RMSEs in the five folds are computed for each model. The quality of the regression model is measured by the AIC score [1]. Smaller values of RMSE or AIC indicates better performance.

The results of the regression task are shown in Table 1. We see that the access patterns with period information led to the two models, namely `Period` and `Period-stepAIC`, with the best prediction performance. Among the two models, the better model is obtained by using `stepAIC` to perform model selection. `Period-stepAIC` also has the lowest AIC score, indicating that it has the highest quality. We further use the paired t-test to compare `Period-stepAIC` with the three baseline models. The test results show that differences in RMSE is statistically significant at a significance level of 5%. They show that the access patterns with period information can be useful for predicting students' performance.

Table 1. Results of the score prediction task by the regression models using different sets of features. The best results are highlighted in bold.

Model	Explanatory variables	stepAIC	AIC	RMSE
Mean	Intercept term only		569	21.52 ± 2.10
Scores	Other assessment scores		546	18.57 ± 1.10
Scores-stepAIC	Other assessment scores	✓	545	18.62 ± 1.36
Basic	Basic access patterns		562	24.18 ± 5.48
Basic-stepAIC	Basic access patterns	✓	537	16.74 ± 2.29
Period	Access patterns with periods		533	16.57 ± 1.78
Period-stepAIC	Access patterns with periods	✓	**527**	**14.90 ± 2.01**

Figure 6 shows the regression coefficients in Period-stepAIC. We see that variables $Z212$, $Z211$, $Z27$, and $Z28$ show a significant relationship with the examination score. The access patterns corresponding to those variables are shown in Table 2. The pattern $Z212$ has the highest coefficient value. The pattern and the coefficient suggests that those students who had exhibited a pattern of access to resources before the lesson usually had a higher examination score. Variables $Z27$ and $Z211$ show two patterns of access on resources during the class. They also led to a higher examination score, but their effects are smaller.

The negative value of the coefficient for variable $Z28$ indicates that students exhibiting that pattern usually had a lower examination score. It may look surprising that more access to the learning resources would have a detrimental effect. From Table 2 we see that the pattern represents reading the solutions after the lesson. This may suggest that some students did not attempt the exercises during the class but try to catch up the progress by reading the solutions quickly outside the class. Thus they may get lowers in examination in general.

```
Coefficients:
            Estimate Std. Error t value Pr(>|t|)
(Intercept)    26.57       8.73    3.04  0.00358 **
Z29            10.56       5.82    1.81  0.07528 .
Z212           35.71      10.09    3.54  0.00081 ***
Z211           25.36       9.22    2.75  0.00796 **
Z22            -9.51       6.16   -1.54  0.12823
Z27            28.19       4.84    5.83  2.9e-07 ***
Z28           -32.66       5.90   -5.53  8.7e-07 ***
---
Signif. codes:  0 '***' 0.001 '**' 0.01 '*' 0.05 '.' 0.1 ' ' 1
```

Fig. 6. Regression coefficients β_i in the model Period-stepAIC. The variables in the first column denote the access patterns discovered. The LTM containing those latent variables is not shown due to space limitation. The coefficient estimates are given in the second column.

Table 2. The access patterns with period information that show significant relationship with the examination scores as shown in Fig. 6.

Z212	0.517	tax-with-variables.html-before runtime-error.html-before compile-error.html-before average4.html-before input-test.html-before runtime-error-debug.html-before runtime-error-println.html-before
Z27	0.500	rocket-no-redundancy.html-during quadratic-equations.html-during rocket-parameterized.html-during rocket-structured.html-during euclidean-distance.html-during runtime-error-debug.html-during runtime-error-println.html-during
Z211	0.474	bmi.html-during square.html-during compile-error.html-during first-steps.html-during operators.html-during locate-file.html-during reorder.html-during
Z28	0.222	solution/Lab04-Distance.html-after solution/Lab05-BmiCalculator2.html-after solution/Lab04-Solver.html-after solution/Lab03-Squarer.html-after solution/Lab04-Rocket3.html-after solution/Lab04-Rocket4.html-after solution/Lab04-Rocket2.html-after

6   Related Work

There are three main approaches to analyze log data in the education context. The first approach uses sequential pattern mining to identify frequent action sequences [11,17,18]. It can be used, for example, to compare between the action sequences frequently taken by more or less successful students [15]. The second approach uses process mining to discover the interactive workflows of the students [4]. The resulting process models are often used to study self-regulated learning strategies [3,13].

The third approach uses cluster analysis to find groups of similar students. It often uses summary statistics aggregated from the raw log data as input [7,12,15]. Some studies cluster students based on usage summary [5] or their background [2] for improving the results of process mining. Our proposed method also belongs to the third approach. However, it clusters students based on access to individual items, requiring no prior knowledge for grouping related items.

HLTA was originally proposed for hierarchical topic detection [6] based on LTMs. LTMs have been mainly used for multidimensional clustering [8,16,19]. Our proposed method can also be considered as using LTMs to perform multi-dimensional clustering on students based on their access. Other applications of LTMs are reviewed in [14,22].

7   Conclusion

In this paper, we propose using HLTA to extract access patterns from log data. We also show how to obtain more insights from the log data by considering the access period as well. The results of the analysis on the log data from an undergraduate programming course show that interesting access patterns can be discovered. Our experimental results on regression further show that the extracted access patterns with period information can be useful for predicting the students' performance in terms of examination scores. Our results demonstrate that the access patterns discovered by HLTA can provide valuable insights into the students' behaviors in a course.

Acknowledgement. Research on this paper was supported by the Education University of Hong Kong under grant RG70/2017-1018R and Dean's Research Fund.

References

1. Akaike, H.: A new look at the statistical model identification. IEEE Trans. Autom. Control **19**(6), 716–723 (1974)
2. Ariouat, H., Cairns, A.H., Barkaoui, K., Akoka, J., Khelifa, N.: A two-step clustering approach for improving educational process model discovery. In: IEEE 25th International Conference on Enabling Technologies: Infrastructure for Collaborative Enterprises (WETICE), pp. 38–43 (2016)
3. Bannert, M., Reimann, P., Sonnenberg, C.: Process mining techniques for analysing patterns and strategies in students' self-regulated learning. Metacognition Learn. **9**(2), 161–185 (2014)
4. Bogarín, A., Cerezo, R., Romero, C.: A survey on educational process mining. Wiley Interdisc. Rev.: Data Min. Knowl. Disc. **8**(1), e1230 (2018)
5. Bogarín, A., Romero, C., Cerezo, R., Sánchez-Santillán, M.: Clustering for improving educational process mining. In: Proceedings of the Fourth International Conference on Learning Analytics and Knowledge, pp. 11–15 (2014)
6. Chen, P., Zhang, N.L., Liu, T., Poon, L.K.M., Chen, Z., Khawar, F.: Latent tree models for hierarchical topic detection. Artif. Intell. **250**, 105–124 (2017)
7. Chen, S.Y., Liu, X.: Mining students' learning patterns and performance in web-based instruction: a cognitive style approach. Interact. Learn. Environ. **19**(2), 179–192 (2011)
8. Chen, T., Zhang, N.L., Liu, T., Poon, K.M., Wang, Y.: Model-based multidimensional clustering of categorical data. Artif. Intell. **176**, 2246–2269 (2012)
9. Cover, T.M., Thomas, J.A.: Elements of Information Theory. Wiley, California (2006)
10. Cowell, R.G., Dawid, A.P., Lauritzen, S.L., Spiegelhalter, D.J.: Probabilistic Networks and Expert Systems. Springer, New York (1999). https://doi.org/10.1007/b97670
11. Kinnebrew, J.S., Segedy, J.R., Biswas, G.: Integrating model-driven and data-driven techniques for analyzing learning behaviors in open-ended learning environments. IEEE Trans. Learn. Technol. **10**(2), 140–153 (2017)
12. Li, L.Y., Tsai, C.C.: Accessing online learning material: quantitative behavior patterns and their effects on motivation and learning performance. Comput. Educ. **114**, 286–297 (2017)
13. Maldonado-Mahauad, J., Pérez-Sanagustín, M., Kizilcec, R., Morales, N., Munoz-Gama, J.: Mining theory-based patterns from big data: identifying self-regulated learning strategies in massive open online courses. Comput. Hum. Behav. **80**, 179–196 (2018)
14. Mourad, R., Sinoquet, C., Zhang, N.L., Liu, T., Leray, P.: A survey on latent tree models and applications. J. Artif. Intell. Res. **47**(1), 157–203 (2013)

15. Perera, D., Kay, J., Koprinska, I., Yacef, K., Zaïane, O.R.: Clustering and sequential pattern mining of online collaborative learning data. IEEE Trans. Knowl. Data Eng. **21**(6), 759–772 (2009)
16. Poon, L.K.M.: Clustering with multidimensional mixture models: analysis on world development indicators. In: Cong, F., Leung, A., Wei, Q. (eds.) ISNN 2017. LNCS, vol. 10261, pp. 153–160. Springer, Cham (2017). https://doi.org/10.1007/978-3-319-59072-1_19
17. Poon, L.K.M., Kong, S.-C., Wong, M.Y.W., Yau, T.S.H.: Mining sequential patterns of students' access on learning management system. In: Tan, Y., Takagi, H., Shi, Y. (eds.) DMBD 2017. LNCS, vol. 10387, pp. 191–198. Springer, Cham (2017). https://doi.org/10.1007/978-3-319-61845-6_20
18. Poon, L.K.M., Kong, S.-C., Yau, T.S.H., Wong, M., Ling, M.H.: Learning analytics for monitoring students participation online: visualizing navigational patterns on learning management system. In: Cheung, S.K.S., Kwok, L., Ma, W.W.K., Lee, L.-K., Yang, H. (eds.) ICBL 2017. LNCS, vol. 10309, pp. 166–176. Springer, Cham (2017). https://doi.org/10.1007/978-3-319-59360-9_15
19. Poon, L.K.M., Zhang, N.L., Liu, T., Liu, A.H.: Model-based clustering of high-dimensional data: variable selection versus facet determination. Int. J. Approximate Reasoning **54**(1), 196–215 (2013)
20. R Core Team: R: A Language and Environment for Statistical Computing. R Foundation for Statistical Computing, Vienna (2018). https://www.R-project.org/
21. Zhang, N.L.: Hierarchical latent class models for cluster analysis. J. Mach. Learn. Res. **5**, 697–723 (2004)
22. Zhang, N.L., Poon, L.K.M.: Latent tree analysis. In: Proceedings of the Thirty-First AAAI Conference on Artificial Intelligence, pp. 4891–4897 (2017)

Measuring Hotel Review Sentiment: An Aspect-Based Sentiment Analysis Approach

Thang Tran[1,2]([✉]), Hung Ba[1,3], and Van-Nam Huynh[1]

[1] School of Knowledge Science, Japan Advanced Institute of Science and Technology, Nomi, Japan
{txthang,hungba,huynh}@jaist.ac.jp
[2] Informatics Department, Tay Nguyen University, Buon Ma Thuot, Vietnam
[3] Faculty of Economics and Business, Hoa Sen University, Ho Chi Minh City, Vietnam

Abstract. Reviews of travelers regarding different characteristics of a hotel are one of the worthiest sources for the managers to enhance their services, facilities, and marketing campaigns. In finding a way to improve the practical experience of both buy-side and sell-side in the hospitality market, we apply sentiment analysis for hospitality data from a user-generated content site named TripAdvisor. Typically, from big data including both quantitative data and qualitative data of customer's reviews, our contributions are first proposing a framework to utilize big data analysis to identify which aspects/features along with their polarities that customers are focusing, and then inferring and grouping them into 11 topics toward different 405 hotels in Ho Chi Minh City. This study adds more contributions to finding the emerging opinions of customers towards the different topics of hotel reviews by providing an annotated dataset in hotel reviews which ultimately benefits for further research in this field.

Keywords: Aspect-based sentiment analysis · Machine learning · Hotel reviews · Annotated hospitality dataset

1 Introduction

User-generated content (UGC) is the online platform where customers reflect their thoughts and place their rating for products and services [16]. In UGC, customer feedback is an integral part of the continuous improvement process implemented in the hotel industry, and yet a comprehensive characterization of the customer experience is difficult to achieve. Nowadays, consumers are participating and spending more time on social media to make friends, co-create, share information, experiences, and opinions [7]. Their purchase as a decision-making process is being influenced by those factors through social media networks. Hence, managers are focusing on online communication platforms to reach online consumers and to take advantage of their feedbacks in the social networks.

© Springer Nature Switzerland AG 2019
H. Seki et al. (Eds.): IUKM 2019, LNAI 11471, pp. 393–405, 2019.
https://doi.org/10.1007/978-3-030-14815-7_33

Simultaneously, to determine the human perceptions, opinions, attitudes towards entities such as services, products [10], researchers have proposed many tasks in sentiment analysis which mainly focus on document-level, sentence-level, and aspect level. Among these studies, aspect-based sentiment analysis (ABSA) is more sophisticated and requires a huge amount of knowledge to discover specific aspects along with sentiment of a product that people were discussing or talking about in reviews. Two subtasks are performed consecutively to accomplish the ABSA goal: aspect term extraction (ATE) extracts the aspect terms and then followed by aspect polarity classification (APC), which determines the polarity expressed towards these aspects.

In this study, we first use an unsupervised learning algorithm to identify the topics mentioned on the hotel reviews and then we measure customer's opinions towards aspects using a deep learning model. The goal of this study is to help hotel managers gain more accurate, useful information and insights from their customers and services.

In the next section, we briefly present a literature review on the sentiment analysis and data mining approach in the hotel industry. Next, the methodological details of data collection-processing and the sentiment analytical approach are described in Sect. 3. Section 4 shows the empirical results based on the online review dataset of hotels in Ho Chi Minh City, Vietnam. Finally, conclusions and future research directions are provided in Sect. 5.

2   Literature Reviews

2.1   Sentiment Analysis

User opinions can be expressed toward anything, such as a product, service, an individual, an organization using a quintuple $(e_i, a_{ij}, o_{ijkl}, h_k, t_l)$ [10], where e_i is an entity's name, a_{ij} is an aspect of e_i, o_{ijkl} is the opinion towards aspect a_{ij} of entity e_i, h_k is the opinion holder, and t_l is the time when the opinion is expressed by h_k. The opinion o_{ijkl} here can be expressed by positive, negative, neutral, or by using some different types such as star ratings or numerical scales.

To extract the quintuple from the text review, Liu et al. proposed a sequence of 6 tasks [10]. However, this research mainly focuses on aspect extraction (Task 2) and aspect sentiment classification (Task 5), also referred as aspect term extraction (ATE) and aspect polarity classification (APC).

2.2   Aspect Extraction and Grouping

Aspect Extraction: The goal of this task has also known as aspect mining or topic detection [11] which identifies the aspects in reviews that users of UGC focused on. This task is usually performed on each sentence of reviews which normally mentions more than one aspect. Mining aspects could be accomplished using supervised or unsupervised learning algorithm.

In supervised learning, a set of aspects is predefined, and a training set of annotated reviews is used to train the algorithms, such as Naïve Bayes, Support

Vector Machine (SVM), or long short-term memory (LSTM) [1,11]. Supervised learning algorithms are crucial in the tasks related to the comparison of aspects for all reviews over time and for all classifiers over the accuracy performances.

In unsupervised learning, there are two basic approaches for discovering aspects in a corpus of text reviews under the unsupervised algorithms:

– Symbolic approaches that rely on syntactic description of terms, namely nouns and/or noun phrases [10]. These approaches basically detect nouns or noun phrases by part-of-speech tagging and add a word to the set of aspects if its frequency is larger than a specific threshold.
– Statistical approaches that exploit the fact that the words composing a term tend to be found close to each other and reoccurring [12].

Aspect Grouping and Refining: In this regard, Zhai et al. proposed a semi-supervised learning method to classify aspect terms into some user-handcrafted aspect groups [18]. First, they assigned a small number of aspects as word seeds for each group; then a semi-supervised learning method was applied to classify the rest aspect terms into suitable categories.

Another method called multilevel latent semantic association was introduced in [5]. At the first stage, they group all the words in aspect terms into a set of topics using Latent Dirichlet Allocation (LDA), which will be explained later in some detail, and then build latent topic structures for aspect expressions based on the previous results. At the second stage, those aspect terms will be grouped by LDA again based on their latent topic structures produced at the first stage and their surrounding words in the reviews to form a document. LDA is applied to such document to get the final result. Guo et al. proposed a similar method to group aspects from different languages into aspect categories [6], which can be used to compare opinions along different aspects from different languages (English and Chinese).

2.3 Aspect Sentiment Classification

In the hotel reviews, dictionary-based approaches use a list of negative and positive words to find the polarity of a text [10]. Neidhardt et al. used a lexical database to measure the sentiment scores of reviews on an online travel forum [13]. The supervised approaches including Naïve Bayes [4], SVM [17], or deep learning techniques require a set of manually annotated reviews to train the classifiers. In many domains, the research of [3,8] showed that deep learning technique such as CNN with proper setup of layers and hyper-parameters archives remarkable results in classifying text polarity of various text corpora.

In general, most of the research on the hotel reviews focus on the relationships of review rating and hotel selection, while lacking the understanding of the review content in the combination of its sentiment and its aspects. Hence, in this research, we combine these two tasks by using BiLSTM-CRF model [9] trained on the top of a new annotated dataset for a new hospitality dataset to identify aspects along with their sentiment opinions simultaneously.

3   Methodology

3.1   Proposed Framework

We propose the framework in Fig. 1 to perform aspect mining and sentiment analysis for the hotel reviews. First, we extract a new hotel review dataset from TripAdvisor[1] and then apply text preprocessing techniques in natural language processing, including language detection, noise removal, lemmatization, to get the processed dataset. This dataset is fed into a well-known BiLSTM-CRF model trained on the top of a new annotated dataset (Sect. 4) to extract the aspect terms along with their sentiments, which explicitly expressed in the review sentences.

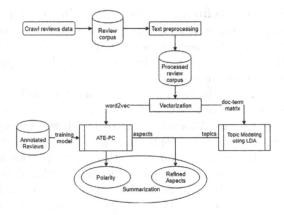

Fig. 1. Aspect-based sentiment analysis on hospitality media

Simultaneously, we implement a topic modeling model using LDA on this dataset to discover the topics and keywords represented for each topic. Those aspect terms will be used as the input for the final stage of LDA model to infer the refined topics for them. Finally, we conduct a summarization for the whole dataset based on the achieved results.

3.2   BiLSTM-CRF for the Combining ATE-PC Task

In this study, we employ the BiLSTM-CRF model, which has been known as the most effective and powerful sequence labeling model for named entity recognition (NER) task, to extract the aspect terms along with their polarities. However, we slightly modify the input and output of the model by combining aspect terms and polarities simultaneously using sequential labeling with new IOB

---

[1] https://www.tripadvisor.com/.

encoding format. The architecture of BiLSTM-CRF model using new sequential labeling [15] is showed in Fig. 2:

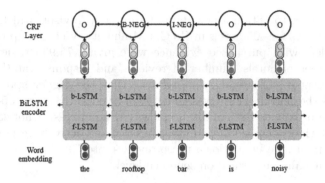

Fig. 2. BiLSTM-CRF using new IOB encoding format.

3.3 Latent Dirichlet Allocation for Topic Modeling

LDA is a statistical generative model developed by Blei et al. [2] to mimic the process of how people compose their documents. This algorithm assumes that each document has its topic distribution and each topic has its word distribution:

- Assume each document has its own distribution of topics, a topic is randomly drawn.
- Assume each topic has its own distribution of words, a word is randomly drawn.

LDA iteratively discovers the topic distribution for each document and word distribution for each topic by fitting this two steps generative process to the observed words in documents until it finds the best set of parameters that describe the topic and word distribution. Based on the advantage that LDA can automatically specify the keywords or rules to identify the topic, we implement LDA for topic modeling to identify the topics and their keywords of our hospitality dataset.

3.4 Grouping and Refining Aspects

We assign the name for each topic based on the keywords' weighted values given from the topic modeling task. Then, we feed the aspect terms extracted from BiLSTM-CRF model into the LDA to infer the topic for them. The final result will be the topics which it represents for the aspect terms, along with their sentiments which were expressed on the review sentences.

4   Empirical Results

4.1   Dataset

We used a crawler to obtain a new dataset from TripAdvisor, which consists of all English reviews for 405 hotels in Ho Chi Minh City, Vietnam up to Feb 2017. 58,381 travelers with purchased evidence were giving 75,933 reviews. Table 1 presents number of hotels, number of reviews and response rate(%) for each hotel star segment. The final column shows that the higher star of a hotel, the higher of the response rate was given. Moreover, the 3-star to 5-star hotels account for approximate 82% of total reviews. Although 5-star hotels cover only 4.2% in total, they occupied 25% reviews in total and actively responded to the customers. Finally, we break down those reviews into 532,164 sentences, as the ABSA tasks usually performed on sentence level.

Table 1. Number of hotels per star ranking category and its reviews.

Hotel star	#Hotels	%Hotel	#Reviews	%Reviews	%Response
Unrated	118	29.4%	4,847	4%	27.7%
1	5	1.2%	529	1%	9.5%
2	71	17.7%	9,923	13%	20.1%
3	147	36.7%	20,440	28%	44.3%
4	47	11.7%	21,077	29%	64.6%
5	17	4.2%	19,117	25%	69.9%

4.2   Annotated Hospitality Dataset

To train the BiLSTM-CRF model for the combining ATE-PC task, we manually build a new annotated dataset[2]. Three equally experienced annotators were organized to provide aspect-level annotation for 3,500 sentences extracted from dataset. We assign 1,000 sentences for each annotator, and keep the remaining 500 sentences to measure the inter-annotator agreement. They only annotate the explicit aspects with polarities mentioned in the sentences using a new IOB encoding format: (*B-POS, I-POS, B-NEU, I-NEU, B-NEG, I-NEG, O*). Where *B* indicates the beginning of a aspect, *I* indicates the inside of a aspect,

Table 2. Sentence with labels in new IOB encoding.

Words:	Staff	are very friendly	,	but the	rooftop	bar	is quite noisy	.
Labels:	**B-POS** O	O	O	O O O	**B-NEG**	**I-NEG** O	O	O O

---

[2] https://github.com/txthang/bilstm-crf-lda-hospitality/tree/master/data

POS, **NEU** and **NEG** represents for Positive, Neutral and Negative sentiment respectively. The character **O** points out the outside of aspect, for instance (Table 2):

Next, we briefly calculate the inter-annotator agreement (IAA) to measure the annotator's consistency rate. The evaluation metrics of precision, recall, and F-measure in SemEval'13[3] will be used, along with slight variations and detailed metrics. These metrics was defined as regarding pairwise comparison among two annotators - one set as golden standard and another set as system's output.

$$Precision = \frac{COR + 1/2 \times PAR}{COR + INC + PAR + SPU} \tag{1}$$

$$Recall = \frac{COR + 1/2 \times PAR}{COR + INC + PAR + MIS} \tag{2}$$

$$F_1 = 2 \times \frac{Precision \times Recall}{Precision + Recall}, \tag{3}$$

where COR, INC, PAR, MIS, SPU indicates the correct, incorrect, partial correct, missing and spurious aspect terms between two annotators.

Table 3 shows the pair-wised inter-annotator agreement between three experienced annotators. The averaged f_1 score of three annotators with 87.80% shows a strong agreement between them. The final annotated dataset consists of 3500 unique sentences, of which 500 shared among all three annotators, and then partitioned into a training set (70%) and a test set (30%).

Table 3. Inter-annotator agreement for three annotators: A1, A2, and A3.

	COR	INC	PAR	MIS	SPU	Precision (%)	Recall (%)	F_1 (%)
A1 vs A2	413	32	8	25	21	87.97	87.24	87.60
A1 vs A3	419	30	7	22	10	90.67	88.39	89.52
A2 vs A3	401	37	9	27	19	87.02	85.55	86.28
Average						88.55	87.06	87.80

4.3   The Combining ATE-PC Task Results

We evaluate the performance of ATE and APC tasks separately by converting the output of the model to the output of ATE and APC tasks as in Fig. 3. We apply 10-fold cross-valuation on the training set and then evaluate the model's performance using the test set. Different evaluation criteria will be used for ATE and APC tasks.

---

[3] https://www.cs.york.ac.uk/semeval-2013/task9/.

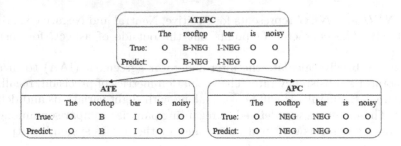

Fig. 3. Converting the ATE-PC output to ATE and APC outputs

ATE Task Evaluation: we defined the Precision, Recall and F_1 score as follow:

$$Precision = \frac{|S \cap G|_{COR} + 1/2 \times |S \cap G|_{PAR}}{|S|} \tag{4}$$

$$Recall = \frac{|S \cap G|_{COR} + 1/2 \times |S \cap G|_{PAR}}{|G|} \tag{5}$$

$$F_1 = \frac{2 \times Precision \times Recall}{Precision + Recall}, \tag{6}$$

where S is a set of predicted aspect terms, G is a set of true aspect terms given by annotated dataset. $|S \cap G|_{COR}$ indicates the number of strict correct aspect terms and $|S \cap G|_{PAR}$ is the number of partial correct aspect terms between S and G. Table 4 shows the averaged result for ATE task:

Table 4. 10-fold cross validation of ATE performance

$|S \cap G|_{COR}$	$|S \cap G|_{PAR}$	Precision	Recall	F_1-score
995	52	0.8693	0.8772	0.8732

With 0.8732 for F_1 score, it is an acceptable result for the complex and complicated task. The result confirms the application of the BiLSTM-CRF model for the ATE task on hospitality media dataset.

APC Task Evaluation: we introduce accuracy to evaluate the performance of the model only for the strict correct aspects extracted by the model. The accuracy is defined as Eq. 7:

$$ACC = \frac{|C|}{|S \cap G|_{COR}}, \tag{7}$$

where C indicates a set of correct aspects where their polarities are correctly detected by the model, and $C \subseteq (S \cap G)_{COR}$. Table 5 shows the performance of the model for APC task.

Table 5. 10-fold cross validation of APC performance

$|S \cap G|_{COR}$	$|C|$	ACC (%)
995	776	80.00

The accuracy score (ACC) indicates that the strict correct aspects terms identified from the model using test data are classified 80% correctly.

ATE-PC for New Hospitality Media Dataset: After training model and test its performance using BiLSTM-CRF, we feed the whole new dataset into the model to extract the aspect terms and their polarities. Given 532,164 review sentences, the model extracted 582,084 aspect terms, including 407,617 (70.03%) aspects with positive sentiment, 84,431 (14.50%) aspects with neutral sentiment, and 90,036 (15.47%) aspects with negative sentiment, as Table 6 showed below:

Table 6. The result for ATE-PC task with new dataset.

#Sentences	#Aspects	#Positive	#Neutral	#Negative
532,164	582,084	407,617	84,431	90,036

4.4   Topic Modeling and Grouping with LDA

For finding the optimal number of topics, we analyze the Coherence score [14], which is a measure used to evaluate topic model, for each number of topics ranges from 5 to 29. From the Fig. 4, we select the number of topics is 11 as the coherence scores are flattened out from it. Table 7 presents the top representative keywords for each topic.

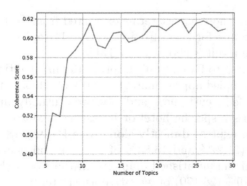

Fig. 4. Coherence scores.

Table 7. Topics and representative keywords

Topic	Keywords	Inferred topic
0	0.153*"breakfast" + 0.062*"food" + 0.037*"buffet"...	Food
1	0.079*"nice" + 0.065*"pool" + 0.053*"bar" + 0....	Hotel Facilities
2	0.120*"good" + 0.091*"service" + 0.047*"price"...	Price & Services
3	0.050*"book" + 0.043*"check" + 0.039*"taxi" + ...	Check in/out & Transportation
4	0.175*"staff" + 0.063*"friendly" + 0.056*"helpful"...	Staff
5	0.210*"stay" + 0.119*"night" + 0.046*"time" + ...	Internet & Devices
6	0.032*"bit" + 0.029*"feel" + 0.028*"work" + 0....	Other
7	0.118*"room" + 0.087*"clean" + 0.052*"bed" + 0...	Room (Internal)
8	0.145*"great" + 0.108*"good" + 0.066*"city" + ...	Restaurant
9	0.276*"room" + 0.073*"floor" + 0.061*"view" + ...	Room (External) & View
10	0.071*"location" + 0.063*"walk" + 0.048*"market"...	Location

We validate the inferred topics of the LDA model by firstly reading the most weighted keywords of each topic and secondly examining the top sentences belong to each topic. For example, topic 0 could be easily inferred with "Food", while the name for topic 5 is set by examining the top sentences as Table 8 follow:

Table 8. Examples for aspect terms assigned to topic 5, where the words (POS, NEG, NEU) after @ indicate the polarities along with aspects.

Sentences	Aspects	Topics
WiFi throughout the hotel is real bonus.	WiFi@POS	5@POS
During one time we needed to use a computer...	computer@NEU	5@NEU
The wifi signal is excellent and one of the best...	wifi signal@POS	5@POS
Air con was efficient and the wi-fi signal was good.	Air con@POS, wi-fi signal@POS	7@POS, 5@POS
..., and it was the best Internet signal...	Internet signal@POS	5@POS

4.5 Inferring Refined Topics and Summarization

In this stage, we feed those aspect terms extracted from Sect. 4.3 into the LDA model to infer the topics for them from the given 11 topics, and then make the summarization. The result is shown as in Table 9.

These topics related to Food, Staff, Internal Room Facilities, External Room Facilities & View occupied approximately 69.60% of all aspect terms mentioned on the reviews. In fact, customers and reviewers usually mention those topics when they discuss a specific hotel or its services. Moreover, topic for Food and its relations was expressed mostly along with Positive (86,698) and Neutral (14,210) sentiment (21.27% and 16.83% of positive and neutral aspect terms), customers expressed the unsatisfied opinions on the Room (External) & View, which occurred 31.29% (28,170) on negative aspect terms.

Figure 5 monitors the trend of positive aspect terms for each topic of 5-star hotels from 2007 to 2016 (by percent). The averaged percentage of positive aspect terms for all topics is larger than 50% except for topic "**Other**" (48.90%).

Table 9. The result for interring topics and summarization

Inferred topic	#Positive	#Neutral	#Negative	%Aspects
Food	**86,698**	**14,210**	9,100	18.90
Hotel Facilities	38,183	5,970	5,844	8.59
Price & Service	43,397	3,289	5,367	8.94
Check in/out & Transportation	6,414	3,323	1,831	1.99
Staff	55,672	7,632	7,120	12.10
Internet & Devices	7,859	365	3,326	1.98
Other	4,144	543	5,714	1.79
Room (Internal)	50,117	8,515	22,753	13.98
Restaurant	8,692	617	468	1.68
Room (External) & View	76,127	3,8997	**28,170**	24.62
Location	30,314	970	343	5.43

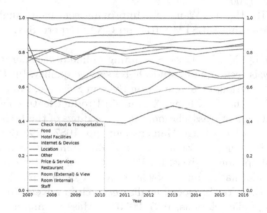

Fig. 5. The trend of positive aspect terms (by %) over 11 topics for 5-star hotels.

Moreover, customers were highly satisfied with those aspect terms related to "**Location**" and "**Restaurant**" of 5-star hotels in Ho Chi Minh City (96.20% and 90% on average).

5   Conclusions

In this study, we proposed a framework to summarize the customer's reviews by combining ATE-PC task with LDA model for the hotel reviews. We aim at improving the service operations and strategies in the hotel industry by automatically identifying the key aspect terms along with their polarities of the hotel services mentioned in the reviews. We is then inferring these aspect terms into 11 topics given from LDA model. This research also contributes to the analytics of hospitality media data for the hotel industry. The aggregate results show the strong performance of 5-star hotels on aspect terms related to 11 topics.

There are some limitations in our current work. Firstly, this study only focuses on the English reviews which might ultimately ignore the significant number of reviews from non-English reviewers. Moreover, the size of the annotated dataset was still limited, that might affect to the performance of ATE-PC task. Additionally, tracking sentiment for a topic over time should also take into consideration the seasonal effects especially in the hospitality industry, so that our study could be extended to monthly, rather than annually opinioned reviews.

References

1. Aggarwal, C.C.: Machine Learning for Text. Springer, New York (2018). https://doi.org/10.1007/978-3-319-73531-3
2. Blei, D.M., Ng, A.Y., Jordan, M.I.: Latent dirichlet allocation. J. Mach. Learn. Res. **3**, 993–1022 (2003)
3. Chen, T., Xu, R., He, Y., Wang, X.: Improving sentiment analysis via sentence type classification using BiLSTM-CRF and CNN. Expert. Syst. Appl. **72**, 221–230 (2017)
4. Duan, W., Cao, Q., Yu, Y., Levy, S.: Mining online user-generated content: using sentiment analysis technique to study hotel service quality. In: Proceedings of the 46th Hawaii International Conference on System Sciences, pp. 3119–3128 (2013)
5. Guo, H., Zhu, H., Guo, Z., Zhang, X., Su, Z.: Product feature categorization with multilevel latent semantic association. In: Proceedings of the 18th ACM Conference on Information and Knowledge Management, pp. 1087–1096 (2009)
6. Guo, H., Zhu, H., Guo, Z., Zhang, X., Su, Z.: Opinionit: a text mining system for cross-lingual opinion analysis. In: Proceedings of the 19th ACM International Conference on Information and Knowledge Management, pp. 1199–1208 (2010)
7. Khang, H., Ki, E.J., Ye, L.: Social media research in advertising, communication, marketing, and public relations, 1997–2010. J. Mass Commun. Q. **89**(2), 279–298 (2012)
8. Kim, Y.: Convolutional neural networks for sentence classification. In: Proceedings of the 2014 Conference on Empirical Methods in Natural Language Processing (EMNLP), pp. 1746–1751 (2014)
9. Lample, G., Ballesteros, M., Subramanian, S., Kawakami, K., Dyer, C.: Neural architectures for named entity recognition. In: Proceedings of NAACL-HLT, pp. 260–270 (2016)
10. Liu, B.: Sentiment Analysis and Opinion Mining. Morgan & Claypool Publishers, San Rafael (2012)
11. Menner, T., Höpken, W., Fuchs, M., Lexhagen, M.: Topic detection: identifying relevant topics in tourism reviews. In: Inversini, A., Schegg, R. (eds.) Information and Communication Technologies in Tourism 2016, pp. 411–423. Springer, Cham (2016). https://doi.org/10.1007/978-3-319-28231-2_30
12. Müller, R.M., Lenz, H.-J.: Business Intelligence. Springer, Heidelberg (2013). https://doi.org/10.1007/978-3-642-35560-8
13. Neidhardt, J., Rümmele, N., Werthner, H.: Predicting happiness: user interactions and sentiment analysis in an online travel forum. Inf. Technol. Tour. **17**(1), 101–119 (2017)

14. Newman, D., Lau, J.H., Grieser, K., Baldwin, T.: Automatic evaluation of topic coherence. In: Human Language Technologies: The 2010 Annual Conference of the North American Chapter of the Association for Computational Linguistics, pp. 100–108 (2010)
15. Nguyen, H., Shirai, K.: A joint model of term extraction and polarity classification for aspect-based sentiment analysis. In: Proceedings of the 10th International Conference on Knowledge and Systems Engineering (KSE), pp. 323–328 (2018)
16. Schmunk, S., Höpken, W., Fuchs, M., Lexhagen, M.: Sentiment analysis: extracting decision-relevant knowledge from UGC. In: Xiang, Z., Tussyadiah, I. (eds.) Information and Communication Technologies in Tourism 2014, pp. 253–265. Springer, Cham (2013). https://doi.org/10.1007/978-3-319-03973-2_19
17. Xiang, Z., Fesenmaier, D.R. (eds.): Analytics in Smart Tourism Design. TV. Springer, Cham (2017). https://doi.org/10.1007/978-3-319-44263-1
18. Zhai, Z., Liu, B., Xu, H., Jia, P.: Grouping product features using semi-supervised learning with soft-constraints. In: Proceedings of the 23rd International Conference on Computational Linguistics, pp. 1272–1280 (2010)

Factors Affecting Carbon Emissons in the G7 and BRICS Countries: Evidence from Quantile Regression

Yefan Zhou[1], Jirakom Sirisrisakulchai[1,3(✉)], Jianxu Liu[2,3], and Songsak Sriboonchitta[1,3]

[1] Faculty of Economics, Chiang Mai University, Chiang Mai 50200, Thailand
sirisrisakulchai@hotmail.com
[2] Faculty of Economics, Shandong University of Finance and Economics, Jinan, China
[3] Puey Ungphakorn Center of Excellence in Econometrics, Chiang Mai University, Chiang Mai 50200, Thailand

Abstract. Over the last three decades, the relationship between growth, foreign direct investment and carbon emissions has become an important issue among environmental economists. The target of this study is to investigate the impact of economic growth and foreign direct investment on carbon emissions in order to provide environmental improvement suggestions. This study tests the validity of the Environmental Kuznets Curve (EKC) hypothesis, including G7 countries (Canada, United State, United Kingdom, Japan, France, Germany, Italy) and BRICS countries (Brazil, Russia, India, China and South Africa) for the period of 1992–2014. We adopt a panel quantile regression model that takes unobserved individual heterogeneity and distributional heterogeneity into consideration. Moreover, to avoid an omitted variable bias, certain related control variables are included in our model. Firstly, our empirical results show that the effect of the independent variables on carbon emissions is heterogeneous across quantiles. Secondly, regarding the impact of FDI on carbon emissions, we find that these results support the pollution halo theory in G7 countries and support the pollution haven hypothesis in BRICS countries. Thirdly, the empirical findings are in support of inverted U-shaped curve of EKC in G7 countries. Finally, the results of the study also provide policymakers with important policy recommendations. In addition, our findings suggest carbon emissions control measures should be tailored differently across low-emissions and high-emissions nations.

Keywords: Carbon emissions · Economic growth · Panel quantile regression · EKC

Supported by Puey Ungphakorn Center of Excellence in Econometrics, Faculty of Economics, Chiang Mai University.

© Springer Nature Switzerland AG 2019
H. Seki et al. (Eds.): IUKM 2019, LNAI 11471, pp. 406–417, 2019.
https://doi.org/10.1007/978-3-030-14815-7_34

1   Introduction

At present, climate changing and environment pollution are consider as one of serious issue for international community. The problem of atmospheric environmental pollution caused by excessive carbon emissions has gradually lead to widespread concern in the world. It is a global environmental pollution problem involved in social production, life and other fields. Therefore, not only will it affect development of the economic in the future, but also impact the choice of the current economic development path. In addition, it will affect distribution of patterns of economic interests and the policy choice of countries around the world. In the view of the world economic development, all countries present the obvious stage characteristics. The main developed countries have finished the industrialization development of high energy consumption and high CO_2 emission. Also, industry structure of developed countries is mainly composed of the tertiary industry which has low energy consumption and low CO_2 emissions. However, developing countries are still at the process of industrialization and the economic development pattern is given priority to the secondary industry, which is high energy consumption with high CO_2 emissions.

Foreign direct investment (FDI) inflows have rapidly increased during the past three decades in almost every region of the world, thus revitalizing the long debate in both academic and policy spheres about their advantages and related costs. Indeed, FDI inflows may provide direct capital financing, generate positive externalities, and consequently stimulate economic growth through technology transfer, spillover effects, productivity gains, and the introduction of new processes and managerial skills. By contrast, they are also considered as one of the major factors that could lead to environmental degradation. FDI mainly influences the environment through technical effect, structural effect and scale effect [1]. But FDI plays an active role in the technical and structural effects, which is beneficial to the reduction of carbon emission intensity, but the increase of FDI plays an important role in the increase of carbon emissions. Therefore, the pollution fees caused by carbon emissions should also be taken into account when evaluating the value of economic growth. On one hand, it is highly possible that consumption of fossil fuels will positively influence economic growth since fossil fuels are inputs of production process. On the other hand, causality can also be expected to run from economic growth to carbon emissions through the income effect.

With the rapid growth of the economy, the industrial structure has changed. Continuous and rapid economic growth generates a series of benefits such as increased income, social stability and increased employment. Yet, emerged with rapid economic growth are woefully some negative phenomena, including excessive energy wastes and environmental degradation. Some researchers proposed the concept of the EKC, which is an inverted U-shape relationship between economic growth and environment quality. This might be taken to suggest future energy. Therefore, the world as a whole is facing the great pressure to control the increase of carbon dioxide emissions.

The target of this study is to investigate the impact of economic growth and foreign direct investment on carbon emissions in order to provide environmental improvement suggestions. This study tests the validity of the EKC hypothesis, including G7 countries (Canada, United State, United Kingdom, Japan, France, Germany and Italy) and BRICS countries (Brazil, Russia, India, China and South Africa) for the period of 1992–2014. We adopt a panel quantile regression model that takes unobserved individual heterogeneity and distributional heterogeneity into consideration.

2   Literature Review and Theoretical Background

2.1   Literature Review

There are two main research groups studying the environment FDI-growth nexus in the environmental economics literature.

Firstly, according to the meaning work of Grossman and Krueger [2], scholars focus on examining the relationship between economic growth and environmental pollution under the validity of the EKC hypothesis. Although the EKC hypothesis illustrates an inverted U-shaped relationship between economic growth and CO_2 emissions, there is some evidence that the EKC hypothesis is a linear relationship [3]. Another evidence of EKC hypothesis on N-shaped relationship was shown by He and Richard [4]. However, some evidences suggest different configurations. For example Zhu et al. [5] found little evidence in support of an inverted U-shaped curve in 5-ASEAN countries. In addition, recent studies appear to present mixed empirical results on the validity of the EKC. For example, Acaravci found a positive long-run elasticity estimate of emissions for nineteen European countries by using autoregressive distributed lag (ARDL) and this studies supported the validity of EKC hypothesis in Denmark and Italy [6].

Secondly, studies focus on FDI-carbon emissions nexus. FDI has become increasingly important around the world. Indeed, the rising FDI flow in developing countries raises an important question regarding whether it has any environmental consequence [7]. Therefore, research on the effect of FDI on carbon emissions is necessary. Although developing countries are active in attracting FDI, previous studies lack an analysis of the complexity correlation of FDI and carbon emissions as well as the causality, which leads to poorer discernment in the pollution haven hypothesis. The conventional view may suggest that, with relaxed environmental standards in developing countries, FDI may promote carbon emissions at large. To attract foreign investment, developing countries have a tendency to ignore environmental concerns through relaxed or non-enforced regulation; in economic theory, this phenomenon is designated the pollution haven hypothesis. However, the effect of FDI can be inverted when low-carbon technologies are introduced to reduce the carbon dioxide emissions by FDI as a whole or when FDI flows to focus on the service industry.

It is believed that foreign companies use better management practices and advanced technologies that are conducive to a clean environment in host countries [8], which is known as the halo effect hypothesis. Similarly, Zeng and Eastin found that overall FDI inflows in less-developing countries promote better environmental awareness [7].

2.2 Theoretical Background

Pollution Haven Hypothesis (PHH) states that natural and environmental resources are scarce endowments attracting polluting industries around the world and has raised diverse research interest and sparked lasting and inconclusive debates. Developing countries uses cheap resources and labor tend to have less stringent environmental regulations. However, the countries with stricter environmental regulations become more expensive for companies as a result of the costs associated with meeting these standards. Therefore, countries and areas may consequently become "pollution havens" for global polluting industries, and the efforts at pollution regulation in developed countries may lead to the transfer of industrial pollution rather than an improvement of global environmental quality [9,10].

Pollution halo Theory states that FDI mainly refers to the transfer of intensive industries to developing countries, and has a positive impact on the host country's environment. The inflow of FDI can not only have a good impact on the environment through technology, but also bring in the good green management system formed in the developed countries, which has a positive impact on the environment of the host country [10].

3 Methodology

3.1 Methodology and Data

The EKC hypothesis is basically used to examine the relationships between economic growth and environmental pollution.

Following Grossman et al. [2] the general form of the EKC hypothesis can be formulated as:

$$E = f\left(Y, Y^2, Z\right) \tag{1}$$

where E is a level of pollution measured as per capita carbon emission, Y is the income measured as real GDP per capita constant USD, and Z are other variables which affect the emissions. The quadratic form of income $\left(Y^2\right)$ will allow the model to capture the inverted U-shape relationship between emissions and income.

In this paper, we will use a fixed effect panel quantile regression model to investigate the impact of economic growth, FDI, and energy consumption and other control variable on carbon emissions. By using a panel quantile regression methodology, we can examine the determinants of carbon emissions throughout the conditional distribution, especially in the countries with the most and least emissions. However, traditional regression techniques focus on the mean effects, which may lead to under- or over-estimating the relevant coefficient or even failing to detect important relationships.

The quantile regression technique was introduced in the seminal paper by Koenker and Bassett [11]. This method is a generalization of median regression analysis to other quantiles. The conditional quantile of y_i given x_i is as follows

$$y_i = (\tau|x_i) = x_{it}\beta_\tau \qquad (2)$$

We studied the effect of economic growth, energy consumption, industrial structure, and financial development on carbon emissions by modifying the specifications of previous studies. But the variable of industrial structure, financial development and trade openness is control variable. We specify the conditional quantiles function for quantile τ as follows:

$$Q_{y_{it}}(\alpha_{it}, \xi_{it}, x_{it}) = \alpha_i + x_t + \beta_{1\tau}FDI_{it} + \beta_{2\tau}GDP_{it} + \beta_{3\tau}GDP_{it}^2 \\ + \beta_{4\tau}EC_{it} + \beta_{5\tau}POP_{it} + \beta_{6\tau}UR_{it} + \beta_{7\tau}Renew_{it} \qquad (3)$$

where the countries are indexed by i and time by time t. y_{it} is the emissions indicator.

3.2   Data

In this study, we used annual data of the G7 countries (Canada, United State, United Kingdom, Japan, France, Germany, Italy) and BRICS (Brazil, Russia, India, and China) countries for the period of 1992–2014 to investigated the impact of economic growth, energy consumption and FDI on carbon emissions. BRICS are five main emerging economics in the world while G7 countries are developed countries.

Table 1. Variable definition

Variable	Definition	Source
CO_2	Carbon dioxide emissions (metric tons per capita)	World Development Indicators
EC	Energy consumption (kg of oil equivalents per capita)	World Development Indicators
GDP	Economic growth (real GDP per capita constant USD at 2010 prices)	World Development Indicators
POP	Population, Total	World Development Indicators
UR	Urban population (% of total)	World Development Indicators
FDI	Foreign direct investment, net inflows (% of GDP)	World Development Indicators
$Renew$	Renewable energy consumption (% of total final energy consumption)	World Development Indicators

The dependent variable is carbon dioxide emissions as a measured of environmental pollution. And the main independent variables are economic growth, energy consumption and foreign direc FDI. Economic growth is expressed by real GDP per capita in constant USD at 2010 prices. And energy consumption is expressed in terms of kg of oil equivalents per capita. FDI is the net inflow as a share of GDP. The definitions and statistical description of all the variables in this study are shown on Table 1. And all variables are transformed into natural logarithms prior to estimation to help eliminate heteroscedasticity.

4 Empirical Findings

4.1 Panel Root Test Results

We applied Levin et al. (LLC), Im et al. (IPS), Maddala and Wu (MW, ADF) [12] panel unit root to examines whether there is a unit root in the sequence of economic variables. Our empirical results observed that the null hypothesis of the existence of a unit root could be rejected for one of the variables in their level form [13]. However, the unit root null hypothesis for one of the variables at the first difference could almost be completely rejected at 1% level. Therefore, we reject the null hypothesis of non-stationary at 1% level of significance and uses the first difference in the panel of G7 group countries and BRICS countries.

4.2 Panel Cointegration Results

As the results of the panel unit root tests indicate that the variables contain a panel unit root at their first difference, we can proceed to examine whether there is a long-run relationship among these variables using the Johansen Fisher panel cointegration test proposed by Maddala and Wu [12]. In the Johansen-type panel cointegration test, results are known to depend heavily on the VAR system lag order. Our results from the use of one lag indicate that six cointegrating vectors exist. Thus, energy consumption, economic growth, foreign direct investment, total population, urbanization level and renewable energy consumption are cointegrated in selected panels G7 group countries and BRICS countries for the period 1992–2014.

4.3 Quantile Regression

The distribution characteristics of carbon emissions can completely on each quantile. And the quantile regression can visually observe the marginal effects of the dependent variables on different quantiles of carbon emissions. Therefore, this paper got quantile regression results of carbon emissions. We chose 11 representative quantiles (from 5th to 95th quantiles) to complete the quantile regression. According to annual average CO_2 emissions metric tons per capita, G7 group countries and BRICS countries are divided into six grades (Table 2).

Regarding FDI for G7 countries, we can observe that the impact of FDI on carbon emissions is clearly heterogeneous. At the 10th, 50th, 75th, 90th quantile,

Table 2. Countries distribution in term of CO_2 emissions metric tons per capita

Quantile	G7 group	Brics
The lower 10th quantile group		India, Brazil
The 10th–25th quantile group	France	China
The 25th–50th quantile group	Italy	
The 50th–75th quantile group	United Kingdom	South Africa
The 275th–90th quantile group	Japan, Germany	Russian Federation
The upper 90th quantile group	Canada, United States	

the coefficient of Δ FDI is negative and significant at the 1% level. The negative coefficient of FDI is sufficient support pollution halo theory in G7 countries. However, the coefficient of FDI on carbon emissions is positive and significant at 1% level from 25th quantile and 90th quantile for BRICS countries. The positive coefficient of FDI for BRICS countries support the pollution heaven hypothesis. We can observe that FDI has an positive influence on carbon emissions for BRICS, implying that most foreign direct investment likely invests in polluted sectors in BRICS countries. However, it may pay more attention to environmental problems for G7 countries. And their environmental regulations are stricter than developing countries. Thus, with the rapidly increase of the economy and the development of technology, the main developed countries have finished the industrialized development of high energy consumption and high carbon emissions. And their industrial structure has changed. So in developed countries, FDI inflowing plays an important role to help to develop environmental and specialized technological skills and technological innovation in production. However, developing countries and areas may become "pollution havens" for

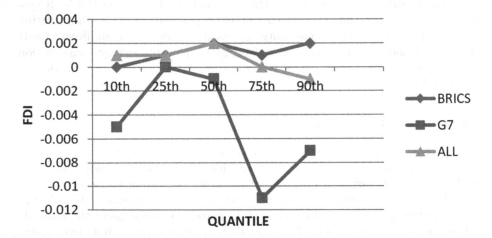

Fig. 1. Change in panel quantile regression coefficient for FDI (1992–2014)

global polluting industries. And because of strict pollution regulation in developed countries, it may lead to the transfer of industrial pollution to developing countries (Fig. 1 and Table 3).

Table 3. Panel regression results for BRICS (1992–2014)

Variable	Quantiles regressions				
	10	25	50	75	90
ΔEC	1.265***	1.089***	1.007***	0.999***	1.317***
	(32.064)	(39.479)	(165.046)	(18.559)	(7.518)
ΔFDI	0.000	0.001***	0.002***	0.001***	0.002***
	(0.032)	(4.575)	(41.822)	(24.677)	(264.638)
ΔGDP	−0.177***	−0.001	0.074***	0.080	0.151
	(−5.1596)	(−0.043)	(31.501)	(0.740)	(1.420)
ΔGDP^2	2.190***	0.869***	0.326***	−0.402	−0.684*
	(14.436)	(3.655)	(6.583)	(−0.542)	(−1.929)
ΔPOP	0.695***	0.435***	0.690***	0.994***	1.107***
	(8.962)	(4.282)	(35.649)	(6.916)	(3.916)
$\Delta RENEW$	−0.146***	0.143***	−0.156***	−0.196***	−0.041
	(−13.866)	(−6.241)	(−47.949)	(−7.615)	(−0.419)
ΔUR	−0.746***	−0.402***	−0.396***	0.176	0.561*
	(−8.981)	(6.006)	(−21.105)	(0.922)	(3.485)

Note: Figures in parentheses are t-values *** Statistical significance at the 1% level. ** Statistical significance at the 5% level. * Statistical significance at the 1% level.

Regarding economic growth, firstly, we can observe that the impact of economic growth for the group of BRICS countries on carbon emissions is positive. There are some significant diffcrent percentiles in the conditional distribution of ΔCO_2. The coefficient of ΔGDP is highly significant and has a positive sign at various quantiles, which increases along with the increase in the ΔCO_2 quantiles. So with the development of economics, carbon emissions is increasing in emerging countries. The result imply that developing countries have to pay more attention to two major issues which are the development of economic and environmental problem. Secondly, the impact of economic growth for G7 countries on carbon emissions is positive. Initially, the coefficient of ΔGDP slightly increases from 10th quantile to 25th quantile then turns to significantly decrease to the lowest point at 50th quantile. Then the coefficient of economic growth increases along the increasing path of carbon emissions from the 50th quantile to the 90th quantile at the peak (Tables 4 and 5).

In terms of ΔGDP^2 for G7 countries, the impact of ΔGDP^2 on carbon emissions is clearly heterogeneous. It is significant and negative at all quantiles. Therefore, the results support EKC hypothesis that the environmental degeneration rises at the first stage with increasing economic growth and then turns

Table 4. Panel regression results for G7 (1992–2014)

Variable	Quantiles regressions				
	10	25	50	75	90
ΔEC	1.177***	1.152***	1.100***	0.945***	0.921***
	(193.173)	(82.045)	(162.465)	(895.737)	(664.025)
ΔFDI	−0.005**	0.000	−0.001 ***	−0.011***	−0.007***
	(−2.455)	(0.452)	(−4.703)	(−3.382)	(−2.681)
ΔGDP	0.162***	0.200***	0.007	0.106***	0.208***
	(21.730)	(15.560)	(0.458)	(36.258)	(940.916)
ΔGDP^2	−1.699***	−5.012***	−1.517***	−2.463***	−2.721***
	(−54.869)	(−17.176)	(−3.847)	(−56.912)	(−70.832)
ΔPOP	−0.431***	−0.109**	−0.133***	0.151***	0.122***
	(−139.789)	(−2.286)	(4.393)	(−22.754)	(−109.049)
$\Delta RENEW$	0.027***	0.015***	−0.006***	−0.016***	−0.017***
	(22.968)	(6.484)	(−3.931)	(−64.540)	(−85.836)
ΔUR	0.351***	0.898***	0.976***	1.676***	1.371***
	(32.264)	(9.052)	(31.058)	(194.925)	(186.539)

Note: Figures in parentheses are t-values *** Statistical significance at the 1% level. ** Statistical significance at the 5% level. * Statistical significance at the 1% level.

Table 5. Panel regression results for all countries (1992–2014)

Variable	Quantiles regressions				
	10	25	50	75	90
ΔEC	1.168***	1.159***	1.110***	1.054***	0.991***
	(177.444)	(143.820)	(238.012)	(107.215)	(95.791)
ΔFDI	0.001**	0.001***	0.002***	0.000	−0.001***
	(10.297)	(2.677)	(8.957)	(0.669)	(−2.808)
ΔGDP	0.151***	−0.020***	0.053***	−0.040***	0.018
	(25.264)	(−2.072)	(3.568)	(−3.195)	(3.018)
ΔGDP^2	−0.247***	−0.086*	−0.304*	−0.069	−0.596***
	(−2.891)	(−0.680)	(−1.742)	(−0.664)	(−4.837)
ΔPOP	−0.202***	0.101*	0.371***	0.164***	0.750***
	(−5.391)	(1.816)	(17.150)	(3.749)	(11.620)
$\Delta RENEW$	0.010***	−0.009***	−0.025***	−0.040***	−0.047***
	(2.934)	(−2.662)	(−11.931)	(−16.738)	(−11.378)
ΔUR	−0.573***	0.023	0.254***	1.085***	0.926***
	(−17.690)	(−0.595)	(3.960)	(6.252)	(0.949)

Note: Figures in parentheses are t-values *** Statistical significance at the 1% level. ** Statistical significance at the 5% level. * Statistical significance at the 1% level.

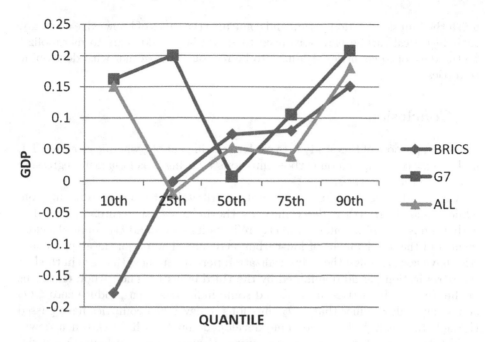

Fig. 2. Change in panel quantile regression coefficient for GDP (1992–2014)

to decrease at the final stage after reaching a threshold level given the level of income. However, regarding as ΔGDP^2 for BRICS countries. We can observe that the coefficient of ΔGDP^2 is positive and strongly significant from 10th quantile to 50th quantile. However, the coefficient of ΔGDP^2 turns to negative and highly significant at 90th quantile. On one hand, the results of BRICS countries imply that economic growth increases lead to the carbon emissions increase and cannot support the EKC hypothesis. On the other hand, the result of G7 countries indicates that when the higher level of economic development can mitigate the increase of carbon cmission for higher income of the countries. Therefore, the results support EKC hypothesis that the environmental degeneration rises at the first stage with increasing economic growth and then turns to decrease at the final stage after reaching a threshold level given the level of income. In addition, the developed countries of the coefficient of ΔGDP are greater than developing countries. One possible explanation of this phenomenon is that developing countries may not have achieved desired level of income at the development stage. Overall, GDP per capita reflects the level of income and social development. The group of G7 countries exhibits an inverted U-shaped curve, meaning that the economic growth level initially increases and then decreases along with the increase of carbon emissions (Fig. 2).

The results illustrate the effects of energy consumption on carbon emissions to be greater in developed countries than in developing countries under the different quantiles. The results indicate that although the developed countries finish the industrialization period and come into the times of knowledge economy, their historical carbon emissions need to spend 50 to 200 years to metabolize. So the developed countries still have to be responsible together with developing countries.

5 Conclusion

The main purpose of this study is to explore the impact of economic growth, FDI, and energy consumption on carbon emissions by using panel quantile regression to achieve the objectives.

Carbon dioxide emissions are mainly attributable to energy consumption, while economic growth is determined by the industrial structure. The level of carbon emissions of a country directly reflects its social and economic development and the development of low-carbon economy. The major developed countries have gone through the industrialization period, and now they are in the last industrialization period dominated by the third industry. The carbon emissions of the Tertiary Industries are low, and some industries even produce zero CO_2 emissions. It determines that CO_2 emissions in developed countries have passed through the peak period, and some developed countries have shown a downward trend in carbon emissions per capita. However, the developing countries are still in the process of industrialization. The mode of economic development in the Second Industries is a high CO_2 emission and a high energy consumption industry model.

Finally, based on the results of the empirical analysis. The following policy implications must be achieve the low carbon economy and low energy consumption development. Firstly, host countries should attempt to assess the environmental impact of FDI before introducing foreign investors into the country. The government should require to multinational companies adopt cleaner technology that will be less harmful to the environment. Secondly, the government of countries have to pay attention to improve environmental economy, specialized technological skills and technical innovative mechanism. Thirdly, in terms of energy consumption, especially developing countries that energy development program needs to shift from fossil fuels, such as oil, to clean and renewable energy [14]. The government and firms should promote development of new energy resources and renewable energy sources. Therefore, carbon emissions control measures and policy recommendations should be tailored differently across the nations with both low and high CO_2 emissions.

References

1. Lee, C.C., Chang, C.P.: FDI, financial development, and economic growth: international evidence. J. Appl. Econ. **12**(2), 249–271 (2009)
2. Grossman, G.H., Krueger, A.B.: Economic growth and the environment. J. Public Econ. **57**(1), 85–101 (1995)

3. Khan, S.A.R., Zaman, K., Zhang, Y.: The relationship between energy-resource depletion, climate change, health resources and the environmental Kuznets curve: evidence from the panel of selected developed countries. Renew. Sustain. Energy Rev. **62**, 468–477 (2016)
4. He, J., Richard, P.: Environmental Kuznets curve for CO_2 in Canada. Ecol. Econ. **69**(5), 1083–1093 (2010)
5. Zhu, H.M., Duan, L.J., Guo, Y.W., Yu, K.M.: The effects of FDI, economic growth and energy consumption on carbon emissions in ASEAN-5: evidence from panel quantile regression. Econ. Model. **58**, 237–248 (2016)
6. Acaravci, A., Ozturk, I.: On the relationship between energy consumption, CO_2 emissions and economic growth in Europe. Energy **35**(12), 5412–5420 (2010)
7. Zeng, K., Eastin, J.: Do developing countries invest up? The environmental effects of foreign direct investment from less-developed countries. World Dev. **40**(11), 2221–2233 (2012)
8. Pao, H.T., Tsai, C.M.: Multivariate Granger causality between CO_2 emissions, energy consumption, FDI (foreign direct investment) and GDP (gross domestic product): evidence from a panel of BRIC (Brazil, Russian Federation, India, and China) countries. Energy **36**(1), 685–693 (2011)
9. Zarsky, L.: Havens, halos and spaghetti: untangling the evidence about foreign direct investment and the environment. Foreign Direct Invest. Environ. **13**, 47–74 (1999)
10. Antonakakis, N., Chatziantoniou, I., Filis, G.: Energy consumption, CO_2 emissions and economic growth: an ethical dilemma. Renew. Sustain. Energy Rev. **68**, 808–824 (2017)
11. Koenker, R., Bassett, G.J.: Regression quantiles. Econometrica **46**, 33–50 (1978)
12. Maddala, G.S., Wu, S.: A comparative study of unit root tests with panel data and a new simple test. Oxf. Bull. Econ. Statement **61**(S1), 631–652 (1999)
13. Pesaran, M.H.: A simple panel unit root test in the presence of cross-section dependence. J. Appl. Econ. **22**(2), 265–312 (2007)
14. Alam, M.M., Murad, M.W., Noman, A.H.M., Ozturk, I.: Relationships among carbon emissions, economic growth energy consumption and population growth: testing environmental Kuznets curve hypothesis for Brazil, China, India and Indonesia. Ecol. Indic. **70**, 466–479 (2016)

Statistical Methods

Construction of Stable Hierarchy Organization from the Perspective of the Maximum Deng Entropy

Bingyi Kang[✉]

College of Information Engineering, Northwest A&F University,
Yangling 712100, Shaanxi, China
bingyi.kang@nwsuaf.edu.cn, bingyi.kang@hotmail.com

Abstract. A method of constructing hierarchy organization from the perspective of the maximum Deng entropy is proposed. The organization generated from the improved method can make the system take the task more stably and flexibly. A method of obtaining the ability of the department of the hierarchy organization through the maximum Deng entropy is proposed, which is compatible with the mechanic of the complex network. Some numerical examples are used to illustrate the effectiveness of the proposed methodologies.

Keywords: Hierarchy organization construction · Deng entropy ·
Maximum Deng entropy · Dempster-Shafer evidence theory

1 Introduction

Hierarchy organization is the common structure of the management organization [3–6,10,12]. Plenty of organizations including the government, enterprise, etc., applied hierarchy structure to ensure their operation workflow. Limited quantitative methods to deal with this problem provide us a possible contribution of this field. We found that Dempster-Shafer evidence framework is an inherent hierarchy structure, at least a possible way to generate the hierarchy structure of the organization although in the real environment the structure of the organization is more simple than the structure generating in the framework of Dempster-Shafer evidence theory. Through investigating the basic structure of hierarchy organization (hierarchy graph) from Dempster-Shafer evidence framework, we found the basic structure is not stable and not flexible.

How to generate a stable and flexible hierarchy construction of an organization is possibly a meaningful issue. We proposed a method of generating the hierarchy structure of an organization from the perspective of the mechanic of the maximum Deng entropy. By this method, the generated hierarchy construction of the organization is more stable and flexible. That means the work can continue in the organization although some departments are in break. We also proposed a method to measure the ability of the different department of the organization.

© Springer Nature Switzerland AG 2019
H. Seki et al. (Eds.): IUKM 2019, LNAI 11471, pp. 421–431, 2019.
https://doi.org/10.1007/978-3-030-14815-7_35

The results show that the proposed method to measure the ability of department is compatible with the mechanic of complex network. Some numerical examples are used to illustrate the effectiveness of the proposed methodology.

The paper is organized as follows. The preliminaries Dempster-Shafer evidence theory framework, Deng entropy are briefly introduced in Sect. 2. In Sect. 3, construction of classical hierarchy organization is discussed. In Sect. 4, some additional explanation about Deng entropy is presented. In Sect. 5, we propose a construction of hierarchy organization from the perspective of maximum Deng entropy. Finally, this paper is concluded in Sect. 6.

2   Preliminaries

In this section, some preliminaries are briefly introduced.

2.1   Framework of Dempster-Shafer Evidence Theory

Dempster-Shafer theory (short for D-S theory) is presented by Dempster and Shafer [1,11]. D-S theory has many advantages to handle uncertain information. First, D-S theory can handle more uncertainty in real world. In contrast to the probability theory in which probability masses can be only assigned to singleton subsets, in D-S theory the belief can be assigned to both singletons and compound sets. Second, in D-S theory, prior distribution is not needed before the combination of information from individual information sources. Third, D-S theory allows one to specify a degree of ignorance in some situations instead of being forced to be assigned for probabilities. D-S theory has been used plenty of application, e.g. prediction [7], data fusion [9]. Some basic concepts in D-S theory are introduced.

Let X be a set of mutually exclusive and collectively exhaustive events, indicated by

$$X = \{\theta_1, \theta_2, \cdots, \theta_i, \cdots, \theta_{|X|}\} \tag{1}$$

where set X is called a frame of discernment (FOD). The power set of X is indicated by 2^X, namely

$$2^X = \{\emptyset, \{\theta_1\}, \cdots, \{\theta_{|X|}\}, \{\theta_1, \theta_2\}, \cdots, \{\theta_1, \theta_2, \cdots, \theta_i\}, \cdots, X\} \tag{2}$$

For a frame of discernment $X = \{\theta_1, \theta_2, \cdots, \theta_{|X|}\}$, a mass function is a mapping m from 2^X to $[0, 1]$, formally defined by:

$$m: \quad 2^X \rightarrow [0, 1] \tag{3}$$

which satisfies the following condition:

$$m(\emptyset) = 0 \quad and \quad \sum_{A \in 2^X} m(A) = 1 \tag{4}$$

where A is a focal element if $m(A)$ is not 0.

2.2 Deng Entropy

With the range of uncertainty mentioned above, Deng entropy [2] can be presented as follows

$$E_d = -\sum_i m(F_i) \log \frac{m(F_i)}{2^{|F_i|} - 1} \tag{5}$$

where, F_i is a proposition in mass function m, and $|F_i|$ is the cardinality of F_i. As shown in the above definition, Deng entropy, formally, is similar with the classical Shannon entropy, but the belief for each proposition F_i is divided by a term $(2^{|F_i|} - 1)$ which represents the potential number of states in F_i (of course, the empty set is not included).

Specially, Deng entropy can definitely degenerate to the Shannon entropy if the belief is only assigned to single elements. Namely,

$$E_d = -\sum_i m(\theta_i) \log \frac{m(\theta_i)}{2^{|\theta_i|} - 1} = -\sum_i m(\theta_i) \log m(\theta_i)$$

3   Construction of Classical Hierarchy Organization

Firstly, we make two simple assumptions about the management departments and executing departments in the hierarchy organization according to the common environment.

Assumption 1. *Assume one can only take one task one time, suppose a* **single set** $\{\theta_i\} \neq \emptyset, i = 1, \dots |X|$ *represents the only executing department, and* $\theta_i, i = 1, \dots |X|$ *represents the basic function of the executing departments.*

For example, a simple construction of classical hierarchy organization is given in Fig. 1, A, B, C are the only executing departments, which can do the functions of a, b, c respectively.

Assumption 2. *If no confusion in the symbols, we assume arbitrary* **multiple set** M *in the organization* $(\emptyset \neq M \in P(X)$ *or* 2^X, $|M| \geq 2$, $P(X)$: *power set of* X, $X : FOD)$ *represents the managing department, which has the functions of management, checking and planing in the field of basic elements of M and* **doesn't** *participate in executing the real tasks.*

For example, a simple construction of classical hierarchy organization is given in Fig. 1, D, E are the management departments, which only consider the management and planing.

In the next section, we introduce the algorithm about the classical construction of hierarchy organization in the Dempster-Shafter evidence framwork.

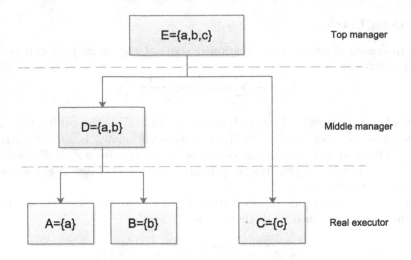

Fig. 1. Construction of classical hierarchy organization

3.1   Generate Classical Construction of Hierarchy Organization

Firstly, we assume the basic functions of the organization is $x_1, x_2, ..., x_n$, namely FOD in Dempster-Shafer evidence framework, the construction of the organization is a graph $G = (V, E)$, V are the nodes representing the different departments with different functions of the organization. E are the edges representing the relations of the departments. The number of nodes V is equal to the power set of FOD $X = \{x_1, x_2, ..., x_n\}$ except empty set. The method of generating the construction of hierarchy organization with basic functions $x_1, x_2, ..., x_n$ is denoted by Algorithm 1.

If we consider an organization with three basic functions a, b, c, the generated construction of hierarchy organization by Algorithm 1 is given in Fig. 2.

3.2   Problems of Construction of Classical Hierarchy Organization

The construction of the classical hierarchy organization is not stable and not flexible. For instance, in the hierarchy organization given as Fig. 2, F_1, F_2, F_3 are the three real executing departments, F_4, F_5, F_6 are the three middle management departments, and F_7 is the top management department according to the Assumptions 1 and 2. According to the two assumptions, F_4, F_5, F_6, F_7 are not responsible for executing the real task. Suppose we want to take a task which needs all the three basic functions a, b, c. Hence, the order is sent by the top management department F_7 to the middle management departments F_4, F_5, F_6, then F_4, F_5, F_6 divide the task to the real executors F_1, F_2, F_3. However, if any of real executors F_1, F_2, F_3 is in trouble (the management department are not responsible for executing real task), the task cannot be completed.

Algorithm 1. Classical construction of hierarchy organization

---

   Input: $FOD = X = \{x_1, x_2, ..., x_n\}$
   Output: $G = (V, E)$
1 **First step** : generate nodes V:
2 $v \in V, v \neq \emptyset, V = 2^X$
3 **Second step** : generate edges E:
4 $i = 1; j = 1; N = |2^X| - 1;$
5 **while** $i < N$ **do**
6   **while** $j < N$ **do**
7   **if** $v_i \subset v_j$ **then**
8 | $E_{i,j} = 1;$
9   **else**
10 | $E_{i,j} = 0;$
11   **end**
12       $i = i + 1$
13   **end**
14    $j = j + 1$
15 **end**

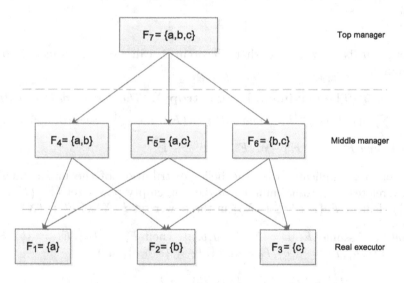

Fig. 2. Construction of classical hierarchy organization

Before we propose the method to improve the problems of construction of classical hierarchy organization using the maximum Deng entropy, we first make some explanations about the maximum Deng entropy.

4   Explanation of the Maximum Deng Entropy

Assume F_i is the focal element and $m(F_i)$ is the basic probability assignment for F_i, then the maximum Deng entropy for a belief function happens when the basic probability assignment satisfy the condition $m(F_i) = \frac{2^{|F_i|}-1}{\sum_i 2^{|F_i|}-1}$, where $i = 1, 2, ..., 2^X - 1$, and X is the scale of the frame of discernment.

Theorem 1 (The maximum Deng entropy). *The maximum Deng entropy:*
$E_d = -\sum_i m(F_i) \log \frac{m(F_i)}{2^{|F_i|}-1}$ *if and only if* $m(F_i) = \frac{2^{|F_i|}-1}{\sum_i 2^{|F_i|}-1}$

Proof. See Ref. [8]

Definition 1. *Non-empty power set S' is denoted as*

$$S' = P(S) - \emptyset \tag{6}$$

where $P(S)$ is the power set (or powerset) of any non-empty set S. It is easy to see that

$$|S'| = 2^{|S|} - 1 \tag{7}$$

Hence, it is easy to see that Theorem 1 can be transformed into the Theorem 2.

Theorem 2 (The maximum Deng entropy). *The maximum Deng entropy:*
$E_d = -\sum_i m(F_i) \log \frac{m(F_i)}{2^{|F_i|}-1}$ *if and only if* $m(F_i) = \frac{|F'_i|}{\sum_i |F'_i|}$ *where* $\emptyset \neq F_i \subseteq X, i = 2^{|X|} - 1$, *and* $F'_i = P(F_i) - \emptyset$, $|F'_i| = 2^{|F_i|} - 1$.

Theorem 2 implicates that the belief distribution of the maximum Deng entropy relates to cardinal number of the non-empty power set $(F', \{F' : F' \in P(F)\})$ element $F, \{F : F \in P(X)\}$ in the power set $P(X)$ of FOD (X).

Example 1. Assume $FOD : X = \{a, b, c\}$, then $F_1 = \{a\}, F_2 = \{b\}, F_3 = \{c\}, F_4 = \{a, b\}, F_5 = \{a, c\}, F_6 = \{b, c\}, F_7 = \{a, b, c\}$, and

$$
\begin{aligned}
F_1' &= \{a\} & F_4' &= \{\{a\}, \{b\}, \{a, b\}\} \\
F_2' &= \{b\} & F_5' &= \{\{a\}, \{c\}, \{a, c\}\} \\
F'_3 &= \{c\} & F_6' &= \{\{b\}, \{c\}, \{b, c\}\} \\
F_7' &= \{\{a\}, \{b\}, \{c\}, \{a, b\}, \{a, c\}, \{b, c\}, \{a, b, c\}\}
\end{aligned}
$$

According to the Theorem 1 or Theorem 2, we can get the basic probability assignment when the maximum Deng entropy can be obtained, the basic probability assignment is denoted as follows

$$m\left(F_1\right) = \frac{2^{|F_1|} - 1}{\sum\limits_{i=1}^{7}\left(2^{|F_i|} - 1\right)} = \frac{|F'_1|}{\sum\limits_{i=1}^{7}|F_i'|} = \frac{1}{1+1+1+3+3+3+7} = \frac{1}{19}$$

$$m\left(F_2\right) = \frac{2^{|F_2|} - 1}{\sum\limits_{i=1}^{7}\left(2^{|F_i|} - 1\right)} = \frac{|F'_2|}{\sum\limits_{i=1}^{7}|F_i'|} = \frac{1}{1+1+1+3+3+3+7} = \frac{1}{19}$$

$$m\left(F_3\right) = \frac{2^{|F_3|} - 1}{\sum\limits_{i=1}^{7}\left(2^{|F_i|} - 1\right)} = \frac{|F'_3|}{\sum\limits_{i=1}^{7}|F_i'|} = \frac{1}{1+1+1+3+3+3+7} = \frac{1}{19}$$

$$m\left(F_4\right) = \frac{2^{|F_4|} - 1}{\sum\limits_{i=1}^{7}\left(2^{|F_i|} - 1\right)} = \frac{|F'_4|}{\sum\limits_{i=1}^{7}|F_i'|} = \frac{3}{1+1+1+3+3+3+7} = \frac{3}{19}$$

$$m\left(F_5\right) = \frac{2^{|F_5|} - 1}{\sum\limits_{i=1}^{7}\left(2^{|F_i|} - 1\right)} = \frac{|F'_5|}{\sum\limits_{i=1}^{7}|F_i'|} = \frac{3}{1+1+1+3+3+3+7} = \frac{3}{19}$$

$$m\left(F_6\right) = \frac{2^{|F_6|} - 1}{\sum\limits_{i=1}^{7}\left(2^{|F_i|} - 1\right)} = \frac{|F'_6|}{\sum\limits_{i=1}^{7}|F_i'|} = \frac{3}{1+1+1+3+3+3+7} = \frac{3}{19}$$

$$m\left(F_7\right) = \frac{2^{|F_7|} - 1}{\sum\limits_{i=1}^{7}\left(2^{|F_i|} - 1\right)} = \frac{|F'_7|}{\sum\limits_{i=1}^{7}|F_i'|} = \frac{7}{1+1+1+3+3+3+7} = \frac{7}{19}$$

The maximum Deng entropy of Example 1 with $FOD : X = \{a, b, c\}$ can by obtained using Eq. (5): max $E_d = 4.2479$.

In the next section, we propose the construction of hierarchy organization from the perspective of maximum Deng entropy, and also propose how to measure the node or position in the hierarchy organization.

5 Proposed Construction of Hierarchy Organization from the Perspective of the Maximum Deng Entropy

In the next section, we propose the construction of hierarchy organization from the perspective of maximum Deng entropy.

5.1 Construction of Hierarchy Organization Using the Maximum Deng Entropy

We assume the basic functions of the organization is $x_1, x_2, ..., x_n$, namely FOD in Dempster-Shafer evidence framework, the construction of the organization is a graph $G = (V, E)$, V are the nodes representing the different departments with different functions of the organization. E are the edges representing the relations of the departments. The node v belongs to the power set of the element of the power set of FOD $X = \{x_1, x_2, ..., x_n\}$ except empty set according to the inherent mechanics of maximum Deng entropy (refer to Theorem 2 and Example 1). The improved method of generating the construction of hierarchy organization with basic functions $x_1, x_2, ..., x_n$ is denoted by Algorithm 2.

Algorithm 2. Construction of hierarchy organization using maximum Deng entropy

---

    Input: $FOD = X = \{x_1, x_2, ..., x_n\}$
    Output: $G = (V, E)$
1 **First step** : generate nodes V:
2 $v \in V, v \neq \emptyset, V = 2^F, F \in 2^X$; (Improved)
3 **Second step** : generate edges E:
4 $i = 1; j = 1; N = |2^X| - 1;$
5 **while** $i < N$ **do**
6   **while** $j < N$ **do**
7   **if** $v_i \subset v_j$ **then**
8           $E_{i,j} = 1;$
9   **else**
10          $E_{i,j} = 0;$
11   **end**
12        $i = i + 1$
13   **end**
14      $j = j + 1$
15 **end**

If we consider an organization with three basic functions a, b, c, the generated construction of hierarchy organization by Algorithm 2 is given in Fig. 3. In Fig. 3, the node is extended to the power set of the element of the power set of FOD $X = \{x_1, x_2, ..., x_n\}$ except empty set. For example, node $F_7 = \{a, b, c\}$ of the classical hierarchy organization in Fig. 2 is extended to $F_7' = \{a, b, c\}' = \{\{a\}, \{b\}, \{c\}, \{a, b\}, \{a, c\}, \{b, c\}, \{a, b, c\}\}$. This change has two advantages. First, node F_7' can directly link to any other node, which means this department can directly send orders to any other node. Second, according to Assumptions 1 and 2, any management department can be as a executor except as only a manager because any one of nodes has the single element. This means if the real executors are in trouble, some other managers can also take the task too. This make the organization more stable and flexible comparing with the classical structure.

How to measure the ability or responsibility of the department of hierarchy organization is a possible meaningful issue, we propose a measurement of the ability of the department of the generated hierarchy organization in the next section.

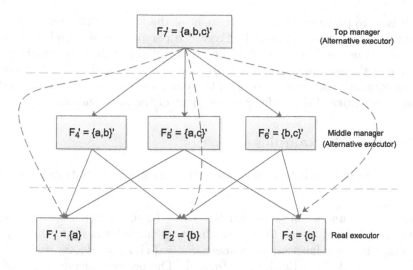

Fig. 3. Proposed construction of organization

5.2 Measuring the Ability of the Department of Hierarchy Organization Using the Maximum Deng Entropy

The ability of the department of the hierarchy organization depends on the numbers of inherent functions of the department. The proposed method is highly related to the ranking method of the field of complex network, which mainly depends on the out-degree of the nodes. (Note: out-degree of the node implicates how many departments of the organization the node can control.) The proposed method of ranking the ability or responsibility of the department considers not only the out-edge but also the inherent function of the node.

We define the ability of the department of hierarchy organization as follow

Definition 2. *The ability of the department of hierarchy organization using the maximum Deng entropy is defined as R, which is denoted as*

$$R\left(F_i'\right) = m\left(F_i\right) = \frac{2^{|F_i|} - 1}{\sum_i \left(2^{|F_i|} - 1\right)} = \frac{|F_i'|}{\sum_i |F_i'|} \tag{8}$$

where F_i', ($F_i' = P\left(F_i\right) - \emptyset$, $F_i \in P(X)$, $P(X)$: power set of X) refers to the ith node or department of the organization.

Definition 3. *If the number of out-edge of the node is given as $N_o(F_i')$, The ability of the department of hierarchy organization can also be defined as R, which is denoted as*

$$R\left(F_i'\right) = \frac{N_o\left(F_i'\right) + 1}{\sum_i \left[N_o\left(F_i'\right) + 1\right]} \tag{9}$$

where $N_o(F_i')$ is the number of out-edge of the node or department of the organization.

It is easy to prove that the measurement of the ability of the organization in Definitions 2 and 3 are compatible. From the perspective of complex network, the proposed method of measuring the ability of the node is reasonable. The mechanic of the maximum Deng entropy, i.e. the generated belief distribution using maximum Deng entropy using Theorem 1 or Theorem 2 can be used to measure the ability of the node or department of the organization.

6   Numerical Examples

We use an example to better understand the relation of Definitions 2 and 3 as follow.

Example 2. Assume an organization has three basic functions a, b, c meaning $FOD : X = \{a, b, c\}$ in the Dempster-Shafer evidence framework, then F_i is a proposition in mass function m, namely, $F_1 = \{a\}, F_2 = \{b\}, F_3 = \{c\}, F_4 = \{a, b\}, F_5 = \{a, c\}, F_6 = \{b, c\}, F_7 = \{a, b, c\}$. The nodes or departments of the organization $F_i, (i = 1, ..., 7)$ is denoted as

$$F_1' = \{a\} \quad F_4' = \{\{a\}, \{b\}, \{a, b\}\}$$
$$F_2' = \{b\} \quad F_5' = \{\{a\}, \{c\}, \{a, c\}\}$$
$$F'_3 = \{c\} \quad F_6' = \{\{b\}, \{c\}, \{b, c\}\}$$
$$F_7' = \{\{a\}, \{b\}, \{c\}, \{a, b\}, \{a, c\}, \{b, c\}, \{a, b, c\}\}$$

Through the Algorithm 2, the generated graph or structure of the organization is given in Fig. 3. Take node F_7' for example, the number of out-edge of node F_7' is $N_o(F_7') = 6$. Some other number of out-edge of nodes can also be obtained given in Table 1. The ability of the department of the organization is given in Table 1 from different ways by mechanic of the maximum Deng entropy, mechanic of transformed maximum Deng entropy, and mechanic of complex network. The three mechanic are inherent compatible.

Table 1. Ability R of generated graph or structure of the organization

Node	F_1'	F_2'	F_3'	F_4'	F_5'	F_6'	F_7'
$N_o(F_i')$	0	0	0	2	2	2	6
$\|F_i'\|$	1	1	1	3	3	3	7
$R(F_i') = \dfrac{2^{\|F_i\|}-1}{\sum_i \left(2^{\|F_i\|}-1\right)}$ [a]	$\frac{1}{19}$	$\frac{1}{19}$	$\frac{1}{19}$	$\frac{3}{19}$	$\frac{3}{19}$	$\frac{3}{19}$	$\frac{7}{19}$
$R(F_i') = \dfrac{N_o(F_i')+1}{\sum_i [N_o(F_i')+1]}$ [b]	$\frac{1}{19}$	$\frac{1}{19}$	$\frac{1}{19}$	$\frac{3}{19}$	$\frac{3}{19}$	$\frac{3}{19}$	$\frac{7}{19}$
$R(F_i') = \dfrac{\|F_i'\|}{\sum_i \|F_i'\|}$ [c]	$\frac{1}{19}$	$\frac{1}{19}$	$\frac{1}{19}$	$\frac{3}{19}$	$\frac{3}{19}$	$\frac{3}{19}$	$\frac{7}{19}$

[a] Mechanic of the maximum Deng entropy (Eq. (5)).
[b] Mechanic of complex network (Eq. (9)).
[c] Mechanic of transformed maximum Deng entropy (Eq. (8)).

7   Conclusion

A new perspective of generating the construction of hierarchy organization using the maximum Deng entropy is proposed. The organization generated from the improved method can make the system take the task more stable and flexible. We also proposed a method of obtaining the ability of the department of the hierarchy organization through the maximum Deng entropy, which is compatible with the mechanic of the complex network. Some numerical examples are used to illustrate the effectiveness of the proposed methodology. In the coming future, we will propose a method of how to optimally assign the task in the proposed structure of the organization if some nodes are in break.

Acknowledgment. The work is partially supported by a startup fund from Northwest A&F University (No. Z109021812).

References

1. Dempster, A.P.: Upper and lower probabilities induced by a multivalued mapping. Ann. Math. Stat. **38**(2), 325–339 (1967)
2. Deng, Y.: Deng entropy. Chaos Solitons Fractals **91**, 549–553 (2016)
3. Diefenbach, T., Sillince, J.A.: Formal and informal hierarchy in different types of organization. Organ. Stud. **32**(11), 1515–1537 (2011)
4. Elsner, W., Hocker, G., Schwardt, H.: Simplistic vs. complex organization: markets, hierarchies, and networks in an organizational triangle - a simple heuristic to analyze real-world organizational forms. J. Econ. Issues **44**(1), 1–29 (2010)
5. Fujie, R., Odagaki, T.: Self organization of social hierarchy and clusters in a challenging society with free random walks. Phys. A Stat. Mech. Appl. **389**(7), 1471–1479 (2010)
6. Fujie, R., Odagaki, T.: Self-organization of social hierarchy on interaction networks. J. Stat. Mech. Theory Exp. **2011**(06), P06011 (2011)
7. Kang, B., Chhipi-Shrestha, G., Deng, Y., Mori, J., Hewage, K., Sadiq, R.: Development of a predictive model for clostridium difficile infection incidence in hospitals using Gaussian mixture model and dempster-shafer theory. Stoch. Environ. Res. Risk Assess. **32**(6), 1743–1758 (2018)
8. Kang, B., Deng, Y.: The maximum deng entropy, p. viXra:1509.0119 (2015)
9. Kang, B., Deng, Y., Sadiq, R., Mahadevan, S.: Evidential cognitive maps. Knowl. Based Syst. **35**, 77–86 (2012)
10. Nguyen, D.T., Ng, D., Yap, P.S.: Instructional leadership structure in singapore: a co-existence of hierarchy and heterarchy. J. Educ. Adm. **55**(2), 147–167 (2017)
11. Shafer, G.: A Mathematical Theory of Evidence. Princeton University Press, Princeton (1976)
12. Valverde, S., Solé, R.V.: Self-organization versus hierarchy in open-source social networks. Phys. Rev. E **76**(4), 046118 (2007)

Restricted Similarity Functions, Distances and Entropies with Intervals Using Total Orders

Zdenko Takáč[1], Mária Minárová[2(✉)], Humberto Bustince[3,4],
and Javier Fernandez[3]

[1] Institute of Information Engineering, Automation and Mathematics,
Slovak University of Technology, Bratislava, Slovakia
`zdenko.takac@stuba.sk`
[2] Department of Mathematics and Descriptive Geometry,
Slovak University of Technology, Bratislava, Slovakia
`maria.minarova@stuba.sk`
[3] Departamento de Estadistica, Informatica y Matematicas,
Institute of Smart Cities, Universidad Publica de Navarra, Pamplona, Spain
`{bustince,fcojavier.fernandez}@unavarra.es`
[4] Navarrabiomed, IdiSNA, Irunlarrea 3, 31008 Pamplona, Spain

Abstract. The paper deals with theoretical investigation targeting in practical utilization in image processing. First the interval valued restricted equivalence functions together with interval valued dissimilarity functions with respect to total order are introduced. The novelty of the approach lays in the using of total orders of intervals and the fact that the outputs of the functions are intervals and not numbers as in previous works. Then both interval valued restricted equivalence functions and interval valued dissimilarity functions are aggregated and similarity measures, distance measures and entropies for interval-valued fuzzy sets are yielded, with respect to total order as well. The investigation is a continuation of previous works of the authors aiming in improvement of existing tools in image processing.

Keywords: Intervals · Total order · Restricted equivalence ·
Dissimilarity · Similarity · Distance · Entropy

1 Introduction

Comparison measures between fuzzy sets which are necessary for those applications where two different fuzzy objects must be compared, as it is the case of problems in image processing [5] or approximate reasoning [2,9–11], among many other fields. Up to now, the different similarity measures which have been

Supported by grant APVV-17-0066, APVV-14-0013, VEGA 1/0614/18 and TIN2016-77356-P(AEI/UE/FEDER) of the Spanish Government.

ⓒ Springer Nature Switzerland AG 2019
H. Seki et al. (Eds.): IUKM 2019, LNAI 11471, pp. 432–442, 2019.
https://doi.org/10.1007/978-3-030-14815-7_36

proposed in the literature to deal with interval-valued fuzzy sets provide as their result a real number in $[0, 1]$. But, the intervals used to provide the membership degree of an element to a given interval-valued fuzzy set can be considered to measure the lack of knowledge or certainty of the expert [6] in order to provide an exact membership degree for that element. For this reason, in this work we propose a new definition which provides as output an interval. The motivation of this investigation is in image processing and the theory is aimed to the practical exploitation, first of all in image processing.

2   Preliminaries

In this section, we introduce several well known notions and results which are necessary for our subsequent developments. We consider closed subintervals of the unit interval $[0, 1]$. In this sense, we denote:

$$L([0,1]) = \{[\underline{X}, \overline{X}] \mid 0 \leq \underline{X} \leq \overline{X} \leq 1\}.$$

We use capital letters to denote elements in $L([0, 1])$.

While fuzzy set in a universe U is a mapping $A : U \rightarrow [0, 1]$, an interval-valued fuzzy set is a mapping $A : U \rightarrow L([0, 1])$. The class of all fuzzy sets in U is denoted by $FS(U)$ and the class of all interval-valued fuzzy sets in U by $IVFS(U)$.

Another key notion in this work is that of order relation. We recall here its definition, adapted for the case of $L([0, 1])$.

Definition 1. *An order relation on $L([0, 1])$ is a binary relation $\leq$ on $L([0, 1])$ such that, for all $X, Y, Z \in L([0, 1])$,*

(i) $X \leq X$, (reflexivity),
(ii) $X \leq Y$ and $Y \leq X$ imply $X = Y$, (antisymmetry),
(iii) $X \leq Y$ and $Y \leq Z$ imply $X \leq Z$, (transitivity).

An order relation on $L([0, 1])$ is called total or linear if any two elements of $L([0, 1])$ are comparable, i.e., if for every $X, Y \in L([0, 1])$, $X \leq Y$ or $Y \leq X$. An order relation on $L([0, 1])$ is partial if it is not total.

Herein $\leq_{TL}$ denotes a total order in $L([0, 1])$ with the minimal element $0_L = [0, 0]$ and maximal element $1_L = [1, 1]$.

Example 1

(i) A total order on $L([0, 1])$ is, for example, the Xu and Yager's order (see [15]):

$$[\underline{X}, \overline{X}] \leq_{XY} [\underline{Y}, \overline{Y}] \quad \text{if} \quad \begin{cases} \underline{X} + \overline{X} < \underline{Y} + \overline{Y} \text{ or} \\ \underline{X} + \overline{X} = \underline{Y} + \overline{Y} \text{ and } \overline{X} - \underline{X} \leq \overline{Y} - \underline{Y}. \end{cases} \tag{1}$$

This definition of Xu and Yager's order was originally provided for Atanassov intuitionistic fuzzy pairs [15].

(ii) Another example of total order is as follows. If, for $\alpha \in [0, 1]$ we define the aggregation function

$$K_\alpha(x, y) = \alpha x + (1 - \alpha)y$$

then, for $\alpha, \beta \in [0, 1]$ with $\alpha \neq \beta$, we can obtain a total order $\leq_{\alpha\beta}$

$$[\underline{X}, \overline{X}] \leq_{\alpha\beta} [\underline{Y}, \overline{Y}] \quad \text{if} \quad \begin{cases} K_\alpha(\underline{X}, \overline{X}) < K_\alpha(\underline{Y}, \overline{Y}) \text{ or} \\ K_\alpha(\underline{X}, \overline{X}) = K_\alpha(\underline{Y}, \overline{Y}) \text{ and } K_\beta(\underline{X}, \overline{X}) \leq K_\beta(\underline{Y}, \overline{Y}). \end{cases} \tag{2}$$

The definition of an interval-valued implication operator with respect to a total order $\leq TL$ reads as follows.

Definition 2. *Let $\leq_{TL}$ be a total order in $L([0, 1])$. An interval-valued (IV) implication function in $L([0, 1])$ with respect to $\leq_{TL}$ is a function $I : (L([0, 1]))^2 \to L([0, 1])$ which verifies the following properties:*

(i) I is a non-increasing function in the first component and an non-decreasing function in the second component with respect to the order $\leq_{TL}$.
(ii) $I(0_L, 0_L) = I(0_L, 1_L) = I(1_L, 1_L) = 1_L$.
(iii) $I(1_L, 0_L) = 0_L$.

We will require the Ordering property (OP) in this paper.

(OP) $I(X, Y) = 1_L \Leftrightarrow X \leq_{TL} Y$.

Definition 3. *Let $n \geq 2$. An (n-dimensional) interval-valued (IV) aggregation function in $L([0, 1])$ with respect to $\leq_{TL}$ is a mapping $M : (L([0, 1]))^n \to L([0, 1])$ which verifies:*

(i) $M(0_L, \cdots, 0_L) = 0_L$.
(ii) $M(1_L, \cdots, 1_L) = 1_L$.
(iii) M is an non-decreasing function with respect to $\leq_{TL}$.

Definition 4. *Let $\leq_{TL}$ be a total order relation in $L([0, 1])$. A function $N : L([0, 1]) \to L([0, 1])$ is an interval-valued negation function (IV negation) if it is a decreasing function with respect to the order $\leq_{TL}$ such that $N(0_L) = 1_L$ and $N(1_L) = 0_L$. A negation N is called strong negation if $N(N(X)) = X$ for every $X \in L([0, 1])$. A negation N is called non-filling if $N(X) = 1_L$ iff $X = 0_L$, while N is called non-vanishing if $N(X) = 0_L$ iff $X = 1_L$.*

Example 2. Let $\leq_{TL}$ be a total order in $L([0, 1])$, and let N be an IV negation function with respect to that order. The function $I : L([0, 1])^2 \to L([0, 1])$ defined by

$$I(X, Y) = \begin{cases} 1_L, & \text{if } X \leq_{TL} Y, \\ \vee(N(X), Y), & \text{if } X >_{TL} Y, \end{cases}$$

is an IV implication function.

It is clear that the function I is a non-decreasing function in the second component and a decreasing function in the first component. Moreover

$$I(0_L, 0_L) = I(0_L, 1_L) = I(1_L, 1_L) = 1_L \text{ and } I(1_L, 0_L) = 0_L.$$

Example 3. Let $\leq_{TL}$ be a total order in $L([0,1])$, and let N be an IV negation function with respect to that order. If $M\colon L([0,1])^2 \to L([0,1])$ is an IV aggregation function, then the function $I\colon L([0,1])^2 \to L([0,1])$ defined by

$$I(X,Y) = \begin{cases} 1_L, & \text{if } X \leq_{TL} Y, \\ M(N(X),Y), & \text{if } X >_{TL} Y, \end{cases}$$

is an IV implication function.

The reasoning is analogous to the previous example.

3   Restricted Equivalence Functions in $L([0,1])$ with Respect to a Total Order

In accordance with the interpretation mentioned in Introduction, up to now any measure which compares two interval-valued fuzzy sets (and which starts with imprecise data) provides an exact value, which is much more precise than the original data.

For this reason, in this section we construct interval-valued equivalence functions [8] and restricted equivalence [3–5] from IV aggregation and negation functions. These functions are the basic elements that we are going to use to build comparison measures with respect to a total order in $L([0,1])$ such that their outcome is an interval and not a single number. We start extending to the case of total orders the definition of equivalence function.

Definition 5. *Let $\leq_{TL}$ be a total order in $L([0,1])$. A map $F\colon L([0,1])^2 \to L([0,1])$ is called an interval-valued (IV) equivalence function (with respect to $\leq_{TL}$)) if F verifies:*

(1) $F(X,Y) = F(Y,X)$ for every $X,Y \in L([0,1])$.
(2) $F(0_L,1_L) = F(1_L,0_L) = 0_L$.
(3) $F(X,X) = 1_L$ for all $X \in L([0,1])$.
(4) If $X \leq_{TL} X' \leq_{TL} Y' \leq_{TL} Y$, then $F(X,Y) \leq_{TL} F(X',Y')$.

Taking possible applications into account, Bustince et al., in [3] introduced in the fuzzy setting the notion of restricted equivalence function. This notion can be extended to the interval-valued setting with respect to a total order in the following way. The justification of using the restricted equivalence function can be found in more details in [3].

Definition 6. *Let $\leq_{TL}$ be a total order in $L([0,1])$. A map $F\colon L([0,1])^2 \to L([0,1])$ is called an interval valued (IV) restricted equivalence function (with respect to $\leq_{TL}$) if F verifies the following properties:*

1. $F(X,Y) = F(Y,X)$ for all $X,Y \in L([0,1])$.
2. $F(X,Y) = 1_L$ if and only if $X = Y$.
3. $F(X,Y) = 0_L$ if and only if $X = 0_L$ and $Y = 1_L$, or, $X = 1_L$ and $Y = 0_L$.
4. If $X \leq_{TL} Y \leq_{TL} Z$, then $F(X,Z) \leq_{TL} F(X,Y)$ and $F(X,Z) \leq_{TL} F(Y,Z)$.

In [3] the condition

$$(5)\ F(X,Y) = F(N(X), N(Y))\ \text{for all}\ X, Y \in L([0,1])$$

where N is a strong negation, was introduced to deal with some specific image processing problems (see [3]). However, since we are not considering such image processing problems in this work, we have preferred not to include it in the definition.

Remark 1. Note that every restricted equivalence function is in particular an equivalence function.

A very interesting aspect of restricted equivalence functions is that they can be related to the well-known notion of bi-implication in order to be characterized. In particular, we have the following result.

Theorem 1. *Let $\leq_{TL}$ be a total order in $L([0,1])$. A function $F: L([0,1])^2 \rightarrow L([0,1])$ is an interval-valued restricted equivalence function with respect to $\leq_{TL}$ if and only if it exists an IV implication function $I : L([0,1])^2 \rightarrow L([0,1])$ with respect to the same order $\leq_{TL}$ which verifies the (OP) property and such that, for every $X, Y \in L([0,1])$, the identity*

$$F(X,Y) = \wedge(I(X,Y), I(Y,X))$$

holds

Proof. The proof is analogous to the one for Theorem 7 in [3].

Example 4. Let $\leq_{TL}$ be a total order in $L([0,1])$. Let I be the IV implication function defined in Proposition 2 with a non-filling and non-vanishing IV negation N. Then, the function
$F: L([0,1])^2 \rightarrow L([0,1])$ defined by

$$F(X,Y) = \wedge(I(X,Y), I(Y,X)),$$

is an IV restricted equivalence function.

We extend the construction method for IV restricted equivalence functions applying a general IV aggregation function instead of the minimum, which generalizes the case of bi-implications.

Theorem 2. *Let $\leq_{TL}$ be a total order in $L([0,1])$. A function $F: L([0,1])^2 \rightarrow L([0,1])$ is an interval-valued restricted equivalence function with respect to $\leq_{TL}$ if and only if there exist an IV aggregation functions $M_1: L([0,1])^2 \rightarrow L([0,1])$ (with respect to $\leq_{TL}$) such that*

(i) $M_1(X,Y) = M_1(Y,X)$ for every $X, Y \in L([0,1])$,
(ii) $M_1(X,Y) = 1_L$ if and only if $X = Y = 1_L$, and
(iii) $M_1(X,Y) = 0_L$ if and only if $X = 0_L$ or $Y = 0_L$,

and an IV implication function $I : L([0,1])^2 \rightarrow L([0,1])$ *which satisfies the (OP) property and such that, for every* $X, Y \in L([0,1])$,

$$F(X,Y) = M_1(I(X,Y), I(Y,X)),$$

Proof. The proof is analogous to that of [13] for the real case.

Example 5. Let $\leq_{TL}$ be a total order in $L([0,1])$. Let $M_1 \colon L([0,1])^2 \rightarrow L([0,1])$ be an IV aggregation function (with respect to $\leq_{TL}$) as in Theorem 2 and let $M_2 \colon L([0,1])^2 \rightarrow L([0,1])$ be an IV aggregation function (with respect to $\leq_{TL}$) such that:

- $M_2(X,Y) = 1_L$ if and only if $X = 1_L$ or $Y = 1_L$, and
- $M_2(X,Y) = 0_L$ if and only if $X = Y = 0_L$.

Then, the function $F(X,Y) = M_1(I(X,Y), I(Y,X))$ is with I the IV implication function defined in the Proposition 3 with a non-filling and non-vanishing IV negation N and taking $M = M_2$, is an IV restricted equivalence function with respect to the order $\leq_{TL}$. Note that, actually:

$$F(X,Y) = \begin{cases} 1_L, & \text{if } X = Y, \\ M_1(M_2(N(Y), X), 1_L), & \text{if } X <_{TL} Y, \\ M_1(M_2(N(X), Y), 1_L), & \text{if } Y <_{TL} X, \end{cases}$$

Remark 2. Note that in Theorem 2 we use a construction method which is more general than the one based in the definition of bi-implication. However, we can use this result to get specific expressions which do not correspond to bi-implication functions. For instance, let's consider Xu and Yager's order. Let M_1 be the IV aggregation function

$$M_1(X,Y) = [\frac{X\overline{Y} + Y\overline{X}}{2}, \frac{XY + \overline{XY}}{2}]$$

and M_2 be the maximum (with respect to Xu and Yager's order). Consider also the IV negation:

$$N(X) = \begin{cases} 1_L & \text{if } X = 0_L \\ 0_L & \text{in other case.} \end{cases}$$

Then the function $F : L([0,1])^2 \rightarrow [0,1]$ given by

$$F(X,Y) = \begin{cases} 1_L & \text{if } X = Y \\ [\frac{X+\overline{X}}{2}, \frac{X+\overline{X}}{2}] & \text{if } X <_{XY} Y \\ [\frac{Y+\overline{Y}}{2}, \frac{Y+\overline{Y}}{2}] & \text{if } Y <_{XY} X, \end{cases}$$

is an example of IV restricted equivalence function which is not obtained using the minimum as it is the case for bi-implications.

As we have already said, the property $F(X,Y) = F(N(X), N(Y))$ is very important for image processing applications when we deal with real valued restricted equivalence functions. In this real case, involutivity, i.e., strong negations, are required. However, as we have seen, getting strong negations for a total order is not an easy task. Nevertheless, it is possible to provide a way of building IV functions F if such a strong negation is provided.

Proposition 1. *Let N be a strictly decreasing IV negation function with respect to a total order $\leq_{TL}$. Let $M_1, M_2 \colon L([0,1])^2 \to L([0,1])$ be two IV aggregation functions (with respect to the same order) as in Example 5. Assume that M_1 and M_2 also satisfy, for every $X, Y \in L([0,1])$:*

1. *If $X <_{TL} Y$ then $M_1(X, 1_L) <_{TL} M_1(Y, 1_L)$.*
2. *If $X <_{TL} Y$ then $M_2(X, 0_L) <_{TL} M_2(Y, 0_L)$.*
3. *$M_2(X, Y) = M_2(Y, X)$.*

Then, the function

$$F(X,Y) = \begin{cases} 1_L, & \text{if } X = Y \\ M_1(M_2(N(Y), X), 1_L), & \text{if } X <_{TL} Y \\ M_1(M_2(N(X), Y), 1_L), & \text{if } Y <_{TL} X \end{cases}$$

and, in particular the function

$$F(X,Y) = \begin{cases} 1_L, & \text{if } X = Y \\ \vee(X, N(Y)), & \text{if } X <_{TL} Y \\ \vee(N(X), Y), & \text{if } Y <_{TL} X \end{cases}$$

verifies that $F(X,Y) = F(N(X), N(Y))$ if and only if N is a strong negation function.

Proof. ($\Leftarrow$) Let $X <_{TL} Y$. Then $N(Y) <_{TL} N(X)$ and

$$F(N(X), N(Y)) = M_1(M_2(N(N(X)), N(Y)), 1_L) = M_1(M_2(X, N(Y)), 1_L)$$
$$= M_1(M_2(N(Y), X), 1_L) = F(X, Y)$$

($\Rightarrow$) Let $X <_{TL} 1_L$. Then $N(X) > N(1_L) = 0_L$,

$$F(N(X), N(1_L)) = M_1(M_2(N(N(X)), 0_L), 1_L)$$

and

$$F(X, 1_L) = M_1(M_2(N(1_L), X), 1_L) = M_1(M_2(0_L, X), 1_L) .$$

So $M_2(N(N(X)), 0_L) = M_2(0_L, X)$ and N is strong.

3.1 Dissimilarity Functions in $L([0,1])$ with Respect to a Total Order

Along this section we also consider that the considered orders between intervals are total.

In the same way as equivalence and restricted equivalence functions measure up to what extent two given (real-valued or interval-valued) data are similar, dissimilarity functions can be used to measure up to what extent they are different. But the same remarks about the necessity of getting an interval-valued result when we are dealing with interval-valued fuzzy sets are true here.

The interval-valued definition of such functions reads as follows [4,5].

Definition 7. *Let $\leq_{TL}$ be a total order in $L([0,1])$. A function $d_L \colon L([0,1])^2 \to L([0,1])$ is called interval valued (IV) restricted dissimilarity function (with respect to $\leq_{TL}$) if d_L satisfies the following conditions:*

1. *$d_L(X,Y) = d_L(Y,X)$ for all $X,Y \in L([0,1])$.*
2. *$d_L(X,Y) = 0_L$ if and only if $X = Y$.*
3. *$d_L(X,Y) = 1_L$ if and only if $X = 0_L$ and $Y = 1_L$, or, $X = 1_L$ and $Y = 0_L$.*
4. *If $X \leq_{TL} Y \leq_{TL} Z$, then $d_L(X,Z) \geq_{TL} d_L(X,Y)$ and $d_L(X,Z) \geq_{TL} d_L(Y,Z)$.*

Proposition 2. *Let N be an IV negation with respect to a total order $\leq_{TL}$. Let $M_1, M_2 \colon L([0,1])^2 \to L([0,1])$ be two IV aggregation function (with respect to the same order) as in Example 5, with M_1 symmetric. Let I be the IV implication operator defined as in Proposition 3 with M_2 and N. Then the function $d_L \colon L([0,1])^2 \to L([0,1])$ defined by*

$$d_L(X,Y) = N(M_1(I(X,Y), I(Y,X)))$$

verifies the properties (1) and (4) in Definition 7 and it also verifies property (2) if N is non-vanishing and property (3) if N is non-filling.

Proof. It follows from Theorem 2 and Example 5, since we are just considering a negation of the function F considered there.

4 Similarity Measures, Distances and Entropy Measures in $L([0,1])$ with Respect to a Total Order

Our constructions in the previous section can be used to build comparison measures between interval-valued fuzzy sets, and, more specifically, to obtain similarity measures, distances in the sense of Fang and entropy measures. We insist that, for the first time in the literature, we are providing measures which are also interval valued, rather than just providing a single number as their result. Along this section, we only deal with a total order $\leq_{TL}$.

First of all, we show how we can build a similarity between interval-valued fuzzy sets defined over the same referential U. We start recalling the definition.

Definition 8. *[3] Let $\leq_{TL}$ be a total order in $L([0,1])$. An interval-valued (IV) similarity measure on $IVFS(U)$ (with respect to $\leq_{TL}$) is a mapping SM : $IVFS(U) \times IVFS(U) \to L([0,1])$ such that, for every $A, B, A', B' \in IVFS(U)$,*

(SM1) SM is symmetric.
(SM2) $SM(A,B) = 1_L$ if and only if $A = B$.
(SM3) $SM(A,B) = 0_L$ if and only if $\{A(u_i), B(u_i)\} = \{0_L, 1_L\}$ for every $u_i \in U$.
(SM4) If $A \subseteq A' \subseteq B' \subseteq B$, then $SM(A,B) \leq_{TL} SM(A', B')$.

Then we have the following result.

Theorem 3. *Let $\leq_{TL}$ be a total order in $L([0,1])$. Let $M : L([0,1])^n \to L([0,1])$ be an IV aggregation function with respect to the total order $\leq_{TL}$ and such that $M(X_1, \ldots, X_n) = 1_L$ if and only if $X_1 = \cdots = X_n = 1_L$ and $M(X_1, \ldots, X_n) = 0_L$ if and only if $X_1 = \cdots = X_n = 0_L$. Then, the function $SM : IVFS(U) \times IVFS(U) \to L([0,1])$ given by*

$$SM(A,B) = M(F(A(u_1), B(u_1)), \ldots, F(A(u_n), B(u_n))) \,,$$

where F is defined as in Example 5 with non-filling and non-vanishing IV negation N, is an IV similarity measure with respect to $\leq_{TL}$.

Proof. It follows from a straightforward calculation.

Example 6. Let us consider Xu and Yager's order. Take as F the function

$$F(X,Y) = \begin{cases} 1_L & \text{if } X = Y \\ [\frac{\underline{X}+\overline{X}}{2}, \frac{\underline{X}+\overline{X}}{2}] & \text{if } X <_{XY} Y \\ [\frac{\underline{Y}+\overline{Y}}{2}, \frac{\underline{Y}+\overline{Y}}{2}] & \text{if } Y <_{XY} X \,, \end{cases}$$

given in Remark 2. Finally, taking M as

$$M(X_1 \ldots, X_n) = [\frac{1}{n}(1\underline{X}_1 + \cdots + \underline{X}_n), \frac{1}{n}(\overline{X}_1 + \cdots + \overline{X}_n)].$$

Then the function

$$SM(A,B) = [\frac{1}{2n}(\underline{\wedge(A(u_1), B(u_1))} + \cdots + \underline{\wedge(A(u_n), B(u_n))}),$$
$$\frac{1}{2n}(\overline{\wedge(A(u_1), B(u_1))} + \cdots + \overline{\wedge(A(u_n), B(u_n))})]$$

is an IV similarity measure.

We can employ the Theorem 3 to recover both IV distances and IV entropy measures with respect to total orders. First of all, let's recall the definition of both concepts.

Definition 9. *[5] Let $\leq_{TL}$ be a total order in $L([0,1])$. A function D : $IVFS(U) \times IVFS(U) \to L([0,1])$ is called an IV distance measure on $IVFS(U)$ if, for every $A, B, A', B' \in IVFS(U)$, D satisfies the following properties:*

(D1) $D(A, B) = D(B, A)$;
(D2) $D(A, B) = 0_L$ if and only if $A = B$;
(D3) $D(A, B) = 1_L$ if and only if A and B are complementary crisp sets;
(D4) If $A \subseteq A' \subseteq B' \subseteq B$, then $D(A, B) \geq_{TL} D(A', B')$.

Definition 10. *[5] Let $\leq_{TL}$ be a total order in $L([0,1])$. Let N be an IV strong negation with respect to $\leq_{TL}$ such that there exists $\varepsilon \in L([0,1])$ with $N(\varepsilon) = \varepsilon$. A function $E : IVFS(U) \to L([0,1])$ is an IV entropy on $IVFS(U)$ with respect to a strong IV negation N if:*

(E1) $E(A) = 0_L$ if and only if A is crisp;
(E2) $E(A) = 1_L$ if and only if $A = \{(u_i, A(u_i) = \varepsilon) | u_i \in U\}$;
(E3) $E(A) \leq_{TL} E(B)$ if A refines B; that is, $A(u_i) \leq_{TL} B(u_i) \leq_{TL} \varepsilon$ or $A(u_i) \geq_{TL} B(u_i) \geq_{TL} \varepsilon$;
(E4) $E(A) = E(N(A))$.

Then the following two results are straight from Theorem 3.

Corollary 1. *Let $\leq_{TL}$ be a total order in $L([0,1])$. Let $M : L([0,1])^n \to L([0,1])$ be an IV aggregation function with respect to the total order $\leq_{TL}$ as in Theorem 3. Then, the function $D : IVFS(U) \times IVFS(U) \to L([0,1])$ given by*

$$D(A, B) = N(M(F(A(u_1), B(u_1)), \ldots, F(A(u_n), B(u_n)))),$$

where F is defined as in Example 5 with a non-filling and non-vanishing negation N, is an IV distance measure.

Proof. It is straight from Theorem 3, since a similarity measure defines a distance in a straightforward way.

Theorem 4. *Let $\leq_{TL}$ be a total order in $L([0,1])$. Let N be a strong IV negation (with respect to the same order $\leq_{TL}$) such that there exists $\varepsilon \in L([0,1])$ with $N(\varepsilon) = \varepsilon$ and M an IV aggregation function (with respect to the same order $\leq_{TL}$) as in Theorem 3 Then, the function $E : IVFS(U) \to L([0,1])$ given by*

$$E(A) = M(F(A(u_1), N(A(u_1))), \ldots, F(A(u_n), N(A(u_n)))),$$

where F is defined as in Example 5 with a non-filling and non-vanishing negation N, is an IV entropy measure.

Proof. It follows from the well known fact that, for a given IV similarity SM, the function $E(A) = SM(A, N(A))$ is an IV entropy measure [5].

5 Conclusions

In this paper we have defined, for the first time in the literature, interval-valued restricted equivalence functions and dissimilarity functions with respect to a total order. Then we have aggregated them receiving similarity measures, distance measures and entropies for interval-valued fuzzy sets. In such a way this investigation can stand as a background for image processing tasks of certain type [1,14].

References

1. Barrenechea, E., Bustince, H., De Baets, B., Lopez-Molina, C.: Construction of interval-valued fuzzy relations with application to the generation of fuzzy edge images. IEEE Trans. Fuzzy Syst. **19**(5), 819–830 (2011)
2. Bustince, H.: Indicator of inclusion grade for interval-valued fuzzy sets. Application to approximate reasoning based on interval-valued fuzzy sets. Int. J. Approx. Reason. **23**(3), 137–209 (2000)
3. Bustince, H., Barrenechea, E., Pagola, M.: Restricted equivalence functions. Fuzzy Sets Syst. **157**(17), 2333–2346 (2006)
4. Bustince, H., Barrenechea, E., Pagola, M.: Image thresholding using restricted equivalence functions and maximizing the measure of similarity. Fuzzy Sets Syst. **128**(5), 496–516 (2007)
5. Bustince, H., Barrenechea, E., Pagola, M.: Relationship between restricted dissimilarity functions, restricted equivalence functions and normal E_N-functions: image thresholding invariant. Pattern Recognit. Lett. **29**(4), 525–536 (2008)
6. Bustince, H., et al.: A historical account of types of fuzzy sets and their relationships. IEEE Trans. Fuzzy Syst. **24**(1), 179–194 (2016)
7. Cornelis, C., Deschrijver, G., Kerre, E.E.: Implication in intuitionistic fuzzy and interval-valued fuzzy set theory: construction, classification, application. Int. J. Approx. Reason. **35**(1), 55–95 (2004)
8. Fodor, J., Roubens, M.: Fuzzy Preference Modelling and Multicriteria Decision Support. Theory and Decision. Kluwer, Dordrecht (1994)
9. Gorzalczany, M.B.: A method of inference in approximate reasoning based on interval-valued fuzzy sets. Fuzzy Sets Syst. **21**, 11–17 (1987)
10. Horanská, L'., Šipošová, A.: A generalization of the discrete Choquet and Sugeno integrals based on a fusion function. Inf. Sci. **451–452**, 83–99 (2018)
11. Horanská, L'., Šipošová, A.: Integration based on fusion functions. Tatra Mt. Math. Publ. **66**, 51–66 (2016)
12. Liu, X.: Entropy, distance measure and similarity measure of fuzzy sets and their relations. Fuzzy Sets Syst. **52**, 305–318 (1992)
13. Paternain, D., et al.: Strong fuzzy subsethood measures and strong equalities via implication functions. J. Mult. Valued Log. Soft Comput. **22**(4–6), 347–371 (2014)
14. Takáč, Z., Minárová, M., Montero, J., Barrenechea, E., Fernandez, J., Bustince, H.: Interval-valued fuzzy strong S-subsethood measures, interval-entropy and P-interval-entropy. Inf. Sci. **432**, 97–115 (2018)
15. Xu, Z.S., Yager, R.R.: Some geometric aggregation operators based on intuitionistic fuzzy sets. Int. J. Gen. Syst. **35**, 417–433 (2006)

Author Index